Technologies and Applications for Big Data Value

Edward Curry • Sören Auer • Arne J. Berre •
Andreas Metzger • Maria S. Perez • Sonja Zillner
Editors

Technologies and Applications for Big Data Value

 Springer

Editors
Edward Curry
Insight SFI Research Centre for Data
Analytics
NUI Galway, Ireland

Sören Auer
Information Centre for Science
and Technology
Leibniz University Hannover
Hannover, Germany

Arne J. Berre
SINTEF Digital
Oslo, Norway

Andreas Metzger
Paluno
University of Duisburg-Essen
Essen, Germany

Maria S. Perez
Universidad Politécnica de Madrid
Boadilla del Monte
Madrid, Spain

Sonja Zillner
Siemens Corporate Technology
München, Germany

ISBN 978-3-030-78309-9   ISBN 978-3-030-78307-5   (eBook)
https://doi.org/10.1007/978-3-030-78307-5

This Springer imprint is published by the registered company Springer Nature Switzerland AG.
The registered company address is: Gewerbestrasse 11, 6330 Cham, Switzerland

Preface

Computer science was created by humankind to solve problems. In 100 BC, early hand-powered computing devices such as the Antikythera mechanism were designed to calculate astronomical positions. In the 1800s, Charles Babbage proposed the Analytical Engine to solve general-purpose computational tasks. In the 1900s, the Bomde by Turing and Welchman was critical to code-breaking. Advances in computer science have been driven by the need for humanity to solve the most pressing challenges of the day. Today, computer science tackles significant societal challenges like organising the world's information, personalised medicine, the search of the Higgs boson, climate change, and weather forecasts.

This book aims to educate the reader on how recent advances in technologies, methods, and processes for big data and data-driven Artificial Intelligence (AI) can deliver value to address problems in real-world applications. The book explores cutting-edge solutions and best practices for big data and data-driven AI and applications for the data-driven economy. It provides the reader with a basis for understanding how technical issues can be overcome to offer real-world solutions to major industrial areas, including health, energy, transport, finance, manufacturing, and public administration.

The book's contributions emanate from the Big Data Value Public-Private Partnership (BDV PPP) and the Big Data Value Association, which have acted as the European data community's nucleus to bring together businesses with leading researchers to harness the value of data to benefit society, business, science, and industry. The technological basis established in the BDV PPP will seamlessly enable the European Partnership on AI, Data, and Robotics.

The book is of interest to two primary audiences: first, undergraduate and postgraduate students and researchers in various fields, including big data, data science, data engineering, machine learning, and AI. Second, practitioners and industry experts engaged in data-driven systems and software design and deployment projects who are interested in employing these advanced methods to address real-world problems.

This book is arranged in two parts. The first part contains "horizontal" contributions of technologies and methods which can be applied in any sector. The second part includes contributions of innovative processes and applications within specific "vertical" sectors. Chapter 1 provides an overview of the book by positioning the chapters in terms of their contributions to technology frameworks, including the Big Data Value Reference Model and the European AI, Data and Robotics Framework, which are key elements of the BDV PPP and the Partnership on AI, Data and Robotics.

Part I: Technologies and Methods details key technical contributions which enable data value chains. Chapter 2 investigates ways to support semantic data enrichment at scale. The trade-offs and challenges of serverless data analytics are examined in Chap. 3. Benchmarking of big data and AI pipelines is the objective of Chap. 4, while Chap. 5 presents an elastic software architecture for extreme-scale big data analytics. Chapter 6 details privacy-preserving technologies for trusted data spaces. Leveraging data-driven infrastructure management to facilitate AIOps is the focus of Chap. 7, and unified big-data workflows over High-Performance-Computing (HPC) and the cloud are tackled in Chap. 8.

Part II: Processes and Applications details experience reports and lessons from using big data and data-driven approaches in processes and applications. The chapters are co-authored with industry experts and cover domains including health, law, finance, retail, manufacturing, mobility, and smart cities. Chapter 9 presents a toolkit for deep learning and computer vision over HPC and cloud architectures. Applying AI to manage acute and chronic clinical conditions is the focus of Chap. 10, while Chap. 11 explores 3D human big data exchange between the health and garment sectors. In Chap. 12, we see how legal knowledge graphs can be used for multilingual compliance services in labour law, contract management, and geothermal energy. Chapter 13 focuses on big data analytics in the banking sector with guidelines and lessons learned from CaixaBank. Chapter 14 explores data-driven AI and predictive analytics for the maintenance of industrial machinery using digital twins. Chapter 15 investigates big data analytics in the manufacturing sector, and Chap. 16 looks at the next generation of data-driven factory operations and optimisation. Large-scale trials of data-driven service engineering are covered in Chap. 17. Chapter 18 describes approaches for model-based engineering and semantic interoperability for digital twins across the product life cycle. In Chap. 19, a data science pipeline for big linked earth observation data is presented, and Chap. 20 looks ahead towards cognitive ports of the future. Distributed big data analytics in a smart city is the focus of Chaps. 21, and 22 looks at system architectures and applications of big data in the maritime domain. The book closes with Chap. 23 exploring knowledge modelling and incident analysis for cargo.

Galway, Ireland   Edward Curry
January 2022

Acknowledgements

The editors thank Ralf Gerstner and all at Springer for their professionalism and assistance throughout the journey of this book. This book was made possible through funding from the European Union's Horizon 2020 research and innovation programme under grant agreement no. 732630 (BDVe). This work was supported by the Science Foundation Ireland, co-funded by the European Regional Development Fund under Grant SFI/12/RC/2289_P2.

We thank all the authors for sharing their work through our book. Very special thanks to all the reviewers who gave their time, effort, and constructive comments that enhanced the overall quality of the chapters. We particularly recognise the dedication and commitments of Amin Anjomshoaa, Felipe Arruda Pontes, Alexandru Costan, Praneet Dhingra, Jaleed Khan, Manolis Marazakis, Ovidiu-Cristian Marcu, Niki Pavlopoulou, Emilio Serrano, Atiya Usmani, Piyush Yadav, and Tarek Zaarour as reviewers.

Finally, we would like to thank our partners at the European Commission, in particular Commissioner Gabriel, Commissioner Kroes, and the Director-General of DG CONNECT Roberto Viola, who had the vision and conviction to develop the European data economy. We thank the current and past members of the European Commission's Unit for Data Policy and Innovation (Unit G.1), Yvo Volman, Márta Nagy-Rothengass, Kimmo Rossi, Beatrice Covassi, Stefano Bertolo, Francesco Barbato, Wolfgang Treinen, Federico Milani, Daniele Rizzi, and Malte Beyer-Katzenberger. Together they have represented the public side of the Big Data Value Partnership and were instrumental in its success.

January 2022

Edward Curry
Sören Auer
Arne J. Berre
Andreas Metzger
Maria S. Perez
Sonja Zillner

Contents

Contents

Editors and Contributors

About the Editors

Edward Curry is Established Professor of Data Science and Director of the Insight SFI Research Centre for Data Analytics and the Data Science Institute at NUI Galway. Edward has made substantial contributions to semantic technologies, incremental data management, event processing middleware, software engineering, as well as distributed systems and information systems. He combines strong theoretical results with high-impact practical applications. The impact of his research have been acknowledged by numerous awards, including best paper awards and the NUIG President's Award for Societal Impact in 2017. The technology he develops with his team enables intelligent systems for smart environments in collaboration with his industrial partners. He is organiser and programme co-chair of major international conferences, including CIKM 2020, ECML 2018, IEEE Big Data Congress, and European Big Data Value Forum. Edward is co-founder and elected Vice President of the Big Data Value Association, an industry-led European big data community.

Sören Auer Following stations at the universities of Dresden, Ekaterinburg, Leipzig, Pennsylvania, Bonn, and the Fraunhofer Society, Prof. Auer was appointed Professor of Data Science and Digital Libraries at Leibniz Universität Hannover and Director of the TIB in 2017. Prof. Auer has made important contributions to semantic technologies, knowledge engineering, and information systems. He is the author (resp. co-author) of over 150 peer-reviewed scientific publications. He has received several awards, including an ERC Consolidator Grant from the European Research Council, a SWSA 10-year award, the ESWC 7-year Best Paper Award, and the OpenCourseware Innovation Award. He has led several large collaborative research projects, such as the EU H2020 flagship project BigDataEurope. He is co-founder of high-potential research and community projects such as the Wikipedia semantification project DBpedia, the OpenCourseWare authoring platform SlideWiki.org, and the innovative technology start-up eccenca.com. Prof. Auer was founding director of the Big Data Value Association, led the semantic data representation in the Industrial/International Data Space, and is an expert for industry, European Commission, and W3C.

Arne J. Berre received his Ph.D. in Computer Science (Dr. Ing) from the Norwegian University of Science and Technology (NTNU) in 1993 on the topic of an Object-oriented Framework for Systems Interoperability. He is Chief Scientist at SINTEF Digital and Innovation Director at the Norwegian Center for AI Innovation (NorwAI), and responsible for the GEMINI centre of Big Data and AI. He is the leader of the BDVA/DAIRO TF6 on Technical priorities, including responsibilities for data technology architectures, data science/AI, data protection, standardisation, benchmarking, and HPC. He has been in the technical lead of more than 20 European projects, including being responsible for the ATHENA Interoperability Framework and the BDVA Reference Model. He has been involved in the development of various standards for Interoperability, including Geographic Information services with OGC, CEN/TC287 and ISO/TC211, ISO 19103 and 19119, and been a lead in the standards development teams of the OMG standards of SoaML, VDML, General Ledger, and Essence. He is the lead of the Norwegian committee for AI and Big Data with ISO SC 42 AI. He has more than 100 scientific publications and has been involved in the organisation of a number of international conferences and workshops. He is the technical coordinator of the DataBench project on Big Data and AI Benchmarking and the COGNITWIN project on Cognitive and Hybrid Digital Twins for the process industry and is responsible for the standardisation framework in the DEMETER project on Digital platform interoperability in the AgriFood domain.

Andreas Metzger received his Ph.D. in Computer Science (Dr.-Ing.) from the University of Kaiserslautern in 2004. He is senior academic councillor at the University of Duisburg-Essen and heads the Adaptive Systems and Big Data Applications group at paluno – The Ruhr Institute for Software Technology. His background and research interests are software engineering and machine learning for adaptive systems. He has co-authored over 120 papers, articles and book chapters. His recent research on online reinforcement learning for self-adaptive software systems received the Best Paper Award at the International Conference on Service-oriented Computing. He is co-organiser of over 15 international workshops and conference tracks and is program committee member for numerous international conferences. Andreas was Technical Coordinator of the European lighthouse project TransformingTransport, which demonstrated in a realistic, measurable, and replicable way the transformations that big data and machine learning can bring to the mobility and logistics sector. In addition, he was member of the Big Data Expert Group of PICASSO, an EU-US collaboration action on ICT topics. Andreas serves as steering committee vice chair of NESSI, the European Technology Platform dedicated to Software, Services and Data, and as deputy secretary general of the Big Data Value Association.

Maria S. Perez received her MS degree in Computer Science in 1998 at the Universidad Politécnica de Madrid (UPM), a Ph.D. in Computer Science in 2003, and the Extraordinary Ph.D. Award at the same university in 2004. She is currently Full Professor at UPM. She is part of the Board of Directors of the Big Data Value Association and also a member of the Research and Innovation Advisory Group of the EuroHPC Joint Undertaking. Her research interests include data science, big data, machine learning, storage, high performance, and large-scale computing. She has co-authored more than 100 articles in international journals and conferences. She has been involved in the organisation of several international workshops and conferences, such as IEEE Cluster, Euro-Par, and International Conference for High Performance Computing, Networking, Storage and Analysis (SC). She has edited several proceedings, books, and special issues in journals, such as FGCS or CCPE. She has participated in a large number of EU projects and Spanish R&D projects. She is currently serving as a program committee member for many relevant conferences, such as ISWC, IEEE Cluster, CCGrid, and CIKM.

Sonja Zillner studied mathematics and psychology at the Albert-Ludwig-University Freiburg, Germany, and accomplished her Ph.D. in computer science, specialising on the topic of Semantics at Technical University in Vienna. Since 2005 she has been working at Siemens AG Technology as Principal Research Scientist focusing on the definition, acquisition, and management of global innovation and research projects in the domain of semantics and artificial intelligence. Since 2020 she has been Lead of Core Company Technology Module "Trustworthy AI" at Siemens Corporate Technology. Before that, from 2016 to 2019 she was invited to consult the Siemens Advisory Board in strategic decisions regarding artificial intelligence. She is chief editor of the Strategic Research Innovation and Deployment Agenda of the European Partnership on AI, Data, and Robotics; leading editor of the Strategic Research and Innovation Agenda of the Big Data Value Association (BDVA); and member of the editing team of the strategic agenda of the European On-Demand Platform AI4EU. Between 2012 and 2018 she was professor at Steinbeis University in Berlin, between 2017 and 2018 guest professor at the Technical University of Berlin, and since 2016 she has been lecturing at Technical University of Munich. She is author of more than 80 publications and holds more than 25 patents in the area of semantics, artificial intelligence, and data-driven innovation.

Contributors

Charilaos Akasiadis NCSR Demokritos, Institute of Informatics & Telecommunications, Greece

Alp Akçay Eindhoven University of Technology, the Netherlands

Özlem Albayrak Teknopar Industrial Automation, Turkey

Marco Aldinucci Università degli studi di Torino, Italy

Ulrich Ahle FIWARE Foundation, Germany

Elias Alevizos NCSR Demokritos, Institute of Informatics & Telecommunications, Greece

Andreas Alexopoulos Aegis IT Research LTD, UK

Danilo Amendola Centro Ricerche Fiat S.C.p.A. (CRF), Italy

Alexander Artikis NCSR Demokritos, Institute of Informatics & Telecommunications, Greece

David Atienza Ecole Polytechnique Fédérale de Lausanne, Switzerland

Sören Auer Leibniz Universität Hannover, Germany

Yolanda Becerra Barcelona Supercomputing Center, Spain

Hanane Becha Traxens, France

Andreu Belsa Pellicer Universitat Politècnica de Valencia, Spain

Konstantina Bereta National and Kapodistrian University of Athens, Greece

Dimitris Bilidas National and Kapodistrian University of Athens, Greece

Gernot Boege FIWARE Foundation, Germany

Omer Boehm IBM, Israel

Pascual Boil Cuatrecasas, Spain

Federico Bolelli Università degli studi di Modena e Reggio Emilia, Italy

Susanna Bonura Engineering Ingegneria Informatica SpA, Italy

Laura Boura P&R Têxteis S.A., Portugal

George Bravos Information Technology for Market Leadership, Greece

David Breitgand IBM Research Haifa, Israel

Mónica Caballero NTT Data Spain, Barcelona, Spain

Santiago Cáceres Instituto Tecnológico de Informática, Spain

Caterina Calefato Domina Srl., Italy

Pablo Calleja-Ibáñez Universidad Politécnica de Madrid, Spain

Alessandro Canepa Fratelli Piacenza, Italy

Antonio Castillo Nieto ATOS, Spain

Roberto Cavicchioli University of Modena (UNIMORE), Italy

Vasilis Chatzigiannakis Information Technology for Market Leadership, Greece

Luca Chiantore Comune di Modena, Italy

Michele Ciavotta University of Milan–Bicocca, Italy

Pietro Cipresso Applied Technology for Neuro-Psychology Lab, Istituto Auxologico, Italy

Calina Ciuhu-Pijlman Philips Research, the Netherlands

Jerome Clavel Agie Charmilles New Technologies, Switzerland

Iacopo Colonnelli Università degli studi di Torino, Italy

Cristovao Cordeiro SIXSQ, Switzerland

Ruben Costa UNINOVA - Instituto de desenvolvimento de novas tecnologías - Associacao., Portugal

Cesare Cugnasco Barcelona Supercomputing Center, Spain

Edward Curry Insight SFI Research Centre for Data Analytics, NUI Galway, Ireland

Vincenzo Cutron University of Milan–Bicocca, Italy

Davide Dalle Carbonare Engineering Ingegneria Informatica SpA, Italy

Giuseppe Danilo Spennacchio Centro Ricerche FIAT, Italy

Eelco de Jong Validaide B.V., the Netherlands

Silvia de la Maza Trimek S.A., Spain

Flavio De Paoli University of Milan–Bicocca, Italy

Aaron Dees Irish Centre for High-End Computing, Ireland

Giorgos Demetriou Circular Economy Research Center, Ecole des Ponts, Business School, Marne-la-Vallée, France

Roberto Díaz-Morales Tree Technology, Spain

Frédéric Donnat Outpost24, France

Elenna Dugundji Centrum Wiskunde & Informatica (CWI), the Netherlands

Juan V. Durá Gil Instituto de Biomecánica de Valencia. Universitat Politècnica de València, Spain

Iliada Eleftheriou University of Manchester, UK

Mauricio Fadel Argerich NEC Laboratories Europe, Heidelberg, Germany

Marcos Fernández-Díaz Tree Technology, Spain

Paulo Figueiras UNINOVA - Instituto de desenvolvimento de novas tecnologías – Associacao., Portugal

José Flich Universitat Politècnica de València, Spain

Lidija Fodor University of Novi Sad - Faculty of Sciences, Serbia

Spiros Fotis Aegis IT Research LTD, UK

Fabiana Fournier IBM, Israel

Chiara Francalanci Politecnico di Milano, Spain

Gerald Fritz TTTech Industrial Automation AG, Austria

Jonathan Fuerst NEC Laboratories Europe, Heidelberg, Germany

Gisela Garcia Volkswagen Autoeuropa, Portugal

Pedro García-López Universitat Rovira i Virgili, Spain

Hugues-Arthur Garious ESI GROUP, France

Gianmarco Genchi Centro Ricerche FIAT, Italy

Nikos Giatrakos Antonios Deligiannakis Athena Research Center, Greece

Christos Gizelis Hellenic Telecommunications Organization S.A., Greece

Martin Golasowski Technical University of Ostrava, Czechia

Jon A. Gómez Universitat Politècnica de València, Spain

Marco Gonzalez Ikerlan Technology Research Centre, Basque Research Technology Alliance (BRTA), Arrasate/Mondragón, Spain

David González TREE Technology, Spain

Thierry Goubier CEA LIST, France

Diogo Graça Volkswagen Autoeuropa, Portugal

Costantino Grana Università degli studi di Modena e Reggio Emilia, Italy

Marco Grangetto Università degli studi di Torino, Italy

Marko Grobelnik Jožef Stefan Institute, Slovenia

Stephan Hachinger Leibniz Supercomputing Centre, Germany

Erez Hadad IBM Research Haifa, Haifa, Israel

Rachael Hagan Queen's University Belfast, UK

Piyush Harsh Cyclops Labs GmbH, Switzerland

Mohamad Haye Leibniz Supercomputing Centre, Germany

Nathan Hazout IBM, Israel

Juanjo Hierro FIWARE Foundation, Germany

Marlene Hildebrand Ecole Polytechnique Federale de Lausanne, Switzerland

Roxana-Maria Holom RISC Software GmbH, Austria

Sotiris Ioannidis Technical University of Crete, Greece

Todor Ivanov Lead Consult, Bulgaria

Arne J. Berre SINTEF Digital, Norway

Charles J. Gillan Queen's University Belfast, UK

Dusan Jakovetic University of Novi Sad - Faculty of Sciences, Serbia

Rubén Jesús García-Hernández Leibniz Supercomputing Centre, Germany

Moez Jomâa SINTEF Industry, Norway

Ana Juan-Ferrer ATOS, Spain

Jan Jürjens Fraunhofer-Institute for Software and Systems Engineering ISST, Germany

Leonidas Kallipolitis Aegis IT Research LTD, UK

Martin Kaltenboeck Semantic Web Company, Austria

Konstantin Karavaev ELSE Corp Srl., Italy

Anthousa Karkoglou National Technical University of Athens, Greece

Vlatka Katusic Circular Economy Research Center, Ecole des Ponts, Business School, Marne-la-Vallée, France

Evangelia Kavakli University of Manchester, UK

Sarah Kerkhove Interuniversitair Microelectronica Centrum, Belgium

Réda Khouani Traxens, France

Dimitris Kiritsis Ecole Polytechnique Federale de Lausanne, Switzerland

Thomas Koch Centrum Wiskunde & Informatica (CWI), the Netherlands

Daniel Köchling BENTELER Automobiltechnik, Germany

Tess Kolkman Eindhoven University of Technology, the Netherlands

Despina Kopanaki Foundation for Research and Technology, Hellas - Institute of Computer Science, Greece

Manolis Koubarakis National and Kapodistrian University of Athens, Greece

George Kousiouris Harokopio University of Athens, Greece

Dimosthenes Krassas Hellenic Telecommunications Organization S.A., Greece

Stefanie Kritzinger RISC Software GmbH, Austria

Dimosthenis Kyriazis University of Piraeus, Greece

Dominika Lekse Philips Electronics, the Netherlands

Simone Leo Center for Advanced Studies, Research and Development in Sardinia, Italy

Christoforos Leventis Foundation for Research and Technology, Hellas - Institute of Computer Science, Greece

Marc Levrier ATOS, France

Antonios Litke National Technical University of Athens, Greece

Pedro López Universitat Politècnica de València, Spain

Tomasz Luniewski CAPVIDIA, USA

Mario Maawad Marcos CaixaBank, Spain

Craig Macdonald University of Glasgow, UK

Donato Magarielli AvioAero, Italy

César Marín Information Catalyst for Enterprise Ltd, Crew, UK

Achilleas Marinakis National Technical University of Athens, Greece

Ramon Martin de Pozuelo CaixaBank, Spain

Miquel Martínez Barcelona Supercomputing Center, Spain

Iván Martínez Rodriguez ATOS Research and Innovation, Madrid, Spain

Jan Martinovic Technical University of Ostrava, Czechia

Julien Mascolo Centro Ricerche FIAT, Italy

Davide Masera Centro Ricerche Fiat, Italy

Richard McCreadie University of Glasgow, UK

Ryan McDonnell Vrije Universiteit Amsterdam, the Netherlands

Ifigeneia Metaxa ATLANTIS Engineering, Greece

Andreas Metzger Paluno, University of Duisburg-Essen, Germany

Luis Miguel Pinho Instituto Superior De Engenharia Do Porto (ISEP), Portugal

Nemanja Milosevic University of Novi Sad - Faculty of Sciences, Serbia

Elena Montiel-Ponsoda Universidad Politécnica de Madrid, Spain

Lucrezia Morabito COMAU Spa, Italy

Vrettos Moulos National Technical University of Athens, Greece

Giuseppe Mulè Siemens, Italy

Luis Muñoz-González Imperial College London, UK

Pietro Pittaro Prima Industrie, Italy

Athanasios Poulakidas Intrasoft International S.A., Luxembourg

Mark Purcell IBM Research Europe, Ireland

Anna Queralt Universitat Politècnica de Catalunya (UPC), Barcelona, Spain
Barcelona Supercomputing Center (BSC), Barcelona, Spain

Bernat Quesada Navidad ATOS, Spain

Eduardo Quiñones Barcelona Supercomputing Center (BSC), Spain

Katharina Rafetseder RISC Software GmbH, Austria

Jose Real INCLIVA Research Institute, University of Valencia, Spain

Josep Redon INCLIVA Research Institute, University of Valencia, Spain

Alfredo Remon Instituto de Biomecánica de Valencia. Universitat Politècnica de València, Spain

Vahideh Reshadat Eindhoven University of Technology, the Netherlands

Gerald Ristow Software AG, Germany

Dumitru Roman SINTEF Digital, Norway

Ricardo Ruiz-Saiz ATOS Research and Innovation, Madrid, Spain

Hernan Ruiz-Ocampo Circular Economy Research Center, Ecole des Ponts, Business School, Marne-la-Vallée, France

Christian Sageder Cybly, Austria

Rizos Sakellariou University of Manchester, UK

Sergio Salmeron-Majadas Atos IT Solutions and Services, USA

Marc Sánchez-Artigas Universitat Rovira i Virgili, Spain

Karel Schorer Vrije Universiteit Amsterdam, the Netherlands

Alberto Scionti LINKS Foundation, Italy

Harald Sehrschön Fill Gesellschaft m.b.H., Austria

Maria Serrano Barcelona Supercomputing Center (BSC), Spain

Amir Shayan Ahmadian Institute for Software Technology - University of Koblenz-, Landau, Germany

Simon Shillaker Imperial College London, England

Murali Shyamsundar Queen's University Belfast, UK

Luísa Silva P&R Têxteis S.A., Portugal

Tatiana Silva TREE Technology, Spain

Konstantinos Sipsas Intrasoft International S.A., Luxembourg

Raül Sirvent Barcelona Supercomputing Center (BSC), Spain

Inna Skarbovsky IBM, Israel

Srdjan Skrbic University of Novi Sad - Faculty of Sciences, Serbia

Katerina Slaninová Technical University of Ostrava, Czechia

Carlijn Snijder Vrije Universiteit Amsterdam, the Netherlands

John Soldatos University of Glasgow, UK

Ilias Spais Aegis IT Research LTD, UK

Dusan Stamenkovic University of Novi Sad - Faculty of Sciences, Serbia

George Stamoulis National and Kapodistrian University of Athens, Greece

Pierre Sutra Université Paris Saclay, Télécom SudParis, France

Enzo Tartaglione Università degli studi di Torino, Italy

Tristan Tarrant Red Hat, Cork, Ireland

Olivier Terzo LINKS Foundation, Italy

Bas Tijsma Phillips Consumer Lifestyle, the Netherlands

Jean-Didier Totow University of Piraeus, Greece

Aphrodite Tsalgatidou SINTEF, Oslo, Norway

Nikos Tzagkarakis National Technical University of Athens, Greece

Fernando Ubis Visual Components Ov., Finland

Perin Unal Teknopar Industrial Automation, Turkey

Marco Vallini Domina Srl., Italy

Francisco Valverde Instituto Tecnológico de Informática, Spain

Giorgos Vasiliadis Foundation for Research and Technology, Hellas - Institute of Computer Science, Greece

Antonio Ventura-Traveset Innovalia Metrology, Spain

Pieter Verhoeven DNV, The Netherlands

Gil Vernik IBM Research Haifa, Haifa, Israel

Michael Vinov IBM, Israel

Giacomo Vitali LINKS Foundation, Italy

Marios Vodas MarineTraffic, Greece

Bruno Volckaert Interuniversitair Microelectronica Centrum, Belgium

Sebastian von Enzberg Fraunhofer Institute for Mechatronic Systems Design, Germany

Casper Wichers Vrije Universiteit Amsterdam, the Netherlands

Alois Wiesinger Fill Gesellschaft m.b.H., Austria

Marina Zapater Ecole Polytechnique Fédérale de Lausanne, Switzerland

Kalliopi Zervanou Eindhoven University of Technology, the Netherlands

Yingqian Zhang Eindhoven University of Technology, the Netherlands

Wojciech Zietak CAPVIDIA, USA

Sonja Zillner Siemens AG, Germany

Dimitris Zissis MarineTraffic, Greece

Dmitry Znamenskiy Philips Research, the Netherlands

Technologies and Applications for Big Data Value

Edward Curry, Sören Auer, Arne J. Berre, Andreas Metzger, Maria S. Perez, and Sonja Zillner

Abstract The continuous and significant growth of data, together with improved access to data and the availability of powerful computing infrastructure, has led to intensified activities around Big Data Value (BDV) and data-driven Artificial Intelligence (AI). Powerful data techniques and tools allow collecting, storing, analysing, processing and visualising vast amounts of data, enabling data-driven disruptive innovation within our work, business, life, industry and society.

The adoption of big data technology within industrial sectors facilitates organisations to gain a competitive advantage. Driving adoption is a two-sided coin. On one side, organisations need to master the technology necessary to extract value from big data. On the other side, they need to use the insights extracted to drive their digital transformation with new applications and processes that deliver real value. This book has been structured to help you understand both sides of this coin and bring together technologies and applications for Big Data Value.

This chapter defines the notion of big data value, introduces the Big Data Value Public-Private Partnership (PPP) and gives some background on the Big Data Value Association (BDVA)—the private side of the PPP. It then moves on to structure the

E. Curry (✉)
Insight SFI Research Centre for Data Analytics, NUI Galway, Ireland
e-mail: edward.curry@nuigalway.ie

S. Auer
Leibniz Universität Hannover, Hanover, Germany

A. J. Berre
SINTEF Digital, Oslo, Norway

A. Metzger
Paluno, University of Duisburg-Essen, Essen, Germany

M. S. Perez
Universidad Politécnica de Madrid, Boadilla del Monte, Madrid, Spain

S. Zillner
Siemens AG, München, Germany

contributions of the book in terms of three key lenses: the BDV Reference Model, the Big Data and AI Pipeline, and the AI, Data and Robotics Framework.

Keywords Data ecosystem · Big data value · Data-driven innovation · Big Data

1   Introduction

The continuous and significant growth of data, together with improved access to data and the availability of powerful computing infrastructure, has led to intensified activities around Big Data Value (BDV) and data-driven Artificial Intelligence (AI). Powerful data techniques and tools allow collecting, storing, analysing, processing and visualising vast amounts of data, enabling data-driven disruptive innovation within our work, business, life, industry and society. The rapidly increasing volumes of diverse data from distributed sources create significant technical challenges for extracting valuable knowledge. Many fundamental, technological and deployment challenges exist in the development and application of big data and data-driven AI to real-world problems. For example, what are the technical foundations of data management for data-driven AI? What are the key characteristics of efficient and effective data processing architectures for real-time data? How do we deal with trust and quality issues in data analysis and data-driven decision-making? What are the appropriate frameworks for data protection? What is the role of DevOps in delivering scalable solutions? How can big data and data-driven AI be used to power digital transformation in various industries?

For many businesses and governments in different parts of the world, the ability to effectively manage information and extract knowledge is now a critical competitive advantage. Many organisations are building their core business to collect and analyse information, to extract business knowledge and insight [3]. The impacts of big data value go beyond the commercial world to significant societal impact, from improving healthcare systems, the energy-efficient operation of cities and transportation infrastructure to increasing the transparency and efficiency of public administration.

The adoption of big data technology within industrial sectors facilitates organisations to gain competitive advantage. Driving adoption is a two-sided coin. On one side, organisations need to master the technology needed to extract value from big data. On the other side, they need to use the insights extracted to drive their digital transformation with new applications and processes that deliver real value. This book has been structured to help you understand both sides of this coin and bring together technologies and applications for Big Data Value.

The chapter is structured as follows: Section 2 defines the notion of Big Data Value. Section 3 explains the Big Data Value Public-Private Partnership (PPP) and Sect. 4 summarises the Big Data Value Association (BDVA). Sections 5, 6 and 7 structure the contributions of the book in terms of three key lenses: BDV Reference

Model (Sect. 5), Big Data and AI Pipeline (Sect. 6) and the AI, Data and Robotics Framework (Sect. 7). Finally, Sect. 8 provides a summary.

2 What Is Big Data Value?

The term "Big Data" has been used by different major players to label data with different attributes [6, 8]. Several definitions of big data have been proposed in the literature; see Table 1.

Big data brings together a set of data management challenges for working with data which exhibits characteristics related to the 3 Vs:

- *Volume (amount of data)*: dealing with large-scale data within data processing (e.g., Global Supply Chains, Global Financial Analysis, Large Hadron Collider).
- *Velocity (speed of data)*: dealing with streams of high-frequency incoming real-time data (e.g., Sensors, Pervasive Environments, Electronic Trading, Internet of Things).
- *Variety (range of data types/sources)*: dealing with data using differing syntactic formats (e.g., Spreadsheets, XML, DBMS), schemas, and meanings (e.g., Enterprise Data Integration).

The Vs of big data challenge the fundamentals of existing technical approaches and require new data processing forms to enable enhanced decision-making, insight discovery and process optimisation. As the big data field matured, other Vs have been added, such as Veracity (documenting quality and uncertainty) and Value [1, 16].

Table 1 Definitions of big data [4]

Big data definition	Source
"Big data is high volume, high velocity, and/or high variety information assets that require new forms of processing to enable enhanced decision making, insight discovery and process optimisation."	[10][13]
"When the size of the data itself becomes part of the problem and traditional techniques for working with data run out of steam."	[12]
Big data is "data whose size forces us to look beyond the tried-and-true methods that are prevalent at that time."	[9]
"Big data is a field that treats ways to analyse, systematically extract information from, or otherwise deal with data sets that are too large or complex to be dealt with by traditional data-processing application software."	[19]
"Big data is a term encompassing the use of techniques to capture, process, analyse and visualise potentially large datasets in a reasonable timeframe not accessible to standard IT technologies. By extension, the platform, tools and software used for this purpose are collectively called "big data technologies"	[15]
"Big data can mean big volume, big velocity, or big variety."	[17]

Table 2 Definitions of big data value [5]

Big data value definition	Source
"Top-performing organisations use analytics five times more than lower performers . . . a widespread belief that analytics offers value."	[11]
"The value of big data isn't the data. It's the narrative."	[7]
"Companies need a strategic plan for collecting and organising data, one that aligns with the business strategy of how they will use that data to create value."	[18]
"We define prescriptive, needle-moving actions and behaviors and start to tap into the fifth V from big data: value."	[1]
"Data value chain recognises the relationship between stages, from raw data to decision making, and how these stages are interdependent."	[14]

The definition of value within the context of big data also varies. A collection of definitions for Big Data Value is provided in Table 2. These definitions clearly show a pattern of common understanding that the Value dimension of big data rests upon successful decision-making and action through analytics [5].

3   The Big Data Value PPP

The European contractual Public-Private Partnership on Big Data Value (BDV PPP) commenced in 2015. It was operationalised with the Leadership in Enabling and Industrial Technologies (LEIT) work programme of Horizon 2020. The BDV PPP activities addressed the development of technology and applications, business model discovery, ecosystem validation, skills profiling, regulatory and IPR environments and many social aspects.

With an initial indicative budget from the European Union of 534 million euros for 2016–2020 and 201 million euros allocated in total by the end of 2018, since its launch, the BDV PPP has mobilised 1570 million euros of private investments (467 million euros for 2018). Forty-two projects were running at the beginning of 2019. The BDV PPP in just 2 years developed 132 innovations of exploitable value (106 delivered in 2018, 35% of which are significant innovations), including technologies, platforms, services, products, methods, systems, components and/or modules, frameworks/architectures, processes, tools/toolkits, spin-offs, datasets, ontologies, patents and knowledge. Ninety-three per cent of the innovations delivered in 2018 had an economic impact, and 48% had a societal impact. By 2020, the BDV PPP had projects covering a spectrum of data-driven innovations in sectors including advanced manufacturing, transport and logistics, health and bioeconomy. These projects have advanced the state of the art in key enabling technologies for big data value and non-technological aspects, such as providing solutions, platforms, tools, frameworks, and best practices for a data-driven economy and future European competitiveness in Data and AI [5].

4   Big Data Value Association

The Big Data Value Association (BDVA) is an industry-driven international non-for-profit organisation that grew over the years to more than 220 members across Europe, with a well-balanced composition of large, small and medium-sized industries as well as research and user organisations. BDVA has over 25 working groups organised in Task Forces and subgroups, tackling all the technical and non-technical challenges of big data value.

BDVA served as the private counterpart to the European Commission to implement the Big Data Value PPP program. BDVA and the Big Data Value PPP pursued a common shared vision of positioning Europe as the world leader in creating big data value. BDVA is also a private member of the EuroHPC Joint Undertaking and one of the leading promoters and driving forces of the European Partnership on AI, Data and Robotics in the framework programme MFF 2021–2027.

The mission of the BDVA is "to develop the Innovation Ecosystem that will enable the data-driven digital transformation in Europe delivering maximum economic and societal benefit, and, to achieve and to sustain Europe's leadership on Big Data Value creation and Artificial Intelligence." BDVA enables existing regional multi-partner cooperation to collaborate at the European-level by providing tools and know-how to support the co-creation, development and experimentation of pan-European data-driven applications and services and know-how exchange. The BDVA developed a joint Strategic Research and Innovation Agenda (SRIA) on Big Data Value [21]. It was initially fed by a collection of technical papers and roadmaps [2] and extended with a public consultation that included hundreds of additional stakeholders representing both the supply and the demand side. The BDV SRIA defined the overall goals, main technical and non-technical priorities, and a research and innovation roadmap for the BDV PPP. The SRIA set out the strategic importance of big data, described the Data Value Chain and the central role of Ecosystems, detailed a vision for big data value in Europe in 2020, analysed the associated strengths, weaknesses, opportunities and threats, and set out the objectives and goals to be accomplished by the BDV PPP within the European research and innovation landscape of Horizon 2020 and at national and regional levels.

5   Big Data Value Reference Model

The BDV Reference Model (see Fig. 1) has been developed by the BDVA, taking into account input from technical experts and stakeholders along the whole big data value chain and interactions with other related PPPs [21]. The BDV Reference Model may serve as a common reference framework to locate big data technologies on the overall IT stack. It addresses the main concerns and aspects to be considered

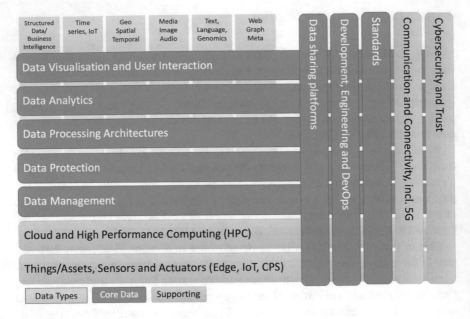

Fig. 1 Big data value reference model

for big data value systems. The model is used to illustrate big data technologies in this book by mapping them to the different topic areas.

The BDV Reference Model is structured into horizontal and vertical concerns.

- Horizontal concerns cover specific aspects along the data processing chain, starting with data collection and ingestion, and extending to data visualisation. It should be noted that the horizontal concerns do not imply a layered architecture. As an example, data visualisation may be applied directly to collected data (the data management aspect) without the need for data processing and analytics.
- Vertical concerns address cross-cutting issues, which may affect all the horizontal concerns. In addition, vertical concerns may also involve non-technical aspects.

The BDV Reference Model has provided input to the ISO SC 42 Reference Architecture, which now is reflected in the ISO 20547-3 Big Data Reference Architecture.

5.1 Chapter Analysis

Table 3 shows how the technical outcomes presented in the different chapters in this book cover the horizontal and vertical concerns of the BDV Reference Model.

Table 3 Coverage of the BDV reference model's core horizontal and vertical concerns by the book's chapters

Ch.	Horizontal concern					Vertical concern		
	Data mgmt.	Data protection	Data processing architectures	Data analytics	Data visualisation and UI	Data sharing	Engineering/DevOps	Standards
Part I. Technologies and methods								
2	+		+					
3								+
4	+	+	+	+		+	+	
5			+	+	+		+	
6		+				+	+	
7	+		+				+	
8	+		+			+	+	
Part II. Processes and applications								
9			+	+			+	
10	+			+				
11	+	+						+
12				+		+		
13		+	+	+	+		+	+
14		+	+	+	+		+	
15		+	+	+	+		+	+
16				+	+	+		
17	+		+				+	+
18	+					+		+
19				+				
20	+			+	+	+		
21			+	+				
22	+		+	+			+	
23	+		+	+				+
Totals	11	6	13	14	6	7	11	7

As this table indicates, the chapters in this book provide a broad coverage of the model's concerns, thereby reinforcing the relevance of these concerns that were spelt out as part of the BDV SRIA.

The majority of the chapters cover the horizontal concerns of data processing architectures and data analytics, followed by data management. This indicates the critical role of big data in delivering value from large-scale data analytics and the need for dedicated data processing architectures to cope with the volume, velocity and variety of data. It also shows that data management is an important basis for delivering value from data and thus is a significant concern.

Many of the chapters cover the vertical concern engineering and DevOps, indicating that sound engineering methodologies for building next-generation Big Data Value systems are relevant and increasingly available.

6 Big Data and AI Pipeline

A Big Data and AI Pipeline model (see Fig. 2) suitable for describing Big Data Applications is harmonised with the Big Data Application layer's steps in ISO 20547-3. This is being used to illustrate Big Data Applications in this book and a mapping to the different topic areas of the BDV Reference Model. Chapter 4 describes the Big Data and AI Pipeline in more detail and relates it to the Big Data Value Reference Model in Fig. 1 and the European AI, Data and Robotics Framework and Enablers in Fig. 3.

6.1   Chapter Analysis

Table 4 gives an overview to which extent the technical contributions described in the different chapters of this book are related to the four Big Data and AI Pipeline steps and in particular any of the six big data types.

The Big Data and AI Pipeline steps are the following:

- *P1:* Data Acquisition/Collection.
- *P2:* Data Storage/Preparation.
- *P3:* Analytics/AI/Machine Learning.
- *P4:* Action/Interaction, Visualisation and Access.

Part I on Technologies and Methods includes chapters that focus on the various technical areas mainly related to the pipeline steps P2 and P3, and mostly independent of the different big data types.

Part II on Processes and Applications includes chapters which typically covers the full big data pipeline, but some chapters also have a more specific focus. With respect to the big data types, a majority of the chapters are related to time series and

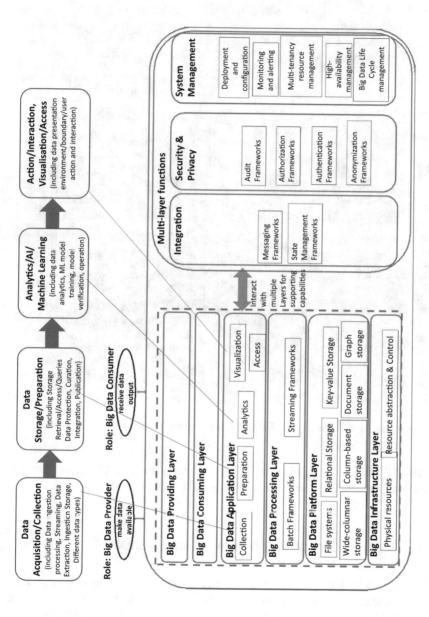

Fig. 2 Big data and AI pipeline on top, related to the ISO 20547-3 Big Data Reference Architecture

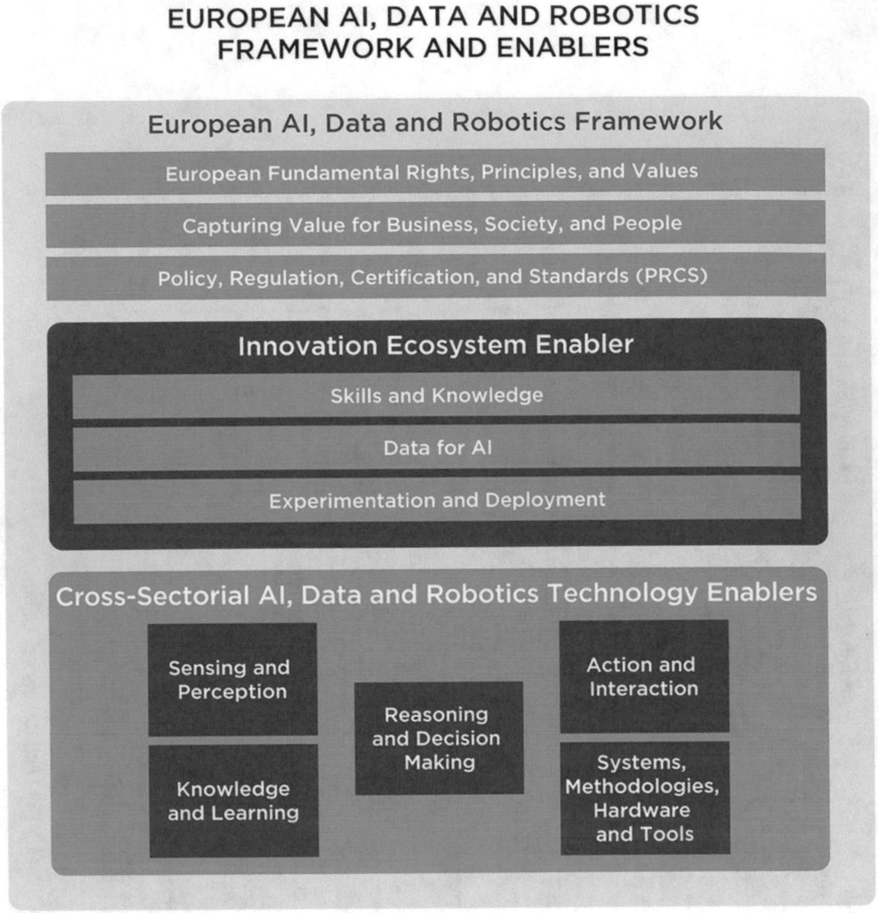

EUROPEAN AI, DATA AND ROBOTICS
FRAMEWORK AND ENABLERS

European AI, Data and Robotics Framework

European Fundamental Rights, Principles, and Values

Capturing Value for Business, Society, and People

Policy, Regulation, Certification, and Standards (PRCS)

Innovation Ecosystem Enabler

Skills and Knowledge

Data for AI

Experimentation and Deployment

Cross-Sectorial AI, Data and Robotics Technology Enablers

Sensing and Perception

Reasoning and Decision Making

Action and Interaction

Knowledge and Learning

Systems, Methodologies, Hardware and Tools

Fig. 3 European AI, data and robotics framework and enablers [20]

IoT data (12), followed by image data (8), spatiotemporal data (7), graph data (6) and also chapters with a focus on text and natural language processing (2).

7 AI, Data and Robotics Framework and Enablers

In September 2020, BDVA, CLAIRE, ELLIS, EurAI and euRobotics announced the official release of the Joint Strategic Research Innovation and Deployment Agenda (SRIDA) for the AI, Data and Robotics Partnership [21], which unifies the strategic focus of each of the three disciplines engaged in creating the Partnership.

Table 4 Coverage of AI pipeline steps and big data types as focused by the book's chapters

Ch.	Big data and AI pipeline steps				Big data types					
	P1	P2	P3	P4	Structured	IoT, Time series	Spatiotemporal	Image, audio	NLP, text	Graph, network
Part I. Technologies and methods										
2	+	+			+	+	+			+
3		+	+							
4	+	+	+	+	+	+	+	+	+	+
5	+	+	+	+		+				
6		+	+							
7		+	+							
8		+	+							
Part II. Processes and applications										
9	+	+	+	+				+		
10	+	+	+			+		+		
11	+	+	+	+			+	+	+	+
12		+	+							
13	+	+	+	+	+	+	+			
14	+	+	+	+		+		+		
15	+	+	+	+		+				
16	+	+	+	+		+				
17	+	+	+	+		+				
18	+	+	+	+		+				
19		+					+	+		
20	+	+								
21	+	+	+	+		+	+	+		+
22	+	+	+	+		+	+	+		+
23	+	+	+						+	+
Total	16	22	19	12	3	12	7	8	3	6

Together, these associations have proposed a vision for an AI, Data and Robotics Partnership: "The Vision of the Partnership is to boost European industrial competitiveness, societal wellbeing and environmental aspects to lead the world in developing and deploying value-driven trustworthy AI, Data and Robotics based on fundamental European rights, principles and values."

To deliver on the vision, it is vital to engage with a broad range of stakeholders. Each collaborative stakeholder brings a vital element to the functioning of the Partnership and injects critical capability into the ecosystem created around AI, Data and Robotics by the Partnership. The mobilisation of the European AI, Data and Robotics Ecosystem is one of the core goals of the Partnership. The Partnership needs to form part of a broader ecosystem of collaborations that cover all aspects of the technology application landscape in Europe. Many of these collaborations will rely on AI, Data and Robotics as critical enablers to their endeavours. Both horizontal (technology) and vertical (application) collaborations will intersect within an AI, Data and Robotics Ecosystem.

Figure 3 sets out the primary areas of importance for AI, Data and Robotics research, innovation and deployment into three overarching areas of interest. The *European AI, Data and Robotics Framework* represents the legal and societal fabric that underpins the impact of AI on stakeholders and users of the products and services that businesses will provide. The *AI, Data and Robotics Innovation Ecosystem Enablers* represent essential ingredients for practical innovation and deployment. Finally, the *Cross Sectorial AI, Data and Robotics Technology Enablers* represent the core technical competencies essential for developing AI, Data and Robotics systems.

7.1   Chapter Analysis

Table 5 gives an overview to which extent the technical contributions described in the different chapters of this book are in line with the three levels of enablers covered in the European AI, Data and Robotics Framework.

Table 5 demonstrates that the European AI, Data and Robotics partnership and framework will be enabled by a seamless continuation of the current technological basis that the BDV PPP has established.

All cross-sectorial AI, Data and Robotics technology enablers are supported through contributions in this book. However, we observe bias towards the topics "Knowledge and Learning," "Reasoning and Decision Making" and "System, Methodologies, Hardware and Tools." This is not surprising as the topics related to the management and analysis of heterogeneous data sources—independently whether they are stored in one or distributed places—is one of the core challenges in the context of extracting value out of data. In addition, tools and methods and processes that integrate AI, Data, HPC or Robotics into systems while ensuring that core system properties and characteristics such as safety, robustness, dependability and trustworthiness can be integrated into the design cycle, tested and validated for

Table 5 Coverage of the European AI, Data and Robotics Framework by the book's chapters

Ch.	AI, Data and Robotics Framework		Innovation Ecosystem Enabler				Cross-Sectorial AI, Data and Robotics Technology Enabler					
	Rights, Principles, and Values	Capturing Value	Policy, Regulation, Certification, and Standards	Skills and Knowledge	Data for AI	Experimentation and deployment	Data Privacy	Sensing and Perception	Knowledge and Learning	Reasoning and Decision-Making	Action and Interaction	System, Methodologies, Hardware and Tools
Part I. Technologies and methods												
2												
3									+			
4					+			+				+
5									+	+	+	+
6					+							
7							+			+		
8												+
Part II. Processes and applications												
9		+				+						+
10		+								+		
11			+				+					
12		+						+	+	+		+
13		+					+		+	+		
14					+				+			+
15		+							+	+		+
16		+							+	+		
17						+						+
18		+			+							
19									+			
20		+	+						+			+
21		+						+	+			
22									+	+	+	+
23								+	+	+	+	
Totals	9	2		4	2	3	4	12	9	3	10	

use, have been in the past and will be in the future core requirements and challenges when implementing and deploying data-driven industrial applications. Finally, many of the chapters describe how data-driven solutions bring value to particular vertical sectors and applications.

8 Summary

The continuous and significant growth of data, together with improved access to data and the availability of powerful computing infrastructure, has led to intensified activities around Big Data Value and data-driven Artificial Intelligence (AI).

The adoption of big data technology within industrial sectors facilitates organisations to gain a competitive advantage. Driving adoption requires organisations to master the technology needed to extract value from big data and use the insights extracted to drive their digital transformation with new applications and processes that deliver real value. This book is your guide to help you understand, design and build technologies and applications that deliver Big Data Value.

References

1. Biehn, N. (2013). *The missing V's in big data: Viability and value.*
2. Cavanillas, J. M., Curry, E., & Wahlster, W. (Eds.). (2016). *New horizons for a data-driven economy: A roadmap for usage and exploitation of big data in Europenull.* https://doi.org/10.1007/978-3-319-21569-3
3. Cavanillas, J. M., Curry, E., & Wahlster, W. (2016). The big data value opportunity. In J. M. Cavanillas, E. Curry, & W. Wahlster (Eds.), *New horizons for a data-driven economy* (pp. 3–11). https://doi.org/10.1007/978-3-319-21569-3_1
4. Curry, E. (2016). The big data value chain: Definitions, concepts, and theoretical approaches. In J. M. Cavanillas, E. Curry, & W. Wahlster (Eds.), *New horizons for a data-driven economy* (pp. 29–37). https://doi.org/10.1007/978-3-319-21569-3_3.
5. Curry, E., Metzger, A., Zillner, S., Pazzaglia, J.-C., Robles, G., & A. (Eds.). (2021). *The elements of big data value: Foundations of the research and innovation ecosystem.* Springer International Publishing.
6. Davenport, T. H., Barth, P., & Bean, R. (2012). How 'big data' is different. *MIT Sloan Management Review, 54,* 22–24.
7. Hammond, K. J. (2013). The value of big data isn't the data. *Harvard Business Review.* Retrieved from https://hbr.org/2013/05/the-value-of-big-data-isnt-the
8. Hey, T., Tansley, S., & Tolle, K. M. (Eds.). (2009). *The fourth paradigm: Data-intensive scientific discovery.* Microsoft Research.
9. Jacobs, A. (2009). The pathologies of big data. *Communications of the ACM, 52*(8), 36. https://doi.org/10.1145/1536616.1536632
10. Laney, D. (2001). 3D data management: Controlling data volume, velocity, and variety. *Application Delivery Strategies, 2.*
11. Lavalle, S., Lesser, E., Shockley, R., Hopkins, M. S., & Kruschwitz, N. (2011). Big data, analytics and the path from insights to value. *MIT Sloan Management Review, 52*(2), 21–32.

12. Loukides, M. (2010, June). What is data science? *O'Reily Radar*. Retrieved from http://radar.oreilly.com/2010/06/what-is-data-science.html
13. Manyika, J., Chui, M., Brown, B., Bughin, J., Dobbs, R., Roxburgh, C., & Byers, A. H. (2011). *Big data: The next frontier for innovation, competition, and productivity*. Retrieved from McKinsey Global Institute website: https://www.mckinsey.com/business-functions/mckinsey-digital/our-insights/big-data-the-next-frontier-for-innovation
14. Miller, H. G., & Mork, P. (2013). *From data to decisions: A value chain for big data*. IT Professional. https://doi.org/10.1109/MITP.2013.11.
15. NESSI. (2012, December). Big data: A new world of opportunities. *NESSI White Paper*.
16. Rayport, J. F., & Sviokla, J. J. (1995). Exploiting the virtual value chain. *Harvard Business Review, 73*, 75–85. https://doi.org/10.1016/S0267-3649(00)88914-1
17. Stonebraker, M. (2012). What does "big data mean." *Communications of the ACM*, BLOG@ACM.
18. Wegener, R., & Velu, S. (2013). The value of big data: How analytics differentiates winners. *Bain Brief*.
19. Wikipedia. (2020). *Big data*. Wikipedia article. Retrieved from http://en.wikipedia.org/wiki/Big_data
20. Zillner, S., Bisset, D., Milano, M., Curry, E., Hahn, T., Lafrenz, R., O'Sullivan, B., et al. (2020). *Strategic research, innovation and deployment agenda – AI, data and robotics partnership. Third release* (third). Brussels: BDVA, euRobotics, ELLIS, EurAI and CLAIRE.
21. Zillner, S., Curry, E., Metzger, A., Auer, S., & Seidl, R. (Eds.). (2017). *European big data value strategic research & innovation agenda*. Retrieved from Big Data Value Association website: https://www.bdva.eu/sites/default/files/BDVA_SRIA_v4_Ed1.1.pdf

Part I
Technologies and Methods

Supporting Semantic Data Enrichment at Scale

Michele Ciavotta, Vincenzo Cutrona, Flavio De Paoli, Nikolay Nikolov, Matteo Palmonari, and Dumitru Roman

Abstract Data enrichment is a critical task in the data preparation process in which a dataset is extended with additional information from various sources to perform analyses or add meaningful context. Facilitating the enrichment process design for data workers and supporting its execution on large datasets are only supported to a limited extent by existing solutions. Harnessing semantics at scale can be a crucial factor in effectively addressing this challenge. This chapter presents a comprehensive approach covering both design- and run-time aspects of tabular data enrichment and discusses our experience in making this process scalable. We illustrate how data enrichment steps of a Big Data pipeline can be implemented via tabular transformations exploiting semantic table annotation methods and discuss techniques devised to support the enactment of the resulting process on large tabular datasets. Furthermore, we present results from experimental evaluations in which we tested the scalability and run-time efficiency of the proposed cloud-based approach, enriching massive datasets with promising performance.

Keywords Big data processing · Data integration · Data enrichment · Data extension · Linked data · Scalability

1 Introduction

Big Data and Business Analytics are among the main value-creating assets for private companies and public institutions—estimates indicate yearly earnings in the order of 274 billion dollars by 2022 [1]. This is made possible by theoretical

M. Ciavotta (✉) · V. Cutrona · F. De Paoli · M. Palmonari
Department of Informatics, Systems and Communication, University of Milan–Bicocca, Milan, Italy
e-mail: michele.ciavotta@unimib.it; vincenzo.cutrona@unimib.it; flavio.paoli@unimib.it; matteo.palmonari@unimib.it

N. Nikolov · D. Roman
Department of Software and Service Innovation, SINTEF AS, Oslo, Norway
e-mail: nikolay.nikolov@sintef.no; dumitru.roman@sintef.no

© The Author(s) 2022
E. Curry et al. (eds.), *Technologies and Applications for Big Data Value*,
https://doi.org/10.1007/978-3-030-78307-5_2

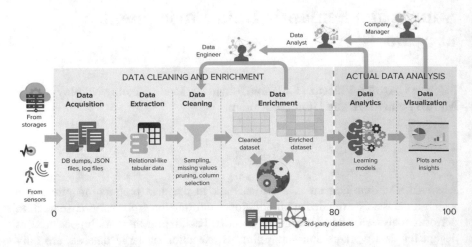

Fig. 1 Infographic representing the main stages of a data project and the related stakeholders

and practical advancements for processing massive amounts of data and developing highly accurate and effective decision-making processes via analytical models. However, very different time frame and effort are required to commission each phase of a data-driven project, which includes, as its primary stages, data acquisition, extraction, cleaning, integration/enrichment, and data analysis and results visualization [2]. Remarkably, the data preparation stage (which encompasses data transformations that also cover cleaning and integration/enrichment) takes up to 80% of the time required by a project. Only the remaining 20% of the time is spent on data analysis and exploration [3]. Such an imbalance (see Fig. 1) poses a problem that gets increasingly more severe with the progressive growth of volume and variability of the involved data [4]. This issue is now widely recognized and needs appropriate tools and methodologies, especially to support the crucial step of data enrichment.

Data enrichment is a specific data integration problem where a dataset that the user (typically a data engineer/scientist) knows is extended with additional information coming from external, possibly unknown, sources. Intuitively, enrichment requires *reconciliation* between values in the main and the external sources to fetch related data from the latter and extend the former. Data enrichment is often pivotal in analytics projects where the model might benefit from features that are not present in the main dataset, e.g., weather-based analysis of digital marketing campaign performance [5]. In recent years, a number of proposals have been presented, both academic and business related, to help data workers in the data preparation phase and, more specifically, in data enrichment tasks; many proposals involve the adoption of semantic techniques.

Semantics play an increasingly important role in Big Data, and, more specifically, in Big Data enrichment, as also acknowledged by the European Big Data Value Strategic Research and Innovation Agenda [2], and dedicated special issues in scientific journals [6–8]. A semantic paradigm that has gained popularity is based on Knowledge Graphs (KGs), which provide graph structures where entities are interconnected and classified. Semantic web technologies support the publication of KGs with standards and shared vocabularies that facilitate access and manipulation of knowledge via web protocols. Several approaches have been proposed to integrate data in different Knowledge Graphs, e.g., using entity reconciliation techniques [6]. However, in most data analytics problems, the user starts with some source legacy dataset that is not structured as a KG. Therefore, approaches have been proposed to transform legacy datasets by giving them a graph structure enriched with shared vocabularies, and, possibly, with background knowledge already available in a graph structure [9]. Such a transformation process is a complex task that can be (partly) sustained by semantic table interpretation [10] and annotation approaches. These approaches aim at mapping an initial relational table to the schema of a reference KG and finally linking values in the source table to entities in the KG [11, 12]. In this case, the focus of the interpretation algorithm is to automatically provide an annotation that enables the fusion of the source table with information in the target KG (e.g., a large and cross-domain information source like Wikidata), aiming at delivering an enriched KG. DAGOBAH [13] is an example of such an algorithm.

In this work, we leverage and take existing work on semantic table interpretation a step forward. We argue that semantic table annotation can provide a valuable paradigm to support data enrichment, modularly and at scale, in a much wider number of scenarios, including when the final objective is to enrich datasets, and not to their transformation into Knowledge Graphs. With *modularly*, we mean that the paradigm can be implemented by an ecosystem of services that provide access to different Knowledge Graphs to support automatic entity linking and data extensions. Automation is a key factor for managing large volumes of data and reaching the *at scale* dimension under certain assumptions that we discuss in this chapter.

We propose a comprehensive approach and a scalable solution to provide data workers with suitable tools to (1) interactively design transformation pipelines on datasets in tabular format, including semantic enrichment using curated knowledge bases (general purpose or domain specific), and (2) deploy and run such pipelines against massive datasets taking full advantage of the potential of scalability offered by modern Cloud services. Most of the related work in this field tackles the problem of automatically inferring the annotations that encode the semantics of a table. However, the primary contribution of this work consists of addressing the issue of implementing reconciliation and extension mechanisms to support both interactive data enrichment on small-size tabular datasets and automatic execution on massive workloads. To this end, we devised a two-phase approach and a service-oriented architecture to support it, whose engine consists of a collection of reconciliation and extension microservices implementing an open interface that can easily be scaled up to manage larger datasets.

To demonstrate the suitability of the proposed approach, we created both general-purpose services for linked data and specialized ones (for instance, for geographical toponyms, weather, and events), which support industry-driven analytic projects that motivated our work and guided a rigorous activity of requirements elicitation [14]. We used these services and business datasets to evaluate the efficiency of the proposed methods, achieving promising results (namely, linear scalability and a performance boost ranging from 4× to 770× over a baseline). Finally, we discuss the current limitations, which point to open issues in making semantic table enrichment approaches applicable at the Big Data scale. In this context, this chapter contributes to a better understanding of the role and challenges of semantics in supporting data enrichment, provides an approach and the corresponding implementation for semantic enrichment of tabular data at scale (thus, contributing to the *Data Management* and *Data Processing Architectures* horizontal concerns of the BDV Technical Reference Model [2] and to the *Knowledge and Learning* cross-sectorial technology enablers of the AI, Data and Robotics Strategic Research, Innovation and Deployment Agenda [15]), reports on the lessons learned in developing the solution, and presents the open problems for future research in the field of scalable semantic enrichment.

Ultimately, the proposed solution aims at filling an existing gap between technologies available today to support data enrichment at scale. This is a process where a natively semantic task like entity reconciliation plays a crucial role and semantics (especially KGs) are a facilitator of the enrichment process. Indeed, some tools provide users with user-friendly functionalities for data preparation, but few offer semantic support. The few solutions that offer such support (e.g., OpenRefine)[1] essentially cover the needs of the exploratory phases of a project by supporting manual transformation and enrichment of datasets. However, they neglect the life-cycle management needed to implement and run production-ready data pipelines and ensure scalability for large volumes of data. On the other hand, tools that provide support for running pipelines on large volumes of data are designed for users who are familiar with programming and process definition and are, therefore, unsuitable for use by data scientists [16]. Furthermore, these solutions, while often offering a wide variety of configurable components to create data pipelines, are poorly designed to incorporate user-specific knowledge, which is often essential to perform data reconciliation tasks effectively.

The rest of the chapter is structured as follows. Section 2 provides a discussion of the main design principles that have driven the definition of the architecture. The components of the platform and the workflow are discussed in Sect. 3. Section 4 illustrates experiments with datasets of different sizes. Finally, a review of the state of the art is reported in Sect. 5, and Sect. 6 concludes the chapter.

2 A Two-Phase Approach

Before discussing the proposed solution for data manipulation and enrichment at scale, we introduce a real-life analytics use case consisting of different data manipulation and enrichment tasks to motivate the principles that guided the formulation of the approach.

---

[1] http://openrefine.org

2.1   Scenario: Weather-Based Digital Marketing Analytics

The JOT Internet Media (JOT)[2] company is analyzing the performance of its digital marketing campaigns using reports from Google AdWords (GAW), and needs to aggregate data on performance (e.g., impressions and clicks) by city, region, or country. Furthermore, JOT, seeking to boost the effectiveness of future advertising campaigns, is interested in investigating the effect of weather on the performance of its campaigns at a regional level, and in training a machine learning model able to predict the most suitable moment to launch a campaign. To train the model and run the analytics, JOT aims to use 3 years of historical data concerning the performance of keywords used in previous campaigns.

The first step might be to enrich the GAW report (the white columns in Table 1) directly with weather data. In this scenario, JOT would access the European Centre for Medium-Range Weather Forecasts (ECMWF)[3] service that provides current forecasts, queryable using geographic bounding boxes and ISO 8601 formatted dates. Since both properties are missing in the original dataset, JOT has to first add them to the dataset. The ISO-formatted date can be easily obtained by applying a data transformation function to the date column (and adding the rightmost column in Table 1). Thus, the next step will be to extend the dataset with GeoNames (GN) identifiers for all mentioned locations. This operation requires to match the region labels adopted by GAW (Google GeoTargets labels) with GN identifiers, which are used in turn to geolocate the regions. Henceforth, this process is referred to as *data reconciliation* and represents a fundamental stage in the enrichment pipeline. Once the working dataset locations have been reconciled against GN (adding the fifth column in Table 1), it is possible to perform the *extension* step where the ECMWF is queried to collect the desired weather-related information (and add the sixth and seventh columns in Table 1). The reconciled and extended data are now suitable for performing the desired analysis.

Table 1 JOT dataset enriched with data from GN and ECMWF

Keyword ID	Clicks	City	Region	Region ID (GN)	Temp. (WS)	Prec. (WS)	Date	Date (ISO)
194906	64	Altenburg	Thuringia	2822542	287.70	0.08	06/09/2017	2017-09-06
517827	50	Ingolstadt	Bavaria	2951839	288.18	0.02	06/09/2017	2017-09-06
459143	42	Berlin	Berlin	2950157	290.48	0.00	06/09/2017	2017-09-06
891139	36	Munich	Bavaria	2951839	288.18	0.02	06/09/2017	2017-09-06
459143	30	Nuremberg	Bavaria	2951839	288.18	0.02	06/09/2017	2017-09-06

Colored columns are appended by different functions: transformation (Date), reconciliation (Region ID), and extension (Temp., Prec.)

---

[2] https://www.jot-im.com

[3] https://www.ecmwf.int

Group	Keyword	City	Region	Clicks	Category	Date
36874389	194906	Altenburg	Thuringia	64	HomeGarden	11/03/2018
36874385	459143	Berlin	Berlin	42	TravelTourism	12/03/2018

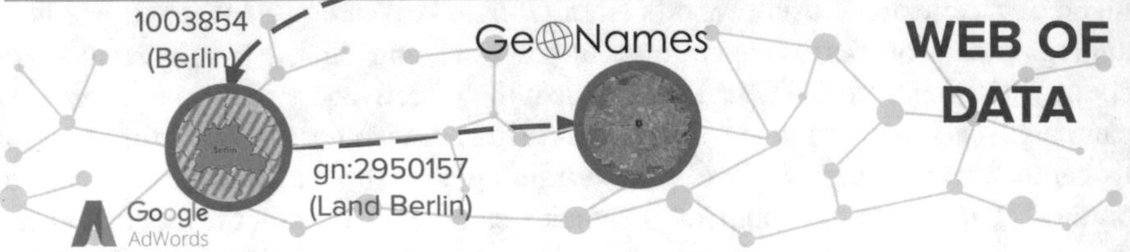

Fig. 2 Semantics for data enrichment

2.2 *Semantics as the Enrichment Enabler*

As demonstrated by the above scenario, data enrichment plays a critical role in the preparation phase of many analytics pipelines, since it can add contextual information to the original dataset to build more effective models.

In the integration of relational datasets, traditional approaches (e.g., record linkage) have proven to be appropriate when the entity values of the involved schemas are compatible (e.g., they feature the same date format). In a more general scenario, however, the user is interested in enriching a dataset (known to the user) by fetching additional information from external datasets (possibly unknown to the user) that only relate semantically with the working dataset. This means that, in general, the terms of the schema and the values of the entities belong to different (possibly implicit) vocabularies or Knowledge Bases (KBs). The role of semantic approaches in such a process is to lift the latent semantics of records and metadata to support the integration of otherwise incompatible data sources.

In our use case, a semantic approach allows JOT to link the company dataset to the ECMWF data source by using a system of identifiers provided by a reference KB (i.e., GN). In this specific example, the reconciliation is performed directly against GN (see Fig. 2). In a more general case, it can happen that the data sources involved refer to different KBs; therefore, the KBs exploited for reconciliation need to be interlinked (e.g., using the *sameAs* predicate) to enable integration.

2.3 *Challenges*

Three main challenges emerge from the above scenario; indeed, the JOT data scientists have to: (1) investigate how to reconcile locations to GN, i.e., they need to look for a service that meets this requirement (suitable for users familiar with programming languages), or to check out the Knowledge Base (KB) that describes GN and build an ad hoc reconciliation service (suitable for users experienced in semantics and the geospatial domain); (2) query a ECMWF endpoint, i.e., they must look for the API documentation (usually geared towards users familiar with programming languages, less applicable to data scientists and domain experts); (3) come up with a scalable architectural solution able to manage and efficiently enrich the whole dataset, meeting possible time constraints. Specifically, since the enrichment process would unavoidably lead to querying external services, efficiency constraints require an effective solution to network latency issues that represent a bottleneck when a large number of API requests have to be issued.

In summary, the key features required to support the design and execution of enrichment-based data transformation at scale can be summarized as follows:

- Column values reconciliation against a reference KB, e.g., matching the spatial references adopted in the source dataset against the system of spatial identifiers adopted by the weather service.
- Data extension based on the reconciliation results, which represents the bridge between the dataset at hand and a reference KB. The extension could add one or more columns to the original dataset.
- An approach that supports the development of a user-friendly environment to design the reconciliation/extension process, and a scalable platform to execute it on massive input datasets.

2.4   Approach Overview

The approach we propose in this work is mainly based on a small-scale design/full-scale execution principle, harnessing semantics to support the reconciliation tasks in data enrichment. A high-level description of the approach is sketched in Fig. 3. The driving principle is to separate the transformation process into two phases: the *design phase*, where the user defines the transformation pipeline by working on a sample and produces a *transformation model* (i.e., an executable representation of the transformation pipeline), and the *processing phase*, where the model is executed against the original dataset to obtain an enriched version of it to feed the analytical activities. Both phases rely on external data sources (e.g., GN and ECMWF) to support reconciliation and extension activities.

A fully automated approach is unsuitable from a user perspective since it would entirely remove the operator control over the process and results. In processes where the matching phase is performed based on semantics, the contribution of the knowledge and experience of a domain expert can impact the final dataset quality. Therefore, complete automation would entail a substantial risk of generating low-quality results and, consequently, be of little use. For this reason, the approach gives full control over the definition of the transformations to the user while automating the enactment of the resulting process. The role of the human in the design phase is to define the pipeline steps over a smaller dataset so that she can control the resulting quality (e.g., they can avoid misinterpretations). In this scenario, the approach

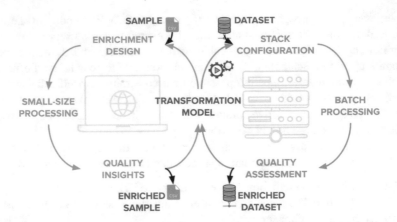

Fig. 3 Summary of design/processing approach

envisions computer-aided support tools offered as a service (the reader is referred to Sect. 3 for more details on the proposed solution) to guide the user in the pipeline composition, facilitating the process and reducing the need for strong programming skills. Once the transformation process has been defined, a full-scale *processing phase*, where the choices of the operator are packed within an executable artifact and run in batch mode over a different (possibly larger) dataset (Fig. 3), takes place.

As for the design phase, the approach proposes a reference implementation for the pipeline execution environment (see Sect. 3 for more details). Notice that featuring independent design and execution improves the efficiency of data management while mitigating confidentiality issues. Since the design phase exploits only a sample, the limited quantity of data to be transferred and manipulated simplifies the support system from an architectural and operational point of view (in practice, data would be handled by browsers on regular hardware) and reduces the risk of leakage of confidential information that might arise whenever full datasets are exposed. The full-scale execution phase can be performed on-premise, thus exploiting corporate infrastructure and tools, without the need for moving the data. Further details on the two phases are summarized as follows.

Design Phase In the design phase, the operator designing the pipeline performs three iterative steps on the working dataset: (1) the *enrichment design*, where the user designs each transformation step for enriching the working table employing a graphical interface that facilitates and automates interactions with reconciliation and extension services; (2) the pipeline execution (*small-size processing*), where each step of the enrichment process is performed over the current dataset; and (3) attaining *quality insights*, i.e., a handful of statistics to enable a general understanding of the overall quality of the result (e.g., the number of missing values). This interactive process is executed every time the user edits the pipeline definition (e.g., adding a new step in the pipeline). The outputs of the design phase are (1) the enriched dataset and (2) an executable transformation model, packaged in an executable that encompasses the steps of the pipeline. It is worth noting that if the user needs to enrich tables with a few thousand rows sporadically, the enriched table produced in the design phase concludes the process. If the user is required to enrich large volumes of data (or at least too large to be interactively managed), our approach assumes that the user carries out the design phase using a representative sample of the original dataset. In that case, the executable with the transformation model can be downloaded and used as the primary step in the processing phase (referred to as Data Flow), presented and discussed in the next section.

Processing Phase This phase aims to execute the transformation pipeline, which has been designed and tested on a smaller (loadable in memory) sample during the previous phase, on a large dataset. As in the previous phase, three steps are implied: (1) data flow definition (*stack configuration*) to support the execution of the enrichment pipeline; (2) *batch execution* (possibly in parallel) of the pipeline; and, finally, (3) *quality assessment* to evaluate the resulting dataset. If the result does not achieve an acceptable quality level (e.g., the number of reconciliation mismatches is above a given threshold), the user could go back to the design phase and modify the pipeline on an updated sample dataset. The new sample could be populated according to an analysis of the log files (e.g., adding a set of rows with values that could not be matched). The goal is to converge after a few iterations and be able to manage the dataset evolution. The *stack configuration* phase defines the pre- and post-processing actions to execute the enrichment pipeline on the dataset. It is composed of standard steps that can be customized according to the project requirements. To serve as an example, the reference data flow that supports the JOT scenario features the following steps: (1) decompress the input dataset and store it in a distributed file system; (2) split data in chunks for parallel processing; (3) execute the pipeline (which includes invoking enrichment services); (4) export the enriched data. The choice of relying on external services implementing the reconciliation and extension functionalities is supported by most available platforms and derives from precise design requirements (not least the requisites of modularity, extensibility, and flexibility). Consequently, the need to perform service invocations to get access to data for enrichment constitutes a fundamental scalability limitation of the entire enrichment process, which is much better performing and predictable for transformations that can be encapsulated within the executable transformation model. In the next section, the issue is discussed in detail.

Finally, on the one hand, these phases have different time requirements—i.e., in the design phase, the system has to be responsive in real time so that the user can provide feedback interactively, while the processing phase can last several hours. On the other hand, the phases deal with datasets of different sizes—the design phase processes only a sample of a dataset, while the processing phase must handle the full data. Therefore, we built an architecture where two main logical components share a set of back-end services to manage these phases. In the following, we provide details about the implementation of those components.

3   Achieving Semantic Enrichment of Tabular Data at Scale

This section outlines the different measures we took to ensure an adequate level of scalability for the enrichment process. The service architecture supporting the described approach is presented in Sect. 3.1. In Sect. 3.2 we discuss the design decisions, strategies, and lessons learned while designing for scalability in this domain, and examine the limitations of the proposed solution as well as possible improvements.

3.1   The Architectural View

The architecture underpinning the execution of our open-source solution, named ASIA (Assisted Semantic Interpretation and Annotation of tabular data) [17] (see Fig. 4), has been designed for modularity and loose coupling resulting in the components specified below:

ASIA User Interface The ASIA front-end is a single-page web application meant to interact with the final user exposing all the services required to support the enrichment. This application is fully integrated within Grafterizer [16] (part of DataGraft [18]), a tool that provides the pipeline abstraction, support for data cleaning and ETL data transformations, a tabular-to-Linked-data generator, and a compiler to produce portable and repeatable data manipulation pipelines. ASIA inherits and extends those features by providing functionalities to streamline the

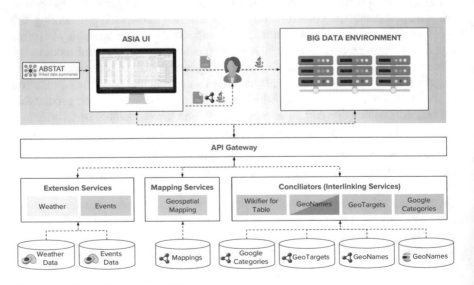

Fig. 4 Detailed architecture of ASIA

mapping between the columns of the working dataset and semantic types and properties (i.e., the schema annotation). This is done by analyzing the table headers and matching them with schema patterns and statistics provided by ABSTAT (Linked Data Summarization with ABstraction and STATistics) [19], an ontology-driven linked data profiling service. Moreover, ASIA UI provides widgets for semantic matching of column values against a shared system of identifiers.

ASIA Back-end This ecosystem consists of an orchestrator (API gateway) and a set of services[4] that are grouped in three categories: Conciliators, Mapping Services, and Extension Services.

- *API Gateway.* This service provides a unified view of the ASIA back-end ecosystem services by isolating the architectural details and relationships of these modules in the background. Moreover, it provides high-level functionalities by orchestrating the execution of the underlying services.
- *Conciliators.* Services for reconciling entity labels to a specific KB. They provide a REST interface compliant with the OpenRefine Reconciliation and Extension APIs.[5] In Fig. 4, conciliator blocks represent reconciliation services, while the GeoNames block also supports *KB-based extensions*, i.e., the possibility to extend the table with information from the reference KB.
- *Mapping Services.* Services in charge of linking KBs to each other, enabling the user to translate seamlessly between different Shared System of Identifiers (SSIs) by identifying URIs that provide suitable inputs to the extension services. The current implementation provides links between GeoTargets (location identifiers used in Google Analytics services) and GeoNames, plus additional *sameAs* links retrieved from DBpedia and Wikidata.
- *Extension Services.* Services for extending the input dataset with information coming from external data sources by using the URIs returned by conciliators and mapping services as inputs. In the JOT use case, we address weather and event datasets. The current implementation relies on data regularly being downloaded and curated (fetched from the ECMWF and EventRegistry[6] services, respectively). This is done to overcome network latency issues.

It is important to note that the ASIA back-end provides functionality to both the front-end (for the definition of the enrichment steps) and the Big Data Environment (presented below). The substantial difference is that in the first case its deployment is designed to provide multi-tenant support in Software-as-a-Service mode,[7] while in the second case, it is deployed in a dedicated way (and often replicating some services) to ensure scalability and efficiency.

---

[4] https://github.com/UNIMIBInside/asia-backend

[5] github.com/OpenRefine/OpenRefine/wiki/Documentation-For-Developers

[6] eventregistry.org

[7] A DataGraft deployment that includes ASIA module is available online at https://datagraft.io

The Big Data Environment This macro-component is mainly responsible for the orchestration and execution of enrichment operations at scale. It provides a high-level interface to configure, process, and monitor data flows. In particular, to handle the scale of data, this component distributes the different replicas of the pipeline steps over a cluster of computational resources (physical machines). For this reason, the operator establishes an appropriate deployment in terms of resources and service replicas to support the parallel execution of each step and leaves the system with the burden of configuring, deploying, and running the flow. More details about this component can be found in the following section.

3.2 Achieving Scalability

The purpose of this section is to present the techniques and strategies employed to achieve a system capable of providing scalable enrichment functionalities.

Stateless, Shared-Nothing Processing The ecosystem of the ASIA back-end is made up of various services and databases capable of serving a number of concurrent invocations. In essence, ASIA services receive a label (in the case of reconciliation) or a URI (in the case of extension) and return one or more corresponding values. They are built to be stateless and thus enable the creation of a platform in which the enrichment pipeline is executed in parallel on non-overlapping segments of the working table (shared-nothing approach [20]).

Distribution and Parallelization The Big Data Environment (Fig. 5) is the component in charge of fostering parallelism and is implemented as a private cloud consisting of a cluster of bare-metal servers running the Docker engine,[8] connected via Gigabit Ethernet and sharing a distributed file system (i.e., GlusterFS).[9] In this environment, data flows are compiled into a chain of Docker containers that are, in turn, deployed and managed by a container Orchestration system (i.e., Rancher).[10] Each of the steps consists of containers working independently and in parallel and scalable on-demand. The communication between two consecutive steps of the chain, i.e., the handover of the partial results, occurs through writing and reading from the file system. For details on the approach for setting up the Big Data Environment, see [21]. The implementation of this container-based solution has several benefits: it makes the data flow deployment independent from the particular stakeholder's hardware infrastructure, also working in heterogeneous distributed environments; it guarantees a flexible deployment, better resource utilization, and seamless horizontal scalability. The GlusterFS distributed file system is fast (as it

---

[8] https://www.docker.com
[9] https://www.gluster.org
[10] https://rancher.com

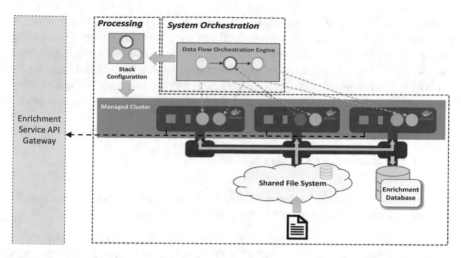

Fig. 5 An overview of the proposed Big Data Environment

lacks a central repository for metadata), linearly scalable, and, therefore, able to support massive amounts of data.

Data and Service Locality One of the primary issues to be addressed for achieving scalability for the enrichment process is the use of remote services. The use of services accessible over the Internet is certainly incompatible with datasets featuring more than a few thousand rows due to the network latency and the high number of invocations. The use of remote, multi-tenant services is generally acceptable in the design phase due to the limited size of managed datasets. However, when large datasets need to be processed, it is imperative to address the issue by making the life-cycle of enrichment data local. In the scenario we adopt, the weather information is downloaded daily from the provider and is treated to enable access using geospatial SSIs. This solution is suitable for datasets that change at a known rate and are thus stable. In the general case, refresh frequency depends on the application domain and the nature of data. The local management of these KBs has the advantage of rightsizing the resources allocated against the incoming workload; moreover, the control over the local network enables reduced and stable round-trip delay times (RDT). Similar considerations have led to deploying the reconciliation services of the ASIA back-end as close as possible (in network terms) to both the reference KBs and the agents (containers) performing the reconciliation pipeline steps.

Scaling up the Enrichment Services In order to manage the workload caused by the simultaneous invocation of reconciliation and extension functions (by the step executing the reconciliation pipeline), the ASIA ecosystem has been implemented to be easily replicable to achieve horizontal scalability of performance. This is achievable as the knowledge bases are used in read-only mode; accordingly, they can be duplicated without consistency issues. A load balancer is used for dispatching requests across the various replicas of ASIA.

Hierarchical Caching Lastly, it should be noted that the same request for reconciliation or extension can be made multiple times by the agents because columns in the processed data can contain repeating values. This, if not mitigated, generates a high number of identical requests. To address this, we implemented a hierarchical caching system in which each agent directly manages the first level of the hierarchy, while the other levels are managed by the stack of ASIA services and databases.

3.3 Discussion on the Limitations

In this section, we discuss the current limitations and aspects that could be improved to bring significant enhancements to the performance of the entire process.

Data Locality In this initial implementation of the Big Data Environment, the data locality principles, intended as one of the chief scalability enablers, are only partially implemented and exploited. Data locality is limited to the life-cycle management of the knowledge bases used by the enrichment services, which are physically brought to the Big Data Enrichment platform to reduce service access times. At the same time, the agents that perform the transformation pipeline may be physically separated from the working dataset they are processing (due to the use of a distributed file system). By deploying the enrichment services in the Big Data Enrichment platform, the speed of the functionalities relying on enrichment services increases dramatically. Similarly, the working dataset is stored in a distributed file system and accessible through the local network. This architectural choice, which enables uniform access times to data, has the disadvantage of raising the average read/write times of a quantity equal to twice the network latency (each agent reads a data chunk and writes a larger one). Nonetheless, by moving the data as close as possible to the agents that have to process it, we can improve the reading and writing performances and affect the whole process positively. This can be done by onboarding partitions of the working dataset on the machines that execute the containers of the agents instead of using the network to transmit partitions.

Distributed Caching The hierarchical caching system that was implemented can be further optimized, mainly because each ASIA replicated deployment has its local memory. Moreover, due to the presence of a load balancer running a round-robin dispatching policy (thus caching unaware), identical requests can be assigned to different replicas of ASIA causing preventable cache misses. The cache used at the agent level is also private, which results in generating much more requests than are strictly necessary. An improved solution to the problem of duplicated requests to the enrichment services can be done through the use of a distributed cache shared among the various instances of ASIA and among the agents that carry out the pipeline in parallel. Such a service (e.g., Ehcache [22]), once deployed on the machines that configure the cluster of the big data environment, would guarantee rapid synchronization of the local caches and would reduce the number of cache misses.

Efficient API Interaction The enrichment service API is another component that may be optimized to provide significant improvement to execution times. In the current implementation, for both the design and processing phases, reconciliation and extension are invoked for each row of the working table. This means that for each line the agent running the pipeline must wait a time equal to the RTD, forcing the system to wait a time roughly equal to twice the network latency at every invocation. A considerable improvement would be obtained by throttling the invocations to the service. The processing times of the input dataset could be further improved if light network protocols (such as Websocket [23]) were used together with improved message serialization (such as Google Protobuf [24]).

4 Evaluation of the Approach

To test the flexibility and scalability of the proposed solution, we performed three experiments of increasing scale involving real datasets. The experiments make use of two enrichment services: the geospatial reconciliation and extension service GN, and the weather extension service W, which is used to enrich the input dataset with weather information. GN takes only one attribute as input (e.g., a toponym), and creates one (reconciliation) or more (extension) columns; W takes two attributes as input—location and date—and appends as many columns as the number of desired weather features.

First, we designed a small-scale experiment reproducing the scenario where a data scientist executes the enrichment pipeline on a commodity machine (the whole cloud-native platform has been installed on a multi-tenant machine with 4 CPUs Intel Xeon Silver 4114 2.20 GHz, and 125GB RAM). The main objective was to assess the performance boost attributable to the introduction different caching levels. We started by testing the reconciliation performance with no caching strategy: 200 K rows (21 columns) from a real company dataset featuring 2227 different toponyms (from Germany and Spain) have been extracted and a pipeline featuring only reconciliation has been created. The measured average time per row was 12.927 ms. The same test was then repeated, enabling the caching level implemented at the reconciliation service level. The cache system improved performance achieving an average processing time of 2.558 ms per row (5 times faster over the baseline). Finally, we enabled the first cache layer, which is implemented locally on the level of the executable of the enrichment pipeline. The objective was to avoid the network latency whenever possible, which is substantial even in a local setup (via the loopback interface). The pipeline, in this case, ran ∼770 times faster than the baseline (0.0168 ms/row on average).

To analyze the behavior of the cache over time, a second experiment was designed, extending the first one as follows: a more complex pipeline (referred to as full-pipeline) was implemented. It reconciles city toponyms to GN, extends reconciled entities with the corresponding first-level administrative division from GN (i.e., regions). After that, it fetches weather information about regions, using the

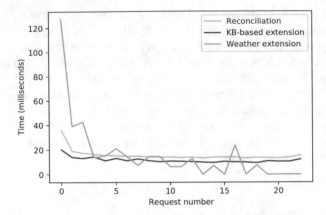

Fig. 6 Request execution time in milliseconds for the second experiment without duplicates

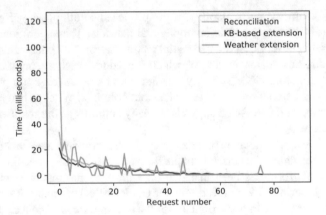

Fig. 7 Request execution time in milliseconds for the second experiment with four duplicates

reconciled administrative level and the date column (i.e., temperature for a specific date and the following one) generating a new dataset with 25 columns. This pipeline was used to enrich a dataset derived from the one used in the first experiment, filtering out duplicates in the reconciliation target column (i.e., each value occurs at most once), resulting in 2227 unique cities (and rows). The outcomes of this experiment, where the cache did not significantly improve the performance (as it was built but never used), are depicted in Fig. 6.[11] Afterwards, a synthetic dataset was built where each line from the previous one is replicated four times to exploit the local cache. As reported in Fig. 7, spikes are still visible due to cache building, but the cache reuse speeds up the process progressively ($4\times$ on average), considerably reducing the execution time (which tends to be purely cache access time).

---

[11] Initial spikes are due to the system startup (e.g., database connectors initialization).

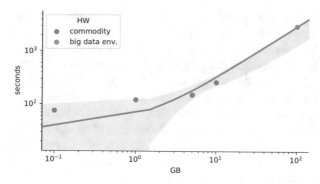

Fig. 8 Total execution time (in seconds) and linear regression curve, for different dataset sizes and two experimental setups

The final experiment was devoted to investigating the system scalability. Firstly, a single physical machine with a single instance of ASIA back-end was used. The full-pipeline was ran to enrich datasets of different sizes: 100MB, 1GB, 5GB, and 10GB. The dataset was split in 10 chunks of equal size and assigned to 10 agents. Performance results (in blue), reported in Fig. 8, measure the total pipeline completion time for different dataset sizes. The implementation achieves a linear trend, which highlights the scalability of the proposed solution. Finally, the enrichment of a ~100GB dataset (~500 million rows, 21 columns) was performed; the pipeline was run on the Big Data Environment deployed on a private cloud infrastructure featuring an 8-node cluster of heterogeneous hosts. Five of the nodes have 4-core CPUs and 15.4GB RAM and three nodes with 12-core CPUs, 64GB RAM, with six 3 TB HDDs holding a GlusterFS distributed file system (shared across the whole cluster). The enrichment agents were deployed on the three 12-core servers.

The transformation accessed a load-balanced (using round-robin load balancing) set of 10 replicas of ASIA back-end services deployed on the same stack.

The linear trend with $R^2 = 0.998$ (please notice Fig. 8 uses a base-10 log scale for the axes) is maintained also for the data point pertaining to the 100GB experiment, despite the different context in which the experiments have been carried out. This is mainly due to similar access and reconciliation times between the two experimental configurations.

5 Related Work

During the past decade, a number of tools devoted to tabular data transformation, reconciliation, and extension have been proposed. Probably the most popular is OpenRefine, an open-source solution that provides reconciliation services (with a focus on Wikidata) to semantic experts. It is a generic tool that cannot be easily customized to use third-party services. OpenRefine provides an interactive user interface with spreadsheet-style interaction in a similar way as Grafterizer does, hence they both encounter memory limitations to handle datasets. We refer the reader to [25] for a detailed discussion on data anomalies in tabular data together with an introduction to the table manipulation approach that was adopted in our work. OpenRefine was designed as a web application without support for batch execution of pipelines, hence the size of data that can be processed is constrained by the available memory. More recently, tools for extending OpenRefine with support to large data processing have been proposed, for example, OpenRefine-HD,[12] which extends OpenRefine to use Hadoop MapReduce jobs on HDFS. One of the few works on transformations of Big Data is discussed in [26], where the authors address how to exploit Apache Spark as a processing engine for iterative data preparation processes. However, all the above proposals require manual preparation of the executions, while in our approach we foresee an automatic deployment of the pipelines. An issue with all such tools is the lack of documentation that explains how the proposed solutions have been implemented, and how they can support scalability when working with distributed workloads, e.g., involving external services.

Semantic table annotation approaches have been proposed for reconciling values in tables; however, most of them cover only schema-level annotations. Approaches, such as [10, 27], are sophisticated and are targeted at Web tables, which are very small (a few hundred rows) and still require considerable computation time, making them inapplicable in Big Data environments. Karma [9] is a tool that provides an interface and a collection of algorithms to interpret tables, maps their schema to an ontology, and learns data transformations. However, Karma does not support external services for value-level reconciliation and data extension. The tool has been used in projects where these processing steps have been applied, but without explicit support by the tool itself [28]. Karma supports the execution of domain-specific scripts for data manipulations, i.e., to implement data cleaning tasks.

One of the commercial tools that are relevant to this work is Trifacta Wrangler,[13] which is a commercial suite of web applications for the exploration and preparation of raw datasets. The toolkit aims to provide data workers with specific smart tools to prepare datasets for different analysis types. Advanced techniques of machine learning, parallel processing, and human–machine interaction are also provided. The suite consists of three software solutions with increasingly advanced features. Large volumes of data can be handled by exploiting Cloud data warehouse deployments. KNIME[14] is a free software for analytics with a modular platform for building and executing workflows using predefined components called nodes. Knime core functionalities are used for standard data mining, analysis, and manipulation, and these features and functionalities can be extended through extensions from various groups and vendors. Big Data extensions allow the user to deploy workflows on Apache

---

[12] https://github.com/rmalla1/OpenRefine-HD

[13] https://www.trifacta.com

[14] https://www.knime.com

Spark and Hadoop clusters. Talend[15] is an Eclipse-based visual programming editor. Similar to KNIME, Talend uses predefined components (called nodes) to set up a pipeline, which can be compiled into executable Java code. Talend also provides an open-source data integration platform with Big Data extensions. None of these tools offer specific functionality for semantic enrichment. However, it is possible, using ad hoc methods (such as downloading data sources locally or using SPARQL queries), to approximate the functionality offered by ASIA. This result, however, can only be obtained by expert users and through specific code implementation. An extensive comparison of these tools with ASIA/DataGraft, where several factors are compared, including the number of correct, incorrect, and missing reconciliations, as well as the number of ambiguous toponyms correctly recognized, is available in [29], where the advantages of our solution in the reconciliation task are discussed in further detail.

6   Conclusions

In this work, we outlined and discussed an approach that addresses the efficient enrichment of massive datasets. The approach was developed as a result of the experience gained by closely working with business partners. We linked it to the practice by identifying the main challenges in the data science field, where actors need to integrate large datasets but often have limited programming expertise. Moreover, we proposed an open-source solution that features several enrichment services, which makes KBs accessible to non-expert users, supporting data enrichment both at the design and run time. Furthermore, repeatability is addressed by packaging the human expert actions within executable models, which can also be exploited to run the user-designed enrichment at a larger scale. The first implementation of our solution was deployed and tested in a real-world scenario. Preliminary experiments highlighted promising performance in terms of scalability; indeed, the prototype system was used to successfully execute a data flow to enrich data in the magnitude of hundreds of GBs continuously. In terms of future work, we plan to further improve the overall performance by addressing the limitations discussed in Sect. 3.2. We also plan to investigate more sophisticated solutions for entity linking in tabular data, which is particularly challenging in large tables [30]. To this end, we plan to develop approaches combining the feedback of possibly more than one users as proposed for analogous tasks [31].

Acknowledgments This work is partly funded by the EC H2020 projects EW-Shopp (Grant nr. 732590), euBusinessGraph (Grant nr. 732003), and TheyBuyForYou (Grant nr. 780247).

---

[15] https://www.talend.com

References

1. IDC. (2019). *Worldwide semiannual big data and analytics spending guide.* https://www.idc.com/getdoc.jsp?containerId=IDC_P33195
2. Zillner, S., Curry, E., Metzger, A., Auer, S., & Seidl, R. (Eds.). (2017). *European big data value strategic research & innovation agenda.*
3. Lohr, S. (2014). For big-data scientists, 'janitor work' is key hurdle to insights. *NY Times, 17.*
4. Furche, T., Gottlob, G., Libkin, L., Orsi, G., & Paton, N. W. (2016). Data wrangling for big data: Challenges and opportunities. In *EDBT* (pp. 473–478).
5. Črešlovnik, D., Košmerlj, A., & Ciavotta, M. (2018). Using historical and weather data for marketing and category management in ecommerce: The experience of EW-shopp. In *Proceedings of ECSA '18* (pp. 31:1–31:5). ACM.
6. Beneventano, D., & Vincini, M. (2019). Foreword to the special issue: "Semantics for big data integration". *Information, 10*, 68.
7. Koutsomitropoulos, D., Likothanassis, S., & Kalnis, P. (2019). Semantics in the deep: Semantic analytics for big data. *Data, 4*, 63.
8. Zhuge, H., & Sun, X. (2019). Semantics, knowledge, and grids at the age of big data and AI. *Concurrency Computation, 31.*
9. Knoblock, C. A., Szekely, P., Ambite, J. L., Goel, A., Gupta, S., Lerman, K., Muslea, M., Taheriyan, M., & Mallick, P. (2012). Semi-automatically mapping structured sources into the semantic web. In *The semantic web: Research and applications* (pp. 375–390).
10. Ritze, D., Lehmberg, O., Bizer, C. (2015). Matching HTML tables to dbpedia. In *Proceedings of the 5th International Conference on Web Intelligence, Mining and Semantics, WIMS 2015*, Larnaca, Cyprus, July 13–15, 2015 (pp. 10:1–10:6).
11. Ermilov, I., & Ngomo, A. C. N. (2016). Taipan: Automatic property mapping for tabular data. In *Knowledge engineering and knowledge management* (pp. 163–179).
12. Kruit, B., Boncz, P., & Urbani, J. (2019). Extracting novel facts from tables for knowledge graph completion. In *The semantic web – ISWC 2019* (pp. 364–381). Springer.
13. Chabot, Y., Labbé, T., Liu, J., & Troncy, R. (2019). DAGOBAH: An end-to-end context-free tabular data semantic annotation system. In *Proceedings of SemTab@ISWC 2019. CEUR Workshop Proceedings* (Vol. 2553, pp. 41–48). CEUR-WS.org.
14. Nikolov, N., Ciavotta, M., & De Paoli, F. (2018). Data wrangling at scale: The experience of ew-shopp. In *Proceedings of the 12th European Conference on Software Architecture: Companion Proceedings* (pp. 32:1–32:4). ECSA '18, ACM.
15. Zillner, S., Bisset, D., Milano, M., Curry, E., Garcìa Robles, A., Hahn, T., Irgens, M., Lafrenz, R., Liepert, B., O'Sullivan, B., & Smeulders, A. (Eds.). (2020). *Strategic research, innovation and deployment agenda – AI, data and robotics partnership.* third release. Brussels. BDVA, EU-Robotics, ELLIS, EurAI and CLAIRE (September 2020).
16. Sukhobok, D., Nikolov, N., Pultier, A., Ye, X., Berre, A., Moynihan, R., Roberts, B., Elvesæter, B., Mahasivam, N., & Roman, D. (2016). Tabular data cleaning and linked data generation with grafterizer. In *ISWC* (pp. 134–139). Springer.
17. Cutrona, V., Ciavotta, M., Paoli, F. D., & Palmonari, M. (2019). ASIA: A tool for assisted semantic interpretation and annotation of tabular data. In *Proceedings of the ISWC 2019 Satellite Tracks. CEUR Workshop Proceedings* (Vol. 2456, pp. 209–212).
18. Roman, D., Nikolov, N., Putlier, A., Sukhobok, D., Elvesæter, B., Berre, A., Ye, X., Dimitrov, M., Simov, A., Zarev, M., Moynihan, R., Roberts, B., Berlocher, I., Kim, S., Lee, T., Smith, A., & Heath, T. (2018). Datagraft: One-stop-shop for open data management. *Semantic Web, 9*(4), 393–411.
19. Palmonari, M., Rula, A., Porrini, R., Maurino, A., Spahiu, B., & Ferme, V. (2015). ABSTAT: Linked data summaries with abstraction and statistics. In *ISWC* (pp. 128–132).
20. Stonebraker, M. (1986). The case for shared nothing. *IEEE Database Engineering Bulletin, 9*(1), 4–9.

21. Dessalk, Y.D., Nikolov, N., Matskin, M., Soylu, A., & Roman, D. (2020). Scalable execution of big data workflows using software containers. In *Proceedings of the 12th International Conference on Management of Digital EcoSystems* (pp. 76–83).
22. Wind, D. (2013). *Instant effective caching with ehcache*. Packt Publishing.
23. Fette, I., & Melnikov, A. (2011). The websocket protocol. Technical Report RFC 6455, IETF.
24. Sumaray, A., & Makki, S. K. (2012). A comparison of data serialization formats for optimal efficiency on a mobile platform. In *Proceedings of ICUIMC '12*.
25. Sukhobok, D., Nikolov, N., & Roman, D. (2017). Tabular data anomaly patterns. In *2017 International Conference on Big Data Innovations and Applications (Innovate-Data)* (pp. 25–34).
26. Wang, H., Li, M., Bu, Y., Li, J., Gao, H., & Zhang, J. (2015). Cleanix: a parallel big data cleaning system. *SIGMOD Record, 44*(4), 35–40.
27. Limaye, G., Sarawagi, S., & Chakrabarti, S. (2010). Annotating and searching web tables using entities, types and relationships. *PVLDB, 3*(1), 1338–1347.
28. Kejriwal, M., Szekely, P. A., & Knoblock, C. A. (2018). Investigative knowledge discovery for combating illicit activities. *IEEE Intelligent Systems, 33*(1), 53–63.
29. Sutton, L., Nikolov, N., Ciavotta, M., & Košmerlj, A. (2019). *D3.5 EW-Shopp components as a service: Final Release*. https://www.ew-shopp.eu/wp-content/uploads/2020/02/EW-Shopp_D3.5_Components-as-a-service_release_v1.1-SUBMITTED_Low.pdf
30. Cutrona, V., Bianchi, F., Jiménez-Ruiz, E., & Palmonari, M. (2020). Tough tables: Carefully evaluating entity linking for tabular data. In *ISWC*.
31. Cruz, I. F., Palmonari, M., Loprete, F., Stroe, C., & Taheri, A. (2016). Quality-based model for effective and robust multi-user pay-as-you-go ontology matching. *Semantic Web, 7*(4), 463–479.

Trade-Offs and Challenges of Serverless Data Analytics

Pedro García-López, Marc Sánchez-Artigas, Simon Shillaker, Peter Pietzuch,
David Breitgand, Gil Vernik, Pierre Sutra, Tristan Tarrant, Ana Juan-Ferrer,
and Gerard París

Abstract Serverless computing has become very popular today since it largely
simplifies cloud programming. Developers do no longer need to worry about
provisioning or operating servers, and they have to pay only for the compute
resources used when their code is run. This new cloud paradigm suits well for many
applications, and researchers have already begun investigating the feasibility of
serverless computing for data analytics. Unfortunately, today's serverless computing
presents important limitations that make it really difficult to support all sorts
of analytics workloads. This chapter first starts by analyzing three fundamental trade-offs of today's serverless computing model and their relationship with
data analytics. It studies how by relaxing disaggregation, isolation, and simple
scheduling, it is possible to increase the overall computing performance, but at the
expense of essential aspects of the model such as elasticity, security, or sub-second
activations, respectively. The consequence of these trade-offs is that analytics

P. García-López (✉) · M. Sánchez-Artigas · G. París
Universitat Rovira i Virgili, Tarragona, Spain
e-mail: pedro.garcia@urv.cat; marc.sanchez@urv.cat; gerard.paris@urv.cat

S. Shillaker · P. Pietzuch
Large Scale Data and Systems Group, Imperial College London, London, England
e-mail: s.shillaker17@imperial.ac.uk; prp@imperial.ac.uk

D. Breitgand · G. Vernik
Cloud Platforms, IBM Research Haifa, Haifa, Israel
e-mail: davidbr@il.ibm.com; gilv@il.ibm.com

P. Sutra
CNRS, Université Paris Saclay, Évry, France
e-mail: pierre.sutra@telecom-sudparis.eu

T. Tarrant
Red Hat, Cork, Ireland
e-mail: ttarrant@redhat.com

A. Juan-Ferrer
ATOS, Barcelona, Spain
e-mail: ana.juanf@atos.net

© The Author(s) 2022
E. Curry et al. (eds.), *Technologies and Applications for Big Data Value*,
https://doi.org/10.1007/978-3-030-78307-5_3

applications may well end up embracing hybrid systems composed of serverless and serverful components, which we call *ServerMix* in this chapter. We will review the existing related work to show that most applications can be actually categorized as *ServerMix*.

Keywords Serverless computing · Data analytics · Cloud computing

1   Introduction

The chapter relates to the technical priority *Data Processing Architectures* of the European Big Data Value Strategic Research & Innovation Agenda [36]. It addresses the horizontal concerns *Data Analytics* and *The Cloud and HPC* of the BDV Technical Reference Model. The chapter relates to the *Systems, Methodologies, Hardware and Tools* cross-sectorial technology enablers of the AI, Data and Robotics Strategic Research, Innovation & Deployment Agenda [37].

With the emergence of serverless computing, the cloud has found a new paradigm that removes much of the complexity of its usage by abstracting away the provisioning of compute resources. This fairly new model was culminated in 2015 by Amazon in its Lambda service. This service offered cloud functions, marketed as FaaS (Function as a Service), and rapidly became the core of serverless computing. We say "core," because cloud platforms usually provide specialized serverless services to meet specific application requirements, packaged as BaaS (Backend as a Service). However, the focus of this chapter will be on the FaaS model, and very often, the words "serverless computing" and "FaaS" will be used interchangeably. The reason why FaaS drew widespread attention is because with FaaS platforms, a user-defined function and its dependencies are deployed to the cloud, where they are managed by the cloud provider and executed on demand. Simply put, users just write cloud functions in a high-level language and the serverless systems manage everything else: instance selection, auto-scaling, deployment, sub-second billing, fault tolerance, and so on. The programming simplicity of functions paves the way to soften the transition to the cloud ecosystem for end users.

Current practice shows that the FaaS model is well suited for many types of applications, provided that they require a small amount of storage and memory (see, for instance, AWS Lambda operational limits [3]). Indeed, this model was originally designed to execute event-driven, stateless functions in response to user actions or changes in the storage tier (e.g., uploading a photo to Amazon S3), which encompasses many common tasks in cloud applications. What was unclear is whether or not this new computing model could also be useful to execute data analytics applications. This question was answered partially in 2017 with the appearance of two relevant research articles: ExCamera [10] and "Occupy the Cloud" [19]. We say "partially," because the workloads that both works handled mostly consisted of "map"-only jobs, just exploiting embarrassingly massive parallelism. In particular, ExCamera proved to be 60% faster and 6x cheaper than using VM instances when encoding videos on the fly over thousands of Lambda functions.

The "Occupy the Cloud" paper showcased simple MapReduce jobs executed over Lambda Functions in their PyWren prototype. In this case, PyWren was 17% slower than PySpark running on r3.xlarge VM instances. The authors claimed that the simplicity of configuration and inherent elasticity of Lambda functions outbalanced the performance penalty. They, however, did not compare the costs between their Lambda experiments against an equivalent execution with virtual machines (VMs).

While both research works showed the enormous potential of serverless data analytics, today's serverless computing offerings importantly restrict the ability to work efficiently with data. In simpler terms, serverless data analytics are way more expensive and less performant than cluster computing systems or even VMs running analytics engines such as Spark. Two recent articles [17, 20] have outlined the major limitations of the serverless model in general. Remarkably, [20] reviews the performance and cost of several data analytics applications and shows that: a MapReduce-like sort of 100 TB was 1% faster than using VMs, but costing 15% higher; linear algebra computations [33] were 3x slower than an MPI implementation in a dedicated cluster, but only valid for large problem sizes; and machine learning (ML) pipelines were 3–5x faster than VM instances, but up to 7x higher total cost.

Furthermore, existing approaches must rely on auxiliary serverful services to circumvent the limitations of the stateless serverless model. For instance, PyWren [19] uses Amazon S3 for storage, coordination, and as indirect communication channel. Locus [28] uses Redis through the ElastiCache service, while ExCamera [10] relies on an external VM-based rendezvous and communication service. Also, Cirrus [7] relies on disaggregated in-memory servers.

The rest of the chapter is structured as follows. Section 1.1 presents the *Server-Mix* model. Trade-offs of serverless architectures are analyzed in Sect. 2, while related work is revisited in Sect. 3. The challenges and advances in CloudButton project are presented in Sect. 4. Finally, Sect. 5 concludes the chapter.

1.1 On the Path to Serverless Data Analytics: The *ServerMix* Model

In the absence of a fully fledged serverless model in today's cloud platforms (e.g., there is no effective solution to the question of serverless storage in the market), current incarnations of serverless data analytics systems are hybrid applications combining serverless and serverful services. In this chapter, we identify them as "*ServerMix*." Actually, we will show how most related work can be classified under the umbrella term of *ServerMix*. We will first describe the existing design trade-offs involved in creating *ServerMix* data analytics systems. We will then show that it is possible to relax core principles such as disaggregation, isolation, and simple scheduling to increase performance, but also how this relaxation of the model may compromise the auto-scaling ability, security, and even the pricing model and fast startup time of serverless functions. For example:

- **Relaxation of disaggregation:** Industry trends show a paradigm shift to dis-aggregated datacenters [12]. By physically decoupling resources and services, datacenter operators can easily customize their infrastructure to maximize the performance-per-dollar ratio. One such example of this trend is serverless computing. That is, FaaS offerings are of little value by themselves and need a vast ecosystem of disaggregated services to build applications. In the case of Amazon, this includes S3 (large object storage), DynamoDB (key-value storage), SQS (queuing services), SNS (notification services), etc. Consequently, departing from a serverless data-shipping model built around these services to a hybrid model where computations can be delegated to the stateful storage tier can easily achieve performance improvements [30]. However, disaggregation is the fundamental pillar of improved performance and elasticity in the cloud.
- **Relaxation of isolation:** Serverless platforms leverage operating system containers such as Docker to deploy and execute cloud functions. In particular, each cloud function is hosted in a separate container. However, functions of the same application may not need such a strong isolation and be co-located in the same container, which improves the performance of the application [1]. Further, cloud functions are not directly network-addressable in any way. Thus, providing direct communication between functions would reduce unnecessary latencies when multiple functions interact with one another, such that one function's output is the input to another one. Leveraging lightweight containers [26], or even using language-level constructs, would also reduce cold starts and boost inter-function communication. However, strong isolation and sandboxing is the basis for multi-tenancy, fault isolation, and security.
- **Flexible QoS and scheduling:** Current FaaS platforms only allow users to provision some amount of RAM and a time slice of CPU resources. In the case of Amazon Lambda, the first determines the other. Actually, there is no way to access specialized hardware or other resources such as the number of CPUs, GPUs, etc. To ensure service level objectives (SLOs), users should be able to specify resource requirements. But, this would lead to implement complex scheduling algorithms that were able to reserve such resources and even execute cloud functions in specialized hardware such as GPUs with different isolation levels. However, this would make it harder for cloud providers to achieve high resource utilization, as more constraints are put on function scheduling. Simple user-agnostic scheduling is the basis for short start-up times and high resource utilization.

It is clear that these approaches would obtain significant performance improvements. But, depending on the changes, such systems would be much closer to a serverful model based on VMs and dedicated resources than to the essence of serverless computing. In fact, we claim in this chapter that the so-called limitations of the serverless model are indeed its defining traits. When applications should require less disaggregation (computation close to the data), relaxation of isolation (co-location, direct communication), or tunable scheduling (predictable performance, hardware acceleration), a suitable solution is to build a *ServerMix* solution. At least for serverless data analytics, we project that in the near future the dependency on serverful computing will increasingly "vanish," for instance, by the appearance of high-throughput, low-latency BaaS storage services, so that many *ServerMix* systems will eventually become 100% serverless. Beyond some technical challenges, we do not see any fundamental reason why pure serverless data analytics would not flourish in the coming years.

In the meantime, we will scrutinize the *ServerMix* model to provide a simplified programming environment, *as much closer as possible to serverless*, for data analytics. To this aim, under the context of the H2020 CloudButton project, we will work on the following three points: (i) Smart scheduling as a mechanism for providing transparent provisioning to applications while optimizing the cost-performance tuple in the cloud; (ii) fine-grained mutable state disaggregation built upon consistent state services; and (iii) lightweight and polyglot serverful isolation-novel lightweight serverful FaaS runtimes based on WebAssembly [15] as universal multi-language substrate.

2 Fundamental Trade-Offs of Serverless Architectures

In this section, we will discuss three fundamental trade-offs underpinning cloud functions architectures—packaged as FaaS offerings. Understand these trade-offs are important, not just for serverless data analytics but to open the minds of designers to a broader range of serverless applications. While prior works such as [17, 20] have already hinted these trade-offs, the contribution of this section is to explain in more detail that the incorrect navigation of these trade-offs can compromise essential aspects of the FaaS model.

The first question to ask is which are the essential aspects of the serverless model. For this endeavor, we will borrow the Amazon's definition of serverless computing, which is an unequivocal reference definition of this new technology. According to Amazon, the four characteristic features of a serverless system are:

- *No server management:* implies that users do not need to provision or maintain any servers
- *Flexible scaling:* entails that the application can be scaled automatically via units of consumption (throughput, memory) rather than units of individual servers
- *Pay for value:* is to pay for the use of consumption units rather than server units
- *Automated high availability:* ensures that the system must provide built-in availability and fault tolerance.

As we argue in this section, these four defining properties can be put in jeopardy but relaxing the tensions among three important architectural aspects that support them. These implementation aspects, which are *disaggregation*, *isolation*, and simple *scheduling*, and their associated trade-offs, have major implications on the success of the FaaS model. In this sense, while a designer can decide to alter one or more of these trade-offs, for example, to improve performance, an oversimplifying or no comprehension of them can lead to hurt the four defining properties of the serverless model. Let us see how the trade-offs affect them.

2.1 Disaggregation

Disaggregation is the idea of decoupling resources over high bandwidth networks, giving us independent resource pools. Disaggregation has many benefits, but importantly, it allows each component to (auto-)scale in an independent manner. In serverless platforms, disaggregation is the standard rather than an option, where applications are run using stateless functions that share state through disaggregated storage (e.g., such Amazon S3) [17, 20, 33]. This concept is backed up by the fact that modern high-speed networks allow for sub-millisecond latencies between the compute and storage layers—even allowing memory disaggregation like in InfiniSwap [14].

Despite the apparent small latencies, several works propose to relax disaggregation to favor performance. The reason is that storage hierarchies, across various storage layers and network delays, make disaggregation a bad design choice for many latency and bandwidth-sensitive applications such as machine learning [7]. Indeed, [17] considers that one of the limitations of serverless computing is its *data-shipping architecture*, where data and state are regularly shipped to functions. To overcome this limitation, the same paper proposes the so-called *fluid code and data placement* concept, where the infrastructure should be able to physically co-locate code and data. In a similar fashion, [2] proposes the notion of *fluid multi-resource disaggregation*, which consists of allowing movement (i.e., fluidity) between physical resources to enhance proximity, and thus performance. Another example of weakening disaggregation is [20]. In this paper, authors suggest to co-locate related functions in the same VM instances for fast data sharing.

Unfortunately, while data locality reduces data movements, it can hinder the elastic scale-out of compute and storage resources. In an effort to scale out wider and more elastically, processing mechanisms near the data (e.g., active storage [29]) have not been put at the forefront of cloud computing, though recently numerous proposals and solutions have emerged (see [18] for details). Further, recent works such as [30] show that active storage computations can introduce resource contention and interferences into the storage service. For example, computations from one user can harm the storage service to other users, thereby increasing the running cost of the application (pay for value). In any case, shipping code to data will interfere with the flexible scaling of serverless architectures due to the lack of fast and elastic datastore in the cloud [7].

Furthermore, ensuring locality for serverless functions would mean, for example, placing related functions in the same server or VM instance, while enabling fast shared memory between them. This would obviously improve performance in applications that require fast access to shared data such as machine learning and PRAM algorithms, OpenMP-like implementations of parallel algorithms, etc. However, as pinpointed in [20], besides going against the spirit of serverless computing, this approach would reduce the flexibility of cloud providers to place functions and consequently reduce the capacity to scale out while increasing the complexity of function scheduling. Importantly, this approach would force developers to think about low-level issues such as server management or whether function locality might lead suboptimal load balancing among server resources.

2.2 Isolation

Isolation is another fundamental pillar of multi-tenant clouds services. Particularly, perfect isolation enables a cloud operator to run many functions (and applications) even on a single host, with low idle memory cost, and high resource efficiency. What cloud providers seek is to reduce the overhead of multi-tenant function isolation and provide high-performance (small startup times), for they leverage a wide variety of isolation technologies such as containers, unikernels, library OSes, or VMs. For instance, Amazon has recently released Firecracker microVMs for AWS Lambda, and Google has adopted gVisor. Other examples of isolation technologies for functions are CloudFlare Workers with WebAssembly or optimized containers such as SOCK [26]. These isolation techniques reduce startup times to the millisecond range, as compared to the second timescale of traditional VMs.

Beyond the list of sandboxing technologies for serverless computing, most of them battled-tested in the industry (e.g., Amazon Firecracker VMs), several research works have proposed to relax isolation in order to improve performance. For instance, [2] proposes the abstraction of a process in serverless computing, with the property that each process can be run across multiple servers. As a consequence of this multi-server vision, the paper introduces a new form of isolation that ensures multi-tenant isolation across multiple servers (where the functions of the same tenant are run). This new concept of isolation is called *coordinated isolation* in the paper. Further, [17] proposes two ways of relaxing isolation. The first one is based on the *fluid code and data placement* approach, and the second way is by allowing direct communication and network addressing between functions. In particular, the paper claims that today's serverless model stymies distributed computing due to its lack of direct communication among functions and advocates for long-running, addressable virtual agents instead.

Another technique to increase performance is to relax isolation and co-locate functions in the same VMs or containers [1, 20]. Co-location may be achieved using language-level constructs that reuse containers when possible. This can make sense for functions belonging to the same tenant [1], since it would heavily reduce cold starts and execution time for function compositions (or workflows). Unfortunately, it is possible that independent sets of sandboxed functions compete for the same server resources and interfere with each other's performance. Or simply, that it becomes impossible to find a single host that have the necessary resources for running a sandbox of multiple functions, affecting important defining properties of serverless computing such as flexible scaling, pay for value, and no server management, among others.

Experiencing similar issues as above, it could be also possible to enable direct communication between functions of the same tenant. In this case, direct communication would permit a variety of distributed communication models, allowing, for example, the construction of replicated shared memory between functions. To put it baldly, each of these forms of relaxing isolation might in the end increase the attack surface, for instance, by opening physical co-residency attacks and network attacks not just to single functions but a collection of them.

2.3 Simple Scheduling

Simple scheduling is another essential pillar of serverless computing. Indeed, cloud providers can ensure Quality of Service (QoS) and Service Level Agreements (SLAs) to different tenants by appropriately scheduling the reserved resources and bill them correspondingly. The goal of cloud scheduling algorithms is to maximize the utilization of the cloud resources while matching the requirements of the different tenants.

In today's FaaS offerings, tenants only specify the cloud function's memory size, while the function execution time is severely limited—for instance, AWS limits the execution time of functions to 15 min. This single constraint simplifies the scheduling of cloud functions and makes it easy to achieve high resource utilization through statistical multiplexing. For many developers, this lack of control on specifying resources, such as the number of CPUs, GPUs, or other types of hardware accelerators, is seen as an obstacle. To overcome this limitation, a clear candidate would be to work on more sophisticated scheduling algorithms that support more constraints on functions scheduling, such as hardware accelerators, GPUs, or the data dependencies between the cloud functions, which can lead to suboptimal function placement. For instance, it is not hard to figure out that a suboptimal placement of functions can result in an excess of communication to exchange data (e.g., for broadcast, aggregation, and shuffle patterns [20]) or in suboptimal performance. Ideally, these constraints should be (semi-)automatically inferred by the platform itself, for instance, from static code analysis, profiling, etc., to not break the property of "no server management," that is, the core principle of serverless. But even in this case, more constraints on function scheduling would make it harder to guarantee flexible scaling.

The literature also proposes ideas to provide predictable performance in serverless environments. For instance, [2] proposes the concept of "fine-grained live orchestration," which involves complex schedulers to allocate resources to serverless processes that run across multiple servers in the datacenter. Hellerstein et al. [17] advocates for heterogeneous hardware support for functions where developers could specify their requirements in DSLs and the cloud providers would then calculate the most cost-effective hardware to meet user SLOs. This would guarantee the use of specialized hardware for functions. In [20], it is supported the claim of harnessing hardware heterogeneity in serverless computing. In particular, it is proposed that serverless systems could embrace multiple instance types (with prices according to hardware specs) or that cloud providers may select the hardware automatically depending on the code (like GPU hardware for CUDA code and TPU hardware for TensorFlow code).

Overall, the general observation is that putting more constraints on function scheduling for performance reasons could be disadvantageous in terms of flexible scaling and elasticity and even hinder high resource utilization. Moreover, it would complicate the pay per use model, as it would make it difficult to pay for the use of consumption units, rather than server units, due to hardware heterogeneity.

2.4 Summary

As a summary, we refer to Fig. 1 as a global view of the overall trade-offs. These trade-offs have serious implications on the serverless computing model and require careful examination. As we have already seen, disaggregation, isolation, and simplified scheduling are pivotal to ensure flexible scaling, multi-tenancy, and millisecond startup times, respectively.

Weakening disaggregation to exploit function and data locality can be useful to improve performance. However, it also means to decrease the scale-out capacity of cloud functions and complicate function scheduling in order to meet user SLOs. The more you move to the left, the closer you are to serverful computing or running VMs or clusters in the datacenter.

With isolation the effect is similar. Since isolation is the key to multi-tenancy, completely relaxing isolation leaves nothing but dedicated resources. In your dedicated VMs, containers, or clusters (serverful), you can run functions very fast

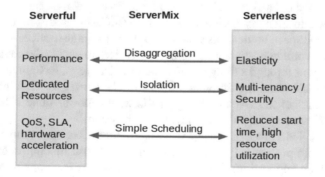

Fig. 1 Trade-offs

without caring about sandboxing and security. But this also entails more complex scheduling and pricing models.

Finally, simple scheduling and agnostic function placement is also inherent to serverless computing. But if you require QoS, SLAs, or specialized hardware, the scheduling and resource allocation gets more complex. Again, moved to the extreme, you end up in serverful settings that already exist (dedicated resources, VMs, or clusters).

Perhaps, the most interesting conclusion of this figure is the region in the middle, which we call *ServerMix* computing. The zone in the middle involves applications that are built combining both serverless and serverful computing models. In fact, as we will review in the related work, many existing serverless applications may be considered *ServerMix* according to our definition.

3   Revisiting Related Work: The *ServerMix* Approach

3.1   Serverless Data Analytics

Despite the stringent constraints of the FaaS model, a number of works have managed to show how this model can be exploited to process and transform large amounts of data [19, 21, 31], encode videos [10], and run large-scale linear algebra computations [33], among other applications. Surprisingly, and contrary to intuition, most of these serverless data analytics systems are indeed good *ServerMix* examples, as they combine both serverless and serverful components.

In general, most of these systems rely on an external, serverful provisioner component [10, 19, 21, 31, 33]. This component is in charge of calling and orchestrating serverless functions using the APIs of the chosen cloud provider. Sometimes the provisioner is called "coordinator" (e.g., as in ExCamera [10]) or "scheduler" (e.g., as in Flint [21]), but its role is the same: orchestrating functions and providing some degree of fault tolerance. But the story does not end here. Many of these systems require additional serverful components to overcome the limitations of the FaaS model. For example, recent works such as [28] use disaggregated in-memory systems such as ElastiCache Redis to overcome the throughput and speed bottlenecks of slow disk-based storage services such as S3. Or even external communication or coordination services to enable the communication among functions through a disaggregated intermediary (e.g., ExCamera [10]).

To fully understand the different variants of *ServerMix* for data analytics, we will review each of the systems one by one in what follows. Table 1 details which components are serverful and serverless for each system.

PyWren [19] is a proof of concept that MapReduce tasks can be run as serverless functions. More precisely, PyWren consists of a serverful function scheduler (i.e., a client Python application) that permits to execute "map" computations as AWS Lambda functions through a simple API. The "map" code to be run in parallel is

Table 1 *ServerMix* applications

Systems	Components	
	Serverful	*Serverless*
PyWren [19]	Scheduler	AWS Lambda, Amazon S3
IBM PyWren [31]	Scheduler	IBM Cloud Functions, IBM COS, RabbitMQ
ExCamera [10]	Coordinator and rendezvous servers (Amazon EC2 VMs)	AWS Lambda, Amazon S3
gg [11]	Coordinator	AW Lambda, Amazon S3, Redis
Flint [21]	Scheduler (Spark context on client machine)	AW Lambda, Amazon S3, Amazon SQS
Numpywren [33]	Provisioner, scheduler (client process)	AWS Lambda, Amazon S3, Amazon SQS
Cirrus [20]	Scheduler, parameter servers (large EC2 VM instances with GPUs)	AWS Lambda, Amazon S3
Locus [28]	Scheduler, Redis service (AWS ElastiCache)	AWS Lambda, Amazon S3

first serialized and then stored in Amazon S3. Next, PyWren invokes a common Lambda function that deserializes the "map" code and executes it on the relevant datum, both extracted from S3. Finally, the results are placed back into S3. The scheduler actively polls S3 to detect that all partial results have been uploaded to S3 before signaling the completion of the job.

IBM-PyWren [31] is a PyWren derived project which adapts and extends PyWren for IBM Cloud services. It includes a number of new features, such as broader MapReduce support, automatic data discovery and partitioning, integration with Jupiter notebooks, and simple function composition, among others. For function coordination, IBM-PyWren uses RabbitMQ to avoid the unnecessary polling to the object storage service (IBM COS), thereby improving job execution times compared with PyWren.

ExCamera [10] performs digital video encoding by leveraging the parallelism of thousands of Lambda functions. Again, ExCamera uses serverless components (AWS Lambda, Amazon S3) and serverful ones (coordinator and rendezvous servers). In this case, apart from a coordinator/scheduler component that starts and coordinates functions, ExCamera also needs a rendezvous service, placed in an EC2 VM instance, to communicate functions among each other.

Stanford's gg [11] is an orchestration framework for building and executing burst-parallel applications over Cloud Functions. gg presents an intermediate representation that abstracts the compute and storage platform, and it provides dependency management and straggler mitigation. Again, gg relies on an external coordinator component, and an external Queue for submitting jobs (gg's thunks) to the execution engine (functions, containers).

Flint [21] implements a serverless version of the PySpark MapReduce framework. In particular, Flint replaces Spark executors by Lambda functions. It is similar to PyWren in two main aspects. On the one hand, it uses an external serverful scheduler for function orchestration. On the other hand, it leverages S3 for input and output data storage. In addition, Flint uses Amazon's SQS service to store intermediate data and perform the necessary data shuffling to implement many of the PySpark's transformations.

Numpywren [33] is a serverless system for executing large-scale dense linear algebra programs. Once again, we observe the *ServerMix* pattern in numpywren. As it is based on PyWren, it relies on an external scheduler and Amazon S3 for input and output data storage. However, it adds an extra serverful component in the system called provisioner. The role of the provisioner is to monitor the length of the task queue and increase the number of Lambda functions (executors) to match the dynamic parallelism during a job execution. The task queue is implemented using Amazon SQS.

Cirrus machine learning (ML) project [20] is another example of a hybrid system that combines serverful components (parameter servers, scheduler) with serverless ones (AWS Lambda, Amazon S3). As with linear algebra algorithms, a fixed cluster size can either lead to severe underutilization or slowdown, since each stage of a workflow can demand significantly different amounts of resources. Cirrus addresses this challenge by enabling every stage to scale to meet its resource demands by using Lambda functions. The main problem with Cirrus is that many ML algorithms require state to be shared between cloud functions, for it uses VM instances to share and store intermediate state. This necessarily converts Cirrus into another example of a *ServerMix* system.

Finally, the most recent example of *ServerMix* is Locus [28]. Locus targets one of the main limitations of the serverless stateless model: data shuffling and sorting. Due to the impossibility of function-to-function communication, shuffling is ill-suited for serverless computing, leaving no other choice but to implement it through an intermediary cloud service, which could be cost-prohibitive to deliver good performance. Indeed, the first attempt to provide an efficient shuffling operation was realized in PyWren [19] using 30 Redis ElastiCache servers, which proved to be a very expensive solution. The major contribution of Locus was the development of a hybrid solution that considers both cost and performance. To achieve an optimal cost-performance trade-off, Locus combined a small number of expensive fast Redis instances with the much cheaper S3 service, achieving comparable performance to running Apache Spark on a provisioned cluster.

We did not include SAND [1] in the list of *ServerMix* systems. Rather, it proposes a new FaaS runtime. In the article, the authors of SAND present it as an alternative high-performance serverless platform. To deliver high performance, SAND introduces two relaxations in the standard serverless model: one in *disaggregation*, via a hierarchical message bus that enables function-to-function communication, and another in *isolation*, through application-level sandboxing that enables packing multiple application-related functions together into the same container. Although SAND was shown to deliver superior performance than Apache OpenWhisk, the paper failed to evaluate how these relaxations can affect the scalability, elasticity, and security of the standard FaaS model.

3.2 Serverless Container Services

Hybrid cloud technologies are also accelerating the combination of serverless and serverful components. For instance, the recent deployment of Kubernetes (k8s) clusters in the big cloud vendors can help overcome the existing application portability issues in the cloud. There exists a plenty of hosted k8s services such as *Amazon Elastic Container Service* (EKS), *Google Kubernetes Engine* (GKE), and *Azure Kubernetes Service* (AKS), which confirm that this trend is gaining momentum. However, none of these services can be considered 100% "serverless." Rather, they should be viewed as a middle ground between cluster computing and serverless computing. That is, while these hosted services offload operational management of k8s, they still require custom configuration by developers. The major similarity to serverless computing is that k8s can provide short-lived computing environments like in the customary FaaS model.

But a very interesting recent trend is the emergence of the so-called serverless container services such as IBM Code Engine, AWS Fargate, Azure Container Instances (ACI), and Google Cloud Run (GCR). These services reduce the complexity of managing and deploying k8s clusters in the cloud. While they offer serverless features such as flexible automated scaling and pay-per-use billing model, these services still require some manual configuration of the right parameters for the containers (e.g., compute, storage, and networking) as well as the scaling limits for a successful deployment.

These alternatives are interesting for long-running jobs such as batch data analytics, while they offer more control over the applications thanks to the use of containers instead of functions. In any case, they can be very suitable for stateless, scalable applications, where the services can scale out by easily adding or removing container instances. In this case, the user establishes a simple CPU or memory threshold and the service is responsible for monitoring, load balancing, and instance creation and removal. It must be noted that if the service or application is more complex (e.g., a stateful storage component), the utility of these approaches is rather small, or they require important user intervention.

An important open source project related to serverless containers is CNCF's KNative. In short, KNative is backed by big vendors such as Google, IBM, and RedHat, among others, and it simplifies the creation of serverless containers over k8s clusters. Knative simplifies the complexity of k8s and Istio service mesh components, and it creates a promising substrate for both PaaS and FaaS applications.

As a final conclusion, we foresee that the simplicity of the serverless model will gain traction among users, so many new offerings may emerge in the next few years, thereby blurring the borders between both serverless and serverful models. Further, container services may become an interesting architecture for *ServerMix* deployments.

4   CloudButton: Towards Serverless Data Analytics

Serverless technologies will become a key enabler for radically simpler, user-friendly data analytics systems in the coming years. However, achieving this goal requires a programmable framework that goes beyond the current FaaS model and has user-adjustable settings to alter the IDS (Isolation-Disaggregation-Scheduling) trade-off (see Sect. 2 for more details)—for example, by weakening function isolation for better performance.

The EU CloudButton project [8] was born out of this need. It has been heavily inspired by "Occupy the Cloud" paper [19] and the statement made by a professor of computer graphics at UC Berkeley quoted in that paper:

> *"Why is there no cloud button?"* He outlined how his students simply wish they could easily *"push a button"* and have their code—existing, optimized, single-machine code—running on the cloud."

Consequently, our primary goal is *to create a serverless data analytics platform, which "democratizes Big Data" by overly simplifying the overall life cycle and cloud programming models* of data analytics. To this end, the 3-year CloudButton research project (2019–2021) will be undertaken as a collaboration between key industrial partners such as IBM, RedHat, and Atos, and academic partners such as Imperial College London, Institut Mines Télécom/Télécom SudParis, and Universitat Rovira i Virgili. To demonstrate the impact of the project, we target two settings with large data volumes: *bioinformatics* (genomics, metabolomics) and *geospatial data* (LiDAR, satellital), through institutions and companies such as EMBL, Pirbright Institute, Answare, and Fundación Matrix.

The project aims to provide full transparency [13] for applications which implies that we will be able to run unmodified single-machine code over effectively unlimited compute, storage, and memory resources thanks to serverless disaggregation and auto-scaling.

As we can see in Fig. 2, the CloudButton's Lithops toolkit [32] will realize this vision of transparency relying on the next building blocks:

- **A high-performance serverless compute engine for Big Data:** The main goal is to support stateful and highly performant execution of serverless tasks. It will also provide efficient QoS management of containers that host serverless functions and a serverless execution framework to support typical dataflow models. As we can see in the Fig. 2, our design includes an extensible backend architecture for compute and storage that covers the major Cloud providers and Kubernetes cluster technologies.
- **Mutable, shared data support in serverless computing:** To simplify the transitioning from sequential to (massively-)parallel code, CloudButton has designed the Crucial [5] middleware on top of RedHat Infinispan that allows

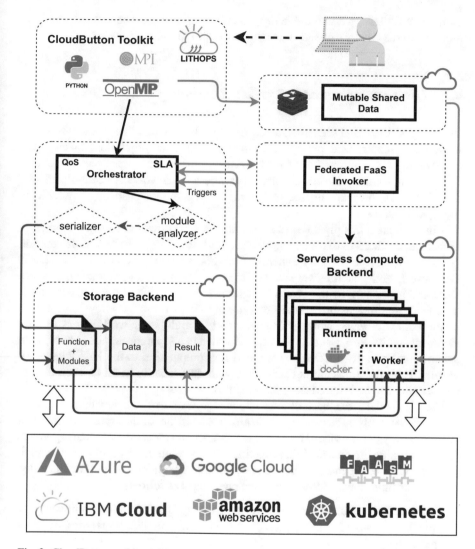

Fig. 2 CloudButton architecture

the quickly spawning and easy sharing of mutable data structures in serverless platforms. This goal will explore the disaggregation area of the IDS trade-off.

- **Novel serverless cloud programming abstractions:** The goal is to express a wide range of existing data-intensive applications with minimal changes. The programming model should at the same time preserve the benefits of a serverless execution model and add explicit support for stateful functions in applications. Thanks to Lithops [32] and FaaSM [34] the toolkit will support almost unmodified data analytics applications in Python [32] but also C/C++ applications [34] using MPI and OpenMP programming models.

In what follows, we will delve deeper into each of these goals, highlighting in more detail the advance of each one with respect to the state of the art.

4.1   High-performance Serverless Runtime

In many real-life cloud scenarios, enterprise workloads cannot be straightforwardly moved to a centralized public cloud due to the cost, regulation, latency and bandwidth, or a combination of these factors. This forces enterprises to adopt a hybrid cloud solution. However, current serverless frameworks are centralized and, out-of-the-box, they are unable to leverage computational capacity available in multiple locations.

Big Data analytics pipelines (a.k.a. analytics workflows) need to be efficiently orchestrated. There exists many serverless workflows orchestration tools (Fission flows, Argo, Apache Airflow), ephemeral serverless composition frameworks (IBM Composer), and stateful composition engines (Amazon Step Functions, Azure Durable Functions). To the best of our knowledge, workflow orchestration tools treat FaaS runtimes as black boxes that are oblivious to the workflow structure. A major issue with FaaS, which is exacerbated in a multi-stage workflow, is its data shipment architecture. Usually, the data is located in a separate storage service, such as Amazon S3 or IBM COS, and shipped for computation to the FaaS cluster. In general, FaaS functions are not scheduled with data locality in mind, even though data locality can be inferred from the workflow structure.

Further, and to the best of our knowledge, none of the existing workflow orchestration tools is serverless in itself. That is, the orchestrator is usually a stateful, always-on service. This is not necessarily the most cost-efficient approach for long running big data analytics pipelines, which might have periods of very high peakedness requiring massive parallelism interleaved with long periods of inactivity.

In CloudButton, we address the above challenges as follows:

- **Federated FaaS Invoker**: CloudButton exploits k8s federation architecture to provide a structured multi-clustered FaaS run time to facilitate analytics pipelines spanning multiple k8s clusters and Cloud Backends.
- **SLA, QoS, and scheduling**: programmers will be enabled to specify desired QoS levels for their functions. These QoS constraints will be enforced by a specialized scheduler (implemented via the k8s custom scheduler framework).
- **Serverless workflow orchestration**: we have constructed a Trigger-based orchestration framework [24] for *ServerMix* analytics pipelines. Tasks in the supported workflows can include massively parallel serverless computations carried out in Lithops.
- **Operational efficiency**: an operations cost-efficiency advisor will track the time utilization of each *ServerMix* component and submit recommendations on its appropriate usage.

4.2 Mutable Shared Data for Serverless Computing

In serverless Big Data applications, thousands of functions run in a short time. From a storage perspective, this requires the ability to scale abruptly the system to be on par with demand. To achieve this, it is necessary to decrease startup times (e.g., with unikernels [25]) and consider new directions for data distribution (e.g., Pocket [22]).

Current serverless computing platforms outsource state management to a dedicated storage tier (e.g., Amazon S3). This tier is agnostic of how data is mutated by functions, requiring serialization. This is cumbersome for complex data types, decreases code modularity and re-usability, and increases the cost of manipulating large objects. In contrast, we advocate that the storage tier should support in-place modifications. Additionally, storage requirements for serverless Big Data include:

- *Fast access (sub-millisecond) to ephemeral mutable data:* to support iterative and stateful computations (e.g., ML algorithms)
- *Fine-grained operations to coordinate concurrent function invocations*
- *Dependability:* to transparently support failures in both storage and compute tiers.

In CloudButton, we tackle these challenges by designing a novel storage layer for stateful serverless computation called Crucial [5]. Our goal is to simplify the transitioning from single-machine to massively parallel code. This requires new advances on data storage and distributed algorithms, such as:

- **Language support for mutable shared data.** The programmer can declare mutable shared data types in a piece of serverless code in a way transparently integrated to the programming language (e.g., with annotations). The storage tier knows the data types, allowing in-place mutations to in-memory shared data.
- **Data consistency.** Shared data objects are distributed and replicated across the storage layer, while maintaining strong consistency. To improve performance, developers can *degrade* data consistency [4, 35] on a per-object basis.
- **Just-right synchronization.** The implementation uses state machine replication atop a consensus layer [23, 27]. This layer self-adjusts to each shared data item, synchronizing replicas only when necessary, which improves performance.

4.3 Novel Serverless Cloud Programming Abstractions: The CloudButton Toolkit

Containers are the foundation of serverless runtimes, but the abstractions and isolation they offer can be restrictive for many applications. A hard barrier between the memory of co-located functions means all data sharing must be done via external storage, precluding data-intensive workloads and introducing an awkward programming model. Instantiating a completely isolated runtime environment for each function is not only inefficient but at odds with how most language runtimes were designed.

This isolation boundary and runtime environment have motivated much prior work. A common theme is optimizing and modifying containers to better suit the task, exemplified by SOCK [26], which makes low-level changes to improve start-up times and efficiency. Others have partly sacrificed isolation to achieve better performance, for example, by co-locating a tenant's functions in the same container [1]. Also, a few frameworks for building serverless applications have emerged [10, 19, 21]. But these systems still require a lot of engineering effort to port existing applications.

Software fault isolation (SFI) has been proposed as an alternative isolation approach, offering memory-safety at low cost [6]. Introducing an intermediate representation (IR) to unify the spectrum of languages used in serverless has also been advocated [17]. WebAssembly is perfectly suited on both counts. It is an IR built on the principles of SFI, designed for executing multi-tenant code [16]. This is evidenced by its use in proprietary serverless technologies such as CloudFlare Workers and Fastly's Terrarium [9].

With the CloudButton toolkit, we build on these ideas and re-examine the serverless programming and execution environment. We have investigated new approaches to isolation and abstraction, focusing on the following areas:

- **Lightweight serverless isolation.** In the Faasm Backend [34], we combine SFI, WebAssembly, and existing OS tooling to build a new isolation mechanism, delivering strong security guarantees at a fraction of the cost of containers.
- **Efficient localized state.** This new isolation approach allows sharing regions of memory between co-located functions, enabling low-latency parallel processing and new opportunities for inter-function communication.
- **Stateful programming abstractions.** To make CloudButton programming seamless, we have created a new set of abstractions [5, 34], allowing users to combine stateful middleware with efficient localized state to easily build high-performance parallel applications.
- **Polyglot libraries and tooling.** By using a shared IR we can reuse abstractions across multiple languages. In this manner we will build a suite of generic tools to ease porting existing applications in multiple languages, including the CloudButton genomics and geospatial use-cases.

5   Conclusions and Future Directions

In this chapter, we have first analyzed three important architectural trade-offs of serverless computing: disaggregation, isolation, and simple scheduling. We have explained that by relaxing those trade-offs, it is possible to achieve higher performance, but also how that loosening can impoverish important serverless traits such as elasticity, multi-tenancy support, and high resource utilization. Moving the trade-offs to the extremes, we have distinguished between serverful and serverless computing, and we have also introduced the new concept of *ServerMix* computing.

ServerMix systems combine serverless and serverful components to accomplish an analytics task. An ideal *ServerMix* system should keep resource provisioning transparent to the user and consider the cost-performance ratio as first citizen.

Finally, we have presented the CloudButton Serverless Data Analytics Platform and explained how it addresses the aforementioned trade-offs. CloudButton has demonstrated different levels of transparency for applications, enabling to run unmodified single-machine code over effectively unlimited compute, storage, and memory resources thanks to serverless disaggregation and auto-scaling. We predict that next-generation Cloud systems will offer a fully Serverless experience to users by combining both Serverless and Serverful infrastructures in a transparent way.

Acknowledgments This work has been partially supported by the EU project H2020 "Cloud-Button: Serverless Data Analytics Platform" (825184) and by the Spanish government (PID2019-106774RB-C22). Thanks also to the Serra Hunter programme from the Catalan government.

References

1. Akkus, I. E., Chen, R., Rimac, I., Stein, M., Satzke, K., Beck, A., Aditya, P., & Hilt, V. (2018). SAND: Towards high-performance serverless computing. In 2018 USENIX annual technical conference (ATC'18), (pp. 923–935).
2. Al-Ali, Z., Goodarzy, S., Hunter, E., Ha, S., Han, R., Keller, E., & Rozner, E. (2018). Making serverless computing more serverless. In IEEE 11th international conference on cloud computing (CLOUD'18), (pp. 456–459).
3. Amazon: AWS lambda limits (2019). https://docs.aws.amazon.com/lambda/latest/dg/limits.html/
4. Attiya, H., & Welch, J. L. (1991). Sequential consistency versus linearizability (extended abstract). In Proceedings of the third annual ACM symposium on parallel algorithms and architectures (SPAA '91), (pp. 304–315).
5. Barcelona-Pons, D., Sánchez-Artigas, M., París, G., Sutra, P., & García-López, P. (2019). On the faas track: Building stateful distributed applications with serverless architectures. In Proceedings of the 20th international middleware conference (pp. 41–54).
6. Boucher, S., Kalia, A., Andersen, D. G., & Kaminsky, M. (2018). Putting the micro back in microservice. In 2018 USENIX annual technical conference (ATC '18) (pp. 645–650).
7. Carreira, J., Fonseca, P., Tumanov, A., Zhang, A. M., & Katz, R. (2018). A case for serverless machine learning. In: Workshop on systems for ML and open source software at NeurIPS.
8. H2020 CloudButton (2019) Serverless data analytics. http://cloudbutton.eu
9. Fastly: Fastly Labs—Terrarium (2019). https://www.fastlylabs.com/
10. Fouladi, S., Wahby, R. S., Shacklett, B., Balasubramaniam, K. V., Zeng, W., Bhalerao, R., Sivaraman, A., Porter, G., & Winstein, K. (2017). Encoding, fast and slow: Low-latency video processing using thousands of tiny threads. In Proceedings of the 14th USENIX symposium on networked systems design and implementation (NSDI'17) (pp. 363–376).
11. Fouladi, S., Romero, F., Iter, D., Li, Q., Chatterjee, S., Kozyrakis, C., Zaharia, M., & Winstein, K. (2019). From laptop to lambda: Outsourcing everyday jobs to thousands of transient functional containers. In 2019 USENIX Annual Technical Conference (ATC'19) (pp. 475–488).

12. Gao, P. X., Narayan, A., Karandikar, S., Carreira, J., Han, S., Agarwal, R., Ratnasamy, S., & Shenker, S. (2016). Network requirements for resource disaggregation. In Proceedings of the 12th USENIX conference on operating systems design and implementation (OSDI'16) (pp. 249–264).

13. García-López, P., Slominski, A., Shillaker, S., Behrendt, M., & Metzler, B. (2020). Serverless end game: Disaggregation enabling transparency. arXiv preprint arXiv:2006.01251.

14. Gu, J., Lee, Y., Zhang, Y., Chowdhury, M., & Shin, K.G. (2017). Efficient memory disaggregation with infiniswap. In 14th USENIX conference on networked systems design and implementation (NSDI'17) (pp. 649–667).

15. Haas, A., Rossberg, A., Schuff, D. L., Titzer, B. L., Holman, M., Gohman, D., Wagner, L., Zakai, A., & Bastien, J. (2017). Bringing the web up to speed with webassembly. In Proceedings of the 38th ACM SIGPLAN conference on programming language design and implementation (PLDI'17) (pp. 185–200).

16. Haas, A., Rossberg, A., Schuff, D. L., Titzer, B. L., Holman, M., Gohman, D., Wagner, L., Zakai, A., & Bastien, J. (2017). Bringing the web up to speed with WebAssembly. In Proceedings of the 38th ACM SIGPLAN conference on programming language design and implementation (PLDI'17) (pp. 185–200).

17. Hellerstein, J. M., Faleiro, J., Gonzalez, J. E., Schleier-Smith, J., Sreekanti, V., Tumanov, A., & Wu, C. (2019). Serverless computing: One step forward, two steps back. In Conference on innovative data systems research (CIDR'19).

18. Istvan, Z., Sidler, D., & Alonso, G. (2018). Active pages 20 years later: Active storage for the cloud. IEEE Internet Computing, 22(4), 6–14.

19. Jonas, E., Pu, Q., Venkataraman, S., Stoica, I., & Recht, B. (2017). Occupy the cloud: Distributed computing for the 99%. In Proceedings of the 2017 symposium on cloud computing (SoCC'17) (pp. 445–451).

20. Jonas, E., et al. (2019). Cloud programming simplified: A Berkeley view on serverless computing. https://arxiv.org/abs/1902.03383

21. Kim, Y., & Lin, J. (2018). Serverless data analytics with Flint. CoRR abs/1803.06354. http://arxiv.org/abs/1803.06354

22. Klimovic, A., Wang, Y., Stuedi, P., Trivedi, A., Pfefferle, J., & Kozyrakis, C. (2018). Pocket: Elastic ephemeral storage for serverless analytics. In Proceedings of the 13th USENIX symposium on operating systems design and implementation (OSDI'18) (pp. 427–444).

23. Lamport, L. (1998). The part-time parliament. ACM Transactions on Computer Systems, 16(2), 133–169. doi:http://doi.acm.org/10.1145/279227.279229

24. López, P. G., Arjona, A., Sampé, J., Slominski, A., Villard, L. (2020). Triggerflow: Trigger-based orchestration of serverless workflows. In: Proceedings of the 14th ACM international conference on distributed and event-based systems (pp. 3–14).

25. Manco, F., Lupu, C., Schmidt, F., Mendes, J., Kuenzer, S., Sati, S., Yasukata, K., Raiciu, C., & Huici, F. (2017). My VM is lighter (and safer) than your container. In Proceedings of the 26th symposium on operating systems principles (SOSP '17) (pp. 218–233).

26. Oakes, E., Yang, L., Zhou, D., Houck, K., Harter, T., Arpaci-Dusseau, A., & Arpaci-Dusseau, R. (2018). SOCK: Rapid task provisioning with serverless-optimized containers. In: 2018 USENIX annual technical conference (ATC'18) (pp. 57–70).

27. Ongaro, D., & Ousterhout, J.K. (2014). In search of an understandable consensus algorithm. In 2014 USENIX conference on USENIX annual technical conference (ATC'14) (pp. 305–319).

28. Pu, Q., Venkataraman, S., & Stoica, I. (2019). Shuffling, fast and slow: Scalable analytics on serverless infrastructure. In: Proceedings of the 16th USENIX symposium on networked systems design and implementation (NSDI'19) (pp. 193–206).

29. Riedel, E., Gibson, G. A., Faloutsos, C. (1998). Active storage for large-scale data mining and multimedia. In Proceedings of the 24rd international conference on very large data bases (VLDB'98) (pp. 62–73).

30. Sampé, J., Sánchez-Artigas, M., García-López, P., & París, G. (2017). Data-driven serverless functions for object storage. In Proceedings of the 18th ACM/IFIP/USENIX middleware conference (pp. 121–133). ACM, New York.

31. Sampé, J., Vernik, G., Sánchez-Artigas, M., & García-López, P. (2018). Serverless data analytics in the IBM Cloud. In Proceedings of the 19th international middleware conference industry, Middleware '18 (pp. 1–8). ACM, New York.
32. Sampé, J., García-López, P., Sánchez-Artigas, M., Vernik, G., Roca-Llaberia, P., & Arjona, A. (2021). Toward multicloud access transparency in serverless computing. *IEEE Software, 38*, 68–74.
33. Shankar, V., Krauth, K., Pu, Q., Jonas, E., Venkataraman, S., Stoica, I., Recht, B., & Ragan-Kelley, J. (2018). Numpywren: Serverless linear algebra. CoRR abs/1810.09679.
34. Shillaker, S., & Pietzuch, P. (2020). Faasm: Lightweight isolation for efficient stateful serverless computing. arXiv preprint arXiv:2002.09344.
35. Wada, H., Fekete, A., Zhao, L., Lee, K., & Liu, A. (2011). Data consistency properties and the trade-offs in commercial cloud storage: the consumers' perspective. In Fifth Biennial conference on innovative data systems research (CIDR'11) (pp. 134–143).
36. Zillner, S., Curry, E., Metzger, A., Auer, S., & Seidl, R. (Eds.) (2017). European big data value strategic research & innovation agenda. Big data value association
37. Zillner, S., et al. (Eds.) (2020). Strategic research, innovation and deployment Agenda—AI, data and robotics partnership. Third Release. BDVA, euRobotics, ELLIS, EurAI and CLAIRE.

Big Data and AI Pipeline Framework: Technology Analysis from a Benchmarking Perspective

Arne J. Berre, Aphrodite Tsalgatidou, Chiara Francalanci, Todor Ivanov, Tomas Pariente-Lobo, Ricardo Ruiz-Saiz, Inna Novalija, and Marko Grobelnik

Abstract Big Data and AI Pipeline patterns provide a good foundation for the analysis and selection of technical architectures for Big Data and AI systems. Experiences from many projects in the Big Data PPP program has shown that a number of projects use similar architectural patterns with variations only in the choice of various technology components in the same pattern. The project DataBench has developed a Big Data and AI Pipeline Framework, which is used for the description of pipeline steps in Big Data and AI projects, and supports the classification of benchmarks. This includes the four pipeline steps of Data Acquisition/Collection and Storage, Data Preparation and Curation, Data Analytics with AI/Machine Learning, and Action and Interaction, including Data Visualization and User Interaction as well as API Access. It has also created a toolbox which supports the identification and use of existing benchmarks according to these steps in addition to all of the different technical areas and different data types in the BDV Reference Model. An observatory, which is a tool, accessed via the toolbox, for observing the popularity,

A. J. Berre
SINTEF Digital, Oslo, Norway
e-mail: Arne.J.Berre@sintef.no

A. Tsalgatidou (✉)
SINTEF, Oslo, Norway

Department of Informatics and Telecommunications, National and Kapodistrian University of Athens, Athens, Greece

C. Francalanci
Politecnico di Milano, Milan, Italy

T. Ivanov
Lead Consult, Sofia, Bulgaria

T. Pariente-Lobo · R. Ruiz-Saiz
ATOS Research and Innovation, Madrid, Spain

I. Novalija · M. Grobelnik
Jožef Stefan Institute, Ljubljana, Slovenia

© The Author(s) 2022
E. Curry et al. (eds.), *Technologies and Applications for Big Data Value*,
https://doi.org/10.1007/978-3-030-78307-5_4

63

importance and the visibility of topic terms related to Artificial Intelligence and Big
Data technologies has also been developed and is described in this chapter.

Keywords Benchmarking · Big Data and AI Pipeline · Blueprint · Toolbox ·
Observatory

1 Introduction

Organizations rely on evidence from the benchmarking domain to provide answers
to how their processes are performing. There is extensive information on why and
how to perform technical benchmarks for the specific management and analytics
processes, but there is a lack of objective, evidence-based methods to measure
the correlation between Big Data Technology (BDT) benchmarks and business
benchmarks of an organization and demonstrate return on investment. When more
than one benchmarking tool exist for a given need, there is even less evidence as
to how these tools compare to each other, and how the results can affect their
business objectives. The DataBench project has addressed this gap by designing
a framework to help European organizations developing BDT to reach for excellence and constantly improve their performance, by measuring their technology
development activity against parameters of high business relevance. It thus bridges
the gap between technical and business benchmarking of Big Data and Analytics
applications.

   The Business Benchmarks of DataBench focus on Quantitative benchmarks
for areas such as Revenue increase, Profit increase and Cost reduction and on
Qualitative benchmarks for areas such as Product and Service quality, Customer
satisfaction, New Products and Services launched and Business model innovation.
The business benchmarks are calculated on the basis of certain business Key
Performance Indicators (KPIs). The business KPIs selected by the project are valid
metrics and can be used as benchmarks for comparative purposes by researchers or
business users for each of the industry and company-size segments measured. These
indicators have been classified in the following four features which group relevant
indicators from different points of views: Business features, Big Data Application
features, Platform and Architecture features, Benchmark-specific features. For
each feature, specific indicators have been defined. Actually, none of the existing
Big Data benchmarks make any attempt to relate the technical measurements
parameters, metrics and KPIs (like Latency, Fault tolerance, CPU utilization, Memory utilization, Price performance, Energy, bandwidth, data access patterns, data
processed per second, data processed per joule, query time, execution time, number
of completed jobs) with the business metrics and KPIs (like operational efficiency,
increased level of transparency, optimized resource consumption, improved process
quality and performance), customer experience (increased customer loyalty and
retention, precise customer segmentation and targeting, optimized customer interaction and service), or new business models (expanded revenue streams from existing
products, creation of new revenue streams from entirely (new) data products).

This chapter presents a Big Data and AI Pipeline Framework that supports technology analysis and benchmarking for both the horizontal and vertical technical priorities of the European Big Data Value Strategic Research and Innovation Agenda [1], and also for the cross-sectorial technology enablers of the AI, Data and Robotics Strategic Research, Innovation and Deployment Agenda [2]. In the following sections, we focus on the DataBench approach for Technical Benchmarks which are using a Big Data and AI Pipeline model as an overall framework, and they are further classified depending on the various areas of the Big Data Value (BDV) Reference Model. Technical benchmarks are also related to the areas of the AI Strategic Research, Innovation and Deployment Agenda (SRIDA) [1] and the ISO SC42 Big Data and AI Reference models [3].

The DataBench Framework is accompanied by a Handbook and a Toolbox, which aim to support industrial users and European technology developers who need to make informed decisions on Big Data Technologies investments by optimizing technical and business performance. The Handbook presents and explains the main reference models used for technical benchmarking analysis. The Toolbox is a software tool that provides access to benchmarking services; it helps stakeholders (1) to identify the use cases where they can achieve the highest possible business benefit and return on investment, so they can prioritize their investments; (2) to select the best technical benchmark to measure the performance of the technical solution of their choice; and (3) to assess their business performance by comparing their business impacts with those of their peers, so they can revise their choices or their organization if they find they are achieving less results than median benchmarks for their industry and company size. Therefore, the services provided by the Toolbox and the Handbook support users in all phases of their journey (before, during and in the ex-post evaluation of their BDT investment) and from both the technical and business viewpoints.

In the following section, we present the Big Data and AI Pipeline Framework, which is used for the description of pipeline steps in Big Data and AI projects, and which supports the classification of benchmarks; the framework also serves as a basis for demonstrating the similarities among Big Data projects such as those in the Big Data Value Public-Private Partnership (BDV PPP) program [4]. We also discuss its relationship with the BDV Reference Model and the Strategic Research and Innovation Agenda (SRIA) [1] and the Strategic Research, Innovation and Deployment Agenda (SRIDA) for a European AI, Data and Robotics Partnership (AI PPP SRIDA) [2]. In Sect. 3, we present Big Data and AI Pipeline examples from the DataBio project [5] for IoT, Graph and SpatioTemporal data. In Sect. 4, we present categorizations of architectural blueprints for realisations of the various steps of the Big Data and AI Pipeline with variations depending on the processing types (batch, real-time, interactive), the main data types involved and on the type of access/interaction (which can be API access action/interaction or a Human interaction). Specializations can also be more complex aggregations/-compositions of multiple specializations/patterns. These blueprints are a basis for selecting specializations of the pipeline that will fit the needs of various projects and instantiations. Section 5 presents how existing Big Data and AI Technical Benchmarks have been classified according to the Big Data and AI Pipeline

Framework that is presented in Sect. 2. These are benchmarks that are suitable for benchmarking of technologies related to the different parts of the pipeline and associated technical areas. Section 6 describes the DataBench Toolbox as well as the DataBench Observatory, which is a tool (accessed via the toolbox) for observing the popularity, importance and the visibility of topic terms related to Artificial Intelligence and Big Data, with particular attention dedicated to the concepts, methods, tools and technologies in the area of Benchmarking. Finally, the conclusions in Sect. 7 present a summary of the contributions and plans for further evolution and usage of the DataBench Toolbox.

2   The Big Data and AI Pipeline Framework

The Big Data and AI Pipeline Framework is based on the elements of the Big Data Value (BDV) Big Data Value Reference Model, developed by the Big Data Value Association (BDVA) [1]. In order to have an overall usage perspective on Big Data and AI systems, a top-level generic pipeline has been introduced to understand the connections between the different parts of a Big Data and AI system in the context of an application flow. Figure 1 depicts this pipeline, following the Big Data and AI Value chain.

The steps of the Big Data and AI Pipeline Framework are also harmonized with the ISO SC42 AI Committee standards [3], in particular the *Collection, Preparation, Analytics* and *Visualization/Access* steps within the Big Data Application Layer of the recent international standard ISO 20547-3 Big Data reference architecture within the functional components of the Big Data Reference Architecture [3, 6]. The following figure shows how the Big Data and AI Pipeline can also be related to the BDV Reference Model and the AI PPP Ecosystem and Enablers (from SRIDA AI). Benchmarks often focus on specialized areas within a total system typically identified by the BDV Reference Model. This is in particular useful for the benchmarking of particular technical components. Benchmarks can also be directly or indirectly linked to the steps of a Big Data and AI Pipeline, which is useful when benchmarks are being considered from a Big Data and AI application perspective, where practical project experiences has shown that these steps can easily be recognized in most application contexts.

Benchmarks are useful both for the evaluation of alternative technical solutions within a Big Data and AI project and for comparing new technology develop-

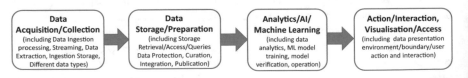

Fig. 1 Top-level, generic Big Data and AI Pipeline pattern

ments/products with alternative offerings. This can be done both from a technical area perspective for selected components and from a pipeline step perspective when seen from the steps of a Big Data and AI application.

As it can be seen in Fig. 1, this pipeline is at a high level of abstraction. Therefore, it can be easily specialized in order to describe more specific pipelines, depending on the type of data and the type of processing (e.g. IoT data and real-time processing). The 3D cube in Fig. 2 depicts the steps of this pipeline in relation to the type of data processing and the type of data being processed. As we can see in this figure, the type of data processing, which has been identified as a separate topic area in the BDV Reference Model, is orthogonal to the pipeline steps and the data types. This is due to the fact that different processing types, like batch/data-at-rest and real-time/data-in-motion and interactive, can span across different pipeline steps and can handle different data types, as the ones identified in the BDV Reference Model, within each of the pipeline steps. Thus, there can be different data types like structured data, times series data, geospatial data, media, image, video and audio data, text data, including natural language data, and graph data, network/web data and metadata, which can all imply differences in terms of storage and analytics techniques.

Other dimensions can similarly be added for a multi-dimensional cube, e.g. for Application domains, and for the different horizontal and vertical technology areas of the BDV Reference Model, and for the technology locations of the Computing Continuum/Trans Continuum—from Edge, through Fog to Cloud and HPC—for the actual location of execution of the four steps, which can happen on all these levels. The same orthogonality can be considered for the area of Data Protection, with Privacy and anonymization mechanisms to facilitate data protection. It also has links to trust mechanisms like Blockchain technologies, smart contracts and various

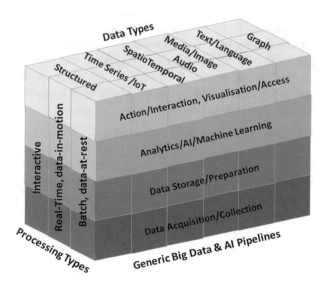

Fig. 2 Top-level, generic Big Data and AI Pipeline cube

forms for encryption. This area is also associated with the area of CyberSecurity, Risk and Trust.

The BDV Reference Model shown in Fig. 3 has been developed by the BDVA [1], taking into account input from technical experts and stakeholders along the whole Big Data Value chain as well as interactions with other related Public–Private Partnerships (PPPs). An explicit aim of the BDV Reference Model in the SRIA 4.0 document is to also include logical relationships to other areas of a digital platform such as Cloud, High Performance Computing (HPC), IoT, Networks/5G, CyberSecurity, etc.

The following describes the steps of the Big Data and AI Pipeline shown on the left of the BDV Reference Model in Fig. 3, with lines connecting them to the typical usage of some of the main technical areas.

Data Acquisition/Collection This step includes acquisition and collection from various sources, including both streaming data and data extraction from relevant external data sources and data spaces. It includes support for handling all relevant data types and also relevant data protection handling for this step. This step is often associated with the use of both real-time and batch data collection, and associated streaming and messaging systems. It uses enabling technologies in the area using data from things/assets, sensors and actuators to collect streaming data-in-motion as well as connecting to existing data sources with data-at-rest. Often, this step also includes the use of relevant communication and messaging technologies.

Data Storage/Preparation This step includes the use of appropriate storage systems and data preparation and curation for further data use and processing. Data storage includes the use of data storage and retrieval in different databases systems—both SQL and NoSQL, like key-value, column-based storage, document storage and graph storage, as well as storage structures such as file systems. This is an area where there historically exist many benchmarks to test and compare various data storage alternatives. Tasks performed in this step also include further data preparation and curation as well as data annotation, publication and presentation of the data in order to be available for discovery, reuse and preservation. Further in this step, there is also interaction with various data platforms and data spaces for broader data management and governance. This step is also linked to handling associated aspects of data protection.

Analytics/AI/Machine Learning This step handles data analytics with relevant methods, including descriptive, predictive and prescriptive analytics and use of AI/Machine Learning methods and algorithms to support decision making and transfer of knowledge. For Machine learning, this step also includes the subtasks for necessary model training and model verification/validation and testing, before actual operation with input data. In this context, the previous step of data storage and preparation will provide data input both for training and validation and test data, as well as operational input data.

Action/Interaction, Visualization and Access This step (including data presentation environment/boundary/user action and interaction) identifies the boundary between the system and the environment for action/interaction, typically through

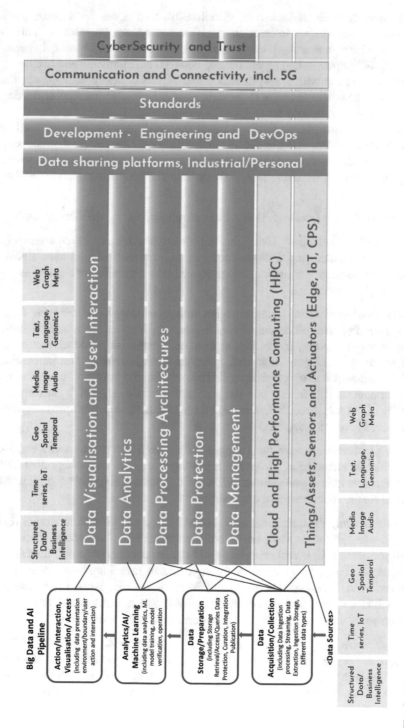

Fig. 3 Big Data and AI Pipeline using technologies from the BDV Reference Model

a visual interface with various data visualization techniques for human users and through an API or an interaction interface for system boundaries. This is a boundary where interactions occur between machines and objects, between machines, between people and machines and between environments and machines. The action/interaction with the system boundaries can typically also impact the environment to be connected back to the data acquisition/collection step, collecting input from the system boundaries.

The above steps can be specialized based on the different data types used in the various applications, and are set up differently based on different processing architectures, such as batch, real-time/streaming or interactive. Also, with Machine Learning there is a cycle starting from training data and later using operational data (Fig. 4).

The steps of the Big Data and AI Pipeline can relate to the AI enablers as follows:

Data Acquisition/Collection Using enablers from *Sensing and Perception* technologies, which includes methods to access, assess, convert and aggregate signals that represent real-world parameters into processable and communicable data assets that embody perception.

Data Storage/Preparation Using enablers from *Knowledge and learning technologies*, including data processing technologies, which cover the transformation, cleaning, storage, sharing, modelling, simulation, synthesising and extracting of insights of all types of data, both that gathered through sensing and perception as well as data acquired by other means. This will handle both training data and operational data. It will further use enablers for *Data for AI*, which handles the availability of the data through data storage through data spaces, platforms and data marketplaces in order to support data-driven AI.

Analytics/AI/Machine Learning Using enablers from *Reasoning and Decision making* which is at the heart of Artificial Intelligence. This technology area also provides enablers to address optimization, search, planning, diagnosis and relies on methods to ensure robustness and trustworthiness.

Action/Interaction, Visualization and Access Using enablers from *Action and Interaction*—where Interactions occur between machines and objects, between machines, between people and machines and between environments and machines. This interaction can take place both through human user interfaces as well as through various APIs and system access and interaction mechanisms. The action/interaction with the system boundaries can typically also be connected back to the data acquisition/collection step, collecting input from the system boundaries.

These steps are also harmonized with the emerging pipeline steps in ISO SC42 AI standard of "Framework for Artificial Intelligence (AI) Systems Using Machine Learning (ML), ISO/IEC 23053, with Machine Learning Pipeline with the related steps of *Data Acquisition, Data Pre-processing, Modeling, Model Deployment and Operation*.

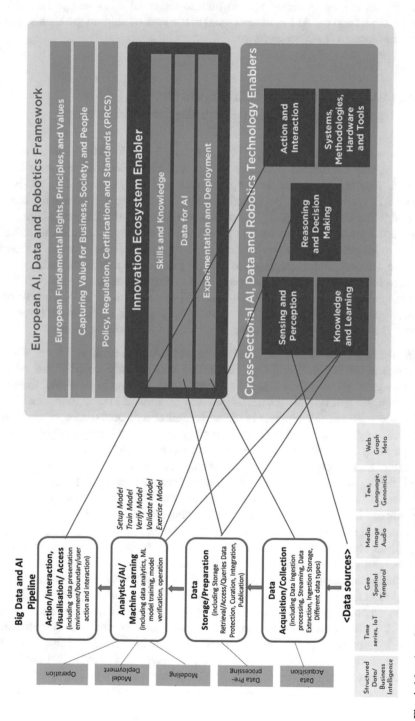

Fig. 4 Mappings between the top-level generic Big Data and AI Pipeline and the European AI and Robotics Framework

Benchmarks can be identified related both to technical areas within the BDV Reference Model and the AI Frameworks and to the various steps in the DataBench Toolbox that supports both perspectives.

3 Big Data and AI Pipeline Examples for IoT, Graph and SpatioTemporal Data: From the DataBio Project

In the following, we present example pipelines which handle different data types. Specifically, they handle IoT data, Graph data and Earth Observation/Geospatial data. Each pipeline is mapped to the four phases of the top-level Generic Big Data and AI Pipeline pattern, presented in Sect. 2. All these pipelines have been developed in the DataBio project [5], which was funded by the European Union's Horizon 2020 research and innovation programme. DataBio focused on utilizing Big Data to contribute to the production of the best possible raw materials from agriculture, forestry, and fishery/aquaculture for the bioeconomy industry in order to produce food, energy and biomaterials, also taking into account responsibility and sustainability issues. The pipelines that are presented below are the result of aggregating Big Data from the three aforementioned sectors (agriculture, forestry and fishery) and intelligently processing, analysing and visualising them.

Pipeline for IoT Data Real-Time Processing and Decision Making
The "Pipeline for IoT data real-time processing and decision making" has been applied to three pilots in the DataBio project from the agriculture and fishery domain, and, since it is quite generic, it can also be applied to other domains. The main characteristic of this pipeline is the collection of real-time data coming from IoT devices to generate insights for operational decision making by applying real-time data analytics on the collected data. Streaming data (a.k.a. events) from IoT sensors (e.g. are collected in real-time, for example: agricultural sensors, machinery sensors, fishing vessels monitoring equipment. These streaming data can then be pre-processed in order to lower the amount of data to be further analysed. Pre-processing can include filtering of the data (filtering out irrelevant data and filtering in only relevant events), performing simple aggregation of the data, and storing the data (e.g. on cloud or other storage model, or even simply as a computer's file system) such that conditional notification on data updates to subscribers can be done. After being pre-processed, data enters the complex event processing (CEP) [7] component for further analysis, which generally means finding patterns in time windows (temporal reasoning) over the incoming data to form new, more complex events (a.k.a. situations or alerts/warnings). These complex events are emitted to assist in decision-making processes either carried out by humans ("human in the loop" [8]) or automatically by actuators, e.g. sensors that start irrigation in a greenhouse as a result of a certain alert. The situations can also be displayed using visualization tools to assist humans in the decision-making process (as, e.g., in [8]).

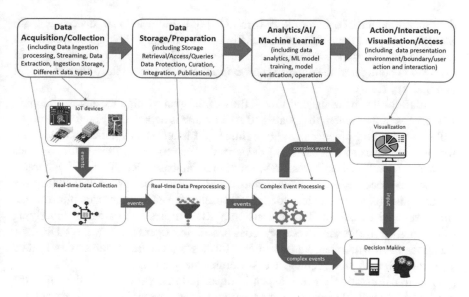

Fig. 5 Mapping of steps of the "Pipeline for IoT data real-time processing and decision making" to the "Generic Big Data and AI Pipeline" steps

The idea is that the detected situations can provide useful real-time insights for operational management (e.g. preventing a possible crop pest or machinery failure).

Figure 5 shows the steps of the pipeline for real-time IoT data processing and decision making that we have just described and their mapping to the steps of top-level Generic Big Data and AI Pipeline pattern that we have analysed in Sect. 2.

Pipeline for Linked Data Integration and Publication

In the DataBio project and some other agri-food projects, Linked Data has been extensively used as a federated layer to support large-scale harmonization and integration of a large variety of data collected from various heterogeneous sources and to provide an integrated view on them. The triplestore populated with Linked Data during the course of DataBio project (and a few other related projects) resulted in creating a repository of over one billion triples, making it one of the largest semantic repositories related to agriculture, as recognized by the EC innovation radar naming it the "Arable Farming Data Integrator for Smart Farming". Additionally, projects like DataBio have also helped in deploying different endpoints providing access to the dynamic data sources in their native format as Linked Data by providing a virtual semantic layer on top of them. This action has been realized in the DataBio project through the implementation of the instantiations of a "Pipeline for the Publication and Integration of Linked Data", which has been applied in different use cases related to the bioeconomy sectors. The main goal of these pipeline instances is to define and deploy *(semi-)automatic processes* to carry out the necessary steps to transform and publish different input datasets for various heterogeneous sources as Linked Data. Hence, they connect different data-processing components to carry out the transformation of data into RDF format [9] or the translation of queries to/from SPARQL [10] and the native data access interface, plus their linking, as well as the mapping specifications to process the input datasets. Each pipeline instance used in DataBio is configured to support specific input dataset types (same format, model and delivery form).

A high-level view of the end-to-end flow of the generic pipeline and its mapping to the steps of the Generic Big Data and AI Pipeline is depicted in Fig. 6. In general, following the best practices and guidelines of Linked Data Publication [11, 12], the pipeline takes as input selected datasets that are collected from heterogeneous sources (shapefiles, GeoJSON, CSV, relational databases, RESTful APIs), curates and/or pre-processes the datasets when needed, selects and/or creates/extends the vocabularies (e.g., ontologies) for the representation of data in semantic format, processes and transforms the datasets into RDF triples according to underlying ontologies, performs any necessary post-processing operations on the RDF data, identifies links with other datasets and publishes the generated datasets as Linked Data, as well as applies required access control mechanisms.

The transformation process depends on different aspects of the data like the format of the available input data, the purpose (target use case) of the transformation and the volatility of the data (how dynamic is the data). Accordingly, the tools and the methods used to carry out the transformation were determined firstly by the format of the input data. Tools like D2RQ [13] were normally used in the case of data coming from relational databases; tools like GeoTriples [14] was chosen mainly for geospatial data in the form of shapefiles; tools like RML Processor [15] for CSV, JSON, XML data formats; and services like Ephedra [16] (within Metaphactory platform) for Restful APIs.

Pipeline for Earth Observation and Geospatial Data Processing
The pipeline for Earth Observation and Geospatial data processing [17], developed in the DataBio project, depicts the common data flow among six project pilots, four of which are from the agricultural domain and two from the fishery domain. To be more specific, from the agricultural domain there are two smart farming pilots, one agricultural insurance pilot and one pilot that provides support to the farmers related to their obligations introduced by the current Common Agriculture Policy [18]. The two pilots from the fishery domain were in the areas of oceanic tuna fisheries immediate operational choice and oceanic tuna fisheries planning.

Some of the characteristics of this pipeline include the following:

- Its initial data input is georeferenced data [19], which might come from a variety of sources such as satellites, drones or even from manual measurements. In general, this will be represented as either in the form of vector or raster data [20]. Vector data usually describes some spatial features in the form of points, lines or polygons. Raster data, on the other hand, is usually generated from image-producing sources such as Landsat or Copernicus satellites.
- Information exchanged among the different participants in the pipeline can be either in raster or vector form. Actually, it is possible and even common that the

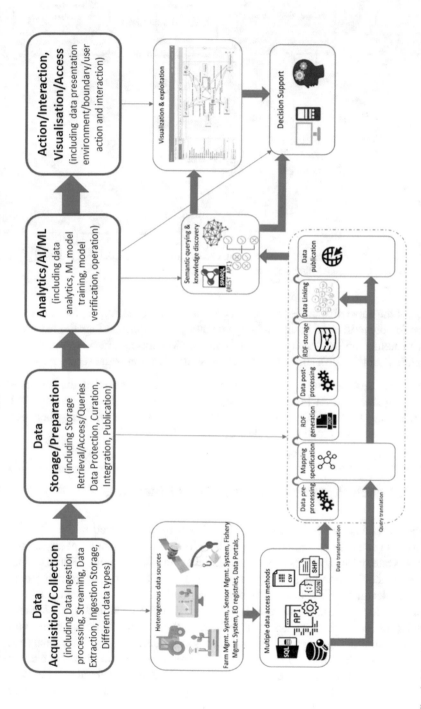

Fig. 6 Mapping of steps of the "Pipeline for linked data integration and publication" to the "Generic big data and AI Pipeline" steps

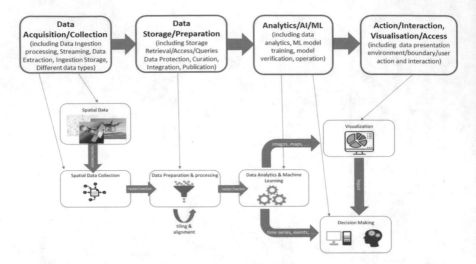

Fig. 7 Mapping of steps of the "Pipeline for earth observation and geospatial data processing" to the "Generic Big Data and AI Pipeline" steps

form of the data will change from one step to another. For example, this can result from feature extraction based on image data or pre-rendering of spatial features.

• For visualization or other types of user interaction options, information can be provided in other forms like: images, maps, spatial features, time series or events.

Therefore, this pipeline can be considered as a specialization of the top-level Generic Big Data and AI Pipeline pattern, presented in Sect. 2, as it concerns the data processing for Earth Observation and Geospatial data. The mapping between the steps of these two pipelines can be seen in Fig. 7.

4 DataBench Pipeline Framework and Blueprints

The top-level Generic Big Data and AI Pipeline pattern discussed in the previous sections has been used as a reference to build architectural blueprints specifying the technical systems/components needed at different stages in a pipeline. For example, in the data acquisition phase of a pipeline, a software broker synchronizing data source and destination is needed. A data acquisition broker will then send data to a lambda function that transforms data in a format that can be stored in a database. In DataBench, we have identified and classified all these technical components with an empirical bottom-up approach as follows: we started from Big Data Analytics (BDA) use cases and then we recognized the commonalities among the technical requirements of the different use cases. Finally, we designed a general architectural blueprint, an overview of which is depicted in Fig. 8. This figure has been detailed in Figs. 9 and 10 for better readability.

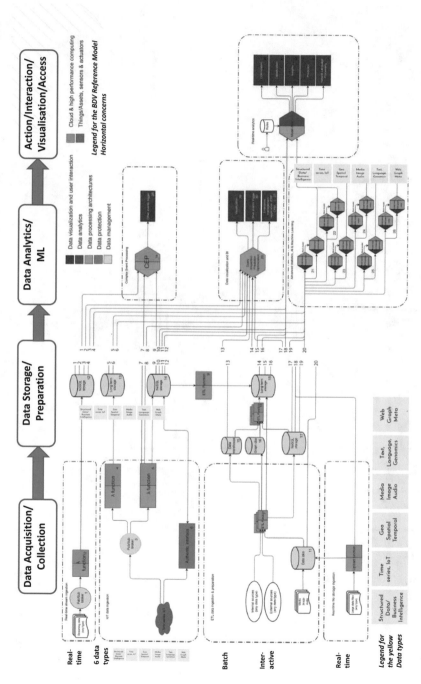

Fig. 8 General architectural blueprint for BDA Pipelines

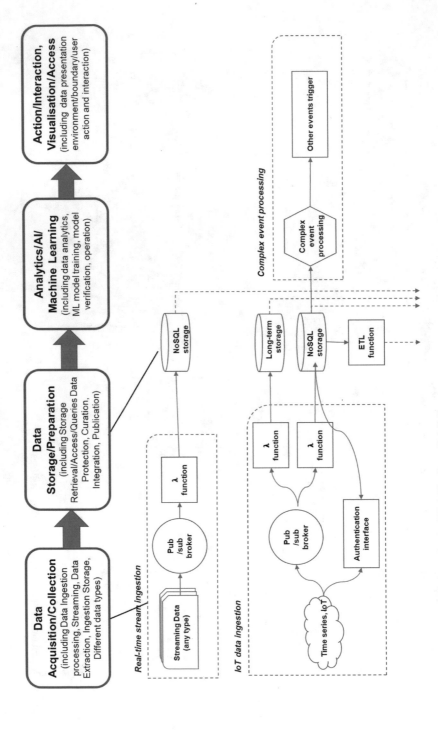

Fig. 9 General architectural blueprint for BDA Pipelines (detail 1)

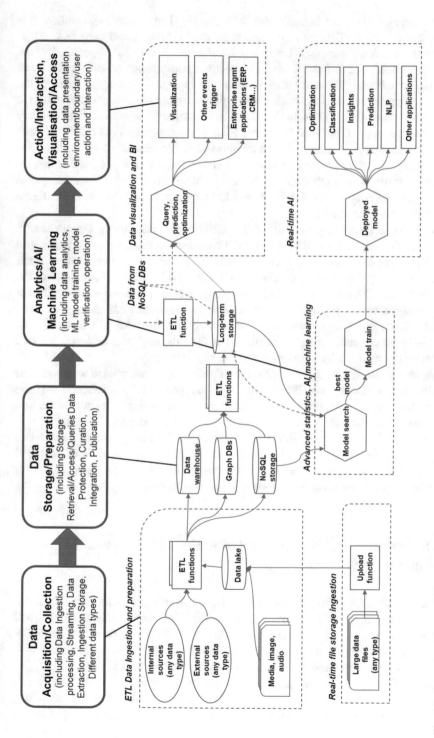

Fig. 10 General architectural blueprint for BDA Pipelines (detail 2)

The general blueprint is consistent with the Big Data Value Association data types classification. This can be seen, e.g., in the "Advanced statistics, AI & Machine Learning" area (see bottom right of Figs. 8 and 10). Specifically, the "Model Search" and "Model Train" architectural components (depicted clearly in Fig. 10) have been replicated for every data type in Fig. 8. The components of the general blueprint can be also seen in the perspective of the horizontal concerns of the BDV Reference Model. Thus, in Fig. 8, we have assigned a specific colour to every horizontal concern of the BDV Reference Model (see *Legend for the BDV Reference Model horizontal concerns*) and each component of the general blueprint has been associated with one or more horizontal concerns by using the respective colours. By mapping the different components of the general blueprint to the horizontal concerns of the BDV Reference Model, we can highlight the interaction among the different Big Data conceptual areas.

We have ensured the generality of this blueprint by addressing the needs of a cross-industry selection of BDA use cases. This selection has been performed based on a European-level large-scale questionnaire (see DataBench deliverables D2.2, D2.3 and D2.4 and desk analyses D4.3, D4.3 and D4.4 in [21]) that have shown the most frequent BDA use cases per industry. We have also conducted an in-depth case study analysis with a restricted sample of companies to understand the functional and technical requirements of each use case. Based on this body of knowledge, we have designed an architectural blueprint for each of the use cases and then inferred the general blueprint which is depicted in the above figure.

We would like to note that the general blueprint can be instantiated to account for the different requirements of different use cases and projects. In DataBench, we have derived use-case-specific blueprints from the general blueprint. The entire collection of use-case-specific blueprints is available from the DataBench Toolbox, as it is discussed in the following section. The Toolbox guides the user from the end-to-end process of the pipelines, the selection of a use case, the specification of technical requirements, down to the selection and benchmarking of specific technologies for the different components of the use-case-specific blueprint.

5   Technical Benchmarks Related to the Big Data and AI Pipeline Framework

As mentioned before, the goal of the DataBench framework is to help practitioners discover and identify the most suitable Big Data and AI technologies and benchmarks for their application architectures and use cases. Based on the BDV Reference Model layers and categories, we initially developed a classification with more than 80 Big Data and AI benchmarks (currently between 1999 and 2020) that we called Benchmark matrix [22]. Then, with the introduction of the DataBench Pipeline Framework, we further extended the benchmark classification to include the pipeline steps and make it easier for practitioners to navigate and search through

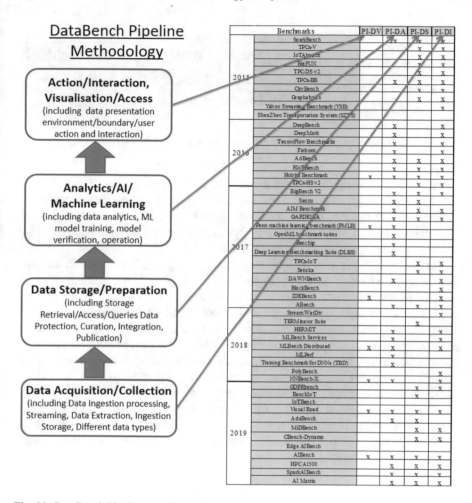

Fig. 11 DataBench Pipeline mapping to benchmarks

the Benchmark matrix. Figure 11 depicts the mapping between the four pipeline steps and the classified benchmarks.

In addition to the mapping of existing technical benchmarks into the four main pipeline steps, there also have been mappings for relevant benchmarks for all of the horizontal and vertical areas of the BDV Reference model. This includes vertical benchmarks following the different data types, such as Structured Data Benchmarks, IoT/Time Series and Stream processing Benchmarks, SpatioTemporal Benchmarks, Media/Image Benchmarks, Text/NLP Benchmarks and Graph/Metadata/Ontology-Based Data Access Benchmarks. It also includes horizontal benchmarks such as benchmarks for Data Visualization (visual analytics), Data Analytics, AI and Machine Learning; Data Protection: Privacy/Security Management Benchmarks

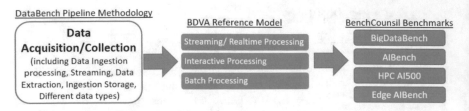

Fig. 12 DataBench Pipeline step mapping to specific category of Benchmarks

related to data management; Data Management: Data Storage and Data Management Benchmarks, Cloud/HPC, Edge and IoT Data Management Benchmarks.

The overall number of technical benchmarks that have been identified and described for these areas are close to 100. All identified benchmarks have been made available through the DataBench Toolbox.

As we can see in Fig. 11, the steps are quite general and map to multiple benchmarks, which is very helpful for beginners that are not familiar with the specific technology types. Similarly, advanced users can go quickly in the pipeline steps and focus on a specific type of technology like batch processing. In this case, focusing on a specific processing category reduces the number of retrieved benchmarks, like in the example in Fig. 12, where only four benchmarks from the BenchCouncil are selected. Then, if further criteria like data type or technology implementation are important, the selection can be quickly reduced to a single benchmark that best suits the practitioner requirements.

The above-described approach for mapping between the DataBench Pipeline Framework and the Benchmark matrix is available in the DataBench Toolbox. The toolbox enables multiple benchmark searches and guidelines via a user-friendly interface that is described in the following sections.

6 DataBench Toolbox

The DataBench Toolbox aims to be a one-stop shop for big data/AI benchmarking; as its name implies, it is not a single tool, but rather a "box of tools". It serves as an entry point of access to tools and resources on big data and AI benchmarking. The Toolbox is based on existing efforts in the community of big data benchmarking and insights gained about technical and business benchmarks in the scope of the DataBench project. From the technical perspective, the Toolbox provides a web-based interface to search, browse and, in specific cases, deploy big data benchmarking tools, or direct to the appropriate documentation and source code to do so. Moreover, it allows to browse information related to big data and AI use cases, lessons learned, business KPIs in different sectors of application, architectural blueprints of reference and many other aspects related to benchmarking big data and

AI from a business perspective. The Toolbox provides access via a web interface to this knowledge base encapsulated in what is called "knowledge nuggets".

The main building blocks of the Toolbox are depicted in Fig. 13 and comprise a front-end DataBench Toolbox Web user interface, Toolbox Catalogues and the Toolbox Benchmarking Automation Framework, which serves as a bridge to the Execution of Benchmarks building block located outside the Toolbox.

The intended users of the Toolbox are technical users, business users, benchmarking providers, and administrators. The support and benefits for each type of user is highlighted in their dedicated user journeys accessible from the Toolbox front-page, except for administrators, who are needed to support all kinds of users and facilitate the aggregation and curation of content to the tool. The Toolbox Catalogues building block shown in Fig. 13 comprises the backend functionality and repositories associated with the management, search and browsing of knowledge nuggets and benchmarking tools. The Toolbox Benchmarking Automation Framework building block serves as a bridge to the Execution of Benchmarks building block located in the infrastructure provided by the user outside the Toolbox (in-house or in the cloud), as the Toolbox does not provide a playground to deploy and execute benchmarks. The automation of the deployment and execution of the benchmarks is achieved via the generation of Ansible Playbooks [23] and enabled by an AWX project [24] for process automation. The steps to be followed by a Benchmark Provider with the help of the Administrator to design and prepare the benchmark with the necessary playbooks for the automation from the Toolbox are described in detail in Sect. 3.1 "Support for adding and configuring benchmarks" of DataBench Deliverable D3.4, which can be found in [21]. Last but not least, the DataBench Toolbox Web building block is the main entry point for the users.

The DataBench Toolbox Web user interface is publicly available for searching and browsing, although some of the options are only available to registered users. Registering to the Toolbox is free and can be done from the front-page. This web user interface is accessible via https://toolbox.databench.eu/. Via this interface, a user can access (1) the Big Data Benchmarking Tools Catalogue (2) the Knowledge Nuggets Catalogue; (3) User journeys which provide a set of tips and advice to different categories of users on how to use and navigate throughout the Toolbox; (4) links to other tools such as the DataBench Observatory explained below in this chapter; and (5) search features. The Toolbox provides several search options, including searching via a clickable representation of the BDV Reference Model and via clickable depiction of the big data architectural blueprints and the generic pipeline presented in Sect. 2. The latter type of searching, depicted in Fig. 14, enables accessing technical benchmarks as well as nuggets related to the clicked elements.

As we mentioned before, one of the tools accessible from the Toolbox is the DataBench Observatory. This is a tool for observing the popularity, importance and the visibility of topic terms related to Artificial Intelligence and Big Data, with particular attention dedicated to the concepts, methods, tools and technologies in the area of Benchmarking.

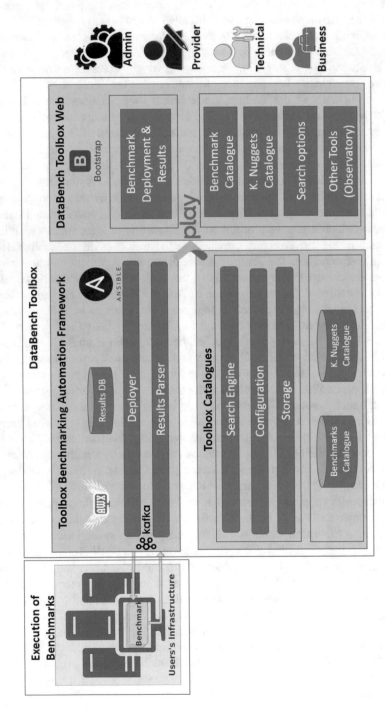

Fig. 13 DataBench Toolbox functional architecture

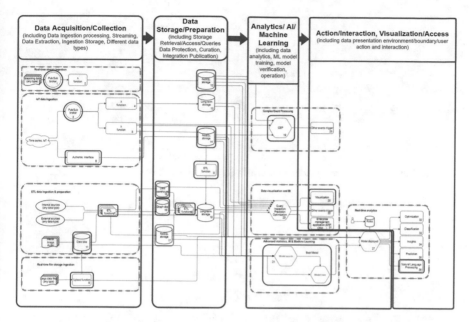

Fig. 14 Search by Pipeline/Blueprint available from the DataBench Toolbox

The DataBench Observatory introduces the popularity index, calculated for ranking the topic terms in time, which is based on the following components: (1) Research component, such as articles from the Microsoft Academic Graph (MAG) [25]; (2) Industry component, such as job advertisements from Adzuna service [26]; (3) Research and Development component, such as EU research projects, e.g. in CORDIS [27] dataset (4) Media component, such as cross-lingual news data from the Event Registry system [28]; (5) Technical Development component, such as projects on GitHub [29]; and (6) General Interest, such as Google Trends. The DataBench observatory provides ranking and trending functionalities, including overall and monthly ranking of topics, tools and technologies, as well as customized trending options. See, e.g., Fig. 15 that demonstrates that Microsoft tools are highly requested in job postings, Python is one of the most popular languages at GitHub (as it is also mentioned in [30, 31]) and users on the web search a lot for Node.js solutions. It is possible here to search for the popularity of various Big Data and AI tools and also for Benchmarks.

See also Fig. 16, which shows time series for topics from the areas of Artificial Intelligence, Big Data and Benchmarking. Users interested in the "Artificial Intelligence" topic can observe its high popularity (score 10 is the maximum normalized popularity) within academic papers.

In the context of the DataBench Toolbox, the DataBench Observatory is targeted at different user groups, such as industrial and business users, academic users, general public, etc. Each type of user can use the Observatory to explore popular topics as well as tools and technologies in areas of their interest.

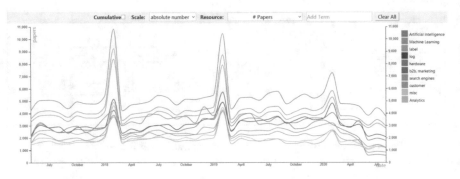

month: All	Topics: All						
Search:							
Topic	Categories	Papers	EU Projects	News	Github	Jobs	Search Volume	Total
---	---	---	---	---	---	---	---	---
Microsoft	infrastructure,computer_science	1.72	1.13	5.24	1.3	10	6.34	4.29
Google	infrastructure,computer_science	3.74	2.23	8.21	2.76	1.27	7.38	4.26
Amplitude	customer,computer_science	5	10	1.03	1.04	1	7.11	4.2
Vector	log,computer_science	10	1	1.19	2.32	1	9.35	4.14
Python	stat_tool,computer_science	1.81	1.06	1.2	10	1	8.96	4.01
Node	b2b_marketing,computer_science	8.71	1	1.17	1.62	1	10	3.92
ARM	hardware,computer_science...	10	1	1.03	1.13	1	9.37	3.92

Fig. 15 DataBench popularity index (Tools and Technologies, accessed in November 2020)

Fig. 16 Time series (Topics, accessed in November 2020)

Furthermore, in order to develop the DataBench observatory tool, we have composed a DataBench ontology based on Artificial Intelligence, Big Data, Benchmarking-related topics from Microsoft Academic Graph and extended/populated the ontology with tools and technologies from the relevant areas, by categories. Microsoft Academic Graph (MAG) taxonomy has been expanded with DataBench terms—over 1700 tools and technologies related to Benchmarking, Big Data, and Artificial Intelligence. New concepts have been aligned with MAG topic, MAG keyword, Wikipedia (for analysis in wikification) and Event Registry concepts. The DataBench ontology is used in the semantic annotation of the unstructured textual information from the available data sources. Figure 17 illustrates the popular tools and technologies in the Graph databases category, sorted by popularity for GitHub data source.

7   Conclusions

This chapter has presented a Big Data and AI Pipeline Framework developed in the DataBench project, supported by the DataBench Toolbox. The Framework contains a number of dimensions, including pipelines steps, data processing types and types

Topic	Papers	EU Projects	News	Github	Jobs	Search Volume	Total
Neo4j	5.95	8.5	10	10	10	10	9.07
Grakn	1	1	1	3.1	1	5	2.02
ArangoDB	1.12	1	1.73	2.5	1	7.43	2.46
OrientDB	1.18	1	1.96	1.6	1	6.41	2.19
GraphDB	1.18	1	1	1.6	1	6.11	1.98
Virtuoso	10	10	1	1	1	8.43	5.24
Microsoft Azure Cosmos DB	1.02	1	8.58	1	1	3.92	2.75
Graph Engine	2.13	1	2.59	1	1	7.51	2.54
TigerGraph	1.07	1	4.01	1	1	4.24	2.05
Dgraph	1	1	1.48	1	1	5.52	1.83
JanusGraph	1.02	1	1	1	1	5.89	1.82

Fig. 17 DataBench popularity index (Tools and Technologies, Category: Graph Database, sorted by GitHub data source, accessed in November 2020)

of different data. The relationship of the Framework is with existing and emerging Big Data and AI reference models such as the BDV Reference Model and the AI PPP, and also the ISO SC42 Big Data Reference Architecture (ISO 20547) [3] and the emerging AI Machine Learning Framework (ISO 23053) [6], with which the pipeline steps also have been harmonized.

Further work is now related to populating the DataBench Toolbox with additional examples of actual Big Data and AI Pipelines realized by different projects, and further updates from existing and emerging technical benchmarks.

The DataBench Toolbox observatory will continuously collect and update popularity indexes for benchmarks and tools. The aim for the DataBench Toolbox is to be helpful for the planning and execution of future Big Data and AI-oriented projects, and to serve as a source for the identification and use of relevant technical benchmarks, also including links to a business perspective for applications through identified business KPIs and business benchmarks.

Acknowledgements The research presented in this chapter was undertaken in the framework of the DataBench project.

"Evidence Based Big Data Benchmarking to Improve Business Performance" [32] funded by the Horizon 2020 Programme under Grant Agreement 780966.

References

1. Zillner, S., Curry, E., Metzger, A., Auer, S., & Seidl, R. (Eds.). (2017). *European big data value strategic research & innovation agenda*. Big Data Value Association.
2. Zillner, S., Bisset, D., Milano, M., Curry, E., García Robles, A., Hahn, T., Irgens, M., Lafrenz, R., Liepert, B., O'Sullivan, B., & Smeulders, A., (Eds.). (2020). Strategic research, innovation and deployment agenda – AI, data and robotics partnership. Third Release. "September 2020, Brussels. BDVA, euRobotics, ELLIS, EurAI and CLAIRE".
3. ISO/IEC 20547–3:2020, Information technology—Big data reference architecture—Part 3: Reference architecture, https://www.iso.org/standard/71277.html
4. https://www.bdva.eu/PPP

5. Södergård, C., Mildorf, T., Habyarimana, E., Berre, A. J., Fernandes, J. A., & Zinke-Wehlmann, C. (Eds.). *Big data in bioeconomy results from the European DataBio project.*
6. ISO/IEC CD 23053.2: 2020 Framework for Artificial Intelligence (AI) Systems Using Machine Learning (ML).
7. https://hazelcast.com/glossary/complex-event-processing/
8. https://www.sciencedirect.com/science/article/pii/S1364815214002618
9. Wood, D., Lanthaler, M., & Cyganiak, R. (2014). RDF 1.1 concepts and abstract syntax [W3C Recommendation]. (Technical report, W3C).
10. Harris, S., & Seaborne, A. (2013). SPARQL 1.1 query language. W3C recommendation. W3C.
11. Hyland, B., Atemezing, G., & Villazón-Terrazas, B. (2014). Best practices for publishing linked data. W3C working group note 09 January 2014. https://www.w3.org/TR/ld-bp/
12. Heath, T., & Bizer, C. (2011). Linked data: Evolving the web into a global data space (1st ed.). *Synthesis lectures on the semantic web: Theory and technology*, 1, 1–136. Morgan & Claypool.
13. http://d2rq.org/
14. http://geotriples.di.uoa.gr/
15. https://github.com/IDLabResearch/RMLProcessor
16. https://www.metaphacts.com/ephedra
17. https://www.europeandataportal.eu/en/highlights/geospatial-and-earth-observation-data
18. https://ec.europa.eu/info/food-farming-fisheries/key-policies/common-agricultural-policy_en
19. https://www.sciencedirect.com/topics/earth-and-planetary-sciences/geo-referenced-data
20. https://gisgeography.com/spatial-data-types-vector-raster/
21. https://www.databench.eu/public-deliverables/
22. http://databench.ijs.si/knowledgeNugget/nugget/53
23. https://docs.ansible.com/ansible/2.3/playbooks.html
24. https://www.ansible.com/products/awx-project
25. https://www.microsoft.com/en-us/research/project/microsoft-academic-graph
26. https://www.adzuna.co.uk
27. https://data.europa.eu/euodp/sl/data/dataset/cordisH2020projects
28. https://eventregistry.org
29. https://github.com
30. https://octoverse.github.com/
31. 2020 StackOverflow Developers Survey. https://insights.stackoverflow.com/survey/2020#technology-programming-scripting-and-markup-languages
32. https://www.databio.eu/en/

An Elastic Software Architecture for Extreme-Scale Big Data Analytics

Maria A. Serrano, César A. Marín, Anna Queralt, Cristovao Cordeiro, Marco Gonzalez, Luis Miguel Pinho, and Eduardo Quiñones

Abstract This chapter describes a software architecture for processing big-data analytics considering the complete compute continuum, from the edge to the cloud. The new generation of smart systems requires processing a vast amount of diverse information from distributed data sources. The software architecture presented in this chapter addresses two main challenges. On the one hand, a new elasticity concept enables smart systems to satisfy the performance requirements of extreme-scale analytics workloads. By extending the elasticity concept (known at cloud side) across the compute continuum in a fog computing environment, combined with the usage of advanced heterogeneous hardware architectures at the edge side, the capabilities of the extreme-scale analytics can significantly increase, integrating both responsive data-in-motion and latent data-at-rest analytics into

M. A. Serrano · E. Quiñones (✉)
Barcelona Supercomputing Center (BSC), Barcelona, Spain
e-mail: eduardo.quinones@bsc.es

C. A. Marín
Information Catalyst for Enterprise Ltd, Crewe, UK
e-mail: cesar.marin@informationcatalyst.com

A. Queralt
Universitat Politècnica de Catalunya (UPC), Barcelona, Spain

Barcelona Supercomputing Center (BSC), Barcelona, Spain
e-mail: anna.queralt@bsc.es

C. Cordeiro
SIXSQ, Meyrin, Switzerland
e-mail: cristovao.cordeiro@sixsq.com

M. Gonzalez
Ikerlan Technology Research Centre, Basque Research Technology Alliance (BRTA), Arrasate/Mondragón, Spain
e-mail: marco.gonzalez@ikerlan.es

L. M. Pinho
Instituto Superior De Engenharia Do Porto (ISEP), Porto, Portugal
e-mail: lmp@isep.ipp.pt

© The Author(s) 2022
E. Curry et al. (eds.), *Technologies and Applications for Big Data Value*,
https://doi.org/10.1007/978-3-030-78307-5_5

a single solution. On the other hand, the software architecture also focuses on the fulfilment of the non-functional properties inherited from smart systems, such as real-time, energy-efficiency, communication quality and security, that are of paramount importance for many application domains such as smart cities, smart mobility and smart manufacturing.

Keywords Smart mobility · Software architecture · Distributed big data analytics · Compute continuum · Fog computing · Edge computing · Cloud computing · Non-functional requirements · Cyber-security · Energy-efficiency · Communications

1 Introduction

The extreme-scale big data analytics challenge refers not only to the heterogeneity and huge amount of data to be processed both on the fly and at rest but also to the geographical dispersion of data sources and the necessity of fulfilling the non-functional properties inherited from the system, such as real-time, energy efficiency, communication quality or security. Examples of smart systems that can exploit the benefits of extreme-scale analytics include production lines, fleets of public transportation and even whole cities. Providing the required computing capacity for absorbing extreme (and geographically dispersed) amounts of collected complex data, while respecting system properties, is of paramount importance to allow converting the data into few concise and relevant facts that can be then consumed and be decided or acted upon.

In a typical smart system (e.g., a smart city), data is collected from (affordable) sensors to gather large volumes of data from distributed sources using Internet of Things (IoT) protocols. The data is then transformed, processed and analysed through a range of hardware and software stages conforming the so-called *compute continuum*, that is from the physical world sensors close to the source of data (commonly referred to as edge computing) to the analytics backbone in the data centres (commonly located in the cloud and therefore referred to as cloud computing). Due to the computing complexity of executing analytics and the limited computing capabilities of the edge side, current approaches forward most of the collected data to the cloud side. There, big data analytics are applied upon large datasets using high-performance computing (HPC) technologies. This complex and heterogeneous layout presents two main challenges when facing extreme-scale big data analytics.

The first challenge refers to the non-functional properties inherited from the application domain:

- *Real-time* big data analytics is becoming a main pillar in industrial and societal ecosystems. The combination of different data sources and prediction models within real-time control loops will have an unprecedented impact in domains such as smart city. Unfortunately, the use of remote cloud technologies makes infeasible to provide real-time guarantees due to the large and unpredictable communication costs on cloud environments.

- *Mobility* shows increased trade-offs and technological difficulties. Mobile devices are largely constrained by the access of *energy*, as well as suffering from *unstable communication*, which may increase random communication delays, unstable data throughput, loss of data and temporal unavailability.
- *Security* is a continuously growing priority for organization of all sizes, as it affects data integrity, confidentiality and potentially impacting on safety. However, strict security policy management may hinder the communication among services and applications, shrinking overall performance and real-time guarantees.

Overall, while processing time and energetic cost of computation is reduced as data analytics is moved to the cloud, the end-to-end communication delay and the performance of the system (in terms of latency) increases and becomes unpredictable, making not possible to derive real-time guarantees. Moreover, as computation is moved to the cloud, the required level of security increases to minimize potential attacks, which may end up affecting the safety assurance levels, hindering the execution and data exchange among edge and cloud resources.

The second challenge refers to the elasticity concept. In recent years, the dramatic growth in both data generation and usage has resulted in the so-called three V's challenges of big data: volume (in terms of data size), variety (in terms of different structure of data, or lack of structure), and velocity (in terms of the time at which data need to be processed). These factors have contributed to the development of the elasticity concept, in which cloud computing resources are orchestrated to provide the right level of service (in terms of system throughput) to big data workloads. The elasticity concept, however, does not match the computing requirements when considering extreme-scale analytics workloads. On the one side, elasticity does not take into account the computing resources located on the edge. The advent of new highly parallel and energy-efficient embedded hardware architectures featuring graphical processing units (GPUs), many-core fabrics or FPGAs, have significantly increased the computing capabilities on the edge side. On the other side, elasticity mainly focuses on system throughput, without taking into account the non-functional properties inherited from the domain.

Addressing together these two important challenges along the compute continuum, that is from the edge to the cloud, is of paramount importance to take full benefit of extreme-scale big data analytics in industrial and societal environments such as smart cities. This chapter describes an end-to-end solution applied along the complete compute continuum to overcome these challenges. Concretely, the ELASTIC project [1], funded by the European Union's Horizon 2020 Programme, faces these challenges and proposes a novel software platform that aims to satisfy the performance requirements of extreme-scale big data analytics through a novel elasticity concept that distributes workloads across the compute continuum. The proposed software framework also considers the non-functional requirements of the system, that is operation with real-time guarantees, enhanced energy efficiency, high communication quality and security against vulnerabilities.

The chapter relates to the technical priority "Data Processing Architectures" of the European Big Data Value Strategic Research & Innovation Agenda [13]. Moreover, the chapter relates to the "Systems, Methodologies, Hardware and Tools" cross-sectorial technology enablers of the AI, Data and Robotics Strategic Research, Innovation & Deployment Agenda [14]. The rest of the chapter is organized as follows: Sect. 2 describes the ELASTIC software architecture. Concretely, Sect. 2.1 motivates the use of such framework in the smart city domain, Sect. 2.2 provides an overview of the layered software framework, and Sects. 2.3–2.6 describe each layer in detail. Finally, Sect. 3 concludes the chapter.

2   Elastic Software Architecture

2.1   Applicability to the Smart City Domain

One of the domains in which extreme-scale big data analytics can have a significant impact on people's day-to-day life is Smart Cities. Big data is increasingly seen as an effective technology capable of controlling the available (and distributed) city resources in a safely, sustainably, and efficiently way to improve the economical and societal outcomes. Cities generate a massive amount of data from heterogeneous and geographically dispersed sources including citizens, public and private vehicles, infrastructures, buildings, etc.

Smart cities can clearly benefit from the proposed software architecture, capable of deploying federated/distributed, powerful and scalable big data systems to extract valuable knowledge, while fulfilling the non-functional properties inherit from the smart cities. This opens the door to a wide range of advanced urban mobility services, including public transportation and traffic management. Therefore, the proposed software architecture is being tested in the city of Florence (Italy), to enhance the tramway public transportation services, as well as its interaction with the private vehicle transportation. The new elasticity concept will enable the efficient processing of multiple and heterogeneous streams of data collected from an extensive deployment of Internet of Things (IoT) sensors, located on board the tram vehicles, along the tramway lines, as well as on specific urban spots around the tram stations (e.g. traffic lights).

Concretely, three specific applications have been carefully identified to assess and highlight the benefits of ELASTIC technology for newly conceived mobility solutions (more details can be found in the ELASTIC project website [1]):

- Next Generation Autonomous Positioning (NGAP) and Advanced Driving Assistant System (ADAS): NGAP enables the accurate and real-time detection of the tram position through data collected and processed from on-board sensors. The positioning information is then sent through a reliable connection to the tram operation control system on the ground. This information also enables the development of ADAS, for obstacle detection and collision avoidance functionalities based on an innovative data fusion algorithm combining the output of multiple sensors (radars, cameras and LIDARs). Data from additional sources, such as fixed sensors placed at strategic positions in the streets (e.g., road crossings), are also integrated to increase the reliability of the system.

- Predictive maintenance: It monitors and profiles the rail track status in real time, enabling the identification of changes in equipment behaviour that foreshadow failure. Furthermore, through offline analytics, potential correlations between unexpected detected obstacles (obtained through the NGAP/ADAS application) and rail track damages are examined. The application also provides recommendations, enabling maintenance teams to carry out remedial work before the asset starts to fail. Finally, the power consumption profile is also monitored and processed in real time, in order to potentially minimize consumption and have an environmentally positive impact.
- Interaction between the public and private transport in the City of Florence: ELASTIC uses the information from the city network of IoT sensors to enhance the quality of the city traffic management, providing valuable outputs for both users and operators that will enable them to: (1) Identify critical situations (e.g. vehicles crossing the intersection with the tram line despite having a red traffic light) (2) Optimize the local traffic regulation strategies (e.g. reduce the waiting time of cars at tram crossings through improved light priority management, or slow down trams to reduce a queue of waiting vehicles, etc.)

2.2 ELASTIC Layered Software Architecture: Overview

In any smart system, large volumes of data are collected from distributed sensors, transformed, processed and analysed, through a range of hardware and software stages conforming the so-called compute continuum, that is from the physical world sensors (commonly referred to as edge computing), to the analytics back-bone in the data centres (commonly referred to as cloud computing). The proposed software architecture to efficiently manage and process this complex data processing scenario is shown in Fig. 1, and it is composed of the following layers:

- *Distributed Data Analytics Platform (DDAP)*: It provides the data accessibility and storage solutions, and the APIs. The data solutions provide the set of mechanisms needed to cope with all data-type variants: formats, syntax, at-rest and in-motion, 3V's (volume, velocity and variety), edge, cloud, etc. The APIs allow to extract valuable knowledge from the connected data sources using distributed and parallel programming models.
- *Computation Orchestrator*: It implements the elasticity concept in which the computing resources will be properly orchestrated across the compute continuum to provide the right level of service to big data analytics workloads. To do so, the orchestrator does not only consider the overall system throughput but also the fulfilment of non-functional properties inherited from the application domain.

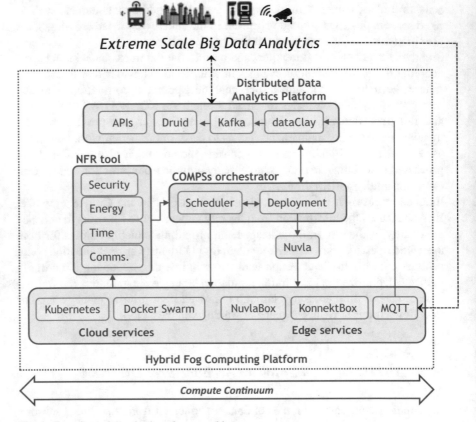

Fig. 1 Overview of the elastic software architecture

This layer supports the APIs exposed to the programmer to efficiently distribute
the execution of the analytics in a transparent way, while exploiting the inherent
parallelism of the system and abstracting the application from the underlying
distributed fog computing architecture.

- *Non-functional Requirements (NFR) Tool*: It provides the required support to
  monitor and manage the behaviour of the system, in order to guarantee some
  level of fulfilment of the non-functional requirements of the supported applications, that is real-time guarantees, energy efficiency, communication quality and
  security properties.
- *Hybrid Fog Computing Platform*: It abstracts the multiple edge and cloud
  computing resources spread across the compute continuum. To do so, this layer
  deploys the application components, that is the computational units distributed
  by the above layer, to virtual resources using container technologies, and
  considering configuration and infrastructural requirements.

Overall, the aim of the elastic software architecture is to enable the design, implementation and efficient execution of extreme-scale big data analytics. To do so, it incorporates a novel elasticity concept across the compute continuum, with the objective of providing the level of performance needed to process the envisioned volume and velocity of data from geographically dispersed sources at an affordable development cost, while guaranteeing the fulfilment of the non-functional properties inherited from the system domain. The following subsections provide a detail description of each software architecture component.

2.3 Distributed Data Analytics Platform

The distributed data analytics platform is developed to cater for domain specific as well as generic needs of analysing data across the compute continuum. Concretely, this layer takes care of two important matters: (1) the actual development of data analytics, providing APIs support (Sect. 2.3.1), and (2) the management, storage and retrieval of data at the time it is needed and at the location where it is needed (Sect. 2.3.2).

2.3.1 Application Programming Interfaces (APIs)

The software architecture provides support for the development of big data analytics methods, capable of analysing all the data collected by IoT sensors and distributed devices. As an example, Deep Neural Networks (DNNs) are used for image processing and predictive modelling, and aggregation and learning methods (based on unsupervised-learning strategies) are used for automatically detecting data patterns, including ant-based clustering, formal concept analysis and frequent pattern mining.

This layer also provides an API to support distributed and parallel computation. This enables the simultaneous use of multiple compute resources to execute software applications. Concretely, the COMPSs [4] task-based programming model is supported to allow developers to simply specify the functions to be executed as asynchronous parallel tasks. At runtime, the system exploits the concurrency of the code, automatically detecting and enforcing the data dependencies between tasks and deploying these tasks to the available resources, which can be edge devices or nodes in a cluster. More details of this component are provided in Sect. 2.4.

One of the key competitive advantages of the DDAP is that these methods are offered to the software developer in a unique development environment. Moreover, big data analytics methods can be optimized to be executed at both, the edge and the cloud side, providing the required flexibility needed to distribute the computation of complex big data analytics workflows across the compute continuum.

2.3.2 Data Accessibility and Storage

One of the main goals of the distributed data analytics platform (DDAP) is to ensure data accessibility across the compute continuum, covering aspects such as data-in-motion and data-at-rest, for data analytic applications. To do so, the DDAP is currently composed of the following main components:

- *dataClay* [8], distributed at the edge/fog side, is responsible for managing the information generated in real time, covering the data-in-motion needs. dataClay is an active object store that can handle arbitrary data structures in the form of objects and collections, as in object-oriented programming, which allows the application programmer to manage data as if it was just in memory. It is highly optimized for accessing and manipulating data at a fine granularity, and it can run in heterogeneous devices, from the edge to the cloud.
- *Druid* [3], distributed across the fog/cloud side, is responsible for collecting all information generated and shared across DDAP; it is a column-based data warehouse tuned to ingest large amounts of time series data such as that generated by a transport infrastructure. Druid is distributed by design and optimized for visual analytics. It contains mechanisms for a fast and easy access to data regardless of its location. This functionality makes Druid a complementing element in DDAP suitable for covering data-at-rest needs.
- *Kafka* [2] is a well-known message queue for streaming data; it functions as a transient message queue to transfer data from dataClay to Druid at each station. In DDAP, Kafka can be seen as the boundary between data-in-motion and data-at-rest.

The combination of dataClay, Druid, and Kafka makes DDAP suitable for real-time and historical big data analytics at the same time, as these solutions complement each other. In particular, Kafka helps enforce a unidirectional data flow from dataClay to Druid, effectively making DDAP operate as a well-known Content Delivery Network. In the latter, data is generated and processed at the edge for real-time needs, then it is collected at the cloud from distributed locations, and finally the content, that is historical big data analytics results, is delivered to interested users.

An example of the DDAP functionality can be seen in Fig. 2. The applications executed in the tram provide different kinds of data, such as the objects detected by its cameras and sensors, or the tram position. In the meantime, the applications executed in the context of a tram stop, which includes the stop itself as well as cabinets with cameras in surrounding intersections, also detect objects that may fall out of the visibility scope of the tram. To provide real-time performance, these applications use COMPSs to distribute the work between the different devices. Both data sources are merged in the tram stop in order to predict possible collisions between the objects detected according to their current trajectories. Simultaneously, the objects detected and their positions at each point in time are pushed to Kafka so that they can be ingested by Druid as they are created, thus immediately enabling them to take part of historical analytics triggered from the cloud.

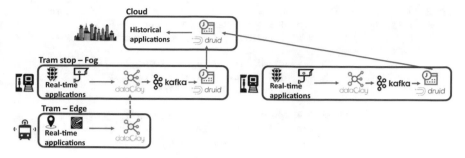

Fig. 2 Example application of DDAP in a transport infrastructure

2.4 Computation Orchestrator

This layer provides the software component in charge of distributing the computation across available computing resources in the hybrid for computing platform. Specifically, it implements the *elasticity* concept to properly orchestrate the computing resources across the compute continuum to provide the right level of service to analytics workloads. Moreover, elasticity will not only consider the overall system throughput but also the fulfilment of non-functional properties inherited from the application domain.

The software component in charge of implementing these features is COMPSs [4]. COMPSs is a distributed framework developed at the Barcelona Supercomputing Center (BSC) mainly composed of a *task-based programming model*, which aims to ease the development of parallel applications for distributed infrastructures, such as Clusters, Clouds and containerized platforms, and a *runtime system* that distributes workloads transparently across multiple computing nodes with regard to the underlying infrastructure. In cloud and big data environments, COMPSs provides scalability and elasticity features allowing the dynamic provision of resources. More specifically, the COMPSs task-based model is offered in the DDAP layer to implement big data analytics methods, and the COMPSs runtime implements the scheduling techniques and deployment capabilities to interact with hybrid resources in a transparent way for the programmer.

2.4.1 Task-based Programming Model

COMPSs offers a portable programming environment based on a task execution model, whose main objective is to facilitate the parallelization of sequential source code (written in Java, C/C++ or Python programming languages) in a distributed and heterogeneous computing environment. One of the main benefits of COMPSs is that the application is agnostic from the underlying distributed infrastructure. Hence, the COMPSs programmer is only responsible for identifying the portions of code, named COMPSs *tasks*, that can be distributed by simply annotating the sequential

Fig. 3 Video analytics
COMPSs example

```
1  @task (camera = IN, returns = numpy.ndarray)
2  def get_video_frame(cameraID):
3      return get_next_frame(cameraID)

5  @task (frame = IN, returns = list)
6  def video_analytics(frame):
7      return process(frame)

9  @task (list_results = IN)
10 def collect_and_display(list_results):
11     update_dashboard(list_results)

13 ### Main function ###
14 while (true):
15     for i,cam in enumerate(cameras_set)
16         frame[i] = get_video_frame(cam)
17         results[i] = video_analytics(frame[i])
18     collect_and_display(results)
```

source code. Data dependencies and their directionality (i.e. in, out or inout) are also identified. Upon them, the COMPSs runtime determines the order in which COMPSs tasks are executed and also the data transfers across the distributed system. A COMPSs task with an in or inout data dependency cannot start its execution until the COMPSs task with an out or inout dependency over the same data element is completed. At run-time, COMPSs tasks are spawned asynchronously and executed in parallel (as soon as all its data dependencies are honoured) on a set of distributed and interconnected computing resources. Moreover, the data elements marked as in and inout are transferred to the compute resource in which the task will execute if needed.

Figure 3 shows a basic example of a Python COMPSs application (PyCOMPSs [12]) that performs video analytics. COMPSs tasks are identified with a standard Python decorator @task, at lines 1, 5 and 9. The returns argument specifies the data type of the value returned by the function (if any), and the *IN* argument defines the data directionality of function parameters. The main code starts at line 14, where the application iterates to process video frames over the time. Then, at line 15 a loop iterates over the available camera video feeds, and first it gets the next frame by instantiating the COMPSs task defined at line 1. At line 17, the COMPSs task that process the video frame (defined at line 5) is instantiated. Finally, all the results are collected at line 18, instantiating the COMPSs task defined at line 9.

2.4.2 Runtime System

The task-based programming model of COMPSs is supported by its runtime system, which manages several aspects of the application execution, keeping the underlying infrastructure transparent to it. The two main aspects are the deployment on the available infrastructure and the scheduling of tasks to available computing resources.

Deployment

One of the main features of COMPSs is that the model abstracts the application from the underlying distributed infrastructure; hence, COMPSs programs do not include any detail that could tie them to a particular platform boosting portability among diverse infrastructures and enabling execution in a fog environment. Instead, it is the COMPSs runtime that features the capabilities to set up the execution environment. The COMPSs runtime is organized as a master-worker structure. The Master is responsible for steering the distribution of the application, as well as for implementing most of the features for initialising the execution environment, processing tasks or data management. The Worker(s) are in charge of responding to task execution requests coming from the Master.

The COMPSs runtime support various scenarios regarding deployment strategy and interoperability between edge/cloud resources. Three different scenarios, compatible between them, are supported:

- Native Linux, monolithic: The big data analytics workload is natively executed in a Linux-like environment. In this configuration, all the nodes available for the execution of a COMPSs workflow require the native installation of COMPSs, and the application.
- Containerized, Docker: A Docker COMPSs application image contains the needed dependencies to launch a COMPSs worker and the user application. In this case, there is no need for setting up the execution environment in advance in all the nodes, but only Docker must be available. Docker image repositories, e.g. Docker Hub, can make the image, and hence the application, available anytime and anywhere. In this deployment, COMPSs takes care of making the image available at the nodes and launching the containers.
- Cloud provider: A cloud infrastructure, in this context, refers to a data centre or cluster with great computing capacity that can be accessed through an API and that can lend some of that computational power, for example in the form of a container. This is the case of a Docker Swarm or Kubernetes cluster. COMPSs also supports the deployment of workers in these infrastructures, using the Nuvla API (see Sect. 2.6).

Scheduling

One key aspect of the COMPSs runtime scheduler is that it maintains the internal representation of a COMPSs application as a Direct Acyclic Graph (DAG) to express the parallelism. Each node corresponds to a COMPSs task instance and edges represent data dependencies. As an example, Fig. 4 shows the DAG representation for three iterations of the COMPSs application presented in Fig. 3, when three camera video feeds are processed. Based on this DAG, the runtime can automatically detect data dependencies between COMPSs tasks.

The COMPSs scheduler is in charge of distributing tasks among the available computing resources and transferring the input parameters before starting the execution, based on different properties of the system such as the non-functional

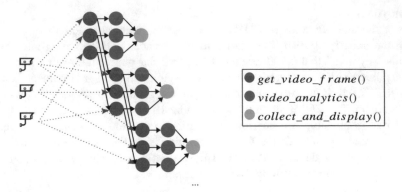

Fig. 4 DAG representation of the COMPSs example in Fig. 3

requirements (real-time, energy, security and communications quality) analysed by the NFR tool (see Sect. 2.5).

2.5   Non-functional Requirements Tool

The software architecture presented in this chapter addresses the challenge of processing extreme-scale analytics, considering the necessity of fulfilling the non-functional properties inherited from the system and its application domain (e.g. smart manufacturing, automotive, smart cities, avionics), such as real time, energy efficiency, communication quality or security.

This task is led by the Non-functional Requirements (NFR) Tool layer (see Fig. 1), in collaboration with the Orchestrator layer, and the Hybrid Fog Computing Platform. The NFR tool continuously monitors and evaluates the extent to which non-functional properties' required levels are guaranteed in the fog computing platform. Moreover, this tool identifies and implements the appropriate mechanisms to deal with the NFRs, monitoring system behaviour and helping taking decisions (such as offloading or reducing performance). *Runtime monitoring* of system status is used to detect NFR violations, while a *Global Resource Manager* guides the evolution towards configurations that are guaranteed to satisfy the system's NFRs.

The NFR monitoring is conceptually constituted by *probes*, i.e. the system tools that provide monitoring data. The probes are in charge of interfacing with the underlying fog platform (OS and/or hardware), to collect the required information, which is used to detect NFR violations. The NFR Monitors are per-property-specific components, which, based on the information from the probes, and the application information, determines if some requirement is not being met. This information is shared with the Orchestrator that may (re)configure the scheduling of a given application to meet its requirements. The Global Resource Manager is the component that considers a holistic approach, providing decisions based on a global

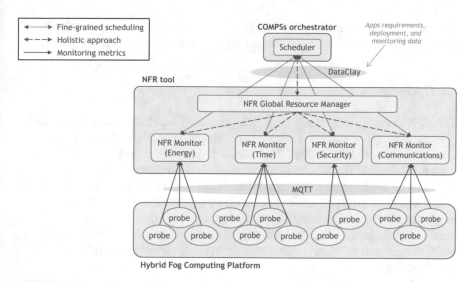

Fig. 5 NFR tool architecture and synergies within the elastic software architecture

view of the system (composed of distributed computing nodes), and considering simultaneously all non-functional properties. This decision is also shared with the Orchestrator to (re)configure the applications accordingly. Figure 5 shows the NFR tool internal structure and the synergies with the Fog platform (see Sect. 2.6) and the Orchestrator layer (see Sect. 2.4).

Next subsections describe the concrete NFR metrics analysis.

2.5.1 Real Time

Coping with real-time computing across the compute continuum requires the ability to specify and manage different timing perspectives. Two main challenges arise: tasks deployed at the edge (e.g. on board the connected car) need to guarantee "hard real-time" responses (e.g. very low latency), while those deployed at the cloud need to guarantee certain QoS levels regarding time: right-time or "soft real-time" guarantees. Closer to the environment, at the edge, tight timing mapping and scheduling approaches can be used, while at the cloud, time is measured in terms of average statistical performance with Quality of Service (QoS) constraints. These perspectives complement each other, and the elastic software architecture provides solutions that try to guarantee the required response time to applications while optimizing energy and communication costs.

To do so, it is necessary to monitor different timing properties, in all nodes of the distributed fog infrastructure. This ranges from monitoring actual CPU utilization and execution time of applications to detection of deadline violations or memory accesses.[1] This monitoring allows to dynamically adjust the system resources to which the application is mapped, depending on the actual load of the system.

2.5.2 Energy

The NFR tool augments the system "introspection" capabilities in terms of power consumption, by means of energy-aware execution models, from the hardware platform to the holistic system. This allows to propagate workload-specific monitoring information from the run-time to the decision-making module, which can be exploited to better adapt to the requirements, as well as to the time predictability and security optimization levels. Furthermore, a richer knowledge of applications' requirements and concurrency structure, coupled with precise energy models for the underlying hardware, combined with the possibility of dynamically switching between edge and cloud deployments, constitutes an enabling factor towards larger energy savings.

Concretely, the NFR tool monitors the power consumption of the different hardware components on edge devices (e.g. System on Chip (SoC), CPU, GPU, etc.). This allows to develop energy-aware execution models and efficiency tune power consumption over the complete continuum.

2.5.3 Security

Verifying that applications correctly comply with security mechanisms and do not contain vulnerabilities is essential. This implies much more than an online analysis and monitoring, e.g. GDPR [6] regulation compliance, secure communication protocols, the use of device certificates and mutual authentication (server and client), etc. Besides these design decisions, in order to guard against security threats, the NFR tool continuously monitors the systems and applications deployed and incorporates security upgrades to software and deploy updates to existing configurations.

The security monitoring component is based on OpenSCAP [9], an open-source tool that simply implements the Security Content Automation Protocol (SCAP), as a vulnerability scanner. OpenSCAP can easily handle the SCAP standards and generate neat, HTML-based reports. The NFR monitor tool and the global resource manager take simple decisions concerning security: The security status of the computing nodes is monitored, providing a security score for each of them. Then, the list of available (secure) nodes is updated for each application, based on its particular requirements.

---

[1] Memory accesses can be used to provide information on contention accessing shared memory, providing a more accurate timing analysis for hard real-time applications.

2.5.4   Communications Quality

In the context of wireless communications, and especially LTE networks, several performance parameters need to be considered to characterize system behaviour. Different types of service prioritize different figures of merit due to the nature of the information to be transmitted and/or received. For instance, packet loss rate plays a paramount role in VoIP services, whereas high throughput is not strictly required, given that VoIP does not generate high volumes of data. On the contrary, video streaming and file transfer services demand a much higher throughput.

The NFR tool considers the following communication monitoring metrics to evaluate the communications quality of the system: active network interfaces, transmitted/received data volume, average throughput, roundtrip time (RTT), packet loss rate (PLR). These metrics provide information that is considered both at the orchestrator, to take fine-grained scheduling decisions (see Sect. 2.4), and at the global resource manager to consider communications quality in the holistic approach.

2.6   Hybrid Fog Computing Platform

Fog computing encompasses the benefits of edge and cloud computing: on the one hand, devices have increased computer capability, on the other hand, Cloud Computing has matured strongly. The main focus of the elastic software architecture is to obtain the best from both approaches (edge and cloud) into fog architecture. While fog computing has recently led to great interest by the research community and industry, it is still a conceptual approach [2]. The elastic software architecture presented in this chapter considers two main concepts for its hybrid fog architecture: (1) a software stack that can be run in (almost) any computing device and (2) the coordination between edge and cloud components to efficiently support elasticity across the compute continuum.

The proposed hybrid fog-computing platform is based on standard open-source reusable components. It follows a microservice-based design, thus decoupling the composing components, which makes the overall solution generic and capable of coping with a wide range of smart edge devices and clouds. The hybrid architecture (see Fig. 6) allows the use of dynamic applications in the form of microservices (containers) or native applications (monolithic). Predictability (native) and flexibility (microservices) can be achieved with this approach. The platform is compatible with both systems offering a high level of flexibility for the use case.

Next subsections provide an overview of each component.

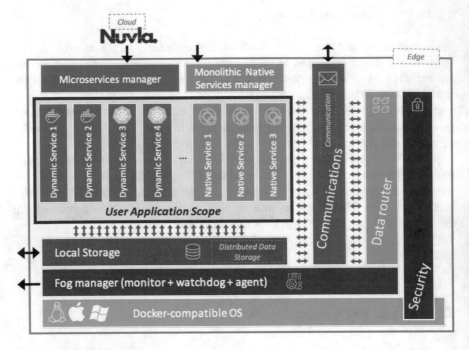

Fig. 6 Hybrid Fog Architecture

2.6.1 Cloud: Nuvla

Nuvla [10] acts both as the orchestration support and deployment engine for all micro-service-based workloads being submitted into both cloud infrastructures and edge devices. As an open-source software stack, Nuvla can be run anywhere. In particular, the elastic software architecture profits from its existing SaaS offering running in the Exoscale cloud,[2] at https://nuvla.io/. Nuvla offers the following services:

- *Application Registration*: users can register Docker Swarm and Kubernetes applications in Nuvla.
- *Infrastructure Registration*: users can register new Docker Swarm and Kubernetes infrastructures in Nuvla (be those at the cloud or at the edge).
- *NuvlaBox Registration*: users can create new NuvlaBoxes via Nuvla. Nuvla will provide users with a "plug-and-play" installation mechanism that can be executed on any Docker compatible device.
- *Resource Sharing*: users can share their Nuvla resources (applications, infrastructures, etc.) with other users.

---

[2] https://www.exoscale.com/.

- *Application deployment*: users can launch their applications into any of their Nuvla infrastructures.
- *Application Monitoring*: all the deployed applications are monitored from Nuvla, giving users an overall view of the deployment status.
- *Edge Monitoring*: all NuvlaBoxes can be monitored and managed from Nuvla. Resource consumption, external peripheral, and lifecycle management options are provided to users from Nuvla.
- *RESTful API*: a standardized and language-agnostic API is available to all Nuvla users, providing full resource management capabilities, plus a comprehensive querying and filtering grammar.

2.6.2 Edge: KonnektBox and NuvlaBox

Two commercial edge solutions are being used as the ground foundation for the edge infrastructure in the elastic software architecture: the KonnektBox [7] and the NuvlaBox [11].

The IKERLAN *KonnektBox* is an industry-oriented digitization solution built over EdgeXFoundry [5], an open-source project backed by Linux Foundation which provides basic edge building blocks. KonnektBox uses a mix between vanilla EdgeXFoundry components and custom services tailored for Industry 4.0 use cases.

The *NuvlaBox* is a secured plug-and-play edge to cloud solution, capable of transforming any Docker compatible device into an edge device. This software solution has been developed by SixSq and is tightly coupled with the application management platform, Nuvla.

2.6.3 Fog Components

As the officially adopted edge software appliances for ELASTIC, both the KonnektBox and NuvlaBox provide their own implementation for each of ELASTIC's Fog Architecture building blocks:

Docker-Compatible OS

In order to comply with the reconfiguration and dynamic fog-cloud service execution requirements of the project, a micro services architecture is required. Docker is the standard open-source micro services software. Docker allows the execution of micro services in the form of Docker containers. Each one of the containers runs in an isolated environment and interfaces with other services via network communications (REST APIs, message brokers, etc.). Docker allows the configuration of priorities and limits for each container. For example, the maximum CPU usage by each container, number of CPUs to use, CPU quota, maximum RAM, etc. If a Linux kernel with the real-time extension is used, the priority of each container can also be defined.

Microservices Manager

The dynamic services refer to the different applications that the platform will run. The applications range from industrial protocol drivers to AI inference, DB manager, etc. These applications shall be independent from one another, and the communication between them should be established via some predefined APIs defined in the data router. The NuvlaBox self-generates TLS credentials to be used on a secure and dedicated endpoint which relays the Docker API via HTTPS, for external orchestration platforms, like Nuvla, to speak with.

Monolithic Native Services Manager

The native services manager is the module in charge of controlling the monolithic native applications run in the system. The NuvlaBox provides the execution of remotely issued operations, via a secured and dedicated HTTPS endpoint, exposing a RESTful API. Such operations include the generation of user-specific SSH keypair for executing Native workflows via SSH. The KonnektBox supports the remote deployment of services via a secured MQTT-over-TLS cloud connection.

Local Storage

The local/distributed storage of the system will be implemented as a software middleware. BSC component dataClay platform will be used as the base component for structured data. The KonnektBox provides an additional local database using Redis and Consul. Apart from the local system storage (in the form of Docker volumes), the NuvlaBox does not provide any dedicated storage for user applications. Such functionality is left entirely to the user's preferences. The selected storage element for ELASTIC (dataClay) is supported, as an additional module, by the NuvlaBox.

Fog Manager

The system manager is the application in charge of starting up the whole fog platform, monitor it and manage it. The system manager will run as a standalone Linux application. This service will send telemetry and statistics data to the cloud in order to update the NFR analysis (QoS, security, etc.). It will implement a watchdog service in order to control that all the microservices are running correctly and the health of the system is good. The manager can stop, start and update each micro service. A local configuration UI can be deployed to allow the local configuration of the platform. Both the NuvlaBox and KonnektBox include several microservices which are responsible for discovering external peripherals, collecting telemetry data, categorizing the host environment and performing regular performance and security scans. All of this information is periodically sent (on a configurable frequency) both to Nuvla and to a local edge dashboard running on the hosting edge device.

Communications

The communications service offers an abstraction layer between platform services and multiple communication protocols. This service allows the platform user to define rules to select automatically the appropriate communication protocol to be used in different use cases or environments. All the protocols would be secured with TLS1.2 (or DTLS1.2 for UDP-based protocols). The KonnektBox uses MQTT-over-TLS for all the connections with the cloud. The NuvlaBox exposes secure and dedicated HTTPS endpoints for configuration application management (separately).

Data Router

The data router abstracts the communication between micro services and serves as a central point for all data communication. A decision algorithm can be implemented to decide where to send the data (other local micro service, the cloud, the edge, etc.). The NuvlaBox together with the Fog Manager's peripheral discovery functionality provides an MQTT-based messaging system, which not only brokers internal application messages but also automatically consumes and serves sensor data from the existing peripherals, to any subscribing user applications.

Security

The security module handles the security credentials of the platform and it checks the device data and binaries for unintended manipulation. All the critical applications of the system should be signed to only allow the execution of trusted and original applications. If the security module detects some anomalies, the device will be restored to factory defaults. The KonnektBox is integrated with OpenSCAP vulnerability scanner. The NuvlaBox on top of the security scans within the Fog Manager has the ability to automatically update its own database of common vulnerabilities. Upon every scan, and for a configurable set of vulnerabilities found, it can proactively take action, halting certain sensitive internal services or even moving the whole edge device into a quarantine state.

2.6.4 Distributed Storage

The distributed storage component in ELASTIC is implemented by dataClay (see Sect. 2.3). Since dataClay runs on different kinds of devices, it can be integrated at any level throughout the edge to cloud continuum. Its function within the elastic architecture is twofold. On the one hand, it is in charge of storing data gathered by the Data Router and making it accessible in other devices. The embedded computing capabilities of dataClay enable the association of a given behaviour to each type of data, such as synchronization policies or filters before handling it to other devices. On the other hand, dataClay is used by other components, such as the NFR tool, or the DDAP.

2.6.5 Communication Middleware

The communication middleware is the software component in charge of the exchange of information between services and other devices. This component offers an abstraction over the communication protocols and physical devices used.

Inter-Service Communication

For communication between services, the middleware offers a standard MQTT broker. MQTT is an IoT-oriented pub-sub communication protocol built over TCP/IP. For example, a service which is getting readings from a temperature sensor can publish data to a topic (e.g. /sensor/temperature/data). Other services can subscribe to that same topic in order to get updates of the temperature in real time.

External Communication

The communication between services and other devices (other edge nodes, cloud, etc.) can be separated in different data streams depending on the desired QoS. Different levels of QoS can be defined with different requirements in order to choose between communication interfaces or protocols. For example, a high priority data stream can be mapped to the 4G/LTE modem. A bulk-data data stream (e.g. high volume of data for offline analysis, etc.) can be transferred when Wi-Fi connectivity is available, because this is not a critical data with real-time constraints.

3 Conclusions

Big data analytics have become a key enabling technology across multiple application domains, to address societal, economic and industrial challenges for safe mobility, well-being and health, sustainable production and smart manufacturing, energy management, etc. A particular challenge for big data analytics in the near future (or even today) is managing large and complex real-world systems, such as production lines, fleets of public transportation and even whole cities, which continuously produce large amounts of data that need to be processed on the fly. Providing the required computational capacity level for absorbing extreme amounts of complex data, while considering non-functional properties, is of paramount importance to allow converting the data into few concise and relevant facts that can be then consumed by humans and be decided or acted upon. The ELASTIC project [1] is facing this challenge by proposing the end-to-end software framework described in this chapter. The final goal is to efficiently distribute extreme-scale big data analytic methods across the compute continuum, to match performance delivered by the different computing resources with the required precision and accuracy.

Acknowledgments The research leading to these results has received funding from the European Union's Horizon 2020 Programme under the ELASTIC Project (www.elastic-project.eu), grant agreement No 825473.

References

1. A software architecture for extreme-scale big-data analytics in fog computing ecosystems (ELASTIC) (2020). https://elastic-project.eu/ [Online; accessed October 2020].
2. Apache Foundation (2017). Apache kafka. https://kafka.apache.org/ [Online; accessed October 2020].
3. Apache Foundation (2020). Apache druid. https://druid.apache.org/ [Online; accessed October 2020].
4. Barcelona Supercomputing Center (BSC). COMP Superscalar (COMPSs). http://compss.bsc.es/ [Online; accessed October 2020].
5. EdgeXFoundry (2020). https://www.edgexfoundry.org/ [Online; accessed November 2020].
6. EU General Data Protection Regulation. https://ec.europa.eu/info/law/law-topic/data-protection/eu-data-protection-rules_en [Online; accessed November 2020].
7. IKERLAN (2020). IKERLAN KONNEKT. https://www.ikerlan.es/en/ikerlankonnekt [Online; accessed October 2020].
8. Martí, J., Queralt, A., Gasull, D., Barceló, A., Costa, J. J., & Cortes, T. (2017). Dataclay: A distributed data store for effective inter-player data sharing. *Journal of Systems and Software, 131*, 129–145.
9. OpenScap (2020). https://www.open-scap.org/. [Online; accessed November 2020].
10. SixSq (2020). Edge and container management software. https://sixsq.com/products-and-services/nuvla/overview. [Online; accessed October 2020].
11. SixSq (2020). Secure and intelligent edge computing software. https://sixsq.com/products-and-services/nuvlabox/overview. [Online; accessed October 2020].

12. Tejedor, E., Becerra, Y., Alomar, G., Queralt, A., Badia, R. M., Torres, J., Cortes, T., & Labarta, J. (2017). Pycompss: Parallel computational workflows in Python. *The International Journal of High Performance Computing Applications, 31*(1), 66–82.
13. Zillner, S., Curry, E., Metzger, A., & Seidl, R. (Eds.) (2017). European big data value strategic research and innovation agenda. https://bdva.eu/sites/default/files/BDVA_SRIA_v4_Ed1.1.pdf. [Online; accessed October 2020].
14. Zillner, S., Bisset, D., Milano, M., Curry, E., Robles, A. G., Hahn, T., Irgens, M., Lafrenz, R., Liepert, B., O'Sullivan, B., & Smeulders, A. (Eds.) (2020). Strategic research and innovation agenda—AI, Data and Robotics Partnership. Third Release. https://ai-data-robotics-partnership.eu/wp-content/uploads/2020/09/AI-Data-Robotics-Partnership-SRIDA-V3.0.pdf. [Online; accessed February 2021].

Privacy-Preserving Technologies for Trusted Data Spaces

Susanna Bonura, Davide Dalle Carbonare, Roberto Díaz-Morales, Marcos Fernández-Díaz, Lucrezia Morabito, Luis Muñoz-González, Chiara Napione, Ángel Navia-Vázquez, and Mark Purcell

Abstract The quality of a machine learning model depends on the volume of data used during the training process. To prevent low accuracy models, one needs to generate more training data or add external data sources of the same kind. If the first option is not feasible, the second one requires the adoption of a federated learning approach, where different devices can collaboratively learn a shared prediction model. However, access to data can be hindered by privacy restrictions. Training machine learning algorithms using data collected from different data providers while mitigating privacy concerns is a challenging problem. In this chapter, we first introduce the general approach of federated machine learning and the H2020 MUSKETEER project, which aims to create a federated, privacy-preserving machine learning Industrial Data Platform. Then, we describe the Privacy

---

All authors have contributed equally and they are listed in alphabetical order.

S. Bonura · D. D. Carbonare
Engineering Ingegneria Informatica SpA—Piazzale dell'Agricoltura, Rome, Italy
e-mail: susanna.bonura@eng.it; davide.dallecarbonare@eng.it

R. Díaz-Morales · M. Fernández-Díaz
Tree Technology—Parque Tecnológico de Asturias, Llanera, Spain
e-mail: roberto.diaz@treetk.com; marcos.fernandez@treetk.com

L. Morabito · C. Napione (✉)
COMAU Spa—Via Rivalta, Grugliasco, Italy
e-mail: lucrezia.morabito@comau.com; chiara.napione@comau.com

L. Muñoz-González
Imperial College London—180 Queen's Gate, London, United Kingdom
e-mail: l.munoz@imperial.ac.uk

Á. Navia-Vázquez
University Carlos III of Madrid, Leganás, Spain
e-mail: angel.navia@uc3m.es

M. Purcell
IBM Research Europe, IBM Campus, Damastown Industrial Estate, Mulhuddart, Ireland
e-mail: markpurcell@ie.ibm.com

111
E. Curry et al. (eds.), *Technologies and Applications for Big Data Value*,
https://doi.org/10.1007/978-3-030-78307-5_6

Operations Modes designed in MUSKETEER as an answer for more privacy before looking at the platform and its operation using these different Privacy Operations Modes. We eventually present an efficiency assessment of the federated approach using the MUSKETEER platform. This chapter concludes with the description of a real use case of MUSKETEER in the manufacturing domain.

Keywords Industrial data platform · Federated learning · Privacy-preserving technologies · Quality assessment

1   Introduction

In recent years, the advancements in Big Data technologies have fostered the penetration of AI and machine learning in many application domains, producing a disruptive change in society and in the way many systems and services are organized. Machine learning technologies provide significant advantages to improve the efficiency, automation, functionality and usability of many services and applications. In some cases, machine learning algorithms are even capable of outperforming humans. It is also clear that machine learning is one of the key pillars for the fourth industrial revolution, and it will have a very significant impact in the future economy. Machine learning algorithms learn patterns, and they are capable of extracting useful information from data [1]. But in some cases, the amount of data needed to achieve a high level of performance is significant. A few years ago, Peter Norvig, Director of Research at Google, recognized: *"we don't have better algorithms, we just have more data."* Thus, apart from the expertise and the computational resources needed to develop and deploy machine learning systems, for many companies, the lack of data can be an important obstacle to participate in this new technological revolution. This can have a profound negative effect on SMEs, which, in many cases, will not be able to compete with the largest companies in this sector. In recent years, this problem has been alleviated by the growth and development of data markets, which enables companies to have access to datasets to develop AI and machine learning models. However, in many sectors, the access to these data markets is very difficult because of the sensitivity of the data or privacy restrictions that impede companies to share or commercialize the data. This can be the case, for example, for healthcare applications, where the privacy of the patients must be preserved and where often the data cannot leave the data owner's facilities. The chapter relates to the technical priority "Mechanisms ensuring data protection and anonymisation, to enable the vast amounts of data which are not open data (and never can be open data) to be part of the Data Value Chain" of the European Big Data Value Strategic Research & Innovation Agenda [27]. It addresses the horizontal concern "Data Protection" of the BDV Technical Reference Model. It addresses the vertical concerns "Data sharing platforms, Industrial/Personal". The chapter relates to the "Systems, Methodologies, Hardware and Tools" cross-sectorial technology enablers of the AI, Data and Robotics Strategic Research, Innovation & Deployment Agenda [28]. In this chapter, we present a solution developed for the MUSKETEER project based on a federated learning approach (Sect. 2) [15, 16, 18, 24], detailing the different Privacy Operations Modes designed to offer higher standards of privacy to the stakeholders of the platform (Sect. 3). From this conceptual level, we then deep dive into the concrete part with an architecture description of the platform (Sect. 4) and an efficiency assessment of the federated approach using the MUSKETEER platform (Sect. 5) before concluding with a real use case description (Sect. 6).

2 Tackling Privacy Concerns: The Solution of Federated Learning

Federated learning is a machine learning technique that allows building collaborative learning models at scale with many participants while preserving the privacy of their datasets [15, 16, 18, 24]. Federated learning aims to train a machine learning algorithm using multiple datasets stored locally in the facilities or devices of the clients participating in the collaborative learning task. Thus, the data is not exchanged between the clients and always remains in their own facilities.

2.1 Building Collaborative Models with Federated Learning

In federated learning, there is a central node (or server) that orchestrates the learning process and aggregates and distributes the information provided by all the participants (or clients). Typically, the server first collects and aggregates the information provided by the different clients, updating the global machine learning model. Then, the server sends the parameters of the updated global model back to the clients. On the other side, the clients get the new parameters sent by the server and train the model locally using their own datasets. This communication process between the server and the clients is repeated for a given number of iterations to produce a high-performance collaborative model. Depending on the application, there are different variants for learning federated machine learning models [16]. For example, in some cases, a global collaborative model is learned from all the clients' data. In other scenarios, only part of the parameters of the machine learning models are shared between participants, whereas the rest of the parameters are local (specific for each client). This allows a level of customization specific to the clients. This approach is also useful in cases where the datasets from the different participants are not completely aligned, that is they do not have exactly the same features.

2.2 Where Can We Use Federated Machine Learning?

To reduce the amount of data to be stored in the cloud, as the computational power of small devices like smartphones and other IoT devices increases, federated machine learning offers new opportunities to build more efficient machine learning models at scale by keeping the data locally on these devices and using their computational power to train the model. This approach was used in one of the first successful use cases of federated learning: *"Gboard"* for Android, a predictive keyboard developed by Google which leverages the smartphones of millions of users to improve the precision of the machine learning model that makes the predictions. In this case Google did not require upload of the text messages from millions of users to the cloud, which can be troublesome from a regulatory perspective, helping to protect the privacy of the users' data. This use case showed that federated machine learning can be applied in settings with millions of participants that have small datasets. Federated learning is also a suitable solution in applications where data sharing or access to data markets is limited. In these cases, the performance of the machine learning models achieved using smaller datasets for training can be unacceptable for their practical application. Thus, federated machine learning boosts performance by training a collaborative model that uses the data from all the participants in the learning task but keeps the data private. The performance of this collaborative model is similar to the one we would obtain by training a standard machine learning model merging the datasets from all the participants, which would require data sharing.

2.3 *MUSKETEER's Vision*

Our mission in MUSKETEER is to develop an industrial data platform with scalable algorithms for federated and privacy-preserving machine learning techniques, including detection and mitigation of adversarial attacks and a rewarding model capable of fairly monetizing datasets according to the real data value. In Fig. 1 we represent the IDP concept.

In MUSKETEER we are addressing specific fundamental problems related to the privacy, scalability, robustness and security of federated machine learning, including the following objectives:

- To create machine learning models over a variety of privacy-preserving scenarios: in MUSKETEER we are defining and developing a platform for federated learning and distributed machine learning with different privacy operation modes to provide compliance with the legal and confidentiality restrictions of most industrial scenarios.
- To ensure security and robustness against external and internal threats: in MUSKETEER we are not only concerned with the security of our software but also with the security and robustness of our algorithms to internal and external threats. We are investigating and developing new mechanisms to make the algorithms robust and resilient to failures, malicious users and external attackers trying to compromise the machine learning algorithms [20].
- To provide a standardized and extendable architecture: the MUSKETEER platform aims to enable interoperability with Big Data frameworks by providing

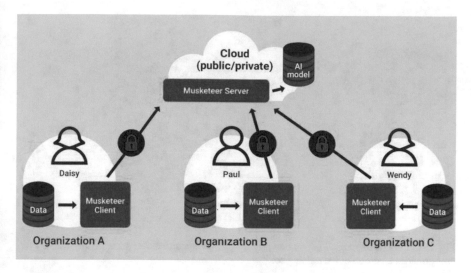

Fig. 1 Musketeer Industrial Data Platform to share data in a federated way

portability mechanisms to load and export the predictive models from/to other platforms. For this, the MUSKETEER design aims to comply with the Reference Architecture model of the International Data Spaces Association (IDSA) [13].

- To enhance the data economy by boosting cross-domain sharing: the MUS-KETEER platform will help to foster and develop data markets, enabling data providers to share their datasets to create predictive models without explicitly disclosing their datasets.
- To demonstrate and validate in two different industrial scenarios: in MUSKE-TEER we want to show that our platform is suitable, providing effective solutions in different industrial scenarios such as manufacturing and healthcare, where federated learning approaches can bring important benefits in terms of cost, performance and efficiency.

3 Privacy Operation Modes (POMs)

Training models with data from different contributors is an appealing approach, since when more and more data is used, the performance of the resulting models is usually better. A centralized solution requires that the data from the different users is gathered in a common location, something that is not always possible due to privacy/confidentiality restrictions. The MUSKETEER platform aims at solving machine learning (ML) problems using data from different contributors while preserving the privacy/confidentiality of the data and/or the resulting models. Essentially, it aims at deploying a distributed ML setup (Fig. 2) such that a model equivalent to the one obtained in the centralized setup is obtained. Nevertheless, even under the

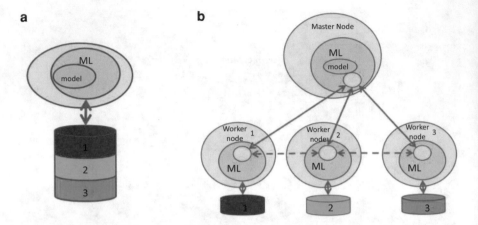

Fig. 2 Centralized approach (**a**) vs our privacy-preserving distributed scenario (**b**) where every user provides a portion of the training set

assumption that the raw data of the users is never directly exchanged or shared among the participants in the training process, any distributed privacy-preserving approach requires the exchange of some information among the participants or at least some interaction among them, otherwise no effective learning is possible.

In the MUSKETEER Machine Learning Library (MMLL) we foresee different possible Privacy Operation Modes (POMs) with different assumptions/characteristics. In what follows, we briefly describe every one of the implemented POMs, each offering a variety of machine learning models/algorithms, but always respecting the privacy restrictions defined in that POM. In what follows, we will name as Master Node (MN) or Aggregator the central object or process that controls the execution of the training procedure and the Workers or participants will be running at the end user side as a part of the MUSKETEER client, and they have direct access to the raw data provided by every user.

3.1 POM 1

This POM is designed for scenarios where the final trained model is not private, since at the end of the training every worker node and also the master node have a copy of the model. This POM implements a federated-learning framework based on the concept introduced by Google in 2016 [15]. Under this paradigm [25], a shared global model is trained under the coordination of the central node, from a federation of participating devices. It enables different devices to collaboratively learn a shared prediction model while keeping all the training data on device, decoupling the ability to perform machine learning from the need to store the data in the cloud. Using this approach, data owners can offer their data to train a predictive model without being exposed to data leakage or data attacks. In addition, since the model updates are specific to improving the current model, there is no reason to store them on the server once they have been applied. Any model trainable by gradient descent or model averaging [14] can be deployed under this scheme. The model is not private, since every worker and the central node will receive it at every iteration. The algorithms developed under this POM conform to the following steps: The aggregator defines the model, which is sent to every worker. Every participant computes the gradient of the model or updates the model (for a model averaging schema) with respect to his data and sends aggregated gradient or the model back to the aggregator, which joins the contributions from all participants and updates the model. This process is repeated until a stopping criterion is met. Under this POM the model is sent to the workers unencrypted and the workers send an averaged gradient vector/updated model to the aggregator.

3.2 POM 2

In some use cases, data owners belong to the same company (e.g. different factories of the same company) and the server that orchestrates the training is in the cloud. The work proposed in [22] shows that having access to the predictive model and to the gradients, it is possible to leak information. Since the orchestrator has access to this information in POM1, if is not completely under our control (e.g. Azure or AWS cloud), POM2 solves the problem by protecting the gradients over the honest-but-curious cloud server. This POM also implements the Federated Machine Learning (FML) paradigm described in POM1 but uses additively homomorphic encryption [26] to preserve model confidentiality from the central node. All gradients are encrypted and stored on the cloud server, and the additive property enables the computation across the gradients. This protection of gradients against the server comes with the cost of increased computational and communication between the learning participants and the server. At the end of the training stage the model is known by all the workers but not by the aggregator. In this POM all the participants use the same pair of public/private keys for encryption/decryption. The aggregator has access to the public key but not to the private one, meaning that it cannot have access to the data encrypted by the workers. It comprises the following steps: every worker has the same pair of public/private keys; the public key is shared with the aggregator, which defines the model and encrypts it. The encrypted model is sent to every participant, which decrypts it with the private key and computes the gradients. Then, it sends back to the master the encrypted computed gradients. The aggregator finally averages the received gradients and updates the local copy of model weights in the encrypted domain thanks to the homomorphic encryption properties. The process is repeated until the end of the training. In this case the model is sent to the workers encrypted, the workers send encrypted gradient vector/updated model to the aggregator, who updates the model in the encrypted domain with no possibility of decryption.

3.3 POM 3

In POM2, every data owner trusts each other, and they can share the private key of the homomorphic encryption (e.g. different servers with data that belongs to the same owner). Using the same key, every data owner uses the same encrypted domain. In many situations it is not possible to transfer the private key in a safe way. POM3 is an extension of POM2 that makes use of a proxy re-encryption [7] protocol to allow that every data owner can handle her/his own private key. This POM is an extension of POM2 that makes use of a proxy re-encryption protocol to allow that every data owner can handle her/his own private key [11]. The aggregator has access to all public keys and is able to transform data encrypted with one public key to a different public key so that all the participants can share the final model. The learning operation under this POM is as follows: every worker generates a different key pair randomly and sends the public key to the aggregator, which defines the initial model, such that it is sent unencrypted to every participant. Here the model is sent to every worker in its own encrypted domain in a sequential communication, the workers send encrypted gradient vector/updated model to the aggregator and the aggregator updates the model in the encrypted domain and uses proxy re-encrypt techniques to translate among different encrypted domains.

3.4 POM 4

This POM uses an additively homomorphic cryptosystem to protect the confidentiality of the data and requires the cooperation of a special node named as Crypto Node (CN), which is an object/process providing support for some of the cryptographic operations not supported by the homomorphism [9]. The CN is able to decrypt, but it only receives encrypted data previously masked by the aggregator. The scheme is cryptographically secure if we guarantee that CN and Master Node (MN) do not cooperate to carry out any operation outside of the established protocols ("honest but curious" assumption). Therefore, the scheme is cryptographically secure if we guarantee that there is no collusion between the MN and the CN. The steps to train a given model are as follows: the MN asks the CN some general public encryption key and distributes it to the workers, which will use it to encrypt the data and send it to the aggregator. The MN starts the training procedure by operating on the (encrypted) model parameters and (encrypted) users' data. The MN is able to perform some operations on the encrypted data (the ones supported by the homomorphism), but for the unsupported ones, it needs to establish a secure protocol with the CN consisting in sending some masked data to the CN, which decrypts the data, computes the unsupported operation in clear text, encrypts the result and sends it back to the MN, which finally removes the masking to obtain the encrypted desired result. As a result of repeating these operations during the training process, the MN never sees the data or the result in clear text, and the CN only sees the clear text of blinded/masked data. The training procedure continues until a stopping criterion is met. This POM is especially useful when the participants cannot actively cooperate during the training process (computational outsourcing) and the crypto node and master node are computationally powerful such that operations on the encrypted domain are highly efficient. Under this POM the MN, CN and Worker Node (WN) are assumed not to collude, and they do not operate outside of the specified protocols ("honest but curious" assumption).

3.5 POM 5

This POM is able to operate only with the aggregator and worker nodes. It uses an additively homomorphic cryptosystem to protect the confidentiality of the model [8]. The data is also protected, since it does not leave the worker facilities, and only some operation results are sent to the aggregator. The MN will help in some of the unsupported operations, that is the MN will play the role of CN. The scheme is cryptographically secure if we guarantee that the workers and aggregator do not operate outside of the protocols. The steps to train a given model are: the MN generates public and private keys and the public keys are distributed to all participants. The initial model parameters are generated at random by the MN. The MN encrypts the model parameters with the secret key and sends the encrypted model to the WNs, which start the training procedure by operating on the (encrypted) model and (un-encrypted) users' data. The WN is able to perform some operations on the encrypted data (the homomorphically supported ones), and for the unsupported ones, it establishes a secure protocol with the MN such that the WN sends some encrypted data with blinding to the MN, which decrypts it, computes the unsupported operation in clear text, encrypts the result and sends it back to the worker, which finally removes the blinding to obtain the result. As a result of this protocol, the MN never sees the data or the result in clear text, and the WN only sees the encrypted model. The procedure is repeated for every needed operation of every algorithm, until a stopping criterion is met. The MN, CN and WN are assumed not to collude and they do not operate outside of the specified protocols ("honest but curious" assumption).

3.6 POM 6

This POM does not use encryption; it relies on Secure Two-Party Computation [4] protocols to solve some operations on distributed data such that both model and data privacy are preserved [2]. Under this POM, raw data is not encrypted, but it is never sent outside the WN. The model trained in the MN can also be kept secret to the WN. Some transformations of the data can be exchanged with the MN, such as aggregated or correlation values, but always guaranteeing that the raw data cannot be derived from them by reverse engineering. Every implemented algorithm will explicitly describe which information in particular is revealed to the MN, such that the participants are aware of this potential partial disclosure before participating in the training process. Some important operations can be directly implemented using protocols for secure dot product or secure matrix multiplication. The security of these operations will be as described in the reference sources of protocol, and in general terms, they prove that the raw data is not exposed. An improved security is achieved during some aggregation operations if a "round robin" or "ring" protocol is used such that global aggregations can be computed without revealing the specific contributions from every participant. POM6 is not a generally applicable procedure, it requires that every algorithm is implemented from scratch, and also it is not guaranteed that any algorithm can be implemented under this POM. The MN, CN and WN are assumed not to collude, and they do not operate outside of the specified protocols ("honest but curious" assumption).

3.7 Algorithms

Over the different POMs, in the MMLL library, there are implementations of several algorithms. The library will contain algorithms capable to infer functions of different nature: Linear models: A simple but widely used class of machine learning models, able to make a prediction by using a linear combination of the input features [8]. MMLL includes alternatives for classification (logistic regression) and regression (linear regression) with different regularization alternatives and cost functions. Kernel Methods: They comprise a very popular family of machine learning models. The main reason of their success is their ability to easily adapt linear models to create non-linear solutions by transforming the input data space onto a high-dimensional one where the inner product between projected vectors can be computed using a kernel function [12]. MMLL provides solutions for classification (SVMs) and regression (SVRs), possibly under model complexity restrictions (budgeted models) [5, 6, 19]. Deep Neural Networks: Deep learning architectures [17] such as multilayer perceptrons or convolutional neural networks are currently the state of the art over a wide variety of fields, including computer vision, speech recognition, natural language processing, audio recognition, machine translation, bioinformatics and drug design, where they have produced results comparable to and in some cases superior to human experts [10]. Clustering: Unsupervised learning is the machine learning task of inferring a function to describe hidden structure from "unlabelled" data. The library will include algorithms for clustering, that is the task of dividing the population or data into a number of groups such that data points in the same groups are more similar to other data points in the same group than those in other groups [21]. In simple words, the aim is to segregate groups with similar characteristics and assign them into clusters. The library will include general purpose clustering algorithms such as k-means [23].

4 Setting Your Own Federated Learning Test Case: Technical Perspective

How to train machine learning algorithms using data collected from different data providers while mitigating privacy concerns is a challenging problem. The concept of a trusted data space serves the need of establishing trusted networks where data can be transferred, accessed and used in a secure mode, to offer secure data access and transfer. This section presents the MUSKETEER platform in relation to how it leverages new paradigms and advancements in machine learning research, such as Federated Machine Learning, Privacy-Preserving Machine Learning and protection against Adversarial Attacks. Moreover, the MUSKETEER platform, following actual industrial standards, implements a trusted and secure data space, to enable scalable privacy-preserving machine learning in a decentralized dataset ownership scenario.

4.1 How to Properly Train Your Machine Learning Model?

That is definitely not an easy question, and it depends on many different aspects. As it will take too much time to make a complete analysis of all of them, we will concentrate only on one of them: the amount of training data. The quality of a machine learning model depends on the volume of training data used during the training process. Small amount of data can produce low accuracy models that cannot be really usable. In this case, we can consider two options to solve the problem: (i) produce more training data by yourself or (ii) increase the training data volume by adding more data sources of the same kind. If the first option is not feasible (e.g. for technical or economic reasons), you can explore the second one by looking for other subjects with the same need. Here is where the concept of federation comes into play. In short, the model that can be produced thanks to the collaboration of the federation participants is better than the one produced by each participant on their own. This paradigm was initially introduced by Google and refers to different devices which collaboratively learn a shared prediction model while keeping all the training data on each device, decoupling the ability to do machine learning from the need to transfer data as well. The collaboration among the federation participants can be implemented with different levels of complexity and has to take into consideration other non-technical aspects. The simplest case is the one that concentrates all data in a single place and the training operation of the model is done using that single data repository. In this case, confidentiality and privacy should not be strong requirements. When the training data cannot be disclosed (e.g. business and/or legal reasons), a more sophisticated configuration has to be adopted. Every single federated participant will train an ML model locally (at their premises and not sending data outside) and will share only the model parameters. All the models produced by the participants are collected by a single subject that aggregates all of them and produces a new one that incorporates all contributions.

4.2   The MUSKETEER Platform

The main result of the MUSKETEER project is the implementation of an industrial data platform with scalable algorithms for federated and privacy-preserving machine learning techniques. The solution is based on the federation of a number of stakeholders contributing together, to build a machine learning model, in a collaborative (or co-operative) way. Different roles are to be assigned: (i) the aggregator, starting the process and taking charge of the computation of the final ML model; (ii) the participants, taking part in a single machine learning model process, built using their own (local) training datasets. From the architectural point of view, the MUSKETEER platform enables the interoperability between a number of distributed big data systems (federation participants) by providing a mechanism to send and retrieve machine learning models. That interoperability mechanism is based on the principles defined in the Reference Architecture Model of the International Data Space Association (IDSA) [13]. The MUSKETEER platform architecture consists of a server side and a client side. The server side is hosted in the cloud (as a number of micro-services), and it makes use of message queues for asynchronous exchange of information among the federation participants, which are often widely distributed geographically. One of the main activities of the server component is to coordinate the exchange of machine learning models between participants and aggregators. Besides the exchange of information for the execution of the actual federated learning tasks, the server side also provides services to manage tasks throughout their lifecycle, such as creating new tasks, browsing created tasks, aggregating tasks, joining tasks as a participant or deleting tasks. The meta-information that is required for task management is stored in a cloud-hosted database (Figs. 3 and 4).

4.3   MUSKETEER Client Connector Components

The client side is represented by the MUSKETEER Client Connector that is a self-contained component that each user has to deploy on-premise in order to work with the MUSKETEER platform. We consider now the case where the training data is stored locally (e.g. in hard drives, Network Attached Storage (NAS) or removable devices that are attached to a single computer), and we want to make use of them to create predictive models without explicitly transferring datasets outside of our system. In this case, the MUSKETEER Client Connector can be deployed in any environment using Docker in order to containerize the Client Connector application itself. Docker containers ensure a lightweight, standalone and executable package

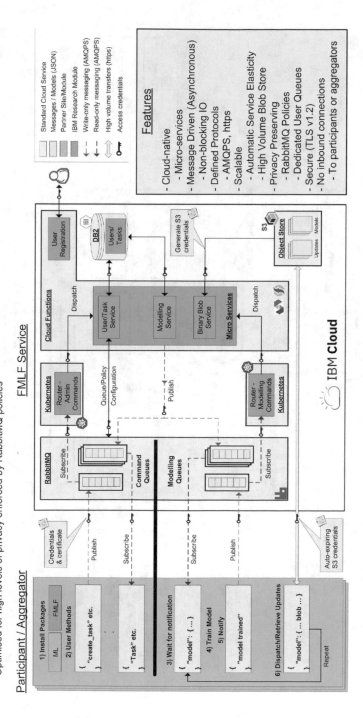

Fig. 3 MUSKETEER Platform Architecture

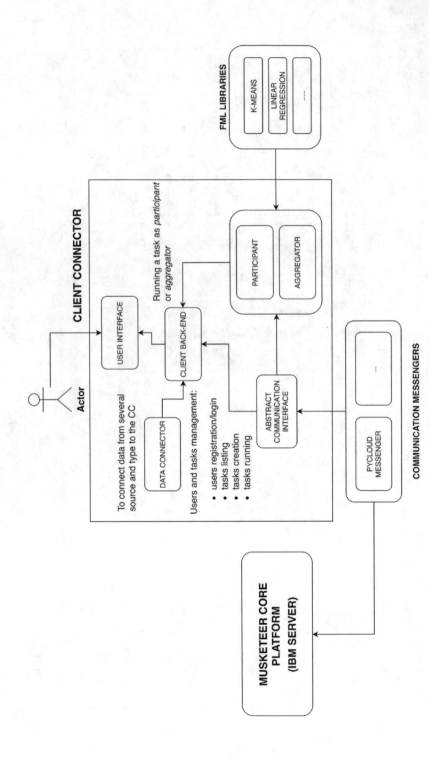

Fig. 4 MUSKETEER client connector architecture

of the software that includes everything needed to run the MUSKETEER Client Connector: operating system, code, runtime, system tools, libraries and settings. In this way the whole application can be easily made available in a sandbox that runs on the host operating system of the user. The MUSKETEER Client Connector consists of five core components and two additional ones that are loaded (as external plug-ins) after the application is up and running: the communication messenger and the federated machine learning library.

1. User Interface is a local web application that performs a set of functionalities where the main ones are: (i) to access the target server platform; (ii) to connect the local data for executing the federated ML model training; (iii) to manage the different tasks for taking part in the federation.
2. Client Back-End acts as a RESTful Web Service that handle all user requests, ranging from local operations (e.g. to connect user data to the Client Connector) to server operations (e.g. tasks and user management); these operations need to use a Communication Messenger library to communicate with a target external server.
3. Data Connector connects user data, which may come from different sources or storage layers, to the Client Connector. In addition, to connect data from different source types, the component can manage and support different kinds of data: in fact, a user can load a *.csv* tabular data from the file system, images files, binary data, a table from a database and so on.
4. Abstract Communication Interface allows the import and use of the communication library. In the MUSKETEER project the Communication Messenger library used is the pycloudmessenger library developed by IBM.[1] After such a library is configured and installed, the MUSKETEER Client Connector can use the APIs to communicate with the MUSKETEER cloud server provided by IBM.
5. Execution component instantiates and runs federated machine learning algorithms according to interfaces defined by the Federated Machine Learning library imported into the MUSKETEER client Connector. In the MUSKETEER project, the FML library imported is provided by the partners TREE Technology and Carlos III de Madrid University.

4.4 Give It a Try

The first prototype of the MUSKETEER Client Connector is available as open-source software from GitHub repositories (Client Connector Backend,[2] Client

---

[1] https://github.com/IBM/pycloudmessenger.

[2] https://github.com/Engineering-Research-and-Development/musketeer-client-connector-backend.

Connector Frontend[3]). Together with the source code you can also find the installation guide. We kindly invite you to download the Client Connector components and try it by setting up your test case of Federated Machine Learning process.

5   Federated Machine Learning in Action: An Efficiency Assessment

This section provides a test use case to show a practical application of Federated Machine Learning in the manufacturing context. The test bed aims to demonstrate, in a easily intuitive way, the potentiality of such algorithms in production environments.

5.1   Robots Learn from Each Other

COMAU is an automation provider, and its robots are installed in dozens of plants, in the automotive domain. In this context, these industrial automation customers are not eager to share their data. Nevertheless, those data are precious for the robot maker with regard to both business (new added value services for customers) and technical aspects (to improve robot performance and quality). Robot joints contain a belt that naturally loses its elasticity over time. With time and usage, the belt tensioning changes, and actually, in order to prevent failures caused by a wrong tension, operators have to regularly check the belt status with a manual maintenance intervention. These operations require time, effort and eventually a production stop. Moreover, these manual tasks bear the risk to be useless if the belt status is still good. The identified solution implemented for this case is a machine learning model able to predict the status of the belt based on the analysis of a set of specific parameters. In order to achieve a solution that provides a reasonable quality of the prediction, it is extremely important to identify the proper training set to use. Given that the same kind of Robot may be operating in different plants and under control of different owners (customers of COMAU), it is possible to assume that: (i) each one of the customers may be interested in running the same predictive ML model for those specific class of robots so as to increase their efficiency; (ii) each one of the customers may face the issue of not having enough training data to produce an effective machine learning model suitable for the production: (iii) customers are not keen to open their data to third parties. In a similar scenario, the MUSKETEER platform comes into play enabling data sharing in a privacy-preserving way for COMAU and its customers and thereafter

---

[3] https://github.com/Engineering-Research-and-Development/musketeer-client-connector-frontend.

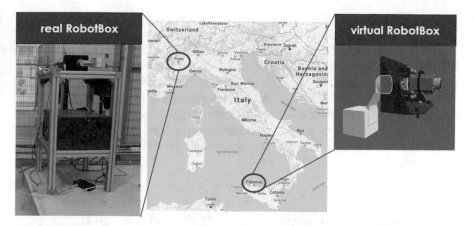

Fig. 5 RobotBox testbeds used for the experiment

with the possibility to use all available training data to build a classification model based on federated machine learning. To test the MUSKETEER platform functionalities and in particular the MUSKETEER Client Connector, two test beds called RobotBox have been setup. The first RobotBox (from now on called "real RobotBox") is located in COMAU headquarters in Turin and the second one (the "virtual RobotBox") is a virtual simulator hosted in an ENGINEERING facility in Palermo. Both test beds are simulating two different customer plants.

5.2 Defining Quality Data

Going a bit more into the details, the RobotBox replicates an axis of a COMAU robot. It is composed of a motor, a belt, a gearbox reducer and a weight (yellow part in Fig. 5). In order to collect data at different belt tensioning levels, we had to space out the motor and gearbox from each other. To do this we installed a slicer to move the motor and a dial gauge to measure the distance between the two parts. We have decided to consider three different belt tensioning levels. The RobotBox always performs the same movement called cycle. After each 24-second cycle, we collected signals from the RobotBox motion controller. For each cycle we calculated 141 features describing the different signals' aspects (e.g. mean value, maximum, minimum, root mean square, skewness, integral, etc.). Those features have been chosen based on the knowledge and literature about belt tensioning. For each level, we have performed around 6000 cycles, for a total of 18,000 samples for each RobotBox, considering the 3 different belt tension levels. For this scenario, we have chosen to train a classification model using an artificial neural network. COMAU plays the role of the aggregator for the federated machine learning task and for this use case also the role of the first participant. On the other side, ENGINEERING plays the role of the second participant. So, COMAU (located in Turin) created the task choosing the MUSKETEER Privacy Operation Mode 1 (POM1), where data cannot leave the facilities of the participants, the predictive models are transferred without encryption and the artificial neural network as algorithm. Moreover, the maximum number of iterations was set to 300 and the learning rate to 0.00015. The architecture of the model shaped as a neural network is a three-layer neural network with 48, 16 and 3 units tuned in the preparatory phase of this evaluation. After the task creation, including validation and test data, COMAU joined the federated process as participant that provides the training data of the real RobotBox. Finally, ENGINEERING joined the same task providing the training data of the virtual RobotBox.

5.3 Federated Data Sharing Is Better than Playing Alone

The final results of the federated task are very promising in comparison with the results of a non-federated approach. Looking at the right side of Fig. 6, it is possible to see how the overall accuracy obtained by the trained model, thanks to the federated task, with both RobotBoxes is 89% and the related confusion matrix is very diagonal. On the other hand, on the left side of Fig. 6, the accuracy of a model trained only with the data of the real RobotBox or only with the data of the virtual RobotBox is lower, respectively 86% and 81%.

6 Use Case Scenario: Improving Welding Quality Assessment Thanks to Federated Learning

The emerging data economy holds the promise of bringing innovation and huge efficiency gains to many established industries. However, the confidentiality and proprietary nature of data is often a barrier as companies are simply not ready to give up their data sovereignty. Solutions are needed to realize the full potential of these new profit pools.

6.1 A Twofold Challenge

MUSKETEER offers to tackle these two dimensions by bringing efficiency while respecting the sovereignty of data providers in industrial assembly lines. Two challenges are presented:

- Improving welding quality assessment to develop predictive maintenance for robots while increasing product safety at the same time
- Training a welding quality assessment algorithm on large datasets from multiple factories

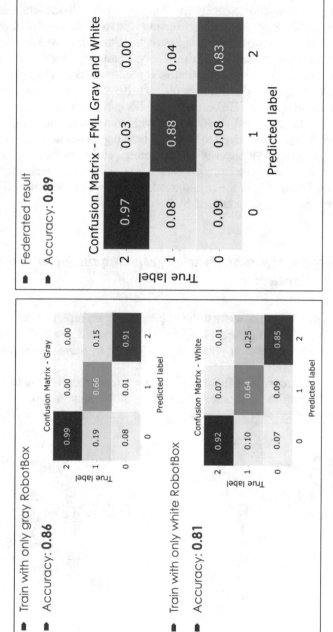

Fig. 6 Comparison of the non-federated and federated approach

The presence of a huge number of machines in industrial automation factories and the elevated cost of downtime result in large expenses for production line maintenance. Getting a more accurate evaluation of robot performance helps to avoid damaging the production capacity contingently (by 5–20% in certain cases) as presented in the report "Predictive maintenance and the smart factory" (Deloitte) [3]. The welding quality assessment can be improved using machine learning algorithms which support the status monitoring of machinery. But a single factory might offer too few data points to create such algorithms. It requires accessing larger datasets from COMAU's robots located in different places to boost the robustness and quality of the machine learning model. However, COMAU's customers can be competitors. Those companies do not intend to share data with competitors and simply waive their data sovereignty. With federated machine learning techniques, COMAU can offer an appropriate level of security for customers and save them costs at the same time. Besides, the aforementioned data might include personal information regarding operators working in the manufacturing plant which can raise additional privacy concerns that have to be tackled by the solution.

6.2 Training an Algorithm While Preserving the Sovereignty of Data Providers

At each piece's welding point, some welding parameters are recorded automatically from the welding robot (represented in Fig. 7), such as the current between the electrodes, resistance, number of points already welded by those electrodes and so on.

Then, a minimum amount of those pieces are sampled from the line to make an ultrasonic non-destructive test to assess the welding spot quality. An operator applies a probe on each welded spot which sends a signal to a computer to be interpreted

Fig. 7 Model of a COMAU robot and welding gun used for the welding of car parts

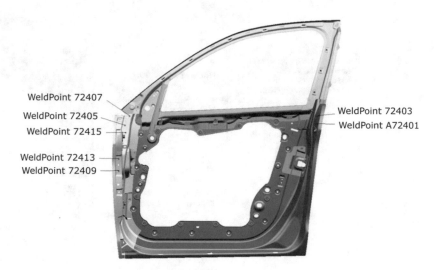

Fig. 8 Presentation of Q+ points on a car door

(graph representing the reflected sound energy versus time). The operator then classifies the welding points into different categories. In a production line only one or two pieces are verified each day whereas the total number of welded pieces goes up to 400 per day. Therefore, only a scarce percentage of pieces are subject to ultrasound testing. Besides, the manual test is limited to a few critical welding points on each piece, called Q+ points, which are always the same. On each car body, there are more than 3000 welding spots, while the Q+ points represent only a very small percentage of them. As an example, Fig. 8 shows which are the few Q+ points in a car door.

Our action here consists of collecting this manual ultrasound testing data and combining it with the welding data from the robot in order to locally train the algorithm. In parallel, this machine learning model is trained on different datasets from other factories. Trained models are eventually merged on the MUSKETEER platform (in different location) to provide the final version of the model. The entire flow is represented in Fig. 9.

As mentioned, collecting data from different factories also raises privacy issues. These data can be sensitive company data but also lead to personal data concerns (e.g. data can include information about operators working at the plant). Using the MUSKETEER platform provides a robust solution mixing a federated machine learning approach (local training) with privacy preserving technologies (highly customized encryption, data poisoning attacks mitigation) while respecting sovereignty of the stakeholders as defined in the reference architecture model of the International Data Spaces Association (IDSA) [13].

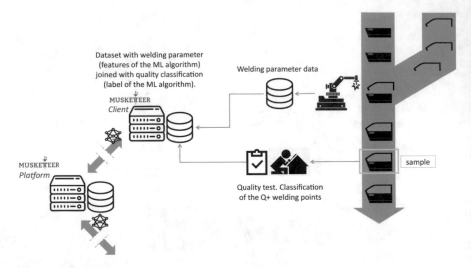

Dataset with welding parameter
(features of the ML algorithm)
joined with quality classification
(label of the ML algorithm).

Welding parameter data

MUSKETEER
Client

MUSKETEER
Platform

sample

Quality test. Classification
of the Q+ welding points

Fig. 9 Data collection and processing in federated learning environment

6.3 *Less Data but More Information*

Based on the combined data coming from robots and ultrasound tests, a robust model is built. In this case the Privacy Operation Mode selected is POM3 (Porthos), where the final trained model is private, the model is hidden to the aggregator and the encryption is implemented with a different private key for each data owner. One of the main reasons to adopt this model is because there is no trust among the data owners, for example in a scenario in which the data providers are competitors. Once the model is trained and has a satisfactory accuracy, thanks to the federated approach, it becomes possible to provide the classification of the welding spot directly from the welding data. This leads to numerous advantages over the limited manual testing:

- Opportunity to estimate the quality of all the welding points (not only the Q+ points) and raise the safety of products accordingly.
- Opportunity to understand if a specific combination of parameters helps to weld with fewer defects.
- Data sharing is allowed while sovereignty of each participant is preserved, and privacy concerns are tackled.

7 Conclusion

In this chapter, we introduced MUSKETEER, a federated machine-learning-based platform and the custom Privacy Operations Modes designed for it, aiming at increasing privacy when sharing data. We presented an efficiency assessment of the federated learning process. The results showed an improved overall accuracy of a model trained in a federated task. The experiment was done using the MUSKETEER platform. The chapter concluded with the presentation of a real use case in the automotive sector to optimize the welding quality of robots in production lines. The data coming from different locations was used to create a federated task.

Acknowledgments This project has received funding from the European Union's Horizon 2020 Research and Innovation Programme under grant agreement No 824988.

References

1. Bishop, C. M. (2006). Pattern recognition and machine learning. Springer, Berlin.
2. Chen, V., Pastro, V., & Raykova, M. (2019). Secure computation for machine learning with SPDZ. arXiv preprint arXiv:1901.00329.
3. Coleman, C., Damodaran, S., & Deuel, E. (2017). Predictive maintenance and the smart factory. Tech. rep., Deloitte.
4. Cramer, R., Damgård, I. B., & Nielsen, J. B. (2015). Secure multiparty computation. Cambridge University, Cambridge.
5. Díaz-Morales, R., & Navia-Vázquez, Á. (2017). LIBIRWLS: A parallel IRWLS library for full and budgeted SVMs. *Knowledge-Based Systems, 136*, 183–186.
6. Díaz-Morales, R., & Navia-Vázquez, Á. (2018). Distributed nonlinear semiparametric support vector machine for big data applications on spark frameworks. *IEEE Transactions on Systems, Man, and Cybernetics: Systems, 50*(11), 4664–4675.
7. Fuchsbauer, G., Kamath, C., Klein, K., & Pietrzak, K. (2019). Adaptively Secure Proxy Re-encryption. In IACR International Workshop on Public Key Cryptography, (pp. 317–346).
8. Giacomelli, I., Jha, S., Joye, M., Page, C. D., & Yoon, K. (2018). Privacy-Preserving ridge regression with only linearly-homomorphic encryption. In International conference on applied cryptography and network security (pp. 243–261).
9. González-Serrano, F. J., Navia-Vázquez, Á., & Amor-Martín, A. (2017). Training support vector machines with privacy-protected data. *Pattern Recognition, 72*, 93–107.
10. Goodfellow, I., Bengio, Y., Courville, A., & Bengio, Y. (2016). Deep learning. MIT Press, Cambridge.
11. Hassan, A., Hamza, R., Yan, H., & Li, P. (2019). An efficient outsourced privacy preserving machine learning scheme with public verifiability. *IEEE Access, 7*, 146322–146330.
12. Hearst, M. A., Dumais, S. T., Osuna, E., Platt, J., & Scholkopf, B. (1998). Support vector machines. *IEEE Intelligent Systems and their Applications, 13*(4), 18–28.
13. IDSA (2019). International Data Spaces Association (IDSA) Reference Architecture Model Version 3.0. Dortmund. https://internationaldataspaces.org/use/reference-architecture/.
14. Kamp, M., Adilova, L., Sicking, J., Hüger, F., Schlicht, P., Wirtz, T., & Wrobel, S. (2018). Efficient decentralized deep learning by dynamic model averaging. In Joint European conference on machine learning and knowledge discovery in databases (pp. 393–409).
15. Konečný, J., McMahan, H. B., Ramage, D., & Richtárik, P. (2016). Federated optimization: Distributed machine learning for on-device intelligence. arXiv preprint arXiv:1610.02527.
16. Konečný, J., McMahan, H. B., Yu, F. X., Richtarik, P., Suresh, A. T., & Bacon, D. (2016). Federated learning: Strategies for improving communication efficiency. In NIPS Workshop on Private Multi-Party Machine Learning.
17. LeCun, Y., Bengio, Y., & Hinton, G. (2015). Deep learning. *Nature, 521*(7553), 436–444.

18. McMahan, B., Moore, E., Ramage, D., Hampson, S., & y Arcas, B. A. (2017). Communication-efficient learning of deep networks from decentralized data. In: Proceedings of the AISTATS (pp. 1273–1282).
19. Morales, R. D., & Vázquez, Á. N.: Improving the efficiency of IRWLS SVMs using parallel Cholesky factorization. *Pattern Recognition Letters, 84*, 91–98 (2016).
20. Muñoz-González, L., Co, K. T., & Lupu, E. C. (2019). Byzantine-Robust federated machine learning through adaptive model averaging. arXiv preprint arXiv:1909.05125.
21. Omari, A., Zevallos, J. J. C., & Morales, R. D. (2017). Nonlinear feature extraction for big data analytics. In Big data analytics: Tools and technology for effective planning, p. 267.
22. Phong, L. T., Aono, Y., Hayashi, T., Wang, L., & Moriai, S. (2018). Privacy-Preserving deep learning via additively homomorphic encryption. *IEEE Transactions on Information Forensics and Security, 13*(5), 1333–1345.
23. Rao, F. Y., Samanthula, B. K., Bertino, E., Yi, X., & Liu, D. (2015). Privacy-Preserving and outsourced multi-user K-means clustering. In Proceedings of the IEEE Conference on Collaboration and Internet Computing (CIC) (pp. 80–89).
24. Shokri, R., & Shmatikov, V. (2015). Privacy-preserving deep learning. In Proceedings of the 22nd ACM SIGSAC conference on computer and communications security, pp. 1310–1321.
25. Yang, Q., Liu, Y., Chen, T., & Tong, Y. (2019). Federated machine learning: Concept and applications. *ACM Transactions on Intelligent Systems and Technology (TIST), 10*(2), 1–19.
26. Yi, X., Paulet, R., & Bertino, E. (2014). Homomorphic encryption. In Homomorphic Encryption and Applications (pp. 27–46). Springer, Berlin.
27. Zillner, S., Curry, E., Metzger, A., Auer, S., & Seidl, R. (Eds.) (2017). European big data value strategic research & innovation agenda. Big Data Value Association.
28. Zillner, S., Bisset, D., Milano, M., Curry, E., García Robles, A., Hahn, T., Irgens, M., Lafrenz, R., Liepert, B., O'Sullivan, B., & Smeulders, A. (Eds.) (2020). Strategic research, innovation and deployment agenda—AI, data and robotics partnership. Third Release. September 2020, Brussels. BDVA, euRobotics, ELLIS, EurAI and CLAIRE.

Leveraging Data-Driven Infrastructure Management to Facilitate AIOps for Big Data Applications and Operations

Richard McCreadie, John Soldatos, Jonathan Fuerst, Mauricio
Fadel Argerich, George Kousiouris, Jean-Didier Totow, Antonio
Castillo Nieto, Bernat Quesada Navidad, Dimosthenis Kyriazis,
Craig Macdonald, and Iadh Ounis

Abstract As institutions increasingly shift to distributed and containerized application deployments on remote heterogeneous cloud/cluster infrastructures, the cost and difficulty of efficiently managing and maintaining data-intensive applications have risen. A new emerging solution to this issue is Data-Driven Infrastructure Management (DDIM), where the decisions regarding the management of resources are taken based on data aspects and operations (both on the infrastructure and on the application levels). This chapter will introduce readers to the core concepts underpinning DDIM, based on experience gained from development of the Kubernetes-based BigDataStack DDIM platform (https://bigdatastack.eu/). This chapter involves multiple important BDV topics, including development, deployment, and operations for cluster/cloud-based big data applications, as well as data-driven analytics and artificial intelligence for smart automated infrastructure self-management. Readers will gain important insights into how next-generation

R. McCreadie (✉) · J. Soldatos · C. Macdonald · I. Ounis
University of Glasgow, Glasgow, United Kingdom
e-mail: richard.mcCreadie@glasgow.ac.uk; john.soldatos@glasgow.ac.uk;
craig.macdonald@glasgow.ac.uk; iadh.ounis@glasgow.ac.uk

J. Fuerst · M. F. Argerich
NEC Laboratories Europe, Heidelberg, Germany
e-mail: jonathan.fuerst@neclab.eu; mauricio.fadel@neclab.eu

J.-D. Totow · D. Kyriazis
University of Piraeus, Pireas, Greece
e-mail: totow@unipi.gr; dimos@unipi.gr

G. Kousiouris
Harokopio University of Athens, Moschato, Greece
e-mail: gkousiou@hua.gr

A. C. Nieto · B. Q. Navidad
ATOS/ATOS Worldline, Chennai, India
e-mail: antonio.castillo@atos.net; bernat.quesada@worldline.com

E. Curry et al. (eds.), *Technologies and Applications for Big Data Value*,
https://doi.org/10.1007/978-3-030-78307-5_7

135

DDIM platforms function, as well as how they can be used in practical deployments to improve quality of service for Big Data Applications.

This chapter relates to the technical priority Data Processing Architectures of the European Big Data Value Strategic Research & Innovation Agenda [33], as well as the Data Processing Architectures horizontal and Engineering and DevOps for building Big Data Value vertical concerns. The chapter relates to the Reasoning and Decision Making cross-sectorial technology enablers of the AI, Data and Robotics Strategic Research, Innovation & Deployment Agenda [34].

Keywords Data processing architectures · Engineering and DevOps for big data

1 Introduction to Data-Driven Infrastructure

For nearly a decade, advances in cloud computing and infrastructure virtualization have revolutionized the development, deployment, and operation of enterprise applications. As a prominent example, the advent of containers and operating system (OS) virtualization facilitates the packaging of complex applications within isolated environments, in ways that raise the abstraction level towards application developers, as well as boosting cost effectiveness and deployment flexibility [32]. Likewise, microservice architectures enable the provision of applications through composite services that can be developed and deployed independently by different IT teams [8]. In this context, modern industrial organizations are realizing a gradual shift from conventional static and fragmented physical systems to more dynamic cloud-based environments that combine resources from different on-premises and cloud environments. As a result of better application isolation and virtualized environments, basic semi-autonomous management of applications has become possible. Indeed, current cloud/cluster management platforms can natively move applications across physical machines in response to hardware failures as well as perform scaling actions based on simple rules. However, while useful, such basic autonomous decision making is insufficient given increasingly prevalent complex big data applications [31]. In particular, such applications rely on complex interdependent service ecosystems and stress the underlying hardware to its limits and whose properties and workloads can vary greatly based on the changing state of the world [4].

Hence, there is an increasing need for smarter infrastructure management solutions, which will be able to collect, process, analyse, and correlate data from different systems, modules, and applications that comprise modern virtualized infrastructures. In this direction, recent works have developed and demonstrated novel Data-Driven Infrastructure Management (DDIM) solutions. For instance, experimental DDIM solutions exist that process infrastructure data streams to detect errors and failures in physical systems (e.g. [16, 29]). In other cases, more advanced data mining techniques like machine learning have been used to detect anomalies in the operation of cloud infrastructures (e.g. [10]). This is complemented by works on data-driven performance management of cloud infrastructures [9] and resources [5]. However, these solutions address individual optimization and resilience concerns, where a more holistic Data-Driven Infrastructure Management approach is required.

This chapter introduces a holistic DDIM approach, based on outcomes of BigDataStack [15], a new end-to-end DDIM solution. The main contributions and innovations of the presented DDIM approach are the following: (i) Data-oriented modelling of applications and operations by analysing and predicting the corresponding data flows and required data services, their interdependencies with the application micro-services and the required underlying resources for the respective data services and operations. Allowing the identification of the applications data-related properties and their data needs, it enables the provision of specific performance and quality guarantees. (ii) Infrastructure management decisions based on the data aspects and the data operations governing and affecting the interdependencies between storage, compute, and network resources, going beyond the consideration of only computational requirements. The proposed DDIM leverages AI and machine learning techniques to enable more efficient and more adaptive management of the infrastructure, known as the AIOps (Artificial Intelligence Operations) paradigm, considering the applications, resources, and data properties across all resource management decisions (e.g. deployment configurations optimizing data operations, orchestration of application and data services, storage and analytics distribution across resources, etc.). (iii) Optimized runtime adaptations for complex data-intensive applications throughout their full lifecycle, from the detection of fault and performance issues to the (re)configuration of the infrastructure towards optimal Quality of Service (QoS) according to data properties. The latter is achieved through techniques that enable monitoring of all aspects (i.e. applications, analytics, data operations, and resources) and enforcement of optimal runtime adaptation actions through dynamic orchestration that addresses not only resources but also data operations and data services adaptations. The proposed DDIM approach has been deployed and validated in the context of three enterprise application environments with pragmatic workloads for different vertical industries, including retail, insurance, and shipping.

The rest of the chapter is structured to incrementally introduce the key building blocks that enable DDIM in BigDataStack. In particular, Sect. 2 introduces the core modelling of user applications, their environment, as well as additional concepts needed to realize DDIM. Section 3 discusses how profiling of user applications can be performed prior to putting them in production. In Sect. 4, we discuss how to monitor applications in production, as well as measure quality of service. Section 5 introduces AIOps decision making within BigDataStack, while Sect. 6 discusses how to operationalize the decisions made. Finally, we provide an illustrative example of DDIM in action within BigDataStack for a real use case in Sect. 7.

2 Modelling Data-Driven Applications

At a fundamental level, DDIM is concerned with altering the deployment of a user's application (to maintain some target state or quality of service objectives) [14, 22]. As such, the first question that needs to be asked is 'what is a user application?' From a practical perspective, a user application is comprised of one or more programs, each needing a compatible environment to run within. However, we cannot assume any language or framework is being used if a general solution is needed. To solve this, the programs comprising the user's application and associated environments need to be encapsulated into packaged units that are readily deployable without additional manual effort. There are two common solutions to this, namely virtual machines and containers [26]. For the purposes of DDIM, containers are generally a better solution, as they have a smaller footprint, have fewer computational overheads and are faster to deploy/alter at runtime [26]. We assume container-based deployment for the remainder of this chapter.

Given container-based deployment, we can now model the user application in terms of containers. It is good practice for a container to be mono-task, that is each container runs only a single program, as this simplifies smarter scheduling on the physical hardware [3]. A user application is then comprised of a series of containers, each performing a different role. DDIM systems then configure, deploy, and maintain these containers over large hardware clusters or clouds. There are a range of commercial and open-source container management solutions currently available, such as Docker Swarm and Kubernetes. The primary function of these solutions is to schedule containers onto cloud or cluster infrastructures. This involves finding machines with sufficient resources for each container, copying the container (image) to those machine(s), mounting any required attached storage resources, setting up networks for communication, starting the container(s), and finally monitoring the container states and restarting them if necessary. At the time of writing, the most popular container management solution is the open-source Kubernetes platform, which is what we will assume is being used moving forward. We discuss the most important Kubernetes concepts for DDIM systems below.

2.1 Application Modelling Concepts

Pods, Deployments, and Jobs For the purposes of modelling the user application in a containerized cloud/cluster, it is reasonable to consider an application to be comprised of a series of 'Pods', where a pod is comprised of one or more containers. A pod abstraction here exists to provide a means to group multiple containers into a single unit that can be deployed and managed together. In our experience, it is useful to distinguish pods along two dimensions: *lifespan* and *statefulness*. First, considering pod lifespan, 'continuous' pods are those that are expected to run indefinitely, representing permanent services which may be user-facing (e.g. a web host). Meanwhile, 'finite' pods are those that are aimed at performing a task, and will end once that task is complete (e.g. a batch learning or analytics task). Continuous pods in effect have an implied Service-Level Objective, that is that they must be kept running regardless of changes to the underlying infrastructure (e.g. due to hardware failures), while finite pods do not. In Kubernetes, continuous pods are managed using 'Deployment' objects, while finite pods are represented as 'Job' objects. Second, considering statefulness, a pod can be stateless meaning that it does not retain any data between requests made to it. This type of pod is the easiest to manage, it holds no critical data that could be lost if the pod needs to be restarted or moved and can often be replicated without issue. On the other hand, stateful pods maintain or build up data over time, which is lost if the pod fails or is killed. As such, the 'cost' of altering the configuration of an application that is comprised of stateful pods can be high, as data is lost when those pods are moved or restarted. In this case, the lost data needs to either be regenerated requiring more time and computational power or may simply be unrecoverable if the underlying input that created the data is no longer available. For this reason, it is recommended that application architects design their system to use only stateless pods where possible.

Services and Routes When scheduling a pod, a machine with the needed resources is only selected at runtime, meaning the network address of that pod cannot be known before then. Furthermore, that address may not be static, as changes in the infrastructure environment may result in the pod being lost/deleted and then a new copy spawned on a different physical node. This complicates the configuration of user applications, as it is commonplace for user programs to expect to be preconfigured with static URLs or IP addresses when two components need to talk to one another. A 'Service' is the solution to this issue, as it is a special entity in Kubernetes in that it has a static URL. Traffic directed at a service will then be forwarded to one or more pods based on a service-to-pod mapping, which is updated over time if changes occur and can also be used for load-balancing requests across multiple copies of a pod. A service can be paired with a 'Route' object to produce an external end-point, enabling requests from the outside world to reach a pod.

Volumes and Volume Claims Containers by their nature are transient, that is their state is lost when they exit. Hence, most pods need some form of persistent storage to write to, for example for writing the final output of a batch operation or as a means to achieve statelessness by reading/writing all state information to an external store. This is handled in Kubernetes using volumes. A volume is simply a definition of a directory within a storage medium that can be mounted to one or more containers, such as an NFS directory or distributed file system like HDFS. However, the available storage amounts and storage types available will vary from cluster to cluster, and as such it is not good practice to directly specify a volume, as this ties the pod to only clusters with that exact volume. Instead, Volume Claims exist, which represent a generic request for a desired amount and type of storage. If a pod specifies a volume claim, Kubernetes will attempt to automatically provide the requested amount and types of storage from its available pool of volumes. In this way, an application can still obtain storage, even if the application owner does not know what exact storage volumes are available on the target cluster.

2.2   Data-Driven Infrastructure Management Concepts

In this section, we will describe additional concepts that are required for DDIM systems to function based on experience from developing BigDataStack, namely Pod Level Objectives, Resource Templates, Workloads, and Actions.

Pod Level Objectives For DDIM systems to meaningfully function, they need a set of objectives to achieve, representing the needs of the application owner. Given the application modelling discussed above, we can consider that an application component, represented by a running pod (and created via a Deployment or Job), could have zero or more objectives associated with it. In the literature, such objectives are typically referred to as Service-Level Objectives (SLOs) [24]. However, this may be somewhat confusing as in containerized environments a 'Service' means something different, as such we will instead refer to these as Pod-Level Objectives (PLOs). A PLO defines a quality of service (QoS) target for a pod to achieve, such as 'cost less than 1.2 U.S. dollars per hour', or 'provide response time less than 300 ms'. A QoS target is comprised of three parts: (1) a metric (e.g. response time), (2) a target value (e.g. 300 ms), and (3) a comparator (e.g. less than). Note an implicit assumption here is that the specified metric is measurable for the pod, either because the pod exports it (e.g. response time), or it can be calculated for the pod by a different component (e.g. cost). If so, a PLO can be checked by comparing the current measured metric value against the QoS target, resulting in a pass or failure. PLO failures are the primary drivers of DDIM systems, as they indicate that changes in the user application or data infrastructure are needed.

Resource Templates To launch a Deployment or Job in a cluster or cloud environment, sufficient computational resources need to be provided, that is CPU capacity, system memory, and potentially GPUs or other specialized hardware [6]. The modelling of resources assigned to a pod is a critical part of DDIM systems, as a lack (or in some cases excess) of such resources is the primary cause of PLO failures. Moreover, the resource allocation for individual pods are often a variable that the DDIM system has control over and hence can manage automatically. In theory, resources, such as allocated system memory, are continuous variables, where any value could be set up to a maximum defined by the target cluster. However, predefined Resource Templates that instead define aggregate 'bundles' of resources for a fixed cost are very common, such as Amazon EC2 Instances. Resource Templates exist as they both simplify the resource selection process for the application owner, while also enabling the cluster owner to divide their available resources in a modular fashion. A basic Resource Template needs to specify CPU capacity and system memory for a pod, as all pods require some amount of these. A Resource Template may optionally list more specialized hardware, such as Nvidia GPUs or Quantum cores based on the requirements of the application and cluster/cloud support available.

Workloads Another factor that DDIM systems often need to consider is the application environment. Continuous applications will typically be serving requests, either from users or other applications. Meanwhile, finite applications most commonly will be concerned with processing very large datasets. We can consider these environmental factors as sources of workload that is placed on pods, that is they quantify the properties of the input to those pods over time. In the case of continuous applications, this might manifest in the number of API calls being made per second and may vary over time (e.g. for user-facing applications distinct day-night cycles are commonly observable). On the other hand, for finite applications, the size of the dataset or database table(s) being processed can be considered to define the workload. Some types of DDIM systems will model such workloads and may even condition their PLOs upon them, for example if the number of requests is less than 500 per second, then response latency should be less than 100 ms.

Actions The final aspect of a user application that is critical to enable DDIM systems is how they can be altered. For a DDIM system to function, it requires a finite set of actions that it can perform for an application. In effect, these actions form a 'toolbox' that the DDIM system can use to rectify PLO failures. Relatively simple actions in containerized environments might include adding/removing replicas or altering the Resource Template for a pod. However, as applications become more complex, associated actions often require multiple steps. For example, scaling a data-intensive API service might involve first replicating an underlying database to provide increased throughput, followed by a reconfiguration step joining the new database instance into the swarm, followed by a scaling action on the API pods to make use of the increased capacity. Moreover, as actions become more complex, the time taken to complete them will grow, hence DDIM systems need to track what actions are currently in progress and their state to enable intelligent decision making.

3   Application Performance Modelling

Much of the emphasis for DDIM systems is on how to manage user applications (semi-)automatically post-deployment. However, some types of DDIM systems, including BigDataStack, support optional functionality to analyse the user application pre-deployment. The core concept here is to gain some understanding of the expected properties of the application, which can be used later to enable more intelligent decision making. From a practical perspective, this is achieved via benchmark tooling, whereby application components can be temporarily deployed with resource templates, subjected to a variety of predefined workloads, in a parameter sweep fashion, and their performance characteristics measured and recorded. This enables the quantification at later stages of the effects of a workload on the QoS metrics of this service.

How Does Benchmarking Function? Fundamentally, benchmarking has three main aspects: (1) the deployment of part or all of the user application with a defined set of resources; (2) the creation of a workload for that application; and (3) measurement of the observed performance characteristics. Deployment in a containerized environment is greatly simplified, as the containers needing tested can be deployed as pods directly upon the infrastructure if correctly configured. On the other hand, generating a workload for the application is more complicated. For continuous pods, typically an external service is needed to generate artificial requests, for example to simulate user traffic patterns. Meanwhile, for finite pods, use of a standard dataset or data sample is used for benchmarking. Finally, measurement of the performance characteristics of an application typically comes in three forms: application exported metrics, infrastructure reported metrics for the application, and metrics about the infrastructure itself. As the name suggests, application exported metrics are metrics directly reported by the application itself based on logging integrated into it by the application engineer, for example request processing time for a website host. Infrastructure metrics about the application represent monitoring the management platform (e.g. Kubernetes) that is passively performing, such as pod-level CPU and system memory consumption. Finally, infrastructure metrics (which are not typically available on public clouds) can provide information about the wider state of the cluster/cloud providing insights into how busy it is. Metrics here might include node-level CPU and memory allocation and information about node-to-node network traffic.

Benchmarking Tools in BigDataStack BigDataStack incorporates a Bench-marking-as-a-Service framework (Flexibench), developed in the context of the project, that exploits baseline tools such as Apache Jmeter and OLTPBench and orchestrates their operation towards a target endpoint (database or otherwise). Flexibench retrieves the necessary setup (type of client to use, workload to launch, desired rate of requests, etc.) via a REST-based interface or a UI-based interface and undertakes their execution. Multiple features are supported such as parallel versus isolated execution of the experiment, trace-driven load injection (based on historical data files), inclusion of the defined data service in the process, or load injection towards an external datapoint. The tool offers REST interfaces for test monitoring and result retrieval and is based on Node-RED, a visual flow programming environment of node.js.

Predictive Benchmarking While undertaking actual benchmarking is the most accurate way to determine the performance characteristics for an application, it can be costly and time consuming to implement as the application needs to be physically deployed and the parameter search space may be extensive. An alternative to this is *predictive benchmarking*. The idea is to use machine learning to estimate what the outcome of a benchmark run would look like, by considering the benchmarking outcomes from other similar application deployments. BigDataStack also supports the creation of predictive benchmarking models via Flexibench through the same UI- or REST-based environment, therefore integrating the two processes (result acquisition and model creation). The creation is performed following a REST- or UI-based request, in which the test series is defined, as well as other parameters that are necessary for result acquisition (such as the identifiers of related services to use for predictions). The framework retrieves the relevant training data and launches a containerized version of an automated model creation algorithm defined in [13].

The resultant model is subsequently validated, and the respective results are made available for the user to examine via the UI. Following, the model can be queried if the relevant input values are supplied (e.g. type and size of workload, resource used, etc.) and the predicted QoS metric (e.g. response time, throughput, etc.) will be returned to the user.

4 Metric Collection and Quality of Service Monitoring

To enable the validation of the application PLOs that act as the triggers for DDIM, constant quantitative performance measurement for pods is required. Therefore, metrics must be collected, stored, and exposed. Indeed, the performance of applications running on platforms like BigDataStack is impacted by many factors, such as infrastructure state, data transaction speeds, and application resourcing. For this reason, the monitoring engine of BigDataStack was developed using a triple monitoring approach. By triple monitoring we mean the collection of performance indicators from: (1) the infrastructure, (2) data transactions, and (3) applications exported metrics.

Metric Collection Strategies Any metric monitoring system can be considered as comprised of multiple agents and a manager. The agents perform measurements and prepare metrics for collection by the manager, which might be a pod within the user application, a database, or Kubernetes itself. Most commonly, the manager periodically requests metrics from the agents (known as polling) via standardized metric end-points. An interval (scraping interval) then defines how frequently the metrics are collected. An alternative approach is to have the agents directly post metric updates to the manager or an intermediate storage location (known as the push method). This can be useful for applications or services that do not/cannot expose a metrics end-point that a manager can connect to. BigDataStack supports both polling and push methods for data collection. The measurement data collected by the manager is then typically held in a time-series database, enabling persistent storage of performance history over time.

Triple Monitoring in BigDataStack BigDataStack's monitoring solution is based on Prometheus, which is the official monitoring tool of Kubernetes. Prometheus requires a definition of target endpoints exposing metrics in its configuration file. However, manually defining this for the many applications that are managed by BigDataStack is infeasible. Instead, BigDataStack exploits the service discovery functionality of Prometheus to automatically configure metric collection from new pods as they are deployed. Under the hood, in this scenario Prometheus periodically communicates with the Kubernetes API to retrieve a list of port end-points exposed by running pods in the BigDataStack managed namespaces, and checks them to see if they export metrics in a format that Prometheus can understand. Some applications may wish to control their own metric collection via their own Prometheus instance. In this context, the triple monitoring engine is able to aggregate metrics collection from multiple Prometheus instances concurrently using Thanos. The monitoring engine of BigDataStack adopts a federated model where several Prometheus instances can be added dynamically for reducing the total scraping duration, thus allowing the collection of very large volumes of data points from a large number of sources. This is coupled with metrics compression by aggregation, minimizing the overhead when working with Big Data applications. For the purposes of DDIM, three different groups of metrics are collected by the triple monitoring engine: infrastructure information, data operations information (i.e. data produced, exchanged, and analysed by applications), and all the data involved in database transactions and object storage operations. Since these metrics are produced by applications with different purposes, specifications, functionalities, and technologies, the triple monitoring engine integrates a data sanitizer to prepare incoming measurements such that they conform with a standard specification. The triple monitoring engine also provides an output REST API for exposing data to all services, as well as a publish/subscription service that enables the streaming consumption of measurements by other applications.

Quality of Service Evaluation Within a DDIM system, the core reason to collect measurements about a user application and infrastructure is to enable evaluation of application Quality of Service (QoS). QoS represents the degree to which the application is meeting the user needs. More precisely, QoS evaluation is concerned with guaranteeing the compliance of a given KPI (Key Performance Indicator) as defined in one or more PLOs (Pod-Level Objectives), typically for a given time period or window. When a user requests a service from BigDataStack, a minimum QoS is agreed between the user and the system, expressed as PLOs. At runtime, metric data is collected by the Triple Monitoring Engine and evaluated against these PLOs to determine whether the application is (or in some cases soon will be) failing to meet the user needs. Such failures can then be used to trigger orchestration actions aimed at rectifying such failures, as discussed in the next section.

5   Automated Decision Making

Having discussed the monitoring of user applications and how to determine when failures have occurred that need to be rectified, we next discuss how DDIM systems can solve these issues through automatic service orchestration.

QoS as an Optimization Problem Service orchestration in the context of Big Data has usually the goal of a Quality of Service (QoS) or Quality of Experience (QoE) sensitive optimization [27]. This optimization problem, under varying contexts, is complex and considered an NP-hard problem [17]. Previous approaches have tackled this optimization problem with heuristics [7, 30] or genetic algorithms [2] that aim to find near-optimal configurations and service compositions, such as configurations respecting overall QoS/QoE constraints, while maximizing a QoS utility function. The composition of services in current cloud-edge Big Data/AI applications usually follows a pipeline pattern in which stream and batch data (potentially recorded at the edge) is processed by multiple components in order to derive the desired results. These new pipelines add new requirements and challenges to service orchestration as they inherently contain complex trade-offs between computation and performance (e.g. resulting accuracy), and errors introduced in early components cascade through the overall pipeline, affecting end-to-end performance and making it impossible to treat the problem as an independent orchestration of components. Initial approaches have addressed the orchestration problem with reasoning across the pipeline components in a probabilistic manner, allowing the user to manually decide the adequate trade-off [25].

We model QoS as a constrained optimization problem. Specifically, (1) we model requirements as constraints, for example to process documents with an end-to-end latency less or equal than 1 s or to run at a cost of less or equal than 10 $ per hour, and (2) we model service performance, such as precision, accuracy, or battery consumption, as objective. There is an important difference between the objective and the constraints: whereas the constraints define a minimum or maximum value for the variable involved (e.g. latency, cost, etc.), the objective does not have a minimum or maximum value expected. In this way, we can define the service requirements as:

$$\underset{\theta}{\text{maximize}} \quad O(\theta)$$

$$\text{subject to} \quad c_i(\theta) \leq C_i, \, i = 1, \ldots, N$$

where:

- θ: is the configuration of parameters used for all of the operators.
- $O(\theta)$: represents the objective of the service, which is determined by the configuration of parameters used.
- $c_i(\theta)$: is a constraint to the service (such as latency), also determined by θ.
- C_i: is the constraint target (e.g. 1 s).
- N: is the total number of constraints.

The developer is in charge of defining the service requirements along with the metrics to monitor them, as well as the parameters that can be adapted and the values they can assume. During runtime, the system is in charge of finding the best configuration of parameter values that maximize (or minimize) the objective while respecting the constraints.

Optimization via Reinforcement Learning Recently, reinforcement learning (RL) has been successfully applied to node selection for execution [19] as well as optimization of overall pipelines using, among others, meta-reasoning techniques to ensure an overall optimization of the pipeline [1, 18, 21]. A key issue with RL-based approaches is the bootstrapping problem, that is how to obtain sufficient performance from the first moment the agent begins to operate. A simple solution is to explore the state space randomly, but this approach is usually time consuming and costly when the state/action space is large, as illustrated in [1]. An alternative approach is to gain experience more cheaply and faster via a sandbox simulation. With enough computational resources, it is possible to produce large volumes of experience data in a short time period, but it is difficult to ensure that the simulated experiences are realistic enough to reflect an actual cloud/cluster environment. To reduce the cost of training RL agents, some works have examined how to leverage external knowledge to improve their exploration efficiency. For example, in [11, 23] prior knowledge like pretrained models and policies are used to bootstrap the exploration phase of an RL agent. However, this type of prior knowledge still originates in previous training and is limited by the availability of such data. In BigDataStack, we leverage Reinforcement Learning in conjunction with expert knowledge to drive service orchestration decisions, as discussed next.

Dynamic Orchestrator in BigDataStack The Dynamic Orchestrator (DO) works alongside the Triple Monitoring Engine (TME) to monitor and trigger the redeployment of BigDataStack applications during runtime to ensure they comply with their Service-Level Objectives (SLOs). The DO receives and manages monitoring requests when a new application or service is deployed into the BigDataStack platform, informing the TME and the Quality of Service (QoS) component what metrics and PLOs should be monitored. When any violation to these PLOs occur, the QoS informs the DO, and the DO is in charge of deciding what redeployment change is necessary, if any. In BigDataStack, we developed flexible orchestration logic that can be applied to any kind of application by applying Reinforcement Learning (RL) that leverages external knowledge to bootstrap the exploration phase of the RL agent. We call this method Tutor4RL. For Tutor4RL, we have modified the RL framework by adding a component we call the Tutor. The tutor possesses external knowledge and helps the agent to improve its decisions, especially in the initial phase of learning when the agent is inexperienced. In each step, the tutor takes as input the state of the environment and outputs the action to take, in a similar way to the agent's policy. However, the tutor is implemented as a series of programmable functions that can be defined by domain experts and interacts with the agent during the training phase. We call these functions knowledge functions and they can be of two types:

- *Constraint functions:* are programmable functions that constrain the selection of actions in a given state, 'disabling' certain options that must not be taken by the agent. For example, if the developer of the application has decided a maximum budget for the application, even if the application load is high and this could be fixed by adding more resources to the deployment, this should not be done if the budget of the user has already reached its maximum.
- *Guide functions:* are programmable functions that express domain heuristics that the agent will use to guide its decisions, especially in moments of high uncertainty, for example at the start of the learning process or when an unseen state is given. Each guide function takes the current RL state and reward as the inputs and then outputs a vector to represent the weight of each preferred action according to the encoded domain knowledge. For example, a developer could

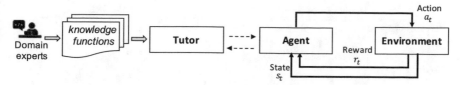

Fig. 1 Overall Working of Tutor4RL

create a guide function that detects the number of current users for an application, and if the number is higher than a certain threshold, more resources might be deployed for the application (Fig. 1).

The benefit coming from using Tutor4RL is twofold. First, during training, the tutor enables a faster bootstrapping to a reasonable performance level. Furthermore, the experience generated by the tutor is important because it provides examples of good behaviour, as it already uses domain knowledge for its decisions. Second, the knowledge the tutor provides does not need to be perfect or extensive. The tutor might have partial knowledge about the environment, that is know what should be done in certain cases only, or might not have a perfectly accurate knowledge about what actions should be taken for a given state. Instead, the tutor provides some 'rules of thumb' the agent can follow during training, and based on experience, the agent can improve upon the decisions of the tutor, achieving a higher reward than it.

Learning What Actions to Take Within BigDataStack, the application engineer (i.e. the domain expert) defines sample guide and constraint functions for the learner. These functions encode domain knowledge of the developer that can guide decision making. Indeed, for some applications these guide functions will be sufficient to manage the application without further effort. On the other hand, for cases where the guide functions are insufficient, reinforcement learning can be enabled. During RF training (i.e. pre-production), the RL agent will experiment with the different available alteration actions that can be performed (discussed in the next section), learning how each affects the metrics tracked by the triple monitoring engine, as well as the downstream PLOs. After a period of exploration, the RF agent can be deployed with the application in production, where it will intelligently manage the application by triggering the optimal alteration action or actions in response to PLO failures.

How Effective is Automatic Service Orchestration? We have compared Tutor4RL performance against vanilla DQN [20] in a scenario where the DO is in charge of controlling two metrics: cost per hour (which varies according to resources used by application) and response time. These are two opposite objectives: if we increase the use of resources, the response time decreases but the cost per hour increases, and if we decrease the use of resources, the opposite is true. However, the SLOs specify thresholds for each metric: cost per hour should be less or equal to \$0.03 and response time should be less than 200 ms. The DO must find the sweet spot that satisfies these two SLOs as long as the application allows it. In

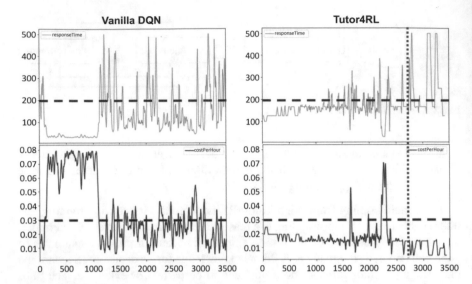

Fig. 2 Tutor4RL Performance compared to Vanilla DQN [20]. DO performance to manage 2 SLOs: costPerHour <0.03 and responseTime <200. Vanilla DQN is shown on the left, while Tutor4RL, with 2 guides and 1 constrain, is shown on the right. The horizontal blue dashed lines show the SLO threshold for the metrics and the pink dotted line show the moment in which guides are not used anymore

fact, it might happen that the application load is too high, and then there is no way of satisfying both SLOs, in these cases the DO behaviour will tend to find the configuration that violates the SLOs proportionally less. However, we believe these are corner cases in which even a human might not be sure what to do, and therefore we have not evaluated the DO's performance in these situations. In Fig. 2, we see the performance of the vanilla DQN agent (left) and the Tutor4RL agent (right) for managing this scenario with two SLOs. Note that on the images we have marked with horizontal lines the thresholds for SLOs and with a vertical line, the moment in which the guide functions from the Tutor are not used anymore, until that point the functions are used on and off with a diminishing frequency from 0.9 to 0. As we can see, the Tutor4RL agent performs better than the vanilla agent by achieving a better satisfaction of SLOs. We still see that once the guides are completely abandoned, the agent commits some mistakes, but it can quickly correct its error. We can avoid this by adding constraints such as not changing the deployment configuration if no SLO is violated, but we wanted to show a case in which the agent is free in its actions and therefore show its learned behaviour better.

6   Operationalizing Application Alterations

Once the decision to alter the user application (as an attempt to rectify a failing PLO) has been made, and the action to perform has been decided, the DDIM needs to operationalize that alteration. In this section we discuss the types of alterations that are possible, as well as how these can be encoded as actions within BigDataStack.

6.1   Types of Alteration Actions

Depending on the type and complexity of the deployed application, as well as the level of permissions that the DDIM has with the infrastructure management system (e.g. Kubernetes), there can be a wide range of alteration actions that might be performed. A useful way to structure actions is based on what problem they aim to solve. In general, there are four common types of problems that can arise as follows:

Insufficient Resources for a Pod Based on unexpectedly poor performance reported by a pod in conjunction with maximized utilization for that pod, the DDIM might decide that a pod needs more resources. This can be solved by the allocation of a different (in terms of resources) pod, which actually refers to a larger Resource Template for that pod. Notably, in Kubernetes this is a destructive action, that is it will involve the pod being killed. For continuous applications, this is typically not an issue, as the pods involved are stateless. However, for Jobs performing this type of operation might result in all progress up to that point being lost. In either scenario, the alteration action involves the launching of a new copy of the target pod with the new larger Resource Template, then halting the previous pod once the new one reaches running status.

Insufficient Application Capacity In this scenario, one or more pods may be reporting unexpectedly poor performance, but the resources are not saturated for any of the application pods. This would indicate that the existing pods are working correctly, but they are unable to keep up with the current workload. The solution here is to scale-up the application to increase its capacity, if supported by the application. For simple applications, this might only involve increasing the number of replicas for one of the pods, which is a non-destructive action (load-balancing across the replicas can be handled by a Service automatically). However, for complex applications this may require multiple steps and can be destructive. For example, consider the scaling of an Apache Spark Streaming application. First, the number of Spark workers needs to be increased, which can be handled by a simple replication factor change on the worker pods. These new workers will be automatically added to the 'Spark Cluster' and will show ready for work. However, the streaming application running on that Spark cluster will not automatically scale to use the additional workers. Instead, the application must be killed and then resubmitted to the Spark cluster before the workers will be allocated to the application.

Insufficient Data Availability Not all issues that might cause PLO failures are necessarily application-related. In this scenario, the DDIM might observe unexpectedly poor performance for the application and at the same time a saturation of resources for one or more data infrastructure components in use by that application (e.g. a database). In this case, the most efficient response would be to scale the data infrastructure components (in the example above, the database) to provide the required additional capacity. This is typically a costly action to perform, in terms of both time and resources. First, appropriate data storage volumes need to be created to hold the replicated data. Second, new infrastructure pods need to be started, and when they are ready, the associated data needs to be imported. Note that this import process is typically performed from some archive service, not via copying from currently running data infrastructure instances, as those are already overloaded by application requests.

Insufficient Network Bandwidth The last case represents failures in the networking/communication infrastructure. Within distributed cluster environments, pod-to-pod as well as external service-to-pod communications are channelled across physical communication links, which can become saturated. A classical example of this is a Denial of Service attack, where external servers attempt to overload an application with requests. While there is no easy fix for this type of failure, specific DDIM setups can control traffic prioritization. In this case, some portion of low-priority in-flight network traffic will be dropped to free up bandwidth for high-priority traffic. This is achieved by re-configuring the cluster's network routing policies based on detected in-bound workloads.

6.2   Considerations When Modelling Alteration Actions

Given the above adaptation scenarios and their possible solutions, it is clear that DDIM solutions require the means to perform complex alteration actions at runtime. There are four key aspects that need to be considered when modelling these alteration actions:

Sequencing and State Dependencies First, a single alteration action can be a complex affair that involves multiple steps. Moreover, the individual steps may have dependencies that require progression to wait until particular application states are achieved. For example, for a scaling action on the data infrastructure, the alteration action is not complete when the new infrastructure pods have started, but rather once the data has finished importing. Hence, an action must be seen as a composite set of lower-level operations that together form the desired action, where both the application and operations have states that can be tracked to determine when new operations can start (as well as when the action as a whole is complete).

Actions Are Application (Type) Dependent Second, it is worth stating that the available alteration action set is not the same across applications. While there are standard operations that are built into management platforms like Kubernetes for controlling factors such as pod replication, just because an operation is technically valid, it does not mean that performing the operation would be efficient for the current application. For example, for deployments where state loss is not an issue, the option to perform destructive runtime resource template changes may be desirable. But such an option might not be effective or cost too much for jobs where progress is reset when the underlying pod is restarted, even if it is possible to do so [28]. Moreover, any reasonably complex application will need support for multistage alteration actions that are not supported natively by existing management platforms. On the other hand, it is notable that applications that are of a similar type, for example those that use a common framework like Apache Spark or Ray, may be able to share actions, enabling common action sets to be shared among similar applications.

Available Actions Can Change Over Time The set of available actions for an application may change depending on that application's state, or the states of associated actions being performed. In the simplest case, once an alteration action is triggered, it makes sense to remove that action from the available set until it completes, as the DDIM system should wait to see the outcome of that alteration before attempting the same action again. Meanwhile, in more complex scenarios, the application engineer might want more control over what actions the DDIM system will consider under different conditions, effectively providing the DDIM system expert knowledge of what actions are reasonable given different application states.

Actions Are Data-Dependent Finally, the potential different actions to be applied heavily depend on the data: Data volumes might highlight the need for altering the data services settings tackling both storage (e.g. dynamic split or dynamic migration of data regions) and analytics (e.g. triggering of scalability of a real-time complex event processing engine). In this context, the DDIM system should monitor and account for the data operations and the related workloads (of data-intensive applications) and trigger the optimum adaptation actions during runtime - including deployment configurations and orchestration decisions.

6.3   Alteration Actions in BigDataStack

BigDataStack delivers a solution for enabling complex alteration actions: the Realization Engine. At its core, the Realization Engine has three roles: (1) to act as a central point of reference by storing all application-related information, (2) to maintain up-to-date state information about the application alteration actions, and (3) to enable triggering and subsequent operationalization of alteration actions for each application. Figure 3 illustrates the application model within BigDataStack. In particular, under this model, the user account or 'owner' owns one or more applications and can also define metrics. A single application has a state, zero or

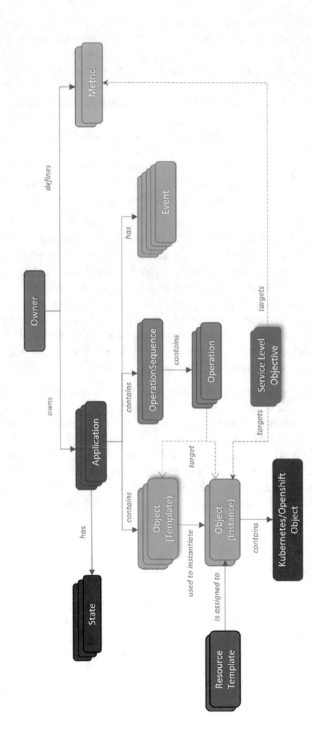

Fig. 3 BigDataStack realization engine application model

more object (templates) representing the different components of the application, zero or more operation sequences representing actions that can be performed for the application, and a series of events generated about the application. An object template (application component) can be instantiated multiple times, producing object instances. Object instances may have an associated resource template describing the resources assigned to that object. An object instance contains a definition of an underlying Kubernetes or OpenShift object that contains the deployment information. Operation sequences represent actions to perform on the application and contain multiple atomic operations. An operation targets either an object template or instance, performing alteration or deployment actions upon it. Service-level objectives can be attached to an object instance, which tracks a metric exported by or about that object.

In this way, alteration actions and their stages have an explicit representation, that is as Operation Sequences and Operations, respectively, which are associated with an application. The application engineer can define any number of operation sequences for their application by specifying the list of operations to perform and their configuration (e.g. the target object(s) for that operation), and can condition the availability of each operation sequence upon the application's current state. An Operation conceptually performs a single change to a BigDataStack Object. Examples of operations include: *Deploy*, *Execute Command On*, *Build*, *Delete*, and *Wait For*. When an operation sequence is triggered within the Realization Engine, internally this first takes the operation sequence template and generates an instance from it that can be run and monitored separately. The Realization Engine then creates a new Pod object on the Kubernetes cluster to run the operation sequence targeting this new instance. Once the Pod object has been created, the responsibility for that operation sequence is passed to the Pod. Once the new Pod reaches running state, it will first load the target operation sequence instance, and subsequently it will process each operation within the sequence in order. State updates are reported to and stored by the Realization Engine within the operation sequence instance itself, enabling BigDataStack mechanisms to track progress for it.

7 Example Use-Case: Live Grocery Recommendation

Finally, having discussed all of the concepts and components of DDIM systems, in this section we summarize how this comes together in BigDataStack to enable a use case: the connected consumer. The connected consumer use case utilizes the BigDataStack environment to implement and offer a recommender system for the grocery market. All of the data that are used for training the analytic algorithms of the use case are corporate data provided by one of the top food retailers companies in Spain. The goal from a DDIM perspective is to host and manage all aspects of the underlying grocery recommendation system.

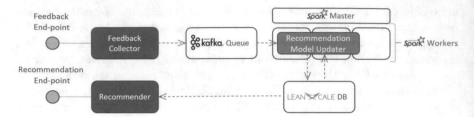

Fig. 4 Overview of the connected consumer grocery recommender

The Grocery Recommendation System Fig. 4 provides an illustration of the grocery recommendation system used by the use case. Each box in this diagram represents a pod. From an external perspective, this system has two end-points: recommendations, which respond with recommended products for a user, and feedback, which receives click and purchase events generated by the user. When a logged-in user opens a homepage on the grocery company store-front, a request is sent to recommendations end-point, which retrieves cached grocery recommendations for that user from a transactional database (in the specific case, LeanXcale [12] has been used as a transactional database). Meanwhile, when a user clicks or purchases a product, an event is sent to the feedback end-point, which reformats the data and sends it via Kafka queue into the main recommendation update component. This is a continuous application component that runs in parallel over multiple Apache Spark workers, which upon receiving item feedback for a user, updates their cached recommendations in real time based on that feedback.

Metrics, Pod Level Objectives, and Workload Within this application, the user cares about three main factors: (1) the response time for recommendations (e.g. less than 100 ms), (2) the delay between feedback being recorded and when the user's recommendations will finish updating (e.g. less than 1 s) and (3) the total cost of the system (e.g. less than 2 US dollars per hour). The volume of both recommendation requests and feedback varies over the course of each day (following the day/night cycle for mainland Europe), with periodic bursts of activity that correspond to flash sales.

Available Actions To achieve these goals, the DDIM system has access to the following actions it can take: (1) increase/decrease replicas for the Feedback Collector, Kafka and/or the Recommender, (2) increase/decrease table replication within the transactional database, and (3) increase the number of Spark workers the recommendation update service has access to.

A Day of Data-Driven Infrastructure Management In the early hours of the morning, the DDIM system will have the application running in a minimal configuration (typically only one instance of each pod) to minimize cost (0.5 USD/hour) when there is little traffic. Around 7am, the workload begins to increase, as online shoppers order groceries before starting work. The triple monitoring engine reports that response times and feedback updates are still within acceptable bounds. At

8:30am, quality of service monitoring reports a POS failure on recommendation updates of 1.1 s on average. The dynamic orchestrator determines based on resource usage that the bottleneck is in the recommendation updater. It triggers an enlargement of the spark cluster, adding an additional worker, and once ready performs a rapid restart of the recommendation updater such that it now leverages both workers. The POS failure is rectified, although cost per hour has increased (0.6 USD/hour). As traffic approaches its peak around mid-day, a second POS failure is reported this time on average recommendation response time (115 ms). In response, the dynamic orchestrator increases the replication factor of the recommender component. After a short delay to collect new measurements, the POS failure is still not resolved, and hence the dynamic orchestrator instructs the database to increase its table replication on the assumption that is where the bottleneck is occurring. This process takes around 10 min to complete, during which the dynamic orchestrator refrains from taking further action for that POS failure. Once the change action has completed, the response time decreases once again, and the POS failure is resolved. Cost per hour is now 1.2 USD/hour. After 6pm, the workloads decrease, and in response, the RL agent within the dynamic orchestrator experiments with decreasing the replication on the recommender to save cost, which results in a POS failure on recommendation response time and so rolls back the change. It tries again a couple of hours later, which does not result in any POS failure. As workloads continue to decrease as the day ends, the dynamic orchestrator instructs the database and recommendation updater to reduce their table replication and number of workers respectively.

8   Conclusions

The future of infrastructure management will be data driven, leveraging recent advances in Big Data Analytics and AI technologies to provide exceptional automation and optimization in the management of diverse virtualized software-defined cloud resources. This chapter has introduced the building blocks of data-driven infrastructure management (DDIM) systems in general and a new DDIM platform, BigDataStack, in particular. Contrary to state-of-the-art DDIM systems that focus on specific optimization aspects (e.g. fault detection or resource allocation), the BigDataStack platform takes a holistic, end-to-end approach to optimizing the configuration of cloud environments. Specifically, BigDataStack-based DDIM includes cutting-edge functionalities in three complementary areas:

- **Deep Application and Action Modelling**: Through deep data-oriented modelling, BigDataStack maintains a more complete view of both data services, their corresponding data flows and the alteration/action space for applications, backed by a metric and state monitoring system that is both efficient and scalable for use with high-parallelism Big Data applications. Indeed, a key take-home message of BigDataStack is that DDIM systems need to not just model applications, but the actions that can be performed on them and have the ability to monitor and interpret the impacts from those actions.

- **Intelligent AIOps Decision Making**: As cluster/cloud infrastructures have become better instrumented and easier to manipulate programmatically, it is now possible to have AI agents learn how to manage such infrastructures as well as humans can (for a fraction of the cost). AI-based decision making enables fast and adaptive management of the infrastructure based on real-time data about the running applications, cluster resources, and data being processed.
- **Complex Multistage Action Management**: BigDataStack provides an atomic operation set from which a wide variety of complex actions can be constructed. Once defined, such actions form templates that can be used to drive fully automated complex runtime adaptations by the AIOps system, enabling end-to-end automated DDIM.

BigDataStack has been validated and evaluated in different real-life environments, including retail, insurance, and shipping, illustrating the efficiency, cost-effectiveness, and flexibility of its DDIM approach. Overall, BigDataStack has provided a novel AIOps showcase, which demonstrates the potential of DDIM for monitoring, analysing, and optimizing the deployment of cloud applications. Moving forward, we aim to extend BigDataStack with support for novel types of cloud computing resources and services, such as the Function-as-a-Service (FaaS) paradigm. Indeed, via FaaS support, we believe big data value chains can be more efficiently enabled, as function-level DDIM is easier to reason about than broader application-level DDIM, increasing precision while also reducing the time-to-convergence to an effective configuration. We also plan to investigate the real-world impact of service performance degradation vs. the cost of maintaining services at different quality of service levels, with the aim of developing future AI models for optimizing this trade-off for the business owner.

Acknowledgments The research leading to these results has received funding from the European Community's Horizon 2020 research and innovation programme under grant agreement n° 779747 (BigDataStack).

References

1. Argerich, M. F., Cheng, B., & Fürst, J. (2019). Reinforcement learning based orchestration for elastic services. In Proceedings of the 2019 IEEE 5th World Forum on Internet of Things (WF-IoT), IEEE (pp. 352–357).
2. Canfora, G., Di Penta, M., Esposito, R., & Villani, M. L. (2005). An approach for qos-aware service composition based on genetic algorithms. In Proceedings of the 7th annual conference on Genetic and evolutionary computation (pp. 1069–1075).
3. Chung, A., Park, J. W., & Ganger, G. R. (2018). Stratus: Cost-aware container scheduling in the public cloud. In Proceedings of the ACM symposium on cloud computing (pp. 121–134).
4. Demchenko, Y., Filiposka, S., Tuminauskas, R., Mishev, A., Baumann, K., Regvart, D., & Breach, T. (2015). Enabling automated network services provisioning for cloud based applications using zero touch provisioning. In 2015 IEEE/ACM 8th international conference on utility and cloud computing (UCC). IEEE (pp. 458–464).

5. Eramo, V., Cianfrani, A., Catena, T., Polverini, M., & Lavacca, F. (2019). Reconfiguration of cloud and bandwidth resources in NFV architectures based on segment routing control/data plane. In Proceedings of the 2019 21st international conference on transparent optical networks (ICTON), IEEE (pp. 1–5).

6. Fard, M. V., Sahafi, A., Rahmani, A. M., & Mashhadi, P. S. (2020). Resource allocation mechanisms in cloud computing: A systematic literature review. *IET Software*.

7. Fürst, J., Argerich, M. F., Cheng, B., & Papageorgiou, A. (2018). Elastic services for edge computing. In Proceedings of the 2018 14th international conference on network and service management (CNSM) , IEEE (pp. 358–362).

8. Gan, Y., & Delimitrou, C. (2018). The architectural implications of cloud microservices. *IEEE Computer Architecture Letters, 17*(2), 155–158.

9. Grabarnik, G. Y., Tortonesi, M., & Shwartz, L. (2016). Data-driven cloud-based it services performance forecasting. In Proceedings of the 2016 IEEE international conference on big data (Big Data), IEEE (pp. 2081–2086).

10. Gulenko, A., Wallschläger, M., Schmidt, F., Kao, O., & Liu, F. (2016). Evaluating machine learning algorithms for anomaly detection in clouds. In Proceedings of the 2016 IEEE international conference on big data (Big Data), IEEE (pp. 2716–2721).

11. Hester, T., Vecerik, M., Pietquin, O., Lanctot, M., Schaul, T., Piot, B., Horgan, D., Quan, J., Sendonaris, A., Dulac-Arnold, G., et al. (2017). Deep q-learning from demonstrations. arXiv preprint arXiv:1704.03732.

12. Kolev, B., Levchenko, O., Pacitti, E., Valduriez, P., Vilaça, R., Gonçalves, R. C., Jiménez-Peris, R., & Kranas, P. (2018). Parallel Polyglot query processing on heterogeneous cloud data stores with LeanXcale. In IEEE BigData, Seattle, United States, IEEE (p. 10).

13. Kousiouris, G., Menychtas, A., Kyriazis, D., Konstanteli, K., Gogouvitis, S. V., Katsaros, G., & Varvarigou, T. A. (2012). Parametric design and performance analysis of a decoupled service-oriented prediction framework based on embedded numerical software. *IEEE Transactions on Services Computing, 6*(4), 511–524.

14. Kraemer, A., Maziero, C., Richard, O., & Trystram, D. (2018). Reducing the number of response time service level objective violations by a cloud-hpc convergence scheduler. *Concurrency and Computation: Practice and Experience, 30*(12), e4352.

15. Kyriazis, D., Doulkeridis, C., Gouvas, P., Jimenez-Peris, R., Ferrer, A. J., Kallipolitis, L., Kranas, P., Kousiouris, G., Macdonald, C., McCreadie, R., et al. (2018). Bigdatastack: A holistic data-driven stack for big data applications and operations. In Proceedings of the 2018 IEEE international congress on big data (BigData Congress), IEEE (pp. 237–241).

16. Lin, Q., Hsieh, K., Dang, Y., Zhang, H., Sui, K., Xu, Y., Lou, J.-G., Li, C., Wu, Y., Yao, R., et al. (2018). Predicting node failure in cloud service systems. In Proceedings of the 2018 26th ACM joint meeting on European software engineering conference and symposium on the foundations of software engineering (pp. 480–490).

17. Mabrouk, N. B., Beauche, S., Kuznetsova, E., Georgantas, N., & Issarny, V. (2009). Qos-aware service composition in dynamic service oriented environments. In ACM/IFIP/USENIX international conference on distributed systems platforms and open distributed processing, Springer (pp. 123–142)

18. Mao, H., Alizadeh, M., Menache, I., & Kandula, S. (2016). Resource management with deep reinforcement learning. In Proceedings of the 15th ACM workshop on hot topics in networks (pp. 50–56).

19. Mirhoseini, A., Pham, H., Le, Q. V., Steiner, B., Larsen, R., Zhou, Y., Kumar, N., Norouzi, M., Bengio, S., & Dean, J. (2017). Device placement optimization with reinforcement learning. arXiv preprint arXiv:1706.04972.

20. Mnih, V., Kavukcuoglu, K., Silver, D., Rusu, A. A., Veness, J., Bellemare, M. G., Graves, A., Riedmiller, M., Fidjeland, A. K., Ostrovski, G., et al. (2015). Human-level control through deep reinforcement learning. *Nature, 518*(7540), 529–533.

21. Modi, A., Dey, D., Agarwal, A., Swaminathan, A., Nushi, B., Andrist, S., & Horvitz, E. (2019). Metareasoning in modular software systems: On-the-fly configuration using reinforcement learning with rich contextual representations. arXiv preprint arXiv:1905.05179.

22. Mohamed, M., Anya, O., Sakairi, T., Tata, S., Mandagere, N., & Ludwig, H. (2016). The RSLA framework: Monitoring and enforcement of service level agreements for cloud services. In Proceedings of the 2016 IEEE international conference on services computing (SCC), IEEE (pp. 625–632).
23. Moreno, D. L., Regueiro, C. V., Iglesias, R., & Barro, S. (2004). Using prior knowledge to improve reinforcement learning in mobile robotics. In Proceedings of the Towards Autonomous Robotics Systems. University of Essex, UK.
24. Nastic, S., Morichetta, A., Pusztai, T., Dustdar, S., Ding, X., Vij, D., & Xiong, Y. (2020). SLOC: Service level objectives for next generation cloud computing. *IEEE Internet Computing, 24*(3), 39–50.
25. Raman, K., Swaminathan, A., Gehrke, J., & Joachims, T. (2013). Beyond myopic inference in big data pipelines. In Proceedings of the 19th ACM SIGKDD international conference on Knowledge discovery and data mining (pp. 86–94).
26. Sharma, P., Chaufournier, L., Shenoy, P., & Tay, Y. (2016). Containers and virtual machines at scale: A comparative study. In Proceedings of the 17th international Middleware conference (pp. 1–13).
27. Syu, Y., Ma, S.-P., Kuo, J.-Y., & FanJiang, Y.-Y. (2012). A survey on automated service composition methods and related techniques. In Proceedings of the 2012 IEEE ninth international conference on services computing, IEEE (pp. 290–297).
28. Voorsluys, W., Broberg, J., Venugopal, S., & Buyya, R. (2009). Cost of virtual machine live migration in clouds: A performance evaluation. In IEEE international conference on cloud computing, Springer (pp. 254–265).
29. Xu, Y., Sui, K., Yao, R., Zhang, H., Lin, Q., Dang, Y., Li, P., Jiang, K., Zhang, W., Lou, J.-G., et al. (2018). Improving service availability of cloud systems by predicting disk error. *2018 {USENIX} Annual Technical Conference* ({USENIX} {ATC}, *18*), 481–494.
30. Yu, T., Zhang, Y., & Lin, K.-J. (2007). Efficient algorithms for web services selection with end-to-end QOS constraints. *ACM Transactions on the Web (TWEB), 1*(1), 6–es.
31. Zhang, D., Han, S., Dang, Y., Lou, J.-G., Zhang, H., & Xie, T. (2013). Software analytics in practice. *IEEE Software, 30*(5), 30–37.
32. Zhu, H., & Bayley, I. (2018). If docker is the answer, what is the question? In *Proceedings of the 2018 IEEE Symposium on Service-Oriented System Engineering (SOSE)*, IEEE (pp. 152–163).
33. Zillner, S., Curry, E., Metzger, A., Auer, S., & Seidl, R. (2017). European big data value strategic research & innovation agenda. In Big Data Value Association.
34. Zillner, S., Bisset, D., Milano, M., Curry, E., Södergård, C., Tuikka, T., et al. (2020). Strategic research, innovation and deployment agenda: AI, data and robotics partnership.

Leveraging High-Performance Computing and Cloud Computing with Unified Big-Data Workflows: The LEXIS Project

Stephan Hachinger, Martin Golasowski, Jan Martinovič, Mohamad Hayek, Rubén Jesús García-Hernández, Kateřina Slaninová, Marc Levrier, Alberto Scionti, Frédéric Donnat, Giacomo Vitali, Donato Magarielli, Thierry Goubier, Antonio Parodi, Andrea Parodi, Piyush Harsh, Aaron Dees, and Olivier Terzo

S. Hachinger · M. Hayek · R. J. García-Hernández
Leibniz Supercomputing Centre (LRZ) of the BAdW, Garching b.M., Germany
e-mail: hachinger@lrz.de; hayek@lrz.de; garcia@lrz.de

M. Golasowski · J. Martinovič · K. Slaninová (✉)
IT4Innovations, VŠB—Technical University of Ostrava, Ostrava, Czechia
e-mail: martin.golasowski@vsb.cz; jan.martinovic@vsb.cz; katerina.slaninova@vsb.cz

M. Levrier
Atos, Campus Teratec, Bruyères-le-Châtel, France
e-mail: marc.levrier@atos.com

A. Scionti · G. Vitali · O. Terzo
LINKS Foundation, Torino, Italy
e-mail: alberto.scionti@linksfoundation.com; giacomo.vitali@linksfoundation.com;
olivier.terzo@linksfoundation.com

F. Donnat
Outpost24, Valbonne, France
e-mail: fdo@outpost24.com

D. Magarielli
AvioAero, Rivalta di Torino (TO), Italy
e-mail: donato.magarielli@avioaero.it

T. Goubier
CEA LIST, Gif-sur-Yvette, France
e-mail: thierry.goubier@cea.fr

A. Parodi · A. Parodi
CIMA Foundation, Savona, Italy
e-mail: antonio.parodi@cimafoundation.org; andrea.parodi@cimafoundation.org

P. Harsh
Cyclops Labs GmbH, Zürich, Switzerland
e-mail: piyush@cyclops-labs.io

A. Dees
Irish Centre for High-End Computing, Technology and Enterprise Campus, Dublin, Ireland
e-mail: aaron.dees@ichec.ie

© The Author(s) 2022
E. Curry et al. (eds.), *Technologies and Applications for Big Data Value*,
https://doi.org/10.1007/978-3-030-78307-5_8

159

Abstract Traditional usage models of Supercomputing centres have been extended by High-Throughput Computing (HTC), High-Performance Data Analytics (HPDA) and Cloud Computing. The complexity of current compute platforms calls for solutions to simplify usage and conveniently orchestrate computing tasks. These enable also non-expert users to efficiently execute Big Data workflows. In this context, the LEXIS project ('Large-scale EXecution for Industry and Society', H2020 GA 825532, https://lexis-project.eu) sets up an orchestration platform for compute- and data-intensive workflows. Its main objective is to implement a front-end and interfaces/APIs for distributed data management and workflow orchestration. The platform uses an open-source Identity and Access Management solution and a custom billing system. The data management API allows data ingestion and staging between various infrastructures. The orchestration API allows execution of workflows specified in extended TOSCA. LEXIS uses innovative technologies like YORC and Alien4Cloud for orchestration or iRODS/EUDAT-B2SAFE for data management, accelerated by Burst Buffers. Three pilot use cases from Aeronautics Engineering, Earthquake/Tsunami Analysis, and Weather and Climate Prediction are used to test the services. On the road towards longer-term sustainability, we are expanding this user base and aiming at the immersion of more Supercomputing centres within the platform.

Keywords High performance computing · Cloud computing · Big data · Workflows · Distributed data management · Orchestration

1 High-Performance Computing, Cloud and Big Data in Science, Research and Industry—and LEXIS

In this chapter, we present how the Horizon-2020 project 'Large-scale EXecution for Industry and Society' (LEXIS[1]) addresses the 'Big Data' theme, establishing automated data analysis and simulation workflows across world-class Supercomputing and Cloud Computing centres in Europe. We relate to the technical priorities 'Data Management' and 'Data Processing Architectures' of the European BDV Strategic Research and Innovation Agenda [1], addressing horizontal ('Cloud and High Performance Computing') and vertical concerns ('Data sharing platforms', 'Cybersecurity and Trust') of the BDV Reference Model (cf. [1]). With respect to the AI, Data and Robotics Strategic Research, Innovation and Deployment Agenda [2], we present 'Systems, Methodologies, Hardware and Tools' as cross-sectorial technology enablers.

The LEXIS collaboration is made up of two of the major European scientific Supercomputing centres (IT4I/CZ and LRZ/DE), scientific and industrial/SME partners with compute- and data-intensive use cases, and industrial/SME technology

---

[1] https://lexis-project.eu, H2020 Grant Agreement No. 825532.

partners. It thus aims to bring the power of scientific Supercomputing to the industrial and enterprise Big Data landscape, but also to equip scientists with industry-standard Big Data tools. In these ways, it seeds knowledge transfer between science, SMEs and industry, and addresses institutions of societal relevance like governmental agencies.

When looking at the history of large-scale computing and Big Data in the last decade, it turns out that science was leading the introduction of powerful distributed computing grids (in particular LCG [3] for CERN's Large Hadron Collider), but industry drove innovations such as Infrastructure-as-a-Service (IaaS) Clouds and Big Data ecosystems. Cloud Computing services (with, e.g. Amazon as one of the pioneers), but also famous frameworks for Big Data analytics like Hadoop (e.g. [4]), were of immediate and sometimes even almost exclusive practical relevance for implementing top-notch data services. Much of the bare compute power, however, remained with scientific High-Performance Computing (HPC), and – in countries with pronounced geopolitical interest – also with the military.

In this situation, bringing ideas from the science and industry/SME worlds together is clearly a key to further development and innovation. The LEXIS project accomplishes this by co-developing an easily-usable platform for the processing of data- and compute-intensive workflows. The LEXIS platform with its ambitious ideas is backed not only by strong computing systems and by an orchestration facility but also by a Distributed Data Infrastructure immersed with the EUDAT [5, 6] (EUropean DATa) system, which will be extensively discussed in this chapter.

From the scientific HPC centres' point of view, our agenda is certainly motivated by practical problems scientists have faced with their computing projects for the last decades. Using a Supercomputer has traditionally involved a steep learning curve and hard work on scripting workflows and submitting them to a job queue. Only researchers with very long-term experience have thus been able to efficiently run extreme-scale simulations.

Clearly, this usage pattern is not practicable for industrial/SME applications and their – often shorter – life cycles. However, it also excluded a major part of all scientists – those without focus on IT – from efficient scientific computing and data analysis. Nowadays, the average scientist appreciates modern low-threshold IT offers, such as (often commercial) IaaS or container platforms, while industry has become aware of the capabilities of supercomputers and of academic developments, for example in quantum computing. Thus, we witness a sort of 'golden age' for projects such as LEXIS which make these worlds converge in order to reach new levels of optimisation.

Below, we present the LEXIS ideas, in the light of collaborations such as the Big Data Value Association (BDVA [7]) and EUDAT, and of the European computing and data landscape in general. We will first cover the vision of HPC-Cloud-Big Data convergence in LEXIS and basics of the LEXIS concept (Sect. 2) and then the LEXIS approach to Authentication/Authorisation as a prerequisite for a secure platform (Sect. 3). Section 4 extensively discusses the European data management approach for Big Data in LEXIS. We close the chapter with a description of the LEXIS Portal as a one-stop shop for our users (Sect. 5) and our conclusions (Sect. 6).

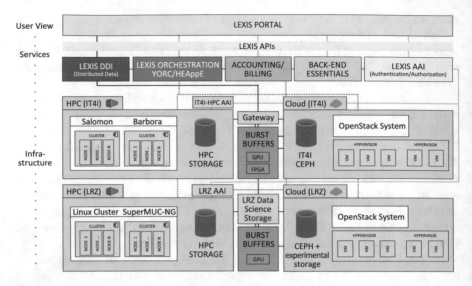

Fig. 1 Simplified scheme of the LEXIS platform: main components as of 2020 (extension to more computing and data centres is planned). The 'back-end essentials' box represents technical components not mentioned in this overview

2 LEXIS Basics: Integrated Data-Heavy Workflows on Cloud/HPC

2.1 LEXIS Vision

The vision of LEXIS is a distributed platform which enables industry, SMEs and scientists to leverage the most powerful European computing and data centres for their simulations, data analytics and visualisation tasks. Via a user-friendly portal with a modern REST-API[2] architecture behind it, the LEXIS Orchestration System for workflows and the LEXIS Distributed Data Infrastructure are addressed. The user uploads necessary data and software containers and specifies workflows with all computing tasks and data flows. From this point on, the Orchestration System takes care of an optimised execution on the LEXIS resources.

Figure 1 gives an overview of the federated systems within the LEXIS platform, from hardware systems (lower part), over service-layer components to APIs and the LEXIS Portal (top part). The architecture can be considered a blueprint for data processing architectures aligned with the BDVA strategy [1, Sect. 3.2]. It federates decentralised, heterogeneous resources to offer data processing services. The platform is the result of a strong initial focus on co-design, identifying available systems to be leveraged and key technologies required. Technological choices were

---

[2] REpresentational State Transfer Application Programming Interface.

made with a preference for state-of-the-art, open-source, extensible and sufficiently mature products. The details of this become clearer in the following parts of this book chapter, which discuss the LEXIS ecosystem from low- to high-level components.

In this section, we describe the LEXIS basics, beginning with hardware systems (Sect. 2.2), which include devices to accelerate computation (GPU and FPGA cards) and data transfer (Burst Buffers). Then, we discuss the Orchestration System (Sect. 2.3), which addresses our hybrid HPC/Cloud Computing facilities, realising a novel processing architecture for Big Data. The section is completed with a glimpse on the LEXIS Pilot use cases (Sect. 2.4), used to co-design and test the platform, and on our billing concept (Sect. 2.5) as part of a future business model.

2.2 LEXIS Hardware Resources

The LEXIS system flexibly utilises computing-time and storage grants on different back-end systems, as specified by the user.

While the LEXIS federation is planned to be constantly extended, currently the compute and data back-end resources (see Fig. 1) are contributed by two flagship Supercomputing centres: the Leibniz Supercomputing Centre (LRZ, Garching near Munich/D) and the IT4Innovations National Supercomputing Centre (IT4I, Ostrava/CZ). Systems available (cf. Fig. 1) include traditional and accelerated Supercomputing resources, on-premises compute clouds, high-end storage resources, and Burst Buffers equipped with GPUs and FPGAs.

At LRZ, 'SuperMUC-NG' (originally one of the world's top-10 HPC machines) provides 311'040 CPU cores (26.9 PFlops) and 719 TB of RAM. Compute time is granted via calls for proposals, which need to devise a promising research agenda. The smaller LRZ Linux Cluster offers (with less bureaucracy) roughly 30'000 CPU cores and 2 PFlops of compute power. This system is also used in LEXIS. Furthermore, a NVIDIA DGX-1 with eight Tesla V-100 GPUs is available for AI workloads.

At IT4I, the 'Barbora' (7'232 compute cores and 45 TB RAM) and 'Salomon' (12-core Xeon, 24'192 cores with 129 TB RAM) HPC systems are available. With some nodes accelerated by NVIDIA Tesla V100-SXM2 and Intel Xeon Phi cards, these systems altogether offer about 3 Pflops. Usage of them is subject to an open-call procedure as for SuperMUC-NG. Some millions of CPU hours have already been allocated to LEXIS in general and can readily be distributed to use cases.

The central LEXIS objective of bringing together HPC and Cloud resources in unified workflows is supported by integrating the LRZ Compute Cloud and IT4I's visualisation nodes into the platform. Altogether, these infrastructures provide more than 3'000 CPU cores for on-demand needs via on-premises OpenStack installations.

Storage resources in the ramp-up phase of LEXIS include access to 50 TB in LRZ's Data Science Storage (DSS, based on IBM Spectrum Scale, formerly known as GPFS [8]) and 150 TB in an 'Experimental Storage System' at LRZ, while IT4I provides 120 TB of Ceph-based [9] storage. These resources serve as back-end for the LEXIS Distributed Data Infrastructure (Sect. 4). They can be extended at any time, allocating more space in the (shared-usage) background storage of the computing centres, which is currently in the 100 PB range.

A typical issue in many data-intensive applications is the slowdown of actual data processing during in- and output of large data sets. LEXIS addresses this by flexibly inserting two 'Burst Buffer' systems per compute site in the data flows. Each of them offers about 10 TB of very fast NVDIMM and NVMe storage. Running the Atos 'Smart Burst Buffer' software, they are able to pre-fetch data or transparently cache output data. Thus, I/O is practically 'instantaneous' for the application, while the Burst Buffer manages the communication with the actual file system (pre-read or delayed write) in the background. In addition, the systems can reprocess data leveraging GPU and FPGA accelerator cards, and their NVDIMM/NVMe storage can be exported via NVMEoF [10] using the Atos 'Smart Bunch of Flash' tools.

2.3 LEXIS Orchestration

Automatising the execution of complex workflows[3] is crucial to enable more users to bring their applications onto efficient computing and data platforms. To address this challenge, the LEXIS Orchestration System uses technologies which minimise the need of users to acquire expertise outside their own domain. It provides the capabilities of composing application workflows in a simple way and of automatically running them on the most suitable set of resources (cf. Sect. 2.2). Moreover, it enables an automated, unified management of workflow steps based on different processing concepts, such as HPDA (High-Performance Data Analytics), HPC and HTC (High-Throughput Computing), fulfilling the respective infrastructural requirements.

The key difference of projects as LEXIS with respect to earlier projects on orchestration is the mixed usage of HPC and IaaS-Cloud (OpenStack) resources, and prospectively also, for example, container platforms to run tasks within one given workflow. Orchestrators that can address all this have been unavailable when designing the LEXIS architecture, and only recently, a few solutions are emerging (cf. [11]). In this context, we decided to use and co-develop Yorc (Ystia Orchestrator, [12]) to orchestrate workflows in LEXIS, and Alien4Cloud (Application LIfecycle ENablement for Cloud, [13]) as a front-end for designing workflows. This open-source system, with which LEXIS partner Atos is experienced, has since been extended to become an HPC-aware meta-orchestrator, addressing all relevant systems via plug-ins.

---

[3] We understand workflows – to give a simplified hint of a definition – as directed acyclic graphs with computational, pre-/post-processing, visualisation or data movement tasks as vertices here.

Alien4Cloud first helps the user to define the so-called topology for his application. This includes the hardware resources (e.g. the amount of CPU cores and RAM) and the software (e.g. frameworks or libraries) needed. Afterwards, it greatly simplifies the specification of the actual workflow (i.e. tasks and their order). Behind its drag-and-drop interface, Alien4Cloud describes all this in an extended version of TOSCA (cf. e.g. [14]). Extensible, generic application templates are provided.

Once the workflow description is generated, the back-end engine will run it on appropriate and available resources. To this end, it considers system characteristics, the user's access rights, locations where data sets reside, and custom constraints in the application template (e.g. deadlines for execution in urgent computing). LEXIS Orchestration is furthermore being equipped with dynamic scheduling capabilities, taking into account, for example, the load and availability of systems in real time.

The actual access to computing resources is mediated by the HEAppE middleware [15, 16], which serves two purposes: (i) security layer for mapping LEXIS users (cf. Sect. 3) to internal Supercomputing-centre accounts, used to actually execute jobs; (ii) sending an appropriately formatted job description (with pointers to the executables, etc.) to the workload managers of LEXIS resources. A wealth of middleware frameworks with this functionality is available on the market since grid times (cf. e.g. [17]). Clearly, HEAppE – besides providing a state-of-the-art implementation with a REST API – has the advantage that it is developed at IT4I as a project partner. Thus, it can easily be adapted, for example to provide usage data for billing (cf. Sect. 2.5).

2.4   LEXIS Pilots

Having described the basics of LEXIS orchestration, we give an impression of the first workflows ('Pilots') the platform is designed to run. These Pilot use cases have been a key for the early requirements analysis and co-design activities to create the LEXIS platform. They have been carefully selected to be sufficiently heterogeneous in order to make LEXIS generically useful.

The three initial LEXIS Pilots cover the following themes: (i) data- and compute-intensive modelling of turbo-machinery and gearbox systems in aeronautics, (ii) earthquake and tsunami data processing and simulations, which are accelerated to enable accurate real-time analysis, and (iii) weather and climate models, where massive amounts of in situ data are assimilated to improve forecasts (also predicting, e.g., flash floods). While Pilot (i) is certainly industry-centric, and (ii) is interesting for public authorities, Pilot (iii) has a broad and mixed range of applications.

Fully orchestrated workflows of the Weather and Climate Large-scale Pilot in LEXIS have already reached considerable intrinsic complexity (Fig. 2; for more details see [18]). Because of their broad range of applications, these are a prime example to elaborate somewhat upon. In the example shown in Fig. 2, the Weather and Research Forecasting (WRF) model [19] as an HPC core drives a fire risk prediction system (RISICO, cf. [18]). Likewise, hydrological risks and air quality

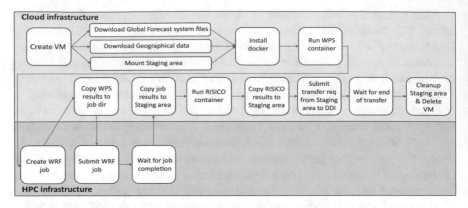

Fig. 2 WRF-RISICO workflow [18] from the Weather and Climate Large-scale Pilot, as presented on a SC20 conference poster (M. Hayek et al.)

can be assessed. However, we strongly target commercial applications here as well. As one example, we aim at predicting optimum sites for wind energy plants with unprecedented accuracy. Also, in collaboration with the SME NUMTECH, highly accurate modelling of agricultural conditions shall be exploited to select, for example, optimum seeding or harvesting times. With respect to previous projects (e.g. [20]), where only limited workflow automation was available, a much broader range of applications is possible without experts having to stand by or even execute steps manually. The available computing systems are optimally leveraged, as the WRF preprocessing system (WPS) and the application models run perfectly as containers on the Cloud infrastructures at LRZ and IT4I, while WRF is a classical HPC job.

Pilots (i) and (ii) will test the LEXIS platform with another variety of different application characteristics. The Earthquake and Tsunami Pilot works with additional database services and urgent computing to feed warning systems before certain deadlines. The Aeronautics Pilot boosts the performance of turbomachinery and gearbox simulations performance to make such computations part of a 'real-time' design process. It thus involves experiments with low-level code optimisation for GPUs attached to HPC nodes, but also quick post-processing and visualisation of simulation snapshots will play a role. Further use cases attracted through a LEXIS 'Open Call' complement all this and contribute to a broad validation of the platform.

2.5 Billing for LEXIS Usage

In order to position LEXIS as a viable innovation platform for SMEs and industry, flexible accounting and billing mechanisms are a must. The accounting process in LEXIS is designed to accommodate resources from both HPC and Cloud systems and to take into account metered consumption of data storage. Abilities to comprehensively support flat-rate or tiered pricing, as well as completely dynamic rating, charging and billing are crucial to support a sustainable LEXIS business model.

Similar to the situation in orchestration, no combined Cloud/HPC/Storage billing framework matching our requirements was known when LEXIS was initiated. Thus, the SME 'Cyclops' (CH) participates in LEXIS and enhances their successful Cyclops cloud billing system to include, for example, HPC usage (compute time) data collectors. The system with its data collectors samples resource usage in near real time. Thus, we will be able to offer paid LEXIS usage with attractive and accurate cost models.

3 Secure Identity and Access Management in LEXIS – LEXIS AAI

The Authentication and Authorisation Infrastructure (AAI) of LEXIS is the actual basis for secure access, and thus a cornerstone for the distributed computing and data platform. All LEXIS systems rely on this AAI and thus offer access control complying with industry standards.

To elaborate a bit more on this, we lay out the motivation (Sect. 3.1) for setting up a LEXIS AAI, give some technical details on our resilient solution and (Sect. 3.2) and describe the role-based access control (RBAC, Sect. 3.3) model thus implemented. The concepts follow current best practices in IT.

3.1 Why an Own LEXIS AAI?

LEXIS as a platform provides unified access to various computing and data facilities, all with their existing, operational user administration and access-rights concepts. As the European identity-provider and AAI federation landscape takes time to consolidate, LEXIS must provide access to its users in a pragmatic and secure way.

Therefore, the LEXIS partners decided, already in the early stages of co-design, to set up a simple, federated and open-standard Identity and Access Management (IAM) solution as a basis for the LEXIS AAI. Thus, a single sign on (SSO) to all parts of the LEXIS platform with the necessary convenience is provided. The LEXIS AAI integrates smoothly with the existing systems for granting access at computing/data centres: Once users authenticate via the LEXIS AAI, they use compute systems via the HEAppE middleware (cf. Sect. 2.3), which addresses the back-end systems using regular, site-specific accounts via secure mechanisms.

3.2   Realisation of LEXIS AAI Following 'Zero-Trust' Security Model

With its central role, the LEXIS AAI had to be built upon a reliable framework with industrial-level backing and widespread usage, but also a rich, state-of-the-art feature set. Based on a requirement analysis [21], considering also experience and prospective maintainability, the LEXIS AAI was decided to rely on the open-source IAM solution 'Keycloak' [22]. Keycloak constitutes the upstream of the 'Red Hat SSO' product. It allows for the implementation of basically any access-policy scheme (role-based or user-based access control, etc.), and for easy integration with applications using OpenID Connect [23], SAML 2.0 [24] and further authentication flows. With its further abilities, for example of delegating authentication to third-party identity providers (also Facebook, Google), it covers almost any imaginable use case.

All components of the LEXIS platform (cf. Fig. 1 in Sect. 2.1) use the central AAI via its APIs, and LEXIS users authenticate preferably via OpenID-Connect tokens. Because LEXIS has opted for a 'zero-trust' model, not a single service on the platform is blindly trusted by any other service. This means, each service checks the provided tokens independently against Keycloak. When tokens are forwarded as needed to back-end services, these will revalidate them.

In order to ensure high availability of the AAI service, a Keycloak cluster is run across IT4I and LRZ in 'Cross-Datacenter-Replication' mode. As all critical traffic between the two centres, the traffic within this cluster is encapsulated in secure channel communication (merely using an encrypted virtual private network).

Keycloak is configured to use one 'realm' for LEXIS identity with several 'clients' (in Keycloak terminology), which are the different components/services of the platform. Authorisations are configured at realm level and are then accessible at client level, allowing reusability, centralisation and simplicity of the configuration and management. If needed, additional client-level settings can be added.

Keycloak's OpenID Connect tokens follow the JSON Web Token (JWT, [25]) standard, which consists of three parts: header, payload and signature. The verification of the signature enables a service to ensure that the token was not modified by a third party and was produced by the expected source. The content of the tokens includes the user identity and information about their granted permissions. We decided to use a unified token for all services in order to minimise load on the Keycloak service. If the need arises for the user's permissions in one resource to be kept hidden from other resources, this concept can, however, easily be modified. Some effort (e.g. [26]) was invested to customise all systems used in LEXIS such that they support the authentication flows of Keycloak.

3.3   RBAC Matrix and Roles in LEXIS

In LEXIS, rights are granted using a role-based access control (RBAC) model. This means that all users are assigned roles, depending on their job and responsibility scope. A fundamental concept in this context is the 'LEXIS Computational Project', where each project corresponds to a group of collaborators who use a set of compute time and storage grants together. Registration of each new LEXIS user and account creation (including the role settings) are subject to an administrative verification process.

A fixed 'LEXIS RBAC matrix' defines all the access policies; that is, which role implies access to which particular LEXIS services, resources or data. When a user tries to access a service or resource, their role attribute key is checked for authorisation. The LEXIS RBAC model not only controls regular user access but also includes elevated-rights roles, for example, for project and organisation management. LEXIS systems already implementing a (more or less sophisticated) access-control mechanism were adapted such that their internal permissions consistently reflect those within the RBAC model.

This being said, all processing, visualisation, data or system-related steps on the LEXIS platform are packaged as workflows, which are being executed on behalf of the user by the LEXIS Orchestration System. These workflows are created by the user via the LEXIS portal. Already for this workflow-building process, the web portal implements a view adapted to the user's roles/rights, comprising, for example, data and systems in the user's scope. For then running the workflow in the back-end, tokens (OpenID Connect/SAML 2.0) issued to the user by the LEXIS AAI are passed through and used to log into the relevant (compute and data) services.

4   Big Data Management for Workflows – LEXIS DDI

LEXIS acts as an infrastructure provider (or 'reseller'), enabling data-heavy workflows on distributed European Supercomputing facilities. This means, it does not directly implement Big Data frameworks such as Spark (e.g. [27]) or Hadoop (e.g. [4]), but leaves it to the users to leverage optimum tools in their workflows. Thus, a prime task in LEXIS is to enable efficient data storage, access and transfer by employing a forefront distributed data management framework in a European context.

For the LEXIS 'Distributed Data Infrastructure' (DDI), we chose to adopt the EUDAT-B2SAFE solution (cf. [6]), based on the Integrated Rule-Oriented Data System (iRODS, [28]). The design is open for federation with new prospective LEXIS partners, for which installation recipes can be provided. The EUDAT ('EUropean DATa') project aims at unifying research data infrastructure, and working with their tools gives us an outstanding opportunity to transfer knowledge from academic data management to the enterprise world.

We expand on system choices, construction plan and features of the DDI below. Section 4.1 gives more details on concept and necessary interfaces of the DDI. The later subsections describe how our system matches the requirements and integrates geographically distributed storage systems (Sect. 4.2), how it handles access rights (Sect. 4.3), how it immerses in the EUDAT context (Sect. 4.4) and how the orchestrator addresses the DDI via APIs (Sect. 4.5).

4.1   Concept of Unified Data Management in LEXIS

The LEXIS DDI must enable the orchestrator and portal, and thus the authenticated users, to retrieve their LEXIS data – no matter where they are physically stored – in a unified, secure and efficient manner at all LEXIS sites. This is ensured by a system for distributed data management fulfilling the following requirements from the early co-design process:

   (i) Unified access on LEXIS data in a file-system-like semantics
  (ii) Reliability and redundancy
 (iii) Support for diverse storage back-end systems
  (iv) Support for the LEXIS AAI
   (v) Support for implementing storage policies, for example selective data mirroring
  (vi) Support for having metadata and persistent identifiers in the system
 (vii) Support for the system to be addressed via REST APIs

With these features, the LEXIS DDI aims to be an academic/industrial data platform (IDP) facilitating data management as envisaged by the BDVA [1, Sect. 3.1]. Basic annotation with metadata shall foster semantic interoperability, and the entire data lifecycle (data taking and processing, internal re-usage with rights management, publication, etc.) is considered in the DDI design.

Within workflows, data access and transfer are to be automatically controlled by the orchestrator (cf. Sect. 2.3), by addressing the DDI via APIs. In order to save precious execution time, the orchestrator may, for example, query the physical location of input data and move compute jobs to the same computing/data centre. Likewise, it may identify and use mirror copies of the data at a proposed computing site, if the user pre-ordered his input data sets to be replicated, for example on IT4I and LRZ.

4.2   Choice of DDI System, Integration of Distributed Storage Systems

Unified access to geographically distributed storage back-ends is probably the most challenging requirement of all discussed above. The back-end systems to be used are of different technological nature and dedicated to various computing clusters, projects or purposes. Often (e.g. in the case of LRZ), the resources are operationally supported and served as a particular file system (e.g. NFS), not as bare storage.

Building the LEXIS DDI on such back-ends, leveraging a distributed file system which needs 'raw disk' access or particular back-end file systems for efficiency (e.g. Ceph [9]) is hard. Various solutions are, however, on the market to integrate existing file systems into one common data management system. Frameworks with a 'file system on file systems' approach (e.g. GlusterFS, or in a way also HDFS, cf. [29]) are usually efficient and scale well but come with a trade-off in terms of flexibility, for example in their storage policies. Also, this approach usually implies a tightly coupled system, whose behaviour in case of high WAN latencies or site failure is certainly not trivial. Thus, we rather went for a looser, middleware-based storage federation approach, whose possible performance penalties [29] are mitigated by the use of burst buffers and HPC-cluster file systems in LEXIS. Excellent open systems in this sector are, for example, iRODS, Onedata, Rucio and dCache (cf. [28, 30]). iRODS stands out through its intuitive file-system-like semantics, flexibility as for storage policies and metadata stored, and most of all through its integration in the feature-rich, European EUDAT [6] data management ecosystem and many other European projects.

Thus, we adopted for the LEXIS DDI an iRODS/EUDAT-B2SAFE based solution, which optimally matches the LEXIS requirements (cf. list/numbering in Sect. 4.1):

(i) Unified LEXIS data access: iRODS has a file-system-like view on all data, which are structured in 'data objects' (similar to files) and 'collections' ($\sim$ folders).

(ii) Reliability: can be boosted with the high-availability setup "HAIRS" [31].

(iii) Support for diverse storage back-end systems: iRODS is extremely flexible and can address any common file system, but also, for example, S3 buckets.

(iv) Support for the LEXIS AAI: here, iRODS has an iRODS-OpenID plugin which we modified to make it work with Keycloak [26].

(v) Support for implementing storage and mirroring policies: this is a traditional strength of the iRODS rule engine.

(vi) iRODS supports storing custom metadata including persistent identifiers.

(vii) The various iRODS clients available (e.g. command-line, Java & Python clients) facilitate the programming of custom 'LEXIS Data System REST APIs'.

Different geographical sites can be loosely bound in iRODS via a 'zone federation' mechanism, which enables transparent data access between the zones, while they are operated independently. Figure 3 gives an overview over the LEXIS iRODS federation, in which each major data/compute site (currently IT4I and LRZ) has its own iRODS zone. To move data between zones/centres, a simple copy command is sufficient, as the zones just appear as different top-level directories. On (transparent) data access, iRODS automatically handles the necessary data transfers with an internal SSL-secured protocol allowing multiple parallel data streams.

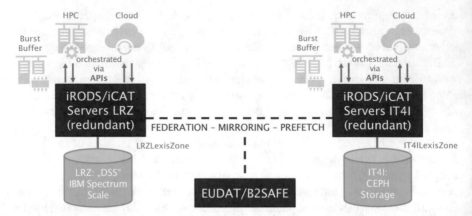

Fig. 3 LEXIS DDI federation. The zone names (IT4ILexisZone and LRZLexisZone) refer to the two computing/data centres federated in the LEXIS DDI by end of 2020. Two main operational back-end storage systems (LRZ's 'Data Science Storage' or 'DSS', and IT4I's Ceph system – cf. Sect. 2.2) are illustrated, as well as the transfer possibilities to Cloud and HPC infrastructure via API calls (cf. Sect. 4.5). Such transfers may leverage the LEXIS Burst Buffers (cf. Sect. 2.2)

Each zone in iRODS has a so-called 'iCAT' or 'provider' iRODS server (cf. [28]) which holds the information on the stored data, permissions, local resources and all other necessary zone-specific information in a database. In case of major problems in one zone (e.g. long-term power outage), the rest of the DDI infrastructure thus remains operational. In addition, the iRODS zones in LEXIS are each set up with a redundant iCAT (cf. [31]), using also a redundant PostgreSQL database back-end with a configuration based on repmgr and pgpool (cf. [32]).

Data mirroring between different zones, as optionally offered in LEXIS to increase resiliency and accelerate immediate data access from different centres, is implemented by the EUDAT-B2SAFE extension (cf. [6]) for iRODS. The DDI thus provides functionality to request replication on different granularity levels, for example by LEXIS Computational Project or by iRODS collection. B2SAFE then helps to set appropriate replication rules for iRODS; that is, it leverages the ability of iRODS to execute rules as a sort of 'event handlers' at so-called policy enforcement points (PEPs). This effectively means that, for example after storing a file in the LEXIS DDI, an arbitrary rule script (written, e.g. in Python and acting on iRODS or also at the operating-system level) can automatically be executed in order to implement data management policies (also beyond B2SAFE-related rules).

4.3   Reflecting AAI Roles in the DDI

Section 3.3 mentioned that LEXIS systems with own access-control mechanisms are set up such as to reflect the LEXIS AAI settings. The iCAT user databases on all sites are thus mirrored from the LEXIS AAI, and iRODS groups are used to implement LEXIS roles, in particular the membership or administrator role of a person in a LEXIS Computational Project (cf. Sect. 3). Actual access rights (for users/groups) are set via the iRODS access control lists.

However, also the directory (or more precisely, iRODS 'collection') structure itself of our DDI reflects project memberships and privacy levels of different data sets. The collection structure of the DDI starts with the iRODS zone (currently '/IT4ILexisZone' and '/LRZLexisZone'). Three collections then exist on the next level, which are named 'user', 'project' and 'public', for data sets which can be accessed only by the user, by all the members of a project or by everybody. At the second level in each of these, collections for each project exist. The third level in 'user' and 'project' contains a collection for each user to write his data sets into. Each data set is automatically stored in a collection named according to a unique identifier. A project administrator can delete project data sets or publish them by moving them to the public collection hierarchy.

All this is implemented by automatically setting up the iRODS access rights (and appropriate inheritance flags) at creation of LEXIS Computational Projects and users as part of the necessary administrative process.

4.4 Immersion in the European 'FAIR' Research Data Landscape

Having taken care of security and privacy where needed, the next most important aspect in modern Research Data Management is controlled data sharing and reuse. In science, this reflects in the 'FAIR' principles [33], also cited by the BDVA [1, Sect. 2.5.2]: Data should be findable, accessible, interoperable and reusable. Even in a context of embargoed or secret enterprise data, companies can strongly profit from a basic 'FAIR' implementation, facilitating internal data sharing and reuse. Such an implementation relies on the assignment of persistent identifiers (PIDs) to data, the availability of a basic description of the data set (i.e. metadata) and clearly actual possibilities of data transfer.

In LEXIS, these prerequisites are largely addressed with the immersion of the DDI in the EUDAT [5, 6] ecosystem. EUDAT calls itself a collaborative data infrastructure ('CDI') for European research and builds its software and services along this line.

Besides EUDAT-B2SAFE (cf. Sect. 4.2), we use the EUDAT-B2HANDLE PID service (cf. [6]) in LEXIS.

Metadata, such as PIDs, but also further basic information (e.g. data set contributors, creation dates or description) are then stored directly in the Attribute-Value-Unit store for each data object and collection in iRODS. Thus, we practically cover the Dublin Core Simple and DataCite metadata schemes (cf. [34, 35]) for each LEXIS data product, enabling us to later make LEXIS data findable via research data search engines (e.g. EUDAT-B2FIND, cf. [6]) on user request.

4.5   APIs of the LEXIS DDI, and Data Transfer Within Workflows

Usage of the LEXIS DDI, be it via the LEXIS Portal (Sect. 5) or other systems of the LEXIS platform, is relying on dedicated REST APIs. These DDI APIs, with standard JSON interfaces, serve to sanitise the DDI usage pattern and thus make data manageable within an automation/orchestration context. Connection to the APIs is secured by the use of HTTPS and Keycloak Bearer tokens. Figure 4 illustrates how the DDI is thus immersed in the LEXIS ecosystem. API specifications and Swagger documentation will be released within the release cycle of the LEXIS platform.

Besides a (meta-)data search API, and many 'smaller' APIs (e.g. for user/rights management), the LEXIS DDI provides a REST API for data staging, which is of immediate importance for our workflows and shall thus be discussed in some detail.

Although data in the LEXIS DDI are available at all participating centres, actual compute jobs on HPC clusters normally require input and output data to be (temporarily) stored on a bare, efficient parallel file system attached to the cluster. Within a given workflow, the orchestrator thus addresses the staging API and automatically manages data movement between the different systems and the DDI as required. This includes moving input, intermediate and output files.

Performing data transfer takes time, necessitating an asynchronous and non-blocking solution for execution of staging API requests. To this purpose, the API uses – behind its front-end – a distributed task queue connected to a broker (cf. the design of [36]). At request submission, the API returns a request ID to the orchestrator. With this ID, the status of the transfer task (in progress, done, failed) can be queried via a separate API endpoint.

On the Orchestration-System side, two TOSCA components have been defined and added to the Ystia/Alien4Cloud catalogue (cf. Sect. 2.3) for transferring data to and from a computing system when executing a task (CopyToJob, CopyFromJob).

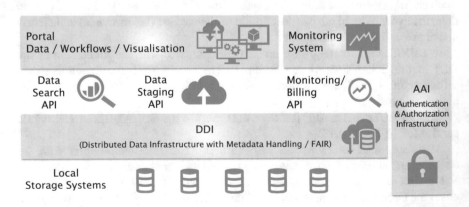

Fig. 4 Immersion of the Distributed Data Infrastructure in the LEXIS ecosystem via its most important APIs (middle part of figure)

These components are associated to a HEAppE job component in the workflow by means of a relationship, which provides attribute values required for the necessary data transfers (e.g., source and target directories). The attribute values are then passed to the staging API to initiate the concrete transfer.

Under the hood, the back-end of the staging API is able to perform transfers using a variety of mechanisms, including those common in the world of scientific computing and HPC. It 'speaks' B2SAFE/GridFTP [6, 37], GLOBUS [38] and SCP/SFTP/SSHFS (e.g. [39]) and chooses the most efficient possibility. In the current optimisation phase of the DDI, we are beginning to regularly benchmark point-to-point speeds between all LEXIS source and target data systems. Thus, we will eliminate bottlenecks and allow the orchestrator to optimise data-transfer paths.

5   The LEXIS Portal: A User-friendly Entry Point to the 'World of HPC/Cloud/Big Data'

The ability to attract a large number of customers to the LEXIS platform mainly depends on the user-friendliness of the system. The LEXIS Portal thus plays a crucial role in lowering the barrier for SMEs to use HPC and cloud resources for solving their Big Data processing needs. It serves as a one-stop-shop and easy entry point to the entire LEXIS platform. Thus, the user does not have to deal with details of our complex infrastructure with its world-class compute and data handling capabilities.

5.1   Portal: Concept and Basic Portal Capabilities

The LEXIS Portal is highly modular by design. It integrates in a plug-and-play manner with the LEXIS DDI, Orchestration System, accounting and billing engine and AAI solution. It gives secure, role-based access to federated resources at multiple computing/data centres. The portal supports (but is not limited to) the following capabilities:

- Registration of users and organisations, and log in
- Creation and management of LEXIS Computational Projects (including addition/deletion of users)
- Requesting access to resources; view of available resources
- View of public and private data sets; data set upload
- Creation of LEXIS workflows, running of workflows as a 'LEXIS job'
- Monitoring and output retrieval for LEXIS jobs
- View of consumption of resources and billing status

In addition, the portal back-end tracks the relationship of centre-specific HPC and cloud computing grants to the organisations and Computational Projects of LEXIS users.

The development strategy of the Portal involves an Agile methodology, however with H2020-compatible project planning for the main directions. Given the unique LEXIS requirements, we are implementing the portal from grounds up, instead of reusing existing portal frameworks. The portal front-end uses React [40], while back-end services are written entirely in Go [41], following API-driven best practices.

5.2 Workflow and Data Management and Visualisation via the Portal

Leveraging the experience with the Alien4Cloud UI [13] for creating workflows, the LEXIS Portal will implement an interface for workflow creation, management and monitoring. A prototype of the latter interface, focusing on part of a Weather and Climate Large-Scale Pilot workflow (cf. Sect. 2.4), is showcased in Fig. 5.

Likewise, the portal provides easy in-browser capabilities to upload new data sets (including resumable uploads based on 'tus' [42]), to find data sets (also by their metadata) and to modify their content. Links to high-bandwidth, out-of-browser data-transfer options for large data sets, such as GridFTP/B2STAGE endpoints of the LEXIS DDI, are provided as well. Output data of user workflows

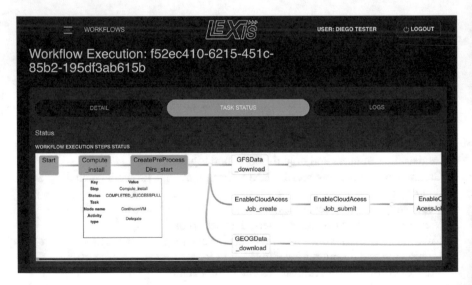

Fig. 5 Prototype of the LEXIS web portal with its workflow-monitoring interface activated. The workflow shown is at the step 'CreatePreProcess Dirs_start'

are automatically registered – with basic metadata – as data products in the LEXIS DDI and can thus be conveniently retrieved via the Portal, or also used as an input for other workflows.

LEXIS will also provide advanced data-visualisation facilities, with the Portal as an entry point. Besides offering resource-friendly in-browser visualisation, the user can also be guided to powerful remote-visualisation systems of the LEXIS centres.

6   Conclusions

In this contribution, we presented LEXIS as a versatile and high-performance Cloud/HPC/Big Data workflow platform, focusing on its EUDAT-based Distributed Data Infrastructure and federation aspects. The LEXIS H2020 project, producing the platform, creates unique opportunities for knowledge transfer between the scientific and industrial IT communities treating Big Data. It enables industrial companies and SMEs to leverage the best European data and Supercomputing infrastructures from academia, while science can profit from applying industrial techniques, be it, for example, in the 'Cloud-Native' or Service Management sectors.

Right from the beginning of the project, the platform was implemented within a co-design framework. We strongly targeted the practical requirements of three representative 'Pilot' use cases in aeronautics engineering, earthquake/tsunami prediction and weather modelling. The pilot simulations, such as the weather models, are of concrete societal and commercial use, for example for the selection of wind energy sites. Despite this practical orientation, the project has managed to consistently follow modern, API-based and secure service design principles.

As more use cases are attracted, the platform is broadly validated, usability will be optimised, and collaboration with European users and projects can be established. This will put all components to a test, including the iRODS systems of our EUDAT-based data infrastructure, which takes care of transparent data transfers within workflows and serves the users for managing their data via the LEXIS Portal. As a result, the LEXIS platform will be optimised to reliably and efficiently execute an entire spectrum of workloads, including visualisation and GPU- or even FPGA-accelerated tasks. Orchestrated LEXIS workflows will thus efficiently combine different computing paradigms (HPC, HTC, Cloud) and analysis methods (classical modelling, AI, HPDA).

We are looking forward to extend the LEXIS federation to more computing and data sites, and software necessary to join the platform will be conveniently packaged. With this open approach, we will continue to push towards a convergence of industrial and academic data science in Europe and towards a convergence of the HPC, Cloud and Big Data ecosystems on the strongest European compute systems. Enlarging the federated LEXIS platform and making it sustainable, besides work on novel functionalities, is the key focus of the project in its second half.

Acknowledgments This work and all contributing authors are funded/co-funded by the EU's Horizon 2020 research and innovation programme (2014–2020) under grant agreement no. 825532 (Project LEXIS – 'Large-scale EXecution for Industry and Society'). The work at IT4I is supported by The Ministry of Education, Youth and Sports from the Large Infrastructures for Research, Experimental Development and Innovations project 'e-Infrastructure CZ – LM2018140'. Likewise, we gratefully acknowledge the use of the computing facilities at LRZ and IT4I, as mentioned in Sect. 2.2. We thank various colleagues from EUDAT for very kind help and good collaboration.

References

1. Zillner, S., Curry, E., Metzger, A., Auer, S., Seidl, R. (Eds.) (2017). European big data value strategic research and innovation agenda. https://bdva.eu/sites/default/files/BDVA_SRIA_v4_Ed1.1.pdf. Cited 6 Nov 2020
2. Zillner, S., et al. (Eds.) (2020). Strategic research, innovation and deployment agenda—AI, data and robotics partnership. Third Release. https://ai-data-robotics-partnership.eu/wp-content/uploads/2020/09/AI-Data-Robotics-Partnership-SRIDA-V3.0.pdf. Cited 1 Feb 2021
3. Shiers, J. (2007). The worldwide LHC computing grid (worldwide LCG). *Computer Physics Communications, 177*(1), 219–223.
4. Venner, J., Wadkar, S., & Siddalingaiah, M. (2014). Pro apache hadoop, 2nd ed. Expert's voice in big data. Apress, Berkeley, CA. doi:10.1007/978-1-4302-4864-4
5. Lecarpentier, D., Wittenburg, P., Elbers, W., Michelini, A., Kanso, R., Coveney, P. V., & Baxter, R. (2013). EUDAT: A new cross-disciplinary data infrastructure for science. *International Journal of Digital Curation, 8*(1), 279–287.
6. EUDAT Collaborative Data Infrastructure (2020). EUDAT—research data services, expertise & technology solutions. https://www.eudat.eu. Cited 6 Nov 2020
7. The Big Data Value Association (2020). BDVA. https://www.bdva.eu/. Cited 6 Nov 2020
8. Schmuck, F. B., & Haskin, R. L. (2002). GPFS: a shared-disk file system for large computing clusters. In Proceedings of the conference on file and storage technologies, FAST '02 (pp. 231–244). USENIX Association, US.
9. Weil, S. A., Brandt, S. A., Miller, E. L., Long, D. D. E., & Maltzahn, C. (2006). Ceph: A scalable, high-performance distributed file system. In Proceedings of the 7th symposium on operating systems design and implementation OSDI '06 (pp. 307–320). USENIX Association, US.
10. NVM Express, Inc. (2016). NVM express over fabrics 1.0. https://nvmexpress.org/wp-content/uploads/NVMe_over_Fabrics_1_0_Gold_20160605.pdf. Cited 6 Nov 2020.
11. Colonnelli, I., Cantalupo, B., Merelli, I., & Aldinucci, M. (2020). Streamflow: Cross-breeding cloud with HPC. *IEEE Transactions on Emerging Topics in Computing*. In print. doi:10.1109/TETC.2020.3019202.
12. Bull/Atos (2020). Ystia Suite. https://ystia.github.io. Cited 6 Nov 2020.
13. FastConnect, Bull/Atos (2020). Alien 4 Cloud. http://alien4cloud.github.io. Cited 6 Nov 2020.
14. Brogi, A., Soldani, J., & Wang, P. (2014). TOSCA in a Nutshell: Promises and perspectives. In Villari, M., Zimmermann, W., Lau, K. K. (Eds.), Service-oriented and cloud computing (pp. 171–186). Springer, Heidelberg. doi:10.1007/978-3-662-44879-3_13
15. Svaton, V. (2020). Home · Wiki · ADAS/HEAppE/Middleware · GitLab. http://heappe.eu. Cited 6 Nov 2020.
16. Svaton, V., Martinovic, J., Krenek, J., Esch, T., & Tomancak, P. (2019). HPC-as-a-service via HEAppE platform. In Barolli, L., Hussain, F. K., Ikeda, M. (Eds.), CISIS 2019, advances in intelligent systems and computing, vol. 993 (pp. 280–293). Springer, Cham. doi:10.1007/978-3-030-22354-0_26

17. Imamagic, E., & Ferrari, T. (2014). EGI grid middleware and distributed computing infrastructures integration. In Proceedings of the international symposium on grids and clouds (ISGC) 2013—PoS, vol. 179 (010). doi:10.22323/1.179.0010.

18. Parodi, A., et al. (2020). LEXIS weather and climate large-scale pilot. In Barolli, L., Poniszewska-Maranda, A., Enokido, T. (Eds.), CISIS 2020, advances in intelligent systems and computing, vol. 1194 (pp. 267–277). Springer, Cham. doi:10.1007/978-3-030-50454-0_25.

19. Powers, J. G., et al. (2017). The weather research and forecasting model: Overview, system efforts, and future directions. *Bulletin of the American Meteorological Society, 98*(8), 1717–1737.

20. Parodi, A., et al. (2017). DRIHM (2US): An e-science environment for hydrometeorological research on high-impact weather events. *Bulletin of the American Meteorological Society, 98*(10), 2149–2166.

21. LEXIS project (2019). Deliverable 4.1: Analysis of mechanisms for securing federated infrastructure. https://cordis.europa.eu/project/id/825532/results. Cited 7 Jan 2021.

22. JBoss (Red Hat Inc.), Keycloak Community (2020). Keycloak. https://www.keycloak.org/. Cited 6 Nov 2020.

23. Sakimura, N., Bradley, J., Jones, M. B., de Medeiros, B., & Mortimore, C. (2014). OpenID connect core 1.0 incorporating errata set 1. https://openid.net/specs/openid-connect-core-1_0.html. Cited 6 Nov 2020.

24. Cantor, S., Kemp, J., Philpott, R., & Maler, E. (2005). Assertions and protocols for the OASIS security assertion markup language (SAML) V2.0. http://docs.oasis-open.org/security/saml/v2.0/saml-core-2.0-os.pdf. Cited 6 Nov 2020.

25. Jones, M. B., Bradley, J., & Sakimura, N. (2015). RFC 7519 – JSON web token (JWT). https://tools.ietf.org/html/rfc7519. Cited 6 Nov 2020.

26. García-Hernández, R. J., & Golasowski, M. (2020). Supporting Keycloak in iRODS systems with OpenID authentication. Presented at CS3—workshop on cloud storage synchronization and sharing services. https://indico.cern.ch/event/854707/contributions/3681126. Cited 6 Nov 2020.

27. Zaharia, M., et al. (2016). Apache spark: A unified engine for big data processing. *Communications of the ACM, 59*(11), 56–65

28. Xu, H., Russell, T., Coposky, J., Rajasekar, A., Moore, R., de Torcy, A., Wan, M., Shroeder, W., & Chen, S. Y. (2017). iRODS primer 2: Integrated rule-oriented data system. In Morgan and Claypool Publishers, Williston, VT. doi:10.2200/S00760ED1V01Y201702ICR057.

29. Depardon, B., Le Mahec, G., & Séguin, C. (2013). Analysis of six distributed file systems. Report hal-00789086. https://hal.inria.fr/hal-00789086. Cited 7 Jan 2021.

30. Fuhrmann, P., Antonacci, M., Donvito, G., Keeble, O., & Millar, P. (2018). Smart policy driven data management and data federations. In Proceedings of the international symposium on grids and clouds (ISGC) 2018 in conjunction with frontiers in computational drug discovery—PoS, vol. 327 (001). doi:10.22323/1.327.0001.

31. Kawai, Y., & Hasan, A. (2010). High-availability iRODS system (HAIRS). In Proceedings of the iRODS user group meeting 2010: Policy-based data management, sharing and preservation. Chapel Hill, NC.

32. Depuydt, J. (2015). Setup a redundant PostgreSQL database with repmgr and pgpool I Jensd's I/O buffer. http://jensd.be/591/linux/setup-a-redundant-postgresql-database-with-repmgr-and-pgpool. Cited 6 Nov 2020.

33. Wilkinson, M. D., et al. (2016). The FAIR guiding principles for scientific data management and stewardship. *Scientific Data, 3*, 1–9.

34. Apps, A., & MacIntyre, R. (2000). Dublin core metadata for electronic journals. In Borbinha, J., Baker, T. (Eds.), Research and advanced technology for digital libraries (pp. 93–102). Springer, Berlin. doi:10.1007/3-540-45268-0_9.

35. Starr, J., & Gastl, A. (2011). isCitedBy: A metadata scheme for DataCite. *D-Lib Magazine, 17*(1–2). doi:10.1045/january2011-starr.

36. Roberts, A. M., Wong, A. K., Fisk, I., & Troyanskaya, O. (2016). GIANT API: An application programming interface for functional genomics. *Nucleic Acids Research, 44*, W587–W592.

37. Allcock, W., Bresnahan, J., Kettimuthu, R., & Link, M. (2005). The globus striped GridFTP framework and server. In SC '05: Proceedings of the 2005 ACM/IEEE conference on supercomputing (pp. 54–54). doi:10.1109/SC.2005.72.
38. Foster, I. (2011). Globus online: Accelerating and democratizing science through cloud-based services. *IEEE Internet Computing, 15*, 70–73.
39. Thelin, J. (2011). Accessing Remote Files Easily and Securely. Linux Journal. https://www.linuxjournal.com/content/accessing-remote-files-easy-and-secure. Cited 6 Nov 2020.
40. Facebook Inc., React contributors (2020). React—A JavaScript library for building user interfaces. http://reactjs.org. Cited 6 Nov 2020.
41. Google, Go Contributors (2020). The go programming language specification—the go programming language. https://golang.org/ref/spec. Cited 6 Nov 2020.
42. Kleidl, M., Transloadit, tus Collaboration (2020). tus—resumable file uploads. https://tus.io/. Cited 6 Nov 2020.

Part II
Processes and Applications

The DeepHealth Toolkit: A Key European Free and Open-Source Software for Deep Learning and Computer Vision Ready to Exploit Heterogeneous HPC and Cloud Architectures

Marco Aldinucci, David Atienza, Federico Bolelli, Mónica Caballero, Iacopo Colonnelli, José Flich, Jon A. Gómez, David González, Costantino Grana, Marco Grangetto, Simone Leo, Pedro López, Dana Oniga, Roberto Paredes, Luca Pireddu, Eduardo Quiñones, Tatiana Silva, Enzo Tartaglione, and Marina Zapater

Abstract At the present time, we are immersed in the convergence between Big Data, High-Performance Computing and Artificial Intelligence. Technological progress in these three areas has accelerated in recent years, forcing different players like software companies and stakeholders to move quickly. The European Union is dedicating a lot of resources to maintain its relevant position in this

M. Aldinucci · I. Colonnelli · M. Grangetto · E. Tartaglione
Università degli studi di Torino, Torino, Italy

D. Atienza · M. Zapater
Ecole Polytechnique Fédérale de Lausanne, Lausanne, Switzerland

F. Bolelli · C. Grana
Università degli studi di Modena e Reggio Emilia, Modena, Italy

M. Caballero
NTT Data Spain, Barcelona, Spain

J. Flich · J. A. Gómez (✉) · P. López · R. Paredes
Universitat Politècnica de València, Valencia, Spain
e-mail: jon@prhlt.upv.es

D. González · T. Silva
TREE Technology, Asturias, Spain

S. Leo · L. Pireddu
Center for Advanced Studies, Research and Development in Sardinia, Sardinia, Italy

D. Oniga
Softwae Imagination and Vision, Bucureşti, Romania

E. Quiñones
Barcelona Supercomputing Center, Barcelona, Spain

© The Author(s) 2022
E. Curry et al. (eds.), *Technologies and Applications for Big Data Value*,
https://doi.org/10.1007/978-3-030-78307-5_9

scenario, funding projects to implement large-scale pilot testbeds that combine the latest advances in Artificial Intelligence, High-Performance Computing, Cloud and Big Data technologies. The DeepHealth project is an example focused on the health sector whose main outcome is the DeepHealth toolkit, a European unified framework that offers deep learning and computer vision capabilities, completely adapted to exploit underlying heterogeneous High-Performance Computing, Big Data and cloud architectures, and ready to be integrated into any software platform to facilitate the development and deployment of new applications for specific problems in any sector. This toolkit is intended to be one of the European contributions to the field of AI. This chapter introduces the toolkit with its main components and complementary tools, providing a clear view to facilitate and encourage its adoption and wide use by the European community of developers of AI-based solutions and data scientists working in the healthcare sector and others.

Keywords Hybrid big data HPC architectures · High performance data analytics · Hardware-specific capabilities for big data GPUs FPGAs · Performance for large-scale processing

1   Context: The European AI and HPC Landscape and the DeepHealth Project

The rapid progress of different technologies is taking place within a virtuous circle that emerged thanks to the synergies between such technologies and has brought us three important advances in recent years, namely, the increase in storage capacity at a reduced price, the increase in data transmission speed and the increase in computing power provided by High-Performance Computing (HPC) and hardware accelerators. These three advances, in combination with the availability of large-enough volumes of data, have considerably boosted the growth and development of Artificial Intelligence (AI) in recent years. Mainly, thanks to the fact that the techniques of Machine Learning (ML), able to learn from data, have reached a good level of maturity and are improving the best results obtained by expert systems at the core of knowledge-based solutions in most application domains. Machine Learning is one of the most important areas of AI, which in turn includes Deep Learning (DL). As such, descriptive/predictive/prescriptive models based on AI/ML/DL techniques are becoming key components of applications and systems deployed in real scenarios for solving problems in a wide variety of sectors (e.g., manufacturing, agriculture and food, Earth sciences, retail, fintech and smart cities, among others). Nevertheless, its use in the health sector is still far from being widely spread (see [1]).

In this scenario, the European Union (EU) is fostering strategic actions to position the EU as a big worldwide player in AI, HPC and Big Data, capable of creating and deploying solutions based on cutting-edge technologies. The

DeepHealth project whose title is "Deep-Learning and HPC to Boost Biomedical Applications for Health" [2], funded by the EC under the topic ICT-11-2018-2019 "HPC and Big Data enabled Large-scale Test-beds and Applications", is one of the innovation actions supported by the EU to boost AI and HPC leadership and promote large-scale pilots. DeepHealth is a 3-year project, kicked off in January 2019 and scheduled to conclude in December 2021. DeepHealth aims to foster the use of technology in the Healthcare sector by reducing the current gap between the availability of mature-enough AI-based medical imaging solutions and their deployment in real scenarios. The main goal of the DeepHealth project is to put HPC power at the service of biomedical applications that require the analysis of large and complex biomedical datasets and apply DL and Computer Vision (CV) techniques to support new and more efficient ways of diagnosis, monitoring and treatment of diseases.

Following this aim, one of the main outcomes of DeepHealth addressing industry needs is the DeepHealth toolkit, a free and open-source software designed to be a European unified framework to offer DL and CV capabilities completely adapted to exploit underlying heterogeneous HPC, Big Data and cloud architectures. The DeepHealth toolkit is aimed at computer and data scientists as well as to developers of AI-based solutions working in any sector. It is a piece of software ready to be integrated into any software platform, designed to facilitate the development and deployment of new applications for specific problems. Within the framework of the DeepHealth project, the toolkit is being developed, tested and validated by using it to implement descriptive/predictive models for 14 healthcare use cases; nevertheless, its usefulness goes beyond the health sector, being applicable, as said, to any application domain or industrial sector. Thanks to all its features, which will be detailed throughout this chapter, the DeepHealth toolkit is technology made in EU that contributes to the development of AI in Europe.

This chapter is aligned with the technical priorities of Data Processing Architectures and Data Analytics of the European Big Data Value Strategic Research and Innovation Agenda [3]. It addresses the vertical concern Engineering and DevOps of the BDV Technical Reference Model, and the horizontal concerns Data Analytics and Data Processing Architectures focusing on the Cloud and HPC. And this chapter also relates to the *Systems, Methodologies, Hardware and Tools* cross-sectorial technology enablers of the AI, Data and Robotics Strategic Research, Innovation and Deployment Agenda [4].

This chapter also introduces the toolkit, its functionalities and its enabling capabilities with the objective to bring it closer to potential users in both industry and academia. To do so, the authors present the toolkit, its components, the adaptations that allow exploiting HPC and cloud computing infrastructures thanks to complementary HPC frameworks, and describe practical aspects to guide on its use and how to effectively integrate it for the development of AI-based applications.

2   A General Overview of the DeepHealth Toolkit

The DeepHealth toolkit is a general-purpose deep learning framework, including image processing and computer vision functionalities, enabled to exploit HPC and cloud infrastructures for running parallel/distributed training and inference processes. All the components of the toolkit are available as free and open-source software under the MIT license [5]. This framework enables data scientists to design and train predictive models based on deep neural networks, and developers to easily integrate the predictive models into existing software applications/platforms in order to quickly build and deploy AI-based solutions (e.g., support decision tools for diagnosis).

The toolkit is specifically designed to cope with big and constantly growing datasets (e.g., medical imaging datasets). Large-enough datasets enable the use of more complex neural networks and drive to improve both the accuracy and robustness of predictive models, but at the cost of dramatically increasing the demand of computing power. To do so, the DeepHealth toolkit incorporates, in a transparent manner, the most advanced parallel programming models to exploit the parallel performance capabilities of HPC and cloud infrastructures, featuring different acceleration technologies such as symmetric multi-processors (SMPs), graphic processing units (GPUs) and field-programmable gate arrays (FPGAs). It also integrates additional frameworks (i.e., COMPSs [6] and StreamFlow [7]) that allow to exploit specialized infrastructures, enabling parallelization mechanisms at different levels. Moreover, the toolkit provides functionalities to be used for both training and inference, addressing the complexity of the different available computational resources and target architectures at both the training and inference stages. Training is performed by AI experts, commonly in research-focused environments, using specialized HPC architectures equipped with FPGAs and GPUs; the goal is to maximize the number of samples processed per second keeping the overall accuracy. Inference is done with trained models in production environments (even using small devices in the edge), where the response time for predicting single samples is crucial.

The core of the toolkit consists of two libraries, namely the European Computer Vision Library (ECVL) and the European Distributed Deep Learning Library (EDDL), that are accompanied by the back end and the front end, two components to allow and facilitate the use of the libraries. The back end is a software-as-a-service module that offers a RESTful API to give access to all the functionalities of both libraries and provides independency from the programming language. The front end is a web-based graphical user interface, mainly oriented to be used by data scientists, for designing, training and testing deep neural networks. Both libraries are implemented in C++ and include a Python API to facilitate the development of client applications and integration with the wide array of Python-based data analysis libraries. Figure 1 depicts the components of the toolkit and highlights the two possible alternatives for developers of domain-specific applications to use the functionalities provided by both libraries: (1) the use of a RESTful API provided

Fig. 1 Components of the DeepHealth toolkit and the two possible alternatives of interacting with the libraries. One through the back end using a RESTful API, and another using the API of both libraries. The execution of training and inference procedures over HPC + cloud infrastructures is performed by the runtime. The runtime includes adaptations to HPC frameworks ready to be executed under the control of resource managers

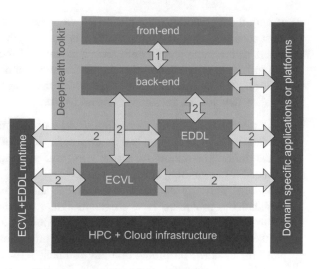

by the back end (represented in Fig. 1 by the arrows labelled 1) or (2) the use of the specific APIs (C++ or Python) of each library (represented in Fig. 1 by the arrows labelled 2). This second option is not independent of the programming language, yet it provides more control and flexibility over the software at the cost of additional programming complexity. The former option requires less effort from platform developers and makes applications independent of specific versions of the libraries. Only versions including changes in the RESTful API will require updating and recompiling applications. It can also be observed from Fig. 1 that all the DL and CV functionalities can be thoroughly used via the front end or via a software application/platform, in this last case with or without the back end. Additionally, Fig. 1 shows the runtime of both libraries, which can be used to launch distributed learning processes to train models on HPC and cloud architectures. Both libraries are designed to run under the control of HPC-specific workflow managers such as COMPSs [6] and StreamFlow [7], presented in Sect. 6. Once the trained models are tested and validated, they are ready to be used in production environments to perform inference from new samples by using the software applications/platforms in which libraries are integrated.

The following describes the typical workflow of the usage of the toolkit by a development team who is requested to address a new use case in a real scenario, considering that the libraries of the DeepHealth toolkit are already integrated in the platform any company developed to deploy AI-based solutions in the health sector. First, (i) data scientists, members of the team, prepare the dataset by splitting it into three subsets, namely training, validation and testing subsets. Next, (ii) the team designs several artificial neural networks and (iii) launches the training processes on HPC and cloud architectures by means of the runtime of the toolkit adapted to HPC frameworks like the ones described in Sect. 6. (iv) The team evaluates the models using the validation subset, and goes back to step (ii) to redesign some models if necessary. Sometimes, the team should come back to step (i) to consider the dataset itself with the knowledge gained from previous iterations. (v) The model that gets the best accuracy using the testing subset is selected; then (vi) computer scientists, members of the same team, configure an instance of the application with the best model and deploy the solution in a production environment.

In itself, the DeepHealth toolkit provides the following features to AI-based application developers, data scientists and ML practitioners in general:

- Increases the productivity of computer and data scientists by decreasing the time needed to design, train and test predictive models throughout the parallelization of the training operations on top of HPC and cloud infrastructures, and without the need for combining numerous tools.
- Facilitates the easy and fast development and deployment of new AI-based applications, providing in a single toolkit, ready to be integrated, the most common DL and CV functionalities with support for different operating systems. Furthermore, it allows to perform training processes outside the application/-platform installed on production environments. To use the resulting predictive models, applications/platforms only need to integrate the libraries following one of the two possible alternatives presented.
- Relaxes the need of having highly skilled AI and HPC/cloud experts. Training processes can be executed in a distributed manner in a transparent way for data/computer scientists, and applications/platforms in production environments do not need to be adapted to run distributed processes on HPC and cloud infrastructures. Therefore, data scientists and developers do not need to have a deep understanding of HPC, DL, CV, Big Data or cloud computing.

3 The European Distributed Deep Learning Library

EDDL is a general-purpose deep learning library initially developed to cover deep learning needs in healthcare use cases within the DeepHealth project. As part of the DeepHealth toolkit, EDDL is a free and open-source software available on a GitHub public repository [5]. Currently, it supports most widely used deep neural network topologies, including convolutional and sequence-to-sequence models, and is being used in different tasks like classification, semantic segmentation of images, image description, event prediction from time-series data, and machine translation. In order to be compatible with existing developments and other deep learning toolkits, the EDDL uses ONNX [8], the standard format for neural network interchange, to import and export neural networks, including both weights and topology.

EDDL provides hardware-agnostic tensor operations to facilitate the development of hardware-accelerated deep learning functionalities and the implementation of the necessary tensor operators, activation functions, regularization functions, optimization methods and all layer types (dense, convolutional and recurrent) to implement state-of-the-art neural network topologies. The EDDL exposes two APIs (C++ and Python) with functionalities belonging to two main groups: neural network manipulation (models, layers, regularization functions, initializers) and tensor operations (creation, serialization, I/O, mathematical transformations). The neural networks section provides both high-level tools, such as fitting and evaluating the whole model, and lower-level ones that allow developers to act on individual epochs and batches, providing finer control albeit with a slight efficiency loss in the case of using the Python API, since a larger part of the program needs to be written in Python to handle loops.

EDDL is implemented in C++, and the Python API, called PyEDDL [5], has been developed to enhance the value of EDDL to the scientific community. The availability of a Python library allows to integrate EDDL functionalities with widely used scientific programming tools such as NumPy/SciPy [9] and Pandas [10]. In particular, PyEDDL supports converting between EDDL tensors and NumPy arrays, which are key to enable interoperability with other Python scientific libraries. Moreover, since PyEDDL is based on a native extension module that wraps the C++ EDDL code, users can take advantage of the simplicity and speed of development of Python without sacrificing performance, using Python as a "glue" language that ties together computationally intensive native routines. PyEDDL allows Python access to the EDDL API and, as mentioned above, adds NumPy interoperability, allowing interaction with a wide array of data sources and tools. Like the rest of the DeepHealth Toolkit, PyEDDL is released as free and open-source software and its source code is available on GitHub [5].

In relation to hardware accelerators, EDDL is ready to run on single computers using either all or a subset of the available cores, all or a subset of the available GPU cards, and coordinating the computation flow on the FPGA cards connected to a single computer. The C++ and Python APIs of the EDDL both include a function to build neural networks that creates all the data structures according to the network topology and allocates all the necessary memory; one of the parameters of the build function is an object for describing the available hardware devices the EDDL will use to run the training and inference processes. EDDL defines the concept of Computing Service to describe hardware devices. Currently, three types of computing services are defined, namely CPU, GPU and FPGA. The number of CPU cores, GPU cards or FPGA cards to be used are indicated by the Computing Service.

Any neural network topology is internally represented by means of two directed and acyclic graphs (DAGs), one for the forward step and another one for the backward step. Each DAG defines the sequence of tensor operations to perform the computation corresponding to the entire network, so that the computations corresponding to a given layer will be performed when all its input dependencies according to the DAG have been satisfied, i.e., when the output of all the layers used as input to a given one are ready. Tensor operations are performed using the hardware devices specified by means of the Computing Service provided as a parameter to the build function. On manycore CPUs, tensor operations are performed by using the Eigen library [11] and parallelized using OpenMP [12]. When using GPU or FPGA cards, the forward and backward algorithms are designed to minimize the number of memory transfers between the CPU and GPU/FPGA cards. In the particular case of GPUs, EDDL has three modes of memory management to address the lack of memory when a given batch size does not fit in the memory of a GPU. The most efficient one tries to allocate the whole batch in the GPU memory to reduce memory transfers at the minimum, the intermediate and least-efficient modes allow to work with larger batch sizes at the cost of increasing the number of memory transfers to perform the forward and backward steps for a given batch of samples.

GPU support in EDDL is done by means of CUDA kernels developed as part of the EDDL code. As mentioned above, the use of different hardware accelerators is completely transparent to developers and programmers who use the EDDL; they only need to create the corresponding Computing Service to use all or a subset of the computational resources. Integrating NVIDIA cuDNN library in the EDDL as an alternative to CUDA kernels is in the work plan of the DeepHealth project.

Table 1 shows the performance in terms of the accuracy obtained with the test set and the time per epoch in seconds during training. The EDDL is compared with TensorFlow [13] and PyTorch [14], the two most popular DL toolkits. The Cifar10 dataset was used. It can be observed that EDDL performs similar to the other toolkits, but EDDL still needs to improve the performance on both CPUs and GPUs when using Batch Normalization and larger topologies like VGG16 and VGG19.

EDDL support for FPGA cards is quite similar to the support for GPU cards. The developer or data scientist using the EDDL simply indicates the target device to run training or inference processes by means of a Computing Service object. Although FPGAs can also be used for training, they are more appealing for inference processes, and, therefore, FPGA support has been optimized for the inference process. Depending on the trained model, FPGA cards can be directly used. This

Table 1 Benchmark to compare EDDL with TensorFlow and PyTorch using Cifar10 with and without Batch Normalization

Model	Accuracy/time	TensorFlow		PyTorch		EDDL	
		No BN	BN	No BN	BN	No BN	BN
VGG16	Test accuracy	77.4%	71.7%	77.9%	/ 76.2%	74.6%	76.4%
	GPU time per epoch	62 s	68 s	72 s	77 s	146 s	204 s
	CPU time per epoch	1313 s	1375 s	887 s	956 s	3107 s	2846 s
VGG19	Test accuracy	66.0%	59.9%	65.5%	59.7%	68.2%	61.0 s
	GPU time per epoch	76 s	81 s	120 s	126 s	190 s	260 s
	CPU time per epoch	1703 s	1809 s	1262 s	1352 s	3872 s	3838 s
RestNet18	Test accuracy	67.6%	64.0%	66.4%	65.7%	67.3%	64.8%
	GPU time per epoch	25 s	26 s	59 s	60 s	36 s	49 s
	CPU time per epoch	1234 s	1244 s	456 s	485 s	932 s	1207 s
ResNet34	Test accuracy	66.6%	66.4%	67.8%	65.5%	66.1%	60.4%
	GPU time per epoch	44 s	46 s	97 s	101 s	65 s	89 s
	CPU time per epoch	2125 s	2140 s	834 s	895 s	1674 s	2119 s
ResNet50	Test accuracy	68.4%	61.3%	68.1%	63.1%	66.4%	61.9%
	GPU time per epoch	47 s	52 s	84 s	92 s	75 s	132 s
	CPU time per epoch	1995 s	2044 s	706 s	835 s	1684 s	2622 s

is the case when the models fit on the typically lower memory resources available on FPGA devices. If models do not fit, then two options can be used. The first one is to iteratively use FPGA cards to run the complete inference process on the model, performing operations on a step-by-step manner driven by the CPU. The FPGA support has been provided to allow this operational mode. However, quantization and compression strategies can be deployed once the model has been trained. In the DeepHealth project, the FPGA kernels mostly used on the Medical sector use cases are being optimized and adapted to quantized and compressed models. In order to deploy a model on FPGAs targeting low resource constraints and high energy efficiency, the EDDL incorporates a strategy to reduce the complexity of a deep neural network. Many techniques have been proposed recently to reduce such complexity [15–17]. These approaches include the so-called pruning techniques, whose aim is to detect and remove the irrelevant parameters from a model [18]. Removing parameters from a model has a huge impact on the deployment of the trained model on FPGA cards, since the overall size of the model reduces as well as the number of operations to generate the outcome decreases and, for instance, the power consumption. This is allowed by the typically high dimensionality of these models, where sparser and more efficient solutions can be found [19]. Towards this end, in order to deploy the model on FPGA cards targeting low resource constraints and high energy efficiency, the approach used in the EDDL is to include a structured sparsity step, where as many neurons as possible are removed from the model with a negligible performance loss.

Regarding distributed learning on HPC/cloud/HPC + cloud architectures, the EDDL includes specific functions to simplify the distribution of batches when training and inference processes are run by means of HPC frameworks like COMPSs or StreamFlow. Concretely, the COMPSs framework allows to accelerate the DL training operations by dividing the training data sets across a large set of computing nodes available on HPC and cloud infrastructures, and upon which partial training operations can then be performed. To do so, EDDL allows to distribute the weights of the network from the master node to worker nodes, and to report gradients from worker nodes to the master node, both synchronously and asynchronously. The EDDL serializes networks using ONNX to transfer weights and gradients between the master node and worker nodes. The serialization includes the network topology, the weights and the bias. To facilitate distributed learning, the serialization functions implemented in the EDDL allow to select whether to include weights or gradients.

EDDL and PyEDDL code is covered by an extensive test suite and complemented by numerous usage examples in Python and C++, including network training and evaluation with different models, ONNX serialization and NumPy compatibility. To facilitate their adoption, EDDL and PyEDDL also provide extensive documentation on installation, tensor and neural network manipulation, API usage and examples [5]. The "getting started" section contains simple examples; the most advanced ones show the use of topologies like VGG16/VGG19 [20] and U-Net [21]. Concerning installation, developers can choose between installing from source code, via conda [22], and via Homebrew for Mac OS X [23]. Additionally, pre-built Docker

images with the DeepHealth toolkit components ready to be used are available on
DockerHub [24] (see Sect. 6.1).

4   The European Computer Vision Library

ECVL is a general-purpose computer vision library developed to support healthcare
use cases within the DeepHealth project, with the aim of facilitating the integration
of existing state-of-the-art libraries such as OpenCV [25]. ECVL currently includes
high-level computer vision functionalities implementing specialized and accelerated
versions of algorithms commonly employed in conjunction with deep learning;
functionalities that are useful for image processing tasks in any sector beyond health.

The design of ECVL is based on the concept of *Image*, which represents
the core element of the entire library. It allows to store raw data, images, and
videos in a multi-dimensional dense numerical single- or multi-channel tensor.
Multiple types of scientific imaging data and data formats (e.g., jpeg, png, bpm,
ppm, pgm, etc.) are natively supported by ECVL. Moreover, the library provides
specific functionalities to handling medical data, such as DICOM, NIfTI and many
proprietary Virtual Slides (VS) formats. In the case of VS, the *Image* object allows
to choose the area and the resolution to be extracted from the file. The availability
of a common software architecture provided by the ECVL Hardware Abstraction
Layer (HAL) allows great flexibility for device differentiation (SMPs, GPUs, and
FPGAs) while keeping the same user interface. This hardware-agnostic API ensures
versatility, flexibility, and extensibility, simplifying the library usage and facilitating
the development of distributed image analysis tasks.

The *Image* class has been designed for representing and manipulating different
types of images with diverse channel configurations, providing both reading
and writing functionalities for all the aforementioned data formats. Arithmetic
operations between images and scalars are performed through the *Image* class.
Obviously, all the classic operations for image manipulation such as rotation,
resizing, mirroring and colour space change are available. Extremely optimized
processing functions, like noising, blurring, contour finding [26], image skeletoniza-
tion [27] and connected components labelling [28] are implemented as well. ECVL
image-processing operations can be applied on-the-fly during deep neural networks
training to implement data augmentation. Given the relevance of data augmentation,
ECVL provides with a simple Domain-Specific Language (DSL) to facilitate the
definition of transformations to be applied and their configuration parameters. A set
of transformations can thus be defined for each split of a dataset (train, validation
and test subsets). Augmentation can be either provided in compiled code or through
the DSL and thus read from file at runtime. More details are available in [29].

Optional modules are supplied with the library and can be activated to enable
additional functionalities, such as the cross-platform GUI based on wxWidgets [30],
which provides simple exploration and visualization of images contained in ECVL

Image objects, and a 3D volumes visualizer to observe different slices of a CT scan from different views.

In order to ensure an efficient and straightforward mechanism to perform distributed model training, ECVL defines the DeepHealth Dataset Format (DDF): a simple and flexible YAML-based syntax [31] that allows to describe a dataset. Regardless of the task being analysed, a DDF file provides all the information required to characterize the dataset and thus performing data loading, image pre- and post-processing and model training. A detailed description of such a format can be found in [29]. Moreover, a specific module to load and parse DDF-defined datasets is implemented and exposed by the library interface.

Like EDDL, ECVL is complemented by a Python API called PyECVL [5]. In addition to simplified programming, its main advantage is the ability to integrate with other scientific programming tools, which are abundant in the Python ecosystem. This interoperability is enabled by supporting the conversion between ECVL images and NumPy arrays. Like PyEDDL, PyECVL is based on a wrapper extension module that reroutes calls to the C++ code, allowing to reap the benefits of Python development without taking a big hit on performance. PyECVL exposes ECVL functionalities to Python, including Image objects, data and colour types, arithmetic operations, image processing, image I/O, augmentations, the DeepHealth dataset parser and the ECVL-EDDL interaction layer. As discussed earlier, its support for to/from array conversion allows to process data with NumPy as well as many other scientific tools based on it.

Regarding hardware accelerators, the ECVL supports the use of GPU and FPGA cards to run the computer vision algorithms needed in training and inference processes. The implementation for GPUs has been done using CUDA kernels, while for FPGA cards it is somewhat more complicated as FPGA cards are reconfigurable devices which allow the designer to fully customize their design and to adapt it to the algorithm they need to run. This enables, for specific application domains, more power- and energy-efficient solutions than, for instance, CPUs and GPUs. The DeepHealth project advocates for the use of FPGAs as accelerator devices for the inference process. In particular, the trained models ready for production can be launched to an FPGA card by using the FPGA support provided in both the ECVL and the EDDL libraries. The use of FPGA cards is totally transparent to data scientists who use the DeepHealth toolkit. Indeed, both libraries enable the developers who use them just to indicate which type of device the application should be using. For the specificities of the ECVL library, the use of FPGAs is appealing as most computer vision algorithms (e.g., image resize, mirror) deal with pixels rather than floating point values. FPGA devices excel at integer operations and offer massive parallelism possibilities within the device.

Like other software packages of the toolkit, ECVL and PyECVL are available as free and open-source software on a public GitHub repository, including documentation, comprehensive tests, and several usage examples of both the C++ and Python APIs [5]. Examples include data augmentation usage, handling of DeepHealth datasets, interaction with EDDL/PyEDDL, image processing and I/O. The documentation includes detailed instructions to install ECVL and PyECVL from different options as in the case of EDDL and PyEDDL. As mentioned in the EDDL section, a set of pre-built Docker images including the components of the DeepHealth toolkit are available in the Docker hub for the DeepHealth project [24].

5   The Back End and the Front End

The four components of the DeepHealth toolkit are the ECVL, the EDDL, the back end and the front end. Figure 1 shows how the back end and the front end are interconnected with the libraries. The back end is a software module where ECVL and EDDL are fully integrated, which offers a RESTful API to allow any software application or platform to access all the functionalities provided by both libraries without the need to use the C++ or Python API. Ready-to-use pre-built Docker images (see Sect. 6.1) are available, including the back end and all the other components of the toolkit, in such a way that the developers of applications/platforms do not have to worry about the installation and configuration of the DeepHealth toolkit, they only need to provision Docker containers and, obviously, programming, using their preferred programming language, the module for their application/platform that will interact with the RESTful API offered by the back end. This way, the back end enables *managed service* usage scenarios, where a potentially complex and powerful computing infrastructure (e.g., high-performance computing, cloud computing or even heterogeneous hardware) could be transparently used to run deep learning jobs without the users needing to directly interface with it.

The front end is a web-based graphical user interface that facilitates the use of all the functionalities of the libraries by interacting with the back end through the RESTful API. The front end is the component of the toolkit visible to any type of user, but it has been mainly designed for data scientists. Without going into implementation details, the main functionalities provided by the front end are: (1) creation/edition of user profiles; (2) creation/edition of projects; (3) dataset uploading; (4) dataset selection; (5) model creation/import/export/edition/selection; (6) definition of tasks (currently supported types are classification and segmentation); (7) definition of data augmentation transformations; (8) launching training/inference processes; (9) monitoring of training processes, including visualization of the evolution of different neural network related KPIs (e.g., accuracy and loss) with respect to both training and validation data subsets; and (10) model evaluation.

In the common usage of the front end, users have the option of loading from the back end any one of the available models in the set of pre-designed models, which can be already trained. Trained models can be used to perform transfer learning tasks or just to reuse the topology by resetting weights and bias before launching a new training process.

6 Complements to Leverage HPC/Cloud Infrastructures

EDDL, ECVL and the back end are designed to be deployed on HPC and cloud infrastructures to distribute the workload of training and inference processes by following the data parallelization programming paradigm. Both libraries include specific functions to enable distributed learning. The distribution of the workload on multiple worker nodes is not directly performed by ECVL and EDDL. Instead, the libraries are complemented with workflow managers like COMPSs [6] and StreamFlow [7], specially designed for HPC/cloud environments, that manage the workload distribution of training and inference processes in combination with resource managers like SLURM [32]. To leverage hybrid HPC + cloud architectures, pre-built Docker images with all the components of the DeepHealth toolkit are ready to be deployed in scalable environments orchestrated by Kubernetes [33].

The DeepHealth toolkit is being tested on multiple HPC, cloud and hybrid HPC + cloud infrastructures to validate its ability to exploit a wide variety of architectures. The infrastructures considered in the DeepHealth project are:

- The Marenostrum supercomputer, composed of 3456 computed nodes based on Intel Xeon Platinum chips, hosted at the Barcelona Supercomputing Center.
- The MANGO cluster, composed of eight interconnected FPGAs, hosted by the Technical University of Valencia (UPV).
- The *OpenDeepHealth* (ODH) platform, implemented by the University of Torino on top of a hybrid HPC + cloud infrastructure. The HPC component is a C3S OCCAM cluster composed of 46 heterogeneous nodes, also including GPU nodes (K40 or V100). The cloud component, serving multi-tenant private Kubernetes instances, is HPC4AI [34], comprising Intel Xeon Gold 80-cores computing nodes (+2000 CPU cores) with 4 GPUs per node (80 CPU cores + V100 or T4 GPUs).
- The hybrid cloud platform, composed of a Kubernetes cluster on premise (private cloud) and another cluster running in Amazon Web Services (public cloud), provided by the company TREE Technology.

The Marenostrum supercomputer and the ODH platform are similar in terms of use; both are HPC infrastructures and both are ready to hold private clouds. The hybrid cloud facilitates vertical and horizontal scalability, providing good adaptability to different situations and uses, the possibility of deploying applications and work with data that can be shared between clouds, improving the performance of the workload. The private part of the hybrid cloud can be deployed on Marenostrum and ODH, as well as in the on-premise computer cluster of any SME. On the other hand, the MANGO cluster is an FPGA-specific computing infrastructure that is being used to evaluate some use cases of the DeepHealth project.

It is worth noting that these infrastructures offer a wide range of computing environments at different levels:

- High number of CPU computing nodes on Marenostrum, multi-GPUs nodes on the ODH platform and FPGAs in the MANGO cluster.

- The private OpenStack cloud implemented in ODH (HPC4AI), and the hybrid private+public cloud provided by TREE Technology.
- Docker containers technology used on top of bare metal layer in ODH (C3S) and in cloud platforms, using orchestration tools like Kubernetes, in TREE Technology platform and ODH (HPC4AI), and StreamFlow, which is described in Sect. 6.3.

6.1 Use of Docker Images Orchestrated by Kubernetes

The hybrid cloud platform provided by TREE Technology is a computing environment that offers the possibility of combining public and private clouds, allowing the deployment of applications and work with data that can be shared between them. This solution was built using Kubernetes technology [33], a distributed container and microservice platform that orchestrates computing, networking and storage infrastructure to support user workloads.

Software containers demonstrated to provide a good way to bundle and deploy applications. However, as system complexity increases (e.g., complex multi-component software applications, multi-node clusters) running deployments become increasingly difficult. Kubernetes supports the automation of much of the work required to maintain and operate such complex services in a distributed environment. The objective of this hybrid environment is to dynamically operate in different Kubernetes clusters running on several public clouds and on-premise infrastructures. Different Kubernetes clusters can have different hardware configurations, that is, they can have different memory and CPU settings, with or without GPUs. Once the different clusters are deployed, both in public and private clouds, it is necessary to orchestrate all the resources. For this ecosystem to work properly and be able to be coupled in the global scheme, two stages must be taken into account:

- Within a multi-cloud or hybrid-cloud context, a tool is needed to facilitate management and security tasks, as this can become a highly error prone and tedious task, while resources and Kubernetes clusters grow.
- A high-level RESTful API helps to abstract the user from the infrastructure itself, simplifying and speeding up the deployment and management of the workflows. It provides functions of varying complexity, which implements functionality abstracting the user from the potentially complex configuration of the clusters (e.g., multi-cloud, hybrid cloud, etc.). The API itself can support the addition of new Kubernetes clusters both on-premise and in the cloud from any provider with the limitation of having a minimum Kubernetes version.

The proposed hybrid cloud based on Kubernetes is a complex system, and its scalability is determined by several factors, like the number and type of nodes in a pool of nodes, the number of Pods available (Pods are the minimum deployable computer units that can be created and managed in Kubernetes), the number of services or back ends behind a service and how resources are allocated. Usually, in a public cloud, the concept of autoscaling is available, which refers to the possibility of scaling the resources of a cluster in a self-managed manner. In the DeepHealth project, the public cloud has been configured with this autoscaling option, while for the private cloud there is no scaling policy in relation to machines.

Concerning the automatic deployment of the DeepHealth toolkit in any cloud configuration, and regardless of the complexity level of the computing infrastructures that any development team of AI-based solutions may have on hand, a set of pre-built CUDA-enabled DeepHealth Docker images, including all the components of the toolkit, are ready to be used on GPU-enabled computing resources to accelerate compute-intensive operations. All the DeepHealth Docker images available in the DockerHub [24] are CUDA-enabled and provide pre-built binaries of the libraries along with all their dependencies, such that these images can be used to create Docker-ready applications. In addition, a *toolkit* image flavour is also provided to support the developers of applications/platforms directly integrating the EDDL and the ECVL, who may prefer the C++ or Python API. These Docker images are built on the *devel* flavour of the NVIDIA/CUDA images, and add a full DeepHealth build configuration to provide a ready-to-use compilation environment for applications.

For simplified scalable deployments on cloud computing resources, a Kubernetes [33] deployment of the DeepHealth toolkit, with the web service configured, has been created and made available. The deployment automatically configures a server for the DeepHealth front end and all the back-end components (i.e., web service, worker, database, job queue, and static content server) in a flexible and scalable way. In fact, once a deployment is created, the available processing capacity can be dynamically scaled using some of the standard features of Kubernetes, such as configuring the required number of worker replicas to achieve the required throughput. The Kubernetes deployment of the DeepHealth toolkit is packaged as a Helm chart for easy deployment [35]. For simpler use cases that do not have particular scalability requirements, a Docker-compose deployment is also available. This configuration cannot distribute work over multiple nodes, but it can be trivially deployed on a single node and thus is well suited for small workloads and exploratory or development work.

6.2 COMPSs

COMPSs [6] offers a portable programming environment based on a task model, whose main objective is to facilitate the parallelization of sequential source code, written in Java or Python programming languages, to run in a distributed and heterogeneous computing environment. In COMPSs, the programmer is responsible for identifying the units of parallelism (named COMPSs tasks) and the synchronization data dependencies existing among them by annotating the sequential source code. The task-based programming model of COMPSs is then supported by its runtime system, which manages several aspects of the application execution and keeps the underlying infrastructure transparent to the programmer. This is a key feature to guarantee the portability of COMPSs applications across a wide range of computing platforms. This will allow the DeepHealth toolkit to be tested and validated within the DeepHealth project in the infrastructures enumerated above. Regarding cloud configurations, the COMPSs runtime is being adapted within the DeepHealth project to support the hybrid cloud infrastructure. COMPSs runtime interacts with the API developed by TREE Technology to deploy workers and distribute the workload on hybrid cloud architectures. The COMPSs runtime is organized as a master-worker structure:

- The Master, executed in the computing resource where the application is launched, is responsible for steering the distribution of the application and data management.
- The worker(s), co-located with the Master or in remote computing resources, are in charge of responding to task execution requests coming from the Master.

One key aspect is that the master maintains the internal representation of a COMPSs application as a Directed Acyclic Graph (DAG) to express the parallelism. Each node corresponds to a COMPSs task and edges represent data dependencies (and so potential data transfers). Based on this DAG, the runtime can automatically detect data dependencies between COMPSs tasks: as soon as a task becomes ready (i.e., when all its data dependencies are resolved), the master is in charge of distributing it among the available workers, transferring the input parameters before starting the execution. When the COMPSs task is completed, the result is either transferred to the worker in which the destination COMPSs task executes (as indicated in the DAG), or transferred to the master if a barrier synchronization call is invoked. The parallelization of the EDDL training operation has been developed with the COMPSs tasking programming model. Due to the fine grain data dependency synchronization mechanisms supported by COMPSs, two parallel training paradigms are supported: *synchronous*, in which weights are collected and aggregated at the end of each epoch, and *asynchronous*, in which weights are increasingly aggregated as soon as a partial training is completed on the corresponding data set.

COMPSs is perfectly adapted to run in environments managed by SLURM [32], an open-source resource manager widely used in High-Performance Computing data centres to manage job queues and job allocation of incoming tasks to servers. The Ecole Polytechnique Fédérale de Lausanne (EPFL) has enhanced the core version of the SLURM resource manager with novel plugins that enable energy- and performance-aware task allocation for CPU- and memory-intensive tasks in order to increase the efficiency (in terms of performance per watt) of multiple tasks when running simultaneously on the same server and cluster. EPFL do so by proposing the use of graph-based techniques and reinforcement learning, which are low overhead and do not impact the execution time of applications. SLURM can interact with COMPSs in order to launch multiple instances of applications in a coordinated way in an HPC infrastructure, creating a separation of concerns between resource managers, while still working in a coordinated way.

6.3 StreamFlow

StreamFlow is a novel Workflow Management System (WMS) explicitly designed in the DeepHealth project, supporting AI pipeline design and execution in different execution environments, including hybrid HPC + cloud and multi-cloud infrastructures. The portability of AI pipelines on critical data across different infrastructures is crucial for the sustainability of DeepHealth foreground technologies. To address this issue, *OpenDeepHealth* embraces StreamFlow. The ability of StreamFlow to handle sequences of computational steps makes it possible to describe a complex application as a workflow and annotate each step with an execution plan potentially targeting different nodes, e.g., selecting GPU nodes when needed, spawning across multiple sites—e.g., allowing transparent access to OCCAM and HPC4AI clusters. The idea behind this approach is that the ability to deal with hybrid workflows (i.e., to coordinate tasks running on different execution environments) can be a crucial aspect for performance optimization when working with massive amounts of input data and different needs in computational steps. Accelerators like GPUs and, in turn, different infrastructures like HPC and clouds, can be used more efficiently by selecting the execution plan that best suits the specific computational needs of each ML application developed in the project.

The StreamFlow framework is a container-native WMS written in Python. It has been designed to explore the potential benefits deriving from waiving two common properties of existing WMSs that can prevent them from fully exploiting the potential of containerization technologies. Instead of forcing a one-to-one mapping between workflow steps and Docker containers, StreamFlow allows the execution of tasks in potentially complex, multi-container environments. This allows support for concurrent execution of multiple communicating tasks in a multi-agent ecosystem, e.g., a SPMD application implemented with MPI or a COMPSs-based distributed training. StreamFlow relaxes the requirement of a single shared data space among all the worker nodes, allowing to spread different steps of a single workflow on multiple, potentially federated architectures without forcing direct communication channels among them. Moreover, StreamFlow clearly separates the definition of the AI pipeline, described as a declarative workflow, from the description of the runtime environment in charge of executing it, enforcing a separation of concerns. This allows taking advantage of using the most efficient infrastructures for the specific purpose of complex AI pipelines without burdening the AI experts with the configuration and management complexity of such infrastructures. At a very high level, an AI pipeline can comprise a training step, which usually requires very high computational power and distributed programming techniques to handle huge datasets, and an inference step, in which a fully trained model should be directly reachable from one or more user applications. StreamFlow can orchestrate the execution of the AI pipeline, targeting the training step on HPC facilities, e.g., by using EDDL with COMPSs distributed runtime, and the inference step on the cloud cluster, e.g., by leveraging EDDL Docker containers deployed on a Kubernetes infrastructure.

7   Conclusions

The DeepHealth toolkit is presented here as a new and emerging software framework that provides European industry and research institutions with deep learning and computer vision functionalities. To cope with huge and constantly growing data sets, the toolkit has been designed to leverage hybrid and heterogeneous HPC + cloud architectures in which either all or some of the worker nodes are equipped with hardware accelerators (e.g., GPUs, FPGAs). The distributed execution of learning and inference processes is done by the runtime of the DeepHealth toolkit in a transparent manner to the common user, i.e., computer and data scientists who do not need a deep understanding of parallel programming, HPC, deep learning or cloud architectures.

The two libraries at the core of the toolkit can be easily integrated into existing software applications/platforms that European companies (SMEs and large industry) have developed to deploy AI-based solutions in any sector (e.g., decision support systems that clinicians can use to diagnose), and can be used to boost the development of new platforms and solutions. All the components of the toolkit are free and open-source software available on public repositories.

In order to foster the use of the DeepHealth toolkit, the authors have introduced the potential user to all the toolkit components and how to integrate the libraries in existing or new software applications. It is worth mentioning that, thanks to pre-built Docker images including all components with all dependencies satisfied, data scientists only need to provision Docker containers according to their needs.

The toolkit constitutes a contribution from Europe in Artificial Intelligence and smart big data analytics. Besides all the features introduced in this chapter (free and open-source framework, easy to use, portable to different architectures, wide application scope), it contributes to reducing the bottlenecks in turning AI into an enabling technology for Science (e.g., provides a way to reduce the complexity of numerical methods used in scientific environments), bringing closer the separate worlds of AI and HPC. Furthermore, it is expected to boost the adoption of AI and HPC technologies by the industry. The toolkit paves the way towards the offering of AI coupled with HPC as a service, which could be a game changer aspect in order to reach a greater number of companies. On the one hand, it offers improvements for companies that only have temporary needs for high-performance computing resources, which will be able to improve their productivity by developing their own AI solutions, and on the other hand, it could unlock the development of novel applications that need to run computationally intensive processes regularly.

Acknowledgments This chapter describes work undertaken in the context of the DeepHealth project, "Deep-Learning and HPC to Boost Biomedical Applications for Health", which has received funding from the European Union's Horizon 2020 research and innovation programme under grant agreement No 825111.

References

1. EIT Health; McKinsey & Company, 2020 [Online]. Accessed November 11, 2020, from https://eithealth.eu/wp-content/uploads/2020/03/EIT-Health-and-McKinsey_Transforming-Healthcare-with-AI.pdf
2. The DeepHealth project [Online]. Accessed October 28, 2020, from https://deephealth-project.eu
3. Zillner, S., Curry, E., Metzger, A., Auer, S., & Seidl, R. (2017). *European big data value strategic research & innovation agenda*. Big Data Value Association.
4. Zillner, S., Bisset, D., Milano, M., Curry, E., García Robles, A., Hahn, T., Irgens, M., Lafrenz, R., Liepert, B., O'Sullivan, B., & Smeulders, A. (2020). Strategic research, innovation and deployment agenda – AI, data and robotics partnership. Third Release, 9 2020. [Online]. Accessed November 30, 2020, from https://ai-data-robotics-partnership.eu/wp-content/uploads/2020/09/AI-Data-Robotics-Partnership-SRIDA-V3.0.pdf
5. The DeepHealth GitHub Organisation [Online]. Accessed January 5, 2021, from https://github.com/deephealthproject
6. Lordan, F., Tejedor, E., Ejarque, J., Rafanell, R., Álvarez, J., Marozzo, F., Lezzi, D., Sirvent, R., Talia, D., & Badia, R. M. (2014). ServiceSs: An interoperable programming framework for the cloud. *Journal of Grid Computing, 12*(1), 67–91.
7. Colonnelli, I., Cantalupo, B., Merelli, I., & Aldinucci, M. (2020). StreamFlow: Cross-breeding cloud with HPC. *IEEE Transactions on Emerging Topics in Computing*.
8. Open Neural Network Exchange. The open standard for machine learning interoperability [Online]. Accessed October 31, 2020, from https://onnx.ai/
9. Harris, C. R., et al. (2020). Array programming with NumPy. *Nature, 585*(7825), 357–362.
10. McKinney, W., et al. (2010). Data structures for statistical computing in python. In *Proceedings of the 9th Python in Science Conference*, Austin, TX.
11. Guennebaud, G., Jacob, B., et al. (2010). Eigen v3 [Online]. Accessed November 2, 2020, from http://eigen.tuxfamily.org
12. Dagum, L., & Menon, R. (1998). OpenMP: An industry standard API for shared-memory programming. *Computational Science & Engineering, 5*(1), 46–55.
13. Abadi, M., et al. (2016). Tensorflow: A system for large-scale machine learning. In *12th USENIX Symposium on Operating Systems Design and Implementation (OSDI 16)*.
14. Paszke, A., et al. (2019). PyTorch: An imperative style, high-performance deep learning library. In *Advances in neural information processing systems*.
15. Tartaglione, E., Lepsøy, S., Fiandrotti, A., & Francini, G. (2018). Learning sparse neural networks via sensitivity-driven regularization. In *Advances in neural information processing systems*.
16. Frankle, J., & Carbin, M. (2019). *The lottery ticket hypothesis: Finding sparse, trainable neural networks*.
17. Molchanov, D., Ashukha, A., & Vetrov, D. (2017). Variational dropout sparsifies deep neural networks. In *34th International Conference on Machine Learning, ICML*.
18. Tartaglione, M., Bragagnolo, A., & Grangetto, M. (2020). Pruning artificial neural networks: A way to find well-generalizing, high-entropy sharp minima. In *Artificial neural networks and machine learning – ICANN2020*.
19. Tartaglione, E., & Grangetto, M. (2019). Take a ramble into solution spaces for classification problems in neural networks. In *International conference on image analysis and processing*.

20. Simonyan, K., & Zisserman, A. (2014). Very deep convolutional networks for large-scale image recognition. arXiv preprint arXiv:1409.1556.
21. Ronneberger, O., Fischer, P., & Brox, T. (2015). U-net: Convolutional networks for biomedical image segmentation. In International conference on medical image computing and computer-assisted intervention.
22. Anaconda Software Distribution. Computer software. [Online]. Accessed November 2, 2020, from https://anaconda.com; https://docs.conda.io
23. Homebrew. The Missing Package Manager for macOS (or Linux) [Online]. Accessed October 31, 2020, from https://brew.sh
24. Docker hub with DeepHealth images [Online]. Accessed October 28, 2020, from https://hub.docker.com/orgs/dhealth
25. Open Source Computer Vision Library (OpenCV) [Online]. Accessed November 3, 2020, from https://github.com/itseez/opencv
26. Allegretti, S., Bolelli, F., & Grana, C. (2020). A warp speed chain-code algorithm based on binary decision trees. In *4th International Conference on Imaging, Vision & Pattern Recognition.*
27. Bolelli, F., & Grana, C. (2019). Improving the performance of thinning algorithms with directed rooted acyclic graphs. In *International conference on image analysis and processing.*
28. Bolelli, F., Allegretti, S., Baraldi, L., & Grana, C. (2019). Spaghetti labeling: Directed acyclic graphs for block-based connected components labeling. *IEEE Transactions on Image Processing, 29*(1), 1999–2012.
29. Cancilla, M., et al. (2021). The DeepHealth toolkit: A unified framework to boost biomedical applications. In *25th IEEE International Conference on Pattern Recognition*, 2021.
30. wxWidgets: Cross-Platform GUI Library [Online]. Accessed November 2, 2020, from https://www.wxwidgets.org
31. YAML Ain't Markup Language [Online]. Accessed November 2, 2020, from https://yaml.org/
32. Yoo, A., Jette, M., & Grondona, M. (2003). Slurm: Simple linux utility for resource management. In *Job scheduling strategies for parallel processing*. Berlin.
33. Kubernetes [Online]. Accessed November 11, 2020, from https://kubernetes.io
34. Aldinucci, M., et al. (2018). HPC4AI, an AI-on-demand federated platform endeavour. In *15th ACM International Conference on Computing Frontiers*, Ischia, 2018.
35. DeepHealth Helm chart repository [Online]. Accessed November 12, 2020, from https://deephealthproject.github.io/helm-charts/index.yaml

Applying AI to Manage Acute and Chronic Clinical Condition

Rachael Hagan, Charles J. Gillan, and Murali Shyamsundar

Abstract Computer systems deployed in hospital environments, particularly physiological and biochemical real-time monitoring of patients in an Intensive Care Unit (ICU) environment, routinely collect a large volume of data that can hold very useful information. However, the vast majority are either not stored and lost forever or are stored in digital archives and seldom re-examined. In recent years, there has been extensive work carried out by researchers utilizing Machine Learning (ML) and Artificial Intelligence (AI) techniques on these data streams, to predict and prevent disease states. Such work aims to improve patient outcomes, to decrease mortality rates and decrease hospital stays, and, more generally, to decrease healthcare costs.

This chapter reviews the state of the art in that field and reports on our own current research, with practicing clinicians, on improving ventilator weaning protocols and lung protective ventilation, using ML and AI methodologies for decision support, including but not limited to Neural Networks and Decision Trees. The chapter considers both the clinical and Computer Science aspects of the field. In addition, we look to the future and report how physiological data holds clinically important information to aid in decision support in the wider hospital environment.

Keywords Healthcare · Predictive analytics · Artificial Intelligence · Clinical decision support

The chapter relates to the technical priorities Data Analytics of the European Big Data Value Strategic Research and Innovation Agenda [1]. It addresses the horizontal concern Data Analytics of the BDV Technical Reference Model and addresses the vertical concerns of Big Data Types and Semantics. The chapter further relates to the Reasoning and Decision Making cross-sectorial technology enablers of the AI, Data and Robotics Strategic Research, Innovation and Deployment Agenda [2].

R. Hagan (✉) · C. J. Gillan · M. Shyamsundar
Queen's University Belfast, Belfast, Northern Ireland, UK
e-mail: rhagan09@qub.ac.uk

© The Author(s) 2022
E. Curry et al. (eds.), *Technologies and Applications for Big Data Value*,
https://doi.org/10.1007/978-3-030-78307-5_10

1 Overview

The chapter begins by discussing Intensive Care medicine and the types of machines and data that is being recorded continuously, and thus producing 'Big Data'. It proceeds to explain some of the challenges that can arise when working with such data, including measurement errors and bias. The subsequent section explains some of the common methodologies used to provide predictive analytics with examples given for both acute and chronic clinical conditions, and discuss our own work for the promotion of lung protective ventilation, to highlight the accuracies that can be achieved when pairing health 'Big Data' with common machine learning methodologies. The chapter concludes by discussing the future of this field and how we, as a society, can provide value to our healthcare systems by utilizing the routinely collected data at our disposal.

2 Intensive Care Medicine and Physiological Data

Intensive Care Units (ICU) offer expensive and labor-intensive treatments for the critically ill and are therefore a costly resource for the health sector around the world. They can also be referred to as Intensive Therapy Units (ITU) or Critical Care Units (CCU). UK estimates in 2007 highlighted that intensive care in the NHS costs £719 million per year [3]. Comparably, American studies reported in 2000 have shown that median costs per ICU stay can range between $10,916 and $70,501 depending on the length of stay [4]. A typical length of stay varies depending on the condition of the patient with studies showing the mean length of stay being 5.04 days [3], while the condition of the patient can change quickly and sometimes unpredictably.

Patients will normally be admitted to intensive care after a serious accident, a serious condition such as a stroke, an infection or for surgical recovery. Throughout their stay in ICU, these patients are monitored closely due to their critical condition, and on average require one nurse for every one or two patients. Many devices and tests may be used to ensure the correct level of care is provided. Patients will be placed on machines to monitor their condition, support organ function and allow for the detection of any improvements or deterioration. The functions of these machines can vary: from the monitoring of vital parameters such as heart rate via the patient monitor to the use of mechanical ventilators that provide respiratory function when a patient is not able to do so themselves.

The workload on the clinical staff in ICU is intense and so utilizing Big Data analytics will allow for healthcare providers to improve their efficiency through better management of resources, detection of decompensation and adverse events, and treatment optimization, among many benefits to both patient outcomes and hospital costs [5].

2.1 *Physiological Data Acquisition*

Health data recorded in ICU can be classified as 'Big Data' due to the volume, complexity, diversity and timeliness of the parameters [6], the aim being to turn these 'Big Data' records into a valuable asset in the healthcare industry.

As highlighted, patients requiring critical care are placed on numerous monitors and machines to help with and provide their care. Equipment can include, but is not limited to [7]:

- Patient monitoring system to monitor clinical parameters such as electrocardiogram (ECG), peripheral oxygen saturation, blood pressure, temperature.
- Organ support systems such as mechanical ventilator, extracorporeal organ support such as continuous renal replacement therapy.
- Syringe pump for the delivery of medicines.

These machines are all monitoring continuously (24×7) and thus representing one example of the emerging field of Big Data. It is important to wean patients off these machines as quickly as possible to avoid dependency and to lower the risk of infection. In addition to the organ support machines, the electronic health record includes laboratory data, imaging reports such as X-rays and CT scans, and daily review record.

2.1.1 Time Series Data

The human physiologic state is a time-dependent picture of the functions and mechanisms that define life and is amenable to mathematical modeling and data analysis. Physiology involves processes operating across a range of time scales resulting in different granularities. Parameters such as heart rate and brain activity are monitored on the millisecond level while others such as breathing have time windows over minutes and blood glucose regulation over hours. Analysis of instantaneous values in these circumstances is rarely of value. On the other hand, application of analytical techniques for time series offers the opportunity to investigate both the trends in individual physiological variables and the temporal correlation between them, thereby enabling the possibility to make predictions.

Features can also be extracted from these time series using packages such as the Python *tsfresh* software toolkit to use as input into Machine Learning models and to gain further insights into the relationship between the parameter and time [8]. By analyzing the time series we can make predictions on the future trajectory and alert care givers of possible issues in order to prevent complications.

Prior research has shown that these streams of data have very useful information buried in them, yet in most medical institutions today, the vast majority of the data collected is either dropped and lost forever, or is stored in digital archives and seldom reexamined [9].

In the past number of years there has been a rapid implementation of Electronic health records, or EHRs, around the world. EHRs are a common practice to record and store real-time, patient health data, enabling authorized users to track a patient from initial admission into the hospital, their deterioration/improvement, diagnoses, and all physiological parameters monitored and drugs given across all healthcare systems.

2.1.2 Publicly Available Datasets

As part of a global effort to improve healthcare, numerous institutions have put together publicly available data sets based on their EHRs for people to use in research; enabling visualization, analysis, and model development.

One of the best-known and commonly used publicly available databases is MIMIC, the Multiparameter Intelligent Monitoring in Intensive Care database. Produced from the critical care units of the Beth Deaconess Medical Centre at Harvard Medical School, Boston, this large, freely available database consists of de-identified health-related data associated with over 40,000 patients who were in critical care between 2001 and 2012 [10]. After being approved access, users are provided with vital signs, medications, laboratory measurements, observations, and charted notes. Furthermore, waveforms are available for use, along with patient demographics and diagnoses.

PhysioNet offers access to a large collection of health data and related open-source software, including the MIMIC databases [11]. All data recorded in these publically available databases are anonymous, ensuring no patient can be identified from their data.

3   Artificial Intelligence in ICU

The broad field of Data Science has emerged within the discipline of Computer Science over the past 10 years, approximately. Its roots arguably lie in the work of Fayyad et al. [12], which defined a pipelined process for the extraction of abstract models from data using an amalgam of techniques, including Statistics and Machine Learning. Artificial Intelligence (AI) is the ability of a computer system to perform tasks that aim to reach the level of intelligent thinking of the human brain through applying machine learning methodologies, further details are discussed in the methodologies section.

3.1   Challenges

As in majority of real-world applications, there comes a series of challenges that arise when applying Big Data-driven analytics to such sensitive and intense data. We must aim to build trust and quality between the decision support tools and the end users. It is always important to question if there is real clinical impact from carrying out such work and exploring AI methodologies for solving particular problems.

3.1.1 Data Integrity

Before building any Big Data-driven analytic system, it is important to ensure data has been collected and preprocessed correctly, with any errors handled accurately. It is crucial to ensure data harmonization, the process of bringing together all data formats from the different machines and tests into one database. Without this step, the algorithms will produce unreliable results, which in turn could result in a critically dangerous implementation and a lack of trust in the system. Errors in the data can be due to a variety of reasons:

- Data input errors: In intensive care, caregivers often have to input readings on the system. Fields will have a required metric in which the data should be entered and these may sometimes be ignored or mishandled. For example, height that should be entered in centimeters but a user may enter the reading in meters, e.g., 1.7 m instead of the 170 cm.
- Sensor errors: With the complexity and multitude of monitors, leads, and machines that a patient can be on at any given time, the sensors can sometimes fail or miss a reading. Sensors can be disconnected for a period of time to deliver medicines or for imaging tests. Patients' movements in the bed can cause a sensor error or unexpected result. These errors will present as outliers in the data and should be dealt with accordingly as to not throw off any predictive modeling.
- Bias in the data: AI methodologies are only as good as the data of which they are trained on. If this data contains, e.g., racial or gender biases, the algorithm will learn from this and produce similar results such as women not being given the same number of tests as men. Similarly, statistical biases can be present due to small sample numbers from underrepresented groups, for instance, an algorithm only being trained with White patients may not pick up the same diagnosis when presented with Hispanic patients [13]. Furthermore, selection biases exist when the selection of study participants is not a true representation of the true population, resulting in both cultural and social differences.
- Bias in the study: Studies can also contain information bias. These include measurement errors for continuous variables and misclassification for categorical variables. For example, a tumor stage being misdiagnosed would lead to algorithms being trained on incorrect data [14].

We must ensure to implement appropriate techniques such as imputation, smoothing and oversampling to prevent errors in our data and build trust with the user.

3.1.2 Alert Fatigue

When building decision support tools, it is crucial to ensure the integrity of alerts raised. Too many can result in alert fatigue, leading to alarms being switched off or ignored. Research has shown that there is on average 10.6 alarms per hour, too many for a single person to handle amidst their already busy work schedule [15].

In addition, 16% of health care professionals will switch off an alarm and 40.4% will ignore or silence the alarm [16].

This influx of alarms resulting in the monitoring system being turned off will lead to vital physiological problems being missed. We need to be confident that our systems are only producing true alerts and that false alerts generated are minimized.

It has further been highlighted that reducing the number of alerts repeated per patient will help reduce override rates and fatigue [17].

3.1.3 Bringing AI Systems to Clinical Trials

While we know these AI models, derived from Big Data-driven predictive analytics, can provide value to the healthcare industry, we must be able to test them in real time in order for them to be implemented and used throughout the world. This requires a clinical trial to be carried out using the system on patients in real time. While there exist the SPRIRT 2013 and CONSORT 2010 checklists for clinical trials, covering checklists for what you intended to do and what you actually did, respectively, neither of these include steps for AI implementation [18].

With the rise of AI, these guidelines have been extended in 2020 to include steps for reporting the quality of data, errors, clinical context, intended use, and any human interactions involved. These additional steps allow reviewers to better evaluate and compare the quality of research and thus systems created in the future [19, 20]. Other authors are further expanding these checklists to include reporting for diagnostic accuracies and prediction models [21, 22].

Researchers have analyzed the current work that has been published in this area [23]. With 93% of papers exploring model development to demonstrate potential for such systems, and only 5% validating these models with data from other centers. A further 1% of the work currently published reports on models that have actually been implemented and tested in real-time hospital systems and the final 1% of research have integrated their models in routine clinical practice which has proven to work. This summary highlights the large potential for the future of AI in healthcare systems, with huge opportunities for bringing forward accurate models to real-world clinical trials.

3.2 AI Methodology

Machine Learning methodologies are commonly used to enable machines to become 'intelligent' and can be classified as being supervised, predicting using a labeled dataset, i.e., a known output or target, or unsupervised, i.e., finding patterns or groupings in unlabeled data [24]. Common methodologies are utilized across the board for predictive and analytical purposes. This chapter focuses on commonly used supervised learning techniques; however, unsupervised methods can further be used to understand data. We can categorize supervised learning techniques into regression and classification models. Regression techniques aim to find relationships between one dependent variable and a series of other independent variables, e.g., a time series as previously discussed is common in physiological data. Classification techniques on the other hand attempt to label outcomes and draw conclusions from observed values, e.g., if patient has disease or not.

The question of whether the numerical models that are generated can actually be understood by humans has become a hot research topic. In the United States, DARPA has conducted significant investigative research in this field [25]. Other research teams have begun to define the concept of explainable AI with respect to several problem domains. They identify three classifications: opaque systems that offer no insight; interpretable systems where mathematical analysis of the algorithm is viable; and comprehensible systems that emit symbols enabling user-driven explanations of how a conclusion is reached [26].

We know from previous research that utilizing Big Data within the ICU can lead to many benefits to both patient and hospital. We can not only greatly improve the care given and thus patient outcomes, see Table 1, but also reduce hospital costs [5] and the stress levels of care givers [27]. McGregor and colleagues have demonstrated the viability of monitoring physiological parameters to detect sleep apnea in neo-natal ICU leading to the software architecture of the IBM InfoSphere product, which has now been extended into a Cloud environment (the Artemis project) making the functionality available at remote hospital sites [28, 29].

Patient deterioration can often be missed due to the multitude of and complicated relational links between physiological parameters. AI-driven multivariate analysis has the potential to ameliorate the work load of ICU staff. Multiple studies have shown AI to be comparable to routine clinical decision making, including ECG analysis, delirium detection, sedation, and identification of septic patients [28, 37–39].

AI-driven predictive analytics within healthcare most commonly uses supervised learning approaches, due to which we aim to base algorithms and decisions on previous examples and train our models with accurate outcomes, in particular regression analysis is used for time series data.

Below we review the more widely adopted Machine Learning methodologies developed [31], and examples are highlighted in Table 1, where these have been used in previous research.

Table 1 Highlighting applications of AI in healthcare

Area	Example	AI methodologies	Limitations
Mortality rates	Prediction of postoperative 30-day mortality [38]	CNN	Generalizability, completeness and accuracy of data
Deterioration	Prediction of patients at risk for deterioration within the next 6 h [39]	Random Forest	False alarms
Disease prediction	Sepsis prediction [40–42]	Gradient boosting Random forest Rules-based system	Generalization, explainability, data quality
Disease prediction	Prediction of kidney failure [35]	RNN	Unbalanced dataset
Data visualization	Display of optimum ventilator parameters to ensure compliance [43]	Regression	Lack of complete adaption
Patient outcomes	Prediction of ICU outcome based on data collected in first 48 h of admission [44]	Expert system ANN	Unbalanced dataset
Patient outcomes	Interpretable deep models for ICU prediction of acute lung injury (ALI) [45]	Gradient boosting	Generalization
Mortality	Prediction of mortality from respiratory distress among long-term mechanically ventilated patients [46]	Regression Decision trees	Limited data
Diagnosis	Comparison of classification models for the diagnosis of asthma [47]	Ensemble models ANN	Small dataset, Noisy waveforms
Risk prediction	System for predicting the risk of heart attacks [48]	Neural network Decision tree	Small dataset, generalization
Risk prediction	Predicting patients at risk for prolonged ventilation [49]	Gradient boosting decision trees	Information bias
Disease prediction	Early prediction of circulatory failure in ICU [50]	Gradient-boosted ensembles	False alarms, generalization
Outcome prediction	ECG analysis to predict atrial fibrillation [36]	CNN	Information bias
Complication prediction	Prediction of several severe complications post open heart surgery [51]	RNN	Generalization, Noisy data

3.2.1 Expert Systems

In a rule-based expert system, also known as a knowledge-based clinical decision support system, knowledge is represented by a series of IF-THEN rules that are created by knowledge acquisition, which involves observing and interviewing human experts and finalizing rules in format 'IF X happens THEN do Y'. These rules cannot be changed, learnt from or adapted to different environments, meaning the human experts must manually monitor and modify the knowledge base through careful management of the rules. Expert systems allow us to view and understand each of the rules applied by the system.

The systems can take over some mundane decision tasks and discussions of health care professionals, saving vital time and money. An example expert system was created for Diabetes patients to provide decision support for insulin adjustment based on simple rules depending on the regimen that patients were placed on [32].

3.2.2 Decision Trees

A decision tree can be used to visually represent a series of decisions used to reach a certain conclusion, useful when exploring medical reasons behind decisions made. The tree starts at the root, asking a specific question to split the data by the given condition. The tree then splits into branches and edges at each node representing the input variables, continually splitting by conditions until the final decision is achieved at the leaf node or the output variable.

They can also be referred to as Classification and Regression Trees (CART) as they can solve both classification and regression problems. A classification tree will arrive at a binary condition, leaf node, i.e., patient survives or not, whereas the regression trees will predict a certain continuous value, i.e., the heart rate of a patient.

The tree will not only explore the conditions used to split the data at each decision but also the features used and which features are the most significant at splitting the data, added as top-level nodes. Researchers have utilized simple decision tree models for the classification of patients with diabetes, among other disease states. Features include age, BMI, and both systolic and diastolic blood pressure of the patient to arrive at a decision whether the patient has diabetes or is healthy. Figure 1 shows how the features are split and decisions are made, in this circumstance [33].

When building decision tree models it is important to monitor the maximum depth of the tree to avoid overfitting and lengthy training times.

3.2.3 Ensemble Methods

To achieve the greatest predictive performance when working with complex problems, ensembles can be created. The decision trees are combined in different ways

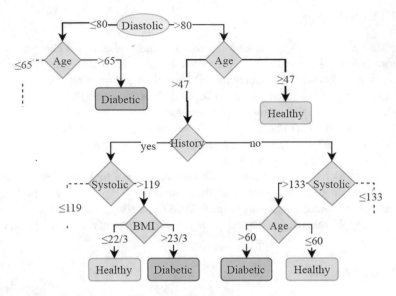

Fig. 1 Decision tree for the classification of diabetes [20]

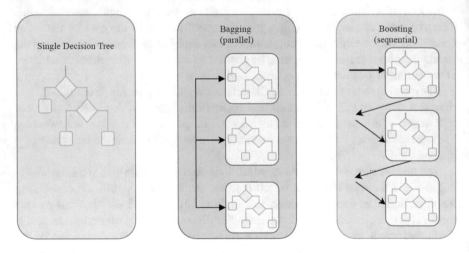

Fig. 2 Decision trees and ensemble methods

depending on the methodology used. The methods can be categorized as bagging or boosting (Fig. 2).

Boosting methods build the trees in a sequential way; for each predicted value multiple models or decision trees are made using different features and combinations of features, then weights are given to these models based on their error so that the final prediction is the most accurate. AdaBoost is an example of this method where very short trees are produced and higher weights are given to more difficult decisions. GradientBoosting further combines boosting with gradient descent, allowing for the optimization of loss functions.

In contrast, for bagging methods, each model is given equal weights, they are combined in parallel and all of the predictions are averaged to get a final, most accurate decision, examples include Bagging and Extra Trees. Bagging can be extended to the RandomForest algorithm by randomly selecting the features used, and decision trees are built to have as many layers as possible.

3.2.4 Neural Networks

A neural network saves a lot of time when working with large amounts of data by combining variables, figuring out which are important and finding patterns that humans might not ever see. The neural network is represented as a set of interconnected nodes, connected by neurons.

They feed the weighted sum of the input values through an activation function, which takes the value and transforms it before returning an output. These activation functions in turn improve the way the neural network learns and allows for more flexibility to model complex relationships between the input and output (Fig. 3).

The neural network can be described as 'Deep Learning' when it has multiple hidden layers between the input and output layers. The neural network learns by

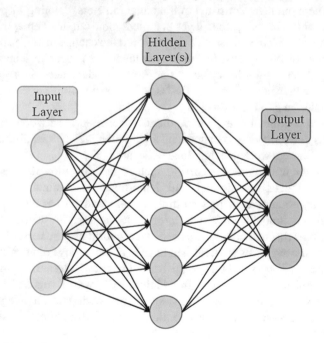

Fig. 3 Neural network structure

figuring out what it got wrong and working backwards to determine what values and connections made the prediction incorrect.

Additionally, there are different types of neural networks, and the list continues to expand as researchers propose new types. There are feedforward neural networks, where all the nodes only feed into the next layer from initial inputs to the output. Recurrent neural networks (RNN) make it possible to feed the output of a node back into the model as an input the next time you run it. Nodes in one layer can be connected to each other and even themselves. Furthermore, they work well with sequential data as they remember previous outputs. The long short-term memory form of the RNN enables the model to store information over longer periods of time, ideal for modeling time series data. Additionally, there are convolutional neural networks (CNN) which look at windows of variables rather than one at a time. Convolution applies a filter, or transformation, to the windows to create features. When working with large databases, pooling is a technique that takes a huge number of variables to create a smaller number of features. The neural network will then use the features generated by convolution and pooling to give its output.

The key parameters to distinguish neural networks are the number of layers and the shape of the input and the output layers.

Researchers have utilized many different formats of neural networks for prediction problems. A medical decision support tool for patient extubation was developed using a multilayer perception artificial neural network, a class of feedforward neural networks. The input layer consisted of 8 input parameters, defining 17 perceptions, and 2 perceptions in the output layer for prediction output. Perceptions can be described as a classifying decision. They explored the change in performance based on the number of perceptions in the hidden layer: 19 producing highly accurate results [34]. Other studies have shown that RNNs produce accurate results for the prediction of kidney failure [35] and CNNs have shown promise in the prediction of atrial fibrillation in ECG analysis [36],

However, it is difficult to explain predictions from neural networks to healthcare professionals due to having to understand a particular weight as a discrete piece of knowledge, although work is being done in the area [37].

With common limitations reoccurring, such as generalization, dataset size, noisy and unbalanced data, we highlight the importance of continuing research and building of larger datasets, across multiple centers, and further the exploration of methodologies to smooth noisy data in order to advance the work of producing accurate systems that can be implemented in real-world healthcare settings.

Imaging and waveforms are a further, huge division of physiological monitoring. Machines such as X-rays and CTs can produce images of internal organs to provide a greater insight into patient state. In addition, ECG waveforms and brain waves can be analyzed for diagnoses. Thus, signal processing is an integral part of understanding physiological data and getting the full picture of patient state.

Furthermore, Machine Learning methodologies can be used for natural language processing to analyze lab notes and patient charts in order to summarize and detect common words and phrases used by care givers, in turn automating the process of flipping through hundreds of pages of records and saving vital time.

4 Use Case: Prediction of Tidal Volume to Promote Lung Protective Ventilation

Around 44.5% of patients in ICU are placed on mechanical ventilation to support respiratory function at any given hour [52]. However, the delivery of high tidal volume values can often lead to lung injury. Tidal volume being the amount of air delivered to the lungs; it is common knowledge amongst critical care providers that tidal volume values should be no greater than 8 ml per kg of ideal body weight. In our recent work we explored regressors for ensemble methods and the long short-term memory (LSTM) form of neural networks to predict tidal volume values to aid in lung protective ventilation [53].

Data acquisition took place at the Regional Intensive Care Unit (RICU), Royal Victoria Hospital, Belfast, over a 3-year period and the VILIAlert system was introduced [54]. The data streams were monitored against the thresholds for lung protective ventilation and if thresholds were breached continuously, an alert was raised. We then turned our attention to predicting these alerts with the aim of preventing violations and protecting the patient's lungs. The VILIAlert system ran for nearly 3 years, recording minutely tidal volume values for almost a thousand patients.

As discussed, noisy signals are common in ICU data. Time series often needs to be filtered to remove noise in the data and produce smooth signals. Methods such as moving average and exponential smoothing can be applied to the data to extract true signals, such as the work we carried out to extract true tidal volume trends [53]. Figure 4 shows how smoothing the time series, shown in blue, removes anomaly in the data, the large jumps, and extracts the true patient trend as shown in red. This work is related to the efforts of the international project known as the Virtual Physiological Human (http://www.vph-institute.org), which seeks to use individualized physiology-based computer simulations in all aspects of prevention, diagnosis, and treatment of diseases [55]. This computational approach has provided immediate insight into the COVID-19 pandemic [56].

We compare multiple regressor ensemble methods for initially predicting 15 min ahead. For each patient, we use the *tsfresh* toolkit to extract features to use as input into the regressor models in order to predict one time bin ahead and report the RMSE between the true observed values and the predicted values from our models. Table 2 reports RMSE calculated for 8 of the patients using each of the ensemble methods. In all models the maximum number of trees is set to 10. We can compare the depth of the bagging method trees: RandomForest being 32 ± 5, ExtraTrees being 37 ± 4, and Bagging being 33 ± 5. In contrast, the boosting methods: AdaBoost and GradientBoosting, set trees of depth four by default. As expected, increasing the number of trees decreases the RMSE.

It is important to take computational time into consideration when choosing algorithms to make predictions in real time. From our experiments we found AdaBoost to give the best trade-off between RMSE and computation time.

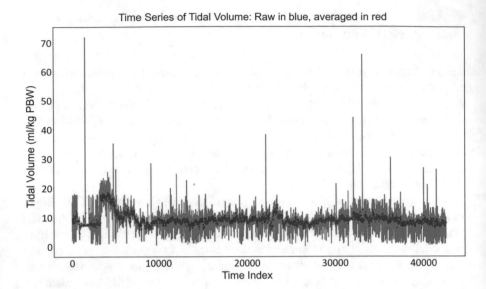

Fig. 4 Time series of a patient's tidal volume profile. Raw minutely values in blue, 15-min averaged bins in red

Table 2 Comparison of regressor ensemble models performance for the prediction of patient's tidal volume one time step ahead

		AdaBoost	RandomForest	Bagging	ExtraTrees	GradientBoosting
Patient	No. data points	RMSE	RMSE	RMSE	RMSE	RMSE
1	517	0.69	0.68	0.70	0.68	0.84
2	150	1.05	1.03	1.03	1.05	1.23
3	1358	0.38	0.38	0.38	0.36	0.39
4	40	1.05	0.99	1.02	1.00	0.95
5	162	0.34	0.32	0.32	0.34	0.41
6	178	1.38	1.28	1.35	1.39	1.32
7	1153	0.62	0.63	0.62	0.62	0.61
8	1245	0.83	0.84	0.84	0.84	0.91

One might expect that predicting further ahead in time would lead to larger RMSE values, however, the change in RMSE is small. We therefore explore using AdaBoost regression for the prediction of tidal volume values up to 1 h ahead, finding very little increase in RMSE values across patients.

As described in this chapter, a benefit of using ensemble models made from decision trees is that we can visualize the features and decisions used to make decisions. Figure 5 is one of the decision trees created when using the AdaBoost method.

The features were extracted by *tsfresh* as the most significant features for this problem. It is interesting here to discuss what these features can mean for our

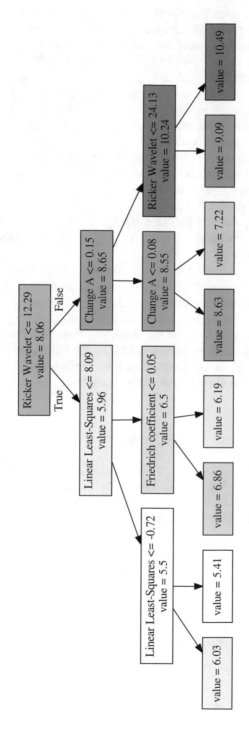

Fig. 5 One of the ten trees created by the AdaBoost method for patient 1 predictions

problem domain. The Ricker Wavelet is used to describe properties of a viscoelastic homogeneous media and the Friedrich coefficient aims at describing the random movement of a particle in a fluid. We can hypothesize from this finding that the amount of fluid in the lungs would be an impacting factor in how a patient's tidal volume can change over time.

A comparison is then made applying long short-term memory neural networks to the same problem: predicting tidal volume 1 h ahead. Two models are created: ModelA has one hidden layer and ModelB has three hidden layers, with a 20% dropout layer between the second and third layers to avoid overfitting. In contrast to the regressor models that work with features extracted from the time series, the LSTM models use the time series values directly; requiring 70% of the time series to train the models. Each layer in our LSTM models has 50 nodes and both models use 20 input points to predict 4 ahead.

The RMSE values for our LSTM models are significantly greater than the AdaBoost method for predicting 1 h ahead, so we deem AdaBoost the better method of the two for this problem.

The VILIAlert system alerted when four consecutive bins were greater than the 8 ml/kg tidal volume threshold. These alert times were stored in the database. We can thus work out the accuracy of our models to predict these alerts, showcasing the possibility of preventing threshold breaches and preventing injury. Table 3 highlights the predictive accuracy of the AdaBoost model for the 8 patients. Total Alerts being the total number of alerts recorded by the VILIAlert system, TP being the true positives: the alerts that would have been predicted by the model, and FN being false negatives: the alerts that would not have been predicted. The accuracy is then calculated using:

$$\text{Accuracy} = \frac{TP}{TP + FN}.$$

For the 84 alerts that were generated for patient 1, 81 would have been predicted using our models and thus those threshold breaches could have been prevented. These results showcase how Machine Learning algorithms, when paired with big

Table 3 Prediction accuracy of alerts using AdaBoost

Patient	Total alerts	TP	FN	Accuracy
1	84	81	3	0.96
2	25	23	2	0.92
3	0	0	0	1.00
4	2	1	1	0.50
5	3	2	1	0.67
6	11	3	8	0.27
7	0	0	0	1.00
8	167	142	25	0.85

data, can provide value in preventing lung injury during mechanical ventilation of intensive care patients.

4.1 The ATTITUDE Study

The ATTITUDE study operated by Queen's University, Belfast, and funded by The Health Foundation UK, aims to develop a clinical decision support tool to improve weaning protocols commonly used in clinical practice, and further understand the barriers in implementing evidence-based care from these tools tested in a proof-of-concept study carried out at the Royal Victoria Hospital ICU, Belfast. Improving patient care, outcomes, and mortality by reducing the duration of weaning can lead to reduced hospital stays and costs, and this study aims to find out if the use of clinical decision support tools can improve the quality of critical care practices.

5 Future of ML and AI in ICU

The methodology discussed in this chapter can, and must be, explored with various and extensive types and volumes of data, to investigate more disease states and clinical conditions. This data is currently being recorded worldwide in what are known as Electronic Health Records and these hold valuable insights which must be utilized to improve healthcare going forward.

From the European Big Data Value Strategic Research and Innovation Agenda [1] we understand the importance of Big Data and utilizing it to benefit both the economy and society. By exploring the already available EHRs we can provide societal benefit by improving patient outcomes and saving lives, and further economic benefits of saving millions in hospital costs, through shorter lengths of stays and disease prediction, among others. Healthcare, being one of the most important and largest sectors, can greatly impact the agenda of a data-driven economy across Europe. Data-driven predictive analytics, built using the methodologies discussed in the chapter, can produce clinical decision support tools, allowing for advanced decision making or automation of procedures. The Machine Learning methodologies can further provide greater insights into patient states and inform healthcare professionals with new information or possible reasoning that would not have been caught by a human. These new insights result in further research questions that can be explored.

This chapter can be aligned with the Big Data Reference model in various ways. The data recorded in ICU is extensive and in various different formats, from structured data and time series to imaging and text inputs. Advanced visualization tools can be used to add value to data by presenting it in user-friendly ways. Data analytics and Machine Learning can be applied for prediction and reasoning of disease states. There exist further protocols and guidelines for the handling of patient information, ensuring efficient data protection and management.

The new AI, data and robotics partnership highlights the value opportunities that exist when transforming the healthcare sector by applying AI to produce value-based and patient-centric care in areas such as pandemic response, disease prevention, diagnosis decision support, and treatment [2].

High-performance computing is an integral part in deploying real-time predictive analytic models in intensive care. We must ensure our machines can process the data efficiently and quickly. Utilizing parallelism will provide speed up data processing and model predictions.

This chapter has explored the use of Big Data-driven analytics for acute and chronic clinical conditions to provide value to healthcare services. While there exists vast research carried out in certain disease states, such as Sepsis [30, 40–42], work is needed to provide greater in-depth analysis and insights into patients' complex physiologic state while in such critical conditions. Recent developments have led to a greater acceptance and excitement in this field, resulting in updated guidelines for testing AI models in real-life clinical trials to promote worldwide acceptance of the use of AI in healthcare.

Acknowledgments Rachael Hagan acknowledges funding for a PhD studentship from the Department for the Economy, Northern Ireland, and Dr. Murali Shyamsundar is an NIHR Clinical Scientist fellow.

References

1. Zillner, S., Curry, E., Metzger, A., et al. (2017). *European big data value strategic research & innovation agenda*. Big Data Value Association.
2. Zillner, S., Bisset, D., Milano, M., Curry, E., García Robles, A., Hahn, T., Irgens, M., Lafrenz, R., Liepert, B., O'Sullivan, B., & Smeulders, A. (Eds.). (2020). *Strategic research, innovation and deployment agenda – AI, data and robotics partnership*. Third release. Brussels. BDVA, euRobotics, ELLIS, EurAI and CLAIRE.
3. Ridley, S., & Morris, S. (2007). Cost effectiveness of adult intensive care in the UK. *Anaesthesia, 62*, 547–554. https://doi.org/10.1111/j.1365-2044.2007.04997.x
4. Teno, J. M., Fisher, E., & Hamel, B. (2000). Decision-making and outcomes of prolonged ICU stays in seriously ill patients. *Journal of the American Geriatrics Society, 48*, 70–74.
5. Bates, D. W., Saria, S., Ohno-Machado, L., et al. (2014). Big data in health care: Using analytics to identify and manage high-risk and high-cost patients. *Health Affairs, 33*, 1123–1131. https://doi.org/10.1377/hlthaff.2014.0041
6. Cavanillas, J. M., Curry, E., & Wahlster, W. (2016). *New horizons for a data-driven economy: A roadmap for usage and exploitation of big data in Europe* (pp. 1–303). https://doi.org/10.1007/978-3-319-21569-3
7. Bhutkar, G., Deshmukh, S., & Detection, D. (2015). Vital medical devices in intensive care unit. https://doi.org/10.13140/RG.2.1.4671.6247
8. Christ, M., Braun, N., Neuffer, J., & Kempa-Liehr, A. W. (2018). Time series FeatuRe extraction on basis of scalable hypothesis tests (tsfresh – A Python package). *Neurocomputing, 307*, 72–77. https://doi.org/10.1016/j.neucom.2018.03.067

9. Drews, F. A. (2008). Patient monitors in critical care: Lessons for improvement. In *Advances in patient safety: New directions and alternative approaches* (pp. 1–13). Rockville, MD: Agency for Healthcare Research and Quality. https://doi.org/NBK43684 [bookaccession].

10. Johnson, A., Pollard, T., Shen, L., et al. (2016). MIMIC-III, a freely accessible critical care database. *Scientific Data*.

11. Goldberger, A. L., Amaral, L. G., et al. (2000). Physiobank, physiotoolkit, and physionet components of a new research resource for complex physiologic signals. *Circulation, 101,* 215–220.

12. Fayyad, U., Piatetsky-shapiro, G., & Smyth, P. (1996). From data mining to knowledge discovery in databases. *American Association for Artificial Intelligence, 17,* 37–54.

13. Parikh, R. B., Teeple, S., & Navathe, A. S. (2019). Addressing bias in artificial intelligence in health care. *JAMA, 322,* 2377–2378. https://doi.org/10.1001/jama.2019.18058

14. Hammer, G. P., Du Prel, J. B., & Blettner, M. (2009). Avoiding Bias in observational studies. *Dtsch Arztebl, 106,* 664–668. https://doi.org/10.3238/arztebl.2009.0664

15. Bridi, A. C., Louro, T. Q., & Da Silva, R. C. L. (2014). Clinical alarms in intensive care: Implications of alarm fatigue for the safety of patients. *Revista Latino-Americana de Enfermagem, 22,* 1034–1040. https://doi.org/10.1590/0104-1169.3488.2513

16. Thangavelu, S., Yunus, J., Ifeachor, E., et al. (2015). Responding to clinical alarms: A challenge to users in ICU/CCU. In *Proceedings – International Conference on Intelligent Systems, Modelling and Simulation, ISMS* (pp. 88–92).

17. Ancker, J. S., Edwards, A., Nosal, S., et al. (2017). Effects of workload, work complexity, and repeated alerts on alert fatigue in a clinical decision support system. *BMC Medical Informatics and Decision Making, 17,* 1–9. https://doi.org/10.1186/s12911-017-0430-8

18. Wicks, P., Liu, X., & Denniston, A. K. (2020). Going on up to the SPIRIT in AI: Will new reporting guidelines for clinical trials of AI interventions improve their rigour? *BMC Medicine, 18,* 4–6. https://doi.org/10.1186/s12916-020-01754-z

19. Statement, C. (2020). The SPIRIT-AI extension. 26. https://doi.org/10.1038/s41591-020-1037-7

20. (2020). Reporting guidelines for clinical trial reports for interventions involving artificial intelligence: The CONSORT-AI extension. 26.

21. Sounderajah, V., Ashrafian, H., Aggarwal, R., et al. (2020). Developing specific reporting guidelines for diagnostic accuracy studies assessing AI interventions: The STARD-AI steering group. *Nature Medicine, 26,* 807. https://doi.org/10.1038/s41591-020-0941-1

22. Collins, G. S., & Moons, K. G. M. (2019). Reporting of artificial intelligence prediction models. *Lancet, 393,* 1577–1579. https://doi.org/10.1016/S0140-6736(19)30037-6

23. Fleuren, L. M., Thoral, P., Shillan, D., et al. (2020). Machine learning in intensive care medicine: Ready for take – off? *Intensive Care Medicine, 46,* 1486–1488. https://doi.org/10.1007/s00134-020-06045-y

24. Deo, R. C. (2015). Basic science for clinicians. *Circulation, 132,* 1920–1930. https://doi.org/10.1161/CIRCULATIONAHA.115.001593

25. Turek, M. (2020). *DARPA Project Explainable Artificial Intelligence (XAI)*. Accessed November 5, 2020, from https://www.darpa.mil/program/explainable-artificial-intelligence%0A

26. Doran, C., Does, T. R. W., & Ai, E. (2018). What does explainable AI really mean? A new conceptualization of perspectives. In *CEUR Workshop Proceedings,* 2071.

27. Bannach-Brown, A., Przybyła, P., Thomas, J., et al. (2019). Machine learning algorithms for systematic review: Reducing workload in a preclinical review of animal studies and reducing human screening error. *Systematic Reviews, 8,* 1–12. https://doi.org/10.1186/s13643-019-0942-7

28. Thommandram, A., Pugh, J. E., Eklund, J. M., et al. (2013). Classifying neonatal spells using real-time temporal analysis of physiological data streams: Algorithm development. In *IEEE EMBS Special Topics Conferenec on Point-of-Care Healthcare Technologies Synerg Towar Better Glob Heal PHT 2013* (pp. 240–243). https://doi.org/10.1109/PHT.2013.6461329

29. Blount, M., Ebling, M., Eklund, J. M., et al. (2010). Real-time analysis for intensive care. *IEEE Engineering in Medicine and Biology Magazine, 29*(2), 110–118.

30. Nemati, S., Holder, A., Razmi, F., et al. (2018). An interpretable machine learning model for accurate prediction of Sepsis in the ICU. *Critical Care Medicine, 46*, 547–553. https://doi.org/10.1097/CCM.0000000000002936
31. Kubat, M. (2017). *An introduction to machine learning* (2nd ed.). Springer.
32. Ambrosiadou, B. V., Goulis, D. G., & Pappasa, C. (1996). Clinical evaluation of the DIABETES expert system for decision support by multiple regimen insulin dose adjustment. *Computer Methods and Programs in Biomedicine, 49*, 105–115.
33. Sayadi, M., Zibaeenezhad, M., Mohammad, S., & Ayatollahi, T. (2017). Simple prediction of type 2 diabetes mellitus via decision tree modeling. *International Cardiovascular Research Journal, 11*, 71–76.
34. Kuo, H. J., Chiu, H. W., Lee, C. N., et al. (2015). Improvement in the prediction of ventilator weaning outcomes by an artificial neural network in a medical ICU. *Respiratory Care, 60*, 1560–1569. https://doi.org/10.4187/respcare.03648
35. Tomašev, N., Glorot, X., Rae, J. W., et al. (2019). A clinically applicable approach to continuous prediction of future acute kidney injury. *Nature, 572*, 116–119. https://doi.org/10.1038/s41586-019-1390-1.A
36. Attia, Z. I., Noseworthy, P. A., Lopez-Jimenez, F., et al. (2019). An artificial intelligence-enabled ECG algorithm for the identification of patients with atrial fibrillation during sinus rhythm: A retrospective analysis of outcome prediction. *Lancet, 394*, 861–867. https://doi.org/10.1016/S0140-6736(19)31721-0
37. Montavon, G., Samek, W., & Müller, K. (2018). Methods for interpreting and understanding deep neural networks. *Digital Signal Processing, 73*, 1–15. https://doi.org/10.1016/j.dsp.2017.10.011
38. Fritz, B. A., Cui, Z., Zhang, M., et al. (2019). Deep-learning model for predicting 30-day postoperative mortality. *British Journal of Anaesthesia, 123*, 688–695. https://doi.org/10.1016/j.bja.2019.07.025
39. Higgins, T., Freeseman-freeman, L., & Henson, K. (2019). Mews ++: Predicting clinical deterioration in admitted patients using a novel inpatient deterioration: Can artificial intelligence predict who will be transferred to the ICU? 48:2019.
40. Li, X., Xu, X., Xie, F., et al. (2020). A time-phased machine learning model for real-time prediction of sepsis in critical care. *Critical Care Medicine*, E884–E888. https://doi.org/10.1097/CCM.0000000000004494
41. Ginestra, J. C., Giannini, H. M., Schweickert, W. D., et al. (2019). Clinician perception of a machine learning-based early warning system designed to predict severe Sepsis and septic shock. *Critical Care Medicine, 47*, 1477–1484. https://doi.org/10.1097/CCM.0000000000003803
42. Komorowski, M., Celi, L. A., Badawi, O., & Gordon, A. C. (2018). The artificial intelligence clinician learns optimal treatment strategies for sepsis in intensive care. *Nature Medicine, 24*. https://doi.org/10.1038/s41591-018-0213-5
43. Eslami, S., de Keizer, N. F., Abu-Hanna, A., et al. (2009). Effect of a clinical decision support system on adherence to a lower tidal volume mechanical ventilation strategy. *Journal of Critical Care, 24*, 523–529. https://doi.org/10.1016/j.jcrc.2008.11.006
44. Jalali, A., Bender, D., Rehman, M., et al. (2016). Advanced analytics for outcome prediction in intensive care units. In *Proceedings of Annual International Conference* of the *IEEE Engineering in Medicine and Biology Society* EMBS (pp. 2520–2524). https://doi.org/10.1109/EMBC.2016.7591243
45. Che, Z., Purushotham, S., Khemani, R., & Liu, Y. (2016). Interpretable deep models for ICU outcome prediction. In *AMIA. Annual Symposium proceedings AMIA Symposium 2016* (pp. 371–380).
46. Boverman, G., & Genc, S. (2014). Prediction of mortality from respiratory distress among long-term mechanically ventilated patients. In *2014 36th Annual International Conference of the IEEE Engineering in Medicine and Biology Society EMBC 2014* (pp. 3464–3467). https://doi.org/10.1109/EMBC.2014.6944368

47. Emanet, N., Öz, H. R., Bayram, N., & Delen, D. (2014). A comparative analysis of machine learning methods for classification type decision problems in healthcare. *Decision Analysis, 1*, 1–20. https://doi.org/10.1186/2193-8636-1-6

48. Florence, S., Amma, N. G. B., Annapoorani, G., & Malathi, K. (2014). Predicting the risk of heart attacks using neural network and decision tree. *International Journal of Innovative Research in Computer and Communication Engineering, 2*, 7025–7030.

49. Parreco, J., Hidalgo, A., Parks, J. J., et al. (2018). Using artificial intelligence to predict prolonged mechanical ventilation and tracheostomy placement. *The Journal of Surgical Research, 228*, 179–187. https://doi.org/10.1016/j.jss.2018.03.028

50. Hyland, S. L., Faltys, M., Hüser, M., et al. (2019). Machine learning for early prediction of circulatory failure in the intensive care unit. *Nature Medicine, 26*. https://doi.org/10.1038/s41591-020-0789-4

51. Meyer, A., Zverinski, D., Pfahringer, B., et al. (2018). Machine learning for real-time prediction of complications in critical care: A retrospective study. *The Lancet Respiratory Medicine, 6*, 905–914. https://doi.org/10.1016/S2213-2600(18)30300-X

52. Harrsion, D. (2014). ICNARC. Number of mechanically ventilated patients during 2012.

53. Hagan, R., Gillan, C. J., Spence, I., et al. (2020). Comparing regression and neural network techniques for personalized predictive analytics to promote lung protective ventilation in intensive care units. *Computers in Biology and Medicine, 126*, 104030. https://doi.org/10.1016/j.compbiomed.2020.104030

54. Gillan, C. J., Novakovic, A., Marshall, A. H., et al. (2018). Expediting assessments of database performance for streams of respiratory parameters. *Computers in Biology and Medicine, 100*, 186–195. https://doi.org/10.1016/j.compbiomed.2018.05.028

55. Viceconti, M., & Hunter, P. (2016). The virtual physiological human: Ten years after. *Annual Review of Biomedical.*

56. Das, A., Saffaran, S., Chikhani, M., et al. (2020). In silico modeling of coronavirus disease 2019 acute respiratory distress syndrome: Pathophysiologic insights and potential management implications. *Critical Care Explorations, 2*, e0202. https://doi.org/10.1097/cce.0000000000000202

3D Human Big Data Exchange Between the Healthcare and Garment Sectors

Juan V. Durá Gil, Alfredo Remon, Iván Martínez Rodriguez,
Tomas Pariente-Lobo, Sergio Salmeron-Majadas, Antonio Perrone,
Calina Ciuhu-Pijlman, Dmitry Znamenskiy, Konstantin Karavaev,
Javier Ordono Codina, Laura Boura, Luísa Silva, Josep Redon, Jose Real,
and Pietro Cipresso

Abstract 3D personal data is a type of data that contains useful information for product design, online sale services, medical research and patient follow-up.

Currently, hospitals store and grow massive collections of 3D data that are not accessible by researchers, professionals or companies. About 2.7 petabytes a year are stored in the EU26.

In parallel to the advances made in the healthcare sector, a new, low-cost 3D body-surface scanning technology has been developed for the goods consumer sector, namely, apparel, animation and art. It is estimated that currently one person is scanned every 15 min in the USA and Europe. And increasing.

The 3D data of the healthcare sector can be used by designers and manufacturers of the consumer goods sector. At the same time, although 3D body-surface scanners

J. V. Durá Gil (✉) · A. Remon
Instituto de Biomecánica de Valencia, Universitat Politècnica de València, Valencia, Spain
e-mail: juan.dura@ibv.org

I. M. Rodriguez · T. Pariente-Lobo
ATOS Research and Innovation, Madrid, Spain

S. Salmeron-Majadas
Atos IT Solutions and Services, Irving, TX, USA

A. Perrone · C. Ciuhu-Pijlman · D. Znamenskiy
Philips Research, Eindhoven, Netherlands

K. Karavaev · J. O. Codina
ELSE Corp Srl, Milan, Italy

L. Boura · L. Silva
P&R Têxteis S.A., Tamel (São Veríssimo), Portugal

J. Redon · J. Real
INCLIVA Research Institute, University of Valencia, Valencia, Spain

CIBERObn ISCIII, Madrid, Madrid, Spain

P. Cipresso
Applied Technology for Neuro-Psychology Lab, Istituto Auxologico Italiano, Milan, Italy

© The Author(s) 2022

E. Curry et al. (eds.), *Technologies and Applications for Big Data Value*,
https://doi.org/10.1007/978-3-030-78307-5_11

225

have been developed primarily for the garment industry, 3D scanners' low cost, non-invasive character and ease of use make them appealing for widespread clinical applications and large-scale epidemiological surveys.

However, companies and professionals of the consumer goods sector cannot easily access the 3D data of the healthcare sector. And vice versa. Even exchanging information between data owners in the same sector is a big problem today. It is necessary to overcome problems related to data privacy and the processing of huge 3D datasets.

To break these silos and foster the exchange of data between the two sectors, the BodyPass project has developed: (1) processes to harmonize 3D databases; (2) tools able to aggregate 3D data from different huge datasets; (3) tools for exchanging data and to assure anonymization and data protection (based on blockchain technology and distributed query engines); (4) services and visualization tools adapted to the necessities of the healthcare sector and the garment sector.

These developments have been applied in practical cases by hospitals and companies of in the garment sector.

Keywords 3D body · Blockchain · Clothing · Obesity

1 Introduction

Three-dimensional (3D) anthropometry refers to the measurement of the human body shape. From 3D data it is possible to obtain more information than from traditional one-dimensional anthropometry, which only uses length, radius or perimeter. Anthropometry plays an important role in industrial design, clothing design and ergonomics where statistical data about the distribution of body dimensions in the population are used to optimize products. Also, changes in lifestyles and nutrition lead to changes in the distribution of body dimensions (e.g. the rise in obesity) and require regular updating of anthropometric data collections.

3D anthropometric data is a key value in scientific and economic sectors like healthcare, consumer goods and professional sports. In the past, the usage of such data was limited by the high price and low precision of 3D scanning technologies. The recent advances in scanning technologies have notoriously widened their usage opportunities. As a result, different actors from different sectors work on the generation of their own dataset of anthropometric data for their own purposes, without taking advantage of similar effort taken by other actors.

In this context, a system to share anthropometric data would allow the different data consumers access to larger datasets and also to reduce the data acquisition costs, thus augmenting their scientific or business opportunities. Besides, data holders would be able to extend the economic benefit reported by their data.

Developing an effective data sharing system is challenging. Success requires a system that gives support to interests from any participant, forming a symbiotic eco-system. Participants in a data-sharing system can be categorized as data providers and data consumers. Regarding this categorization, the participants will have, at least, the following requirements:

- A data provider will want to preserve their data, preventing other users from replicating their dataset. Also, due to legal restrictions, a data provider may need to anonymize any data before sharing it.
- A data consumer needs to know, in detail, the type of data available at any dataset (a sort of data dictionary). The availability of tools to query for specific data (i.e. data filtering tools) is another important requirement.

The chapter relates to the technical priorities *data management* and *data protection* of the European Big Data Value Strategic Research and Innovation Agenda [1]. The BodyPass project focuses on 3D anthropometric data management and data protection. In this context BodyPass has generated tools for:

- Semantic annotation of data. BodyPass has defined an anthropometric data dictionary, or data catalogue, to accomplish seamless integration with and smart access to the various heterogeneous data sources (see Sect. 2.1).
- Templates to harmonize 3D datasets and facilitate 3D data exchange (Sect. 2.2).
- 3D anonymization tools for individual body scans (Sect. 3.1) and rules to protect privacy of health data (Sect. 3).
- Data lifecycle management with control of the data provenance with blockchain (see Sect. 4). BodyPass applies distributed ledger/blockchain technology to enforce consistency in transactions and data management.

The complete BodyPass ecosystem has been tested in practical cases described in Sects. 5 and 6.

2   The Process to Harmonize 3D Datasets

Anthropometric data has key value in many economic and social areas, among others healthcare, apparel and furniture design. The relevance of anthropometric data motivated a number of research efforts to standardize and effectively measure a human body [2–5]. But measuring a human body presents important challenges that are difficult to meet: important features such as repeatability, extrapolation of metrics or accuracy [6–9]. Some problems rely on the fact that the human body is not a rigid body, but a soft, articulated and never fully static body. Some other problems rely on the measuring tools, methodologies or different interpretations of the metric. In this context, the standardization efforts taken up to now seem to be insufficient.

The BodyPass project aims to develop a system where different actors share anthropometric data. Data sources in BodyPass belong to different sectors, currently healthcare and consumer goods sectors. As a matter of fact, this means that the measuring methodologies and measuring tools employed may differ for every dataset in the global federated database of the project. In particular, the healthcare sector data is produced by a commercial solution from Philips and is, in essence, an automatic digital measuring tape. Data from the consumer goods sector is provided by IBV and is obtained from its proprietary automatic digital measuring tape [10, 11]. As a result, all measurements in the federated BodyPass database come from digital measuring tools, meaning that there is a high risk of incompatible measurements due to the use different measuring algorithms.

In this context, the project success has required:

(a) The design and creation of an anthropometric data dictionary that clearly defines every metric. This dictionary must be open to all users and data providers, so they can validate that metrics in their databases are compatible with that in the dictionary (e.g. same metric units are employed in all databases).
(b) To agree on a common parameterization of a reference mesh so that any partner can conduct a template/model fitting process using its proprietary template and afterwards, compute the mapping between both templates. Such reference mesh can act as a kind of 'Rosetta stone' to translate body semantics between different mesh topologies and allow to build compatible body surfaces coming from different sources and parameterised with different template topologies.

2.1 The Anthropometric Data Dictionary

The anthropometric data dictionary developed in the BodyPass project is based on previous international standardization efforts from organizations such as ISO[1] or ISAK,[2] and it considers the project data consumers' requirements as well. Every metric is defined by the following fields:

- Mandatory fields:

  - ID: code that identifies the metric. It is defined as a string that partly includes the metric designation.
  - Designation: It is a self-descriptive name of the metric.
  - Definition: Unambiguous description of the metric.
  - Source: Reference to the company or organization that defined the metric. It usually refers to a standard definition.
  - Other std.: Other standard definitions that can be compatible with the metric as defined.
  - Units: Metrics units employed.

- Optional fields:

---

[1] International Organization for Standardization (http://iso.org).

[2] International Society for the Advancement of Kinanthropometry (https://www.isak.global/Home/Index).

ID:	WaistHeight
Designation:	Waist height (abbr Waist ht)
Definition:	Vertical distance from the waist level to the ground (virtual anthropometer). Waist level (ISO 8559-1:2017 3.1.22) is the midway level between the lowest rib point and the highest point of the hip bone at the side of the body. Lowest rib point (ISO 8559-1:2017 3.1.15) is the inferior point of the bottom of the rib cage (tenth rib) projected horizontally, 45 degrees from the mid-sagittal plane, to the surface of the skin. The highest point of the hip bone (ISO 8559-1:2017 3.1.16) is the highest point at the side of the upper border of the iliac crest.
Source:	ISO 8559-1:2017 5.1.10
Other std.:	ASTM D5219-15 Waist height, ISO 18825-2:2016 2.2.3
Units:	mm

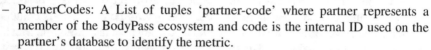

Fig. 1 Details of the definition of a metric in the BodyPass dictionary

- PartnerCodes: A List of tuples 'partner-code' where partner represents a member of the BodyPass ecosystem and code is the internal ID used on the partner's database to identify the metric.
- Media: URLs to media files (images, 3D data or videos) that facilitate metric understanding.

The BodyPass anthropometric data dictionary includes about 100 body metrics and it is available via the project's API and its web interface.[3] The web interface includes resources to access all the BodyPass API services. Among them, the resource DataResourceCatalogue [12] permits to visit the list of metrics in the project's anthropometric data dictionary. BodyPass members can use this resource to append, edit or delete metrics to/from the dictionary. In addition, the resource DataProviderMetrics lists the metrics in the dictionary supported by a given data provider. These two resources allow data consumers to consult which metrics are available in the project and also to check which data provider supports a given metric.

To facilitate the dictionary consultation to data providers, IBV lists all the metrics available in their BodyPass database, about 90 metrics, on a webpage.[4] This webpage includes an index and is far more readable than the BodyPass services deployed to this end. For every metric, the web displays the content of all the mandatory fields and also the related images in the dictionary.

Figure 1 shows the description of a metric in the aforementioned webpage. This particular metric is defined in three standardization international publications, two from ISO and one from ASTM.[5] Its definition includes two images that facilitates its understanding.

---

[3] http://145.239.67.20:3001/explorer/

[4] http://personales.upv.es/alrego/body/BodyPass_DigitalMeasuringTape.html

[5] American Society for Testing and Materials (http://astm.org).

2.2   3D Templates

(a) The initial approach for sharing 3D data was the use of static 3D templates agreed between BodyPass members. However, we discovered that some companies that process 3D data did not accept this approach. They thought that reverse engineering could be used for disclosing its proprietary algorithms. This problem has been solved with an innovative approach: random templates (on-the-fly templates). For this reason, BodyPass supports *standardized templates*: agreed definitions of bodies (or body parts). Standardized templates are being defined and registered in the system once and being reused in the processing of several queries. The pre-computed mapping based on one example could be used to change the 'skin' of any human surface registered with its own mesh to the reference one (or vice-versa) in a very efficient way. Upon demand by any BodyPass member, the set of agreed body part definitions can also be extended in a similar fashion as body metric definitions. In order to protect the proprietary background of BodyPass members, the topology of the released 3D content to the clients will differ from its own developed body template structure. Meaning that, prior to its delivery, data nodes will remesh or resample all 3D content hiding their own topology and parameterization.

(b) *On-the-fly templates.* The BodyPass platform defines a new template for every single data request, herein called 'on-the-fly templates'. This means to create a unique template every time the system processes a new query requiring 3D body surfaces. This strategy provides a higher protection level compared to using standardized templates by making it more difficult to conduct reverse engineering of individual data.

The procedure to manage on-the-fly templates is the following: anytime a data query arrives to the BodyPass system, a template is created on-the-fly and distributed among the processing nodes. In the next query, a new template is created and used. The template creation will provide it a random parameterization and geometry, making it unique. All the subsequent outputs from the nodes must be compliant with the template's features (namely parameterization and pose). Thus, they all can be aggregated in simple operation, i.e. weighted average of average body surfaces created by different nodes (where the weight assigned to a particular average is related to the amount of individual data used to obtain it). In summary, on-the-fly templates will be unique in terms of parameterization and geometry and will determine the parameterization, rigid alignment and pose of the query outputs, i.e., the 3D contents delivered to the client will match the template's features (Fig. 2).

3   Anonymization and Privacy

In this section, we will briefly address the solutions adopted to protect personal data. We have applied two solutions:

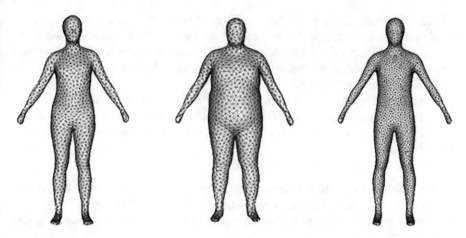

Fig. 2 Examples of on-the-fly templates with random geometry and topology (A-Pose)

(a) Anonymization of individual body scans.
(b) Architectural solution ensuring data security and privacy at hospitals with on-premise data node.

The General Data Protection Regulation (GDPR) in Europe classifies health data as sensible data, and it asks for special protection. For this reason, it was necessary to implement the architectural solution (on-premise data nodes) for processing medical images and to assure that only aggregated data is obtained from the hospitals.

3.1 The Anonymization of Body Scans

The data node of the Institute of Biomechanics (IBV) stores a database with more than 30,000 3D anonymous avatars and their related metrics. BodyPass provides tools to add new data to the database and to query and retrieve data from the database as well. This means that BodyPass's users are allowed to incorporate their own avatars to the database. To this end, the user just needs a 3D scan and basic data from the subject. During the registration of a new avatar in the database, the avatar is measured using the IBV's automatic digital measuring tape, which provides about 80 metrics from the avatar that are also stored in the database.

A basic requirement of the database is to assure the anonymization of the models. The data stored from every model includes:

- Gender, age, country, weight and height.
- The 3D avatar.
- List of metrics automatically extracted from the avatar.

From this data, the only one that can be used to identify the model is the avatar id. Given an avatar id, it would be possible to identify the model using marks on the skin (e.g. tattoos) or facial features. For this reason, avatars are stored without texture, so no skin features are stored. In this context, the face represents the only identifiable feature of the avatar.

A straightforward solution would be to remove or blur the face of the avatar, but this would end in a database full of non-human-faced avatars. This is why we considered a more sophisticated solution that incorporates a synthetic human face to every avatar.

This is performed in a three-step process:

1. The first step identifies the vertices in the raw scan that lay on the model's face.
2. Then, the identified vertices are removed, obtaining a de-faced scan.
3. Finally, the avatar registration process will add a synthetic human-looking face to the avatar.

This process presents two main challenges: The first one is to identify the vertices in the raw scan that belong to the model's face and the second is provide the avatar a human-like face. To identify the vertices which lay on the face, we use AI, in particular a convolutional neural network. The development of such a network required the exploitation of a large dataset generated in previous projects. The dataset was used to train the net and to improve its effectivity.

The computational kernel developed presents a high performance. Although the time-to-solution highly depends on the number of vertices of the input 3D object, a regular mesh of about 50,000 vertices can be processed in just few seconds. In addition, the algorithm presents a high degree of parallelism, allowing the use of massive parallel architectures like GPUs. This brings the possibility to further optimize the kernel if needed.

Once the vertices are identified, they are removed from the 3D object, obtaining a de-faced version of the raw scan.

Finally, the last step is again a challenging process that removes artifacts from the input data, including tasks as hole-filling and noise removal. This is performed via a template-fitting approach [13], which provides a realistic 3D closed avatar. This process is capable of replacing missing data, i.e. holes in the original mesh, with realistic data. In the case of the face, this means that the final avatar will have a face that will perfectly fit to the rest of the avatar and also present a human flavour (Fig. 3).

3.2 Architectural Solution Ensuring Data Security and Privacy at Hospitals

The purpose of the software installed in hospitals is twofold: extract body measurements by processing CT images and serve queries for aggregated data coming

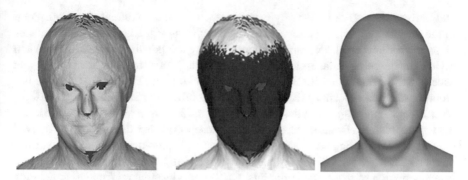

Fig. 3 Detail of the head in a raw scan (left), identification of vertices in the face (middle), synthetic face (right)

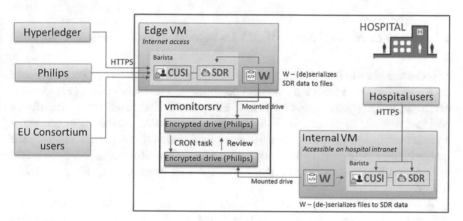

Fig. 4 Software architecture

from the Consortium users via Hyperledger. The software was created to meet the following privacy and security requirements: (1) no personal data should leave the hospital; clients outside the hospital can only receive aggregated data, resistant to de-anonymization efforts via differential privacy; (2) the server holding the personal data cannot be connected to the internet; (3) hospitals should be able to review all outgoing data; (4) no data is sent through the Hyperledger; Consortium users should be able to pick up their data directly at the hospital endpoint; (5) the derived individuals' data (e.g. measurements) is IP sensitive and therefore must be kept out of the hospital's reach, whilst also being unavailable to Philips due to privacy restrictions. Our software solution addressing the above requirements is shown in Fig. 4.

The solution requires the use of two virtual machines (VMs): 'Edge' and 'Internal'. The Edge VM has software that enables it to communicate with BodyPass members. It receives data queries from Hyperledger and makes the results available for download by users. The Internal machine is intended for the processing of raw and derived data. Due to security constraints, it operates without any connection to the internet. Every 20 s data is exchanged between Edge and Internal by means of secure bidirectional file transfers controlled by the hospital, where hospital can review all outgoing data. Both VMs use 'Barista'[6] infrastructure for storage and access to the data. Internally Barista consists of two modules. The first module is a cloud-based data repository (SDR) organized in different 'Studies' and 'Datasets', with access control capabilities. The second module is a web GUI (CUSI) for data manipulation. Datasets in the SDR are used both for the storage of derived data as well as acting as a sort of basic message passing interface for various data processing agents. This way the data processing agents are released from the implementation of their own endpoints. CUSI provides general (image) annotation and 3D visualization capabilities for the data stored in SDR.

4   The Secure Exchange of 3D Data with Blockchain

This chapter aims to describe in detail how the BodyPass approach solves the secure exchange of 3D data from a blockchain perspective. The main objective is to foster exchange, linking and re-use, as well as to integrate 3D data assets from the different business sectors. To cover this, BodyPass has adapted and created tools that allow a secure exchange of information between data owners, companies and subjects (patients and customers). In the BodyPass context, 3D personal data contains useful information for product design, online sale services, medical research and patient follow-up.

A conceptual view of the BodyPass solution is shown in Fig. 5, including the different stakeholders and main building blocks. The building blocks represent aggregated functionality, giving an idea of the flows of usage and retrieval tools.

One of the main drawbacks of the blockchain technology is the incompatibility with managing large amounts of data because of the size limitation in the transactions and the need to replicate the information in the different nodes for the consensus. To overcome this issue, we divided the BodyPass architecture into two different planes as shown in Fig. 6. The first plane depicts the data sharing itself (the Data Sharing Plane), while the second (the Data Management Plane) focuses on the management of the large data elements not so suitable to the blockchain-based plane. In addition, this modular architecture provides two key advantages: It allows scaling the number of data providers easily and facilitates additional control over the data made available by these data providers in the network and who accesses them.

---

[6] Barista is an integrated suite of tools, developed in Philips Research, enabling study-oriented data collection, AI algorithm creation and rapid implementation in user-facing workflows, see https://barista.eu1.phsdp.com

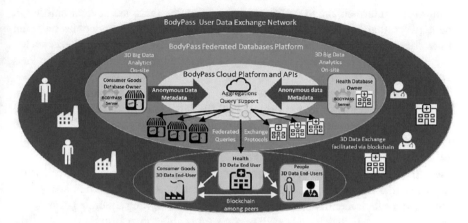

Fig. 5 High level abstraction of the BodyPass ecosystem

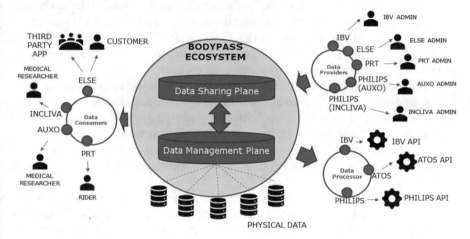

Fig. 6 BodyPass conceptual architecture

There are three different profiles that interact in the BodyPass ecosystem considering their relationship with the data: (1) data providers, (2) data processors and (3) data consumers.

Figure 6 shows these three different profiles, including examples of users of each profile with members of BodyPass.[7] BodyPass attempts to break data silos from several data providers from different sectors and foster exchange and reutilization of data assets that may help data consumers of the network to get external data. On the other hand, trust is a key aspect when devising a data sharing platform, especially if this platform is decentralized and should avoid the intervention of a central

---

[7] https://www.bodypass.eu/partners

authority or intermediary to certify that trust. Finally, there are several entities in the network that may generate interactions, transactions or dependencies between transactions that must also be shared in the repository. This set of characteristics led to the selection of blockchain technology [14] as the main driver of the BodyPass architecture.

The BodyPass ecosystem makes use of blockchain as a distributed ledger. Given the characteristics of the project and the business objectives, the approach selected is a permissioned blockchain, utilized by the members of BodyPass. Public blockchain networks are completely open to interact with the network and require self-governance, and private blockchain networks only allow the participation of selected entities. However, permissioned blockchain networks can adapt to hybrid scenarios like the one in BodyPass, not only letting the participants access the network once their identity has been verified but also assigning concrete permissions that can restrict which activities each participant is able to perform on the network.

Data providers usually have storage solutions for their existing assets. Therefore, the storage of the assets is not part of the scope of the blockchain in the BodyPass ecosystem. Considering these non-transactional data, which occasionally could be dynamically changed or too large, the BodyPass architecture has adopted the design pattern of off-chain storage [15], composed consequently of off-chain data (big chunks of data managed outside the blockchain network).

This approach carries some benefits, like saving bandwidth and storage capacity in the BodyPass ecosystem nodes and avoiding potential confidentiality issues derived from data being distributed out of a designated storage center. General Data Protection Regulation (GDPR) in Europe and companies doing business in Europe drive the need for new off-chain storage in blockchain applications. It is recommended to store sensitive information as off-chain data so that you can delete it if need be [16].

The *Data Sharing Plane* shown in Fig. 6 represents the implementation of the blockchain and provides the features to be a flexible trust model. It is built upon a modular architecture, configurable to choose the most suitable consensus mechanism or certification authority. There is a *Membership Service Provider* to deal with identity management and authentication. Inside the *Data Sharing Plane*, members can participate as if they were private groups by means of channels. Each member can be included in more than one channel, each of them with their own policies. All transactions are stored in the distributed ledger and, therefore, audit efficiency and quality are improved.

Due to the limitation regarding the handling of big data volumes of blockchain networks, this Data Sharing Plane is where the metadata and the permissions of the network components are managed. This has been implemented in BodyPass using Hyperledger Fabric [17] following the logic specified in the chaincode (i.e. a set of rules that govern the blockchain network) developed through the Hyperledger tool Composer [18]. Hyperledger Fabric is a permissioned blockchain network.

Considering that a building block is an asset or software piece from an architectural point of view, the BodyPass functional building blocks have interfaces to access the functionality that they provide. The green boxes shown in Fig. 7 represent

Fig. 7 BodyPass ecosystem
interfaces

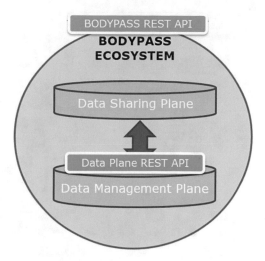

the implementation of these interfaces. This means both planes represented in Fig. 8 have their own API (BODYPASS REST API for the interactions with the Data Sharing Plane and a Data Plane REST API called from the Data Sharing Plane to perform the operations in the Data Management Plane).

The *Data Management Plane* exposes a REST API that will only be consumed by the *Data Sharing Plane*, and therefore the BodyPass actors do not have direct access to the Data Management plane, thus reinforcing the sense of security. Any data-related query will be executed via this API (after it has been 'authorized' and initiated in the Data Sharing Plane). The functionality of the whole ecosystem is exposed also as a REST API, in such a way the interactions with network assets, participants and transactions are available through standard HTTP operations, following the REST architecture (Representational State Transfer). This way, each HTTP request contains all the information necessary to execute it, which allows neither client nor server to remember any previous status. The interface is uniform, only specific actions (POST, GET, PUT and DELETE) are applied on the resources. As benefits, the protocol increases the scalability of the project and allows the internal components to evolve independently.

The *Data Management Plane*, as shown in Fig. 7, is the component that manages access to all off-chain data sources and orchestrates queries to be executed over the distributed storage. The Data Management Plane provides the security and required constraints for the blockchain members. Hash values are stored in order to verify the data when objects are accessed subsequently.

The Data Management Plane contains all the information supplied by the different data providers, which may be accessed for certain users of the BodyPass system who have previously reached an agreement with the data owners.

A detailed view of the Data Management Plane is shown in Fig. 8.

To access the Data Sharing plane REST API, it will be necessary to login in the system, since it is protected with an OAuth Keycloak server. The ATOS Data Hub

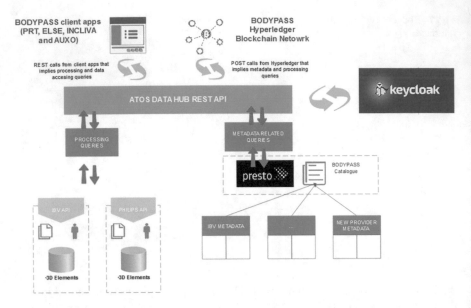

Fig. 8 Data storage plane architecture

REST API will consult the different catalogues supplied by the data providers and will obtain the necessary access information so that the user can retrieve these data. This access information is returned to the third-party application that will make the necessary call to the data providers' APIs, obtaining as a response the required information.

For the access of the blockchain network to the different data catalogs supplied by the data providers through the data plane REST API, PrestoDB [19] will be used. PrestoDB allows to perform federated queries among multiple (relational and NoSQL) databases such as Cassandra, Redis, MongoDB or PostgreSQL with a reasonable performance.

Although it has a slightly higher response time than other query engines, such as Apache Impala, it has a much more complete SQL syntax and has a longer list of database connectors (widening the scope of technologies to be used by the data providers) (Fig. 9).

The BodyPass system will generate a query plan from every request made by the user in the blockchain network. Using this query plan, PrestoDB will be able to consult the necessary data catalogs in order to obtain the necessary access information to retrieve the final data of the supplier. Figure 10 depicts the different steps taken in the application architecture in order to execute a query, showing the interaction between the different data planes and the data providers.

The query planner is the component of the BodyPass system that will be responsible for translating the requirements of the end users in queries to those catalog databases that contain the desired information, considering possible combinations between different data sources. This will be invoked through the POST

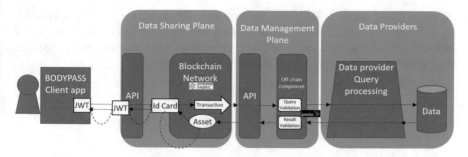

Fig. 9 Transaction data pipeline in a high-level representation of the BodyPass architecture

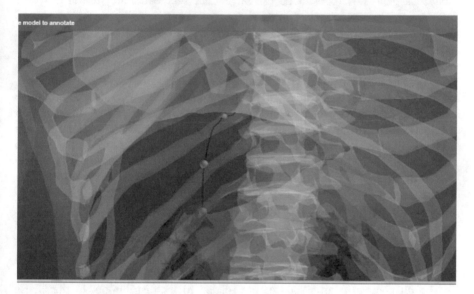

Fig. 10 Internal organ visualization and measurement

method *getQueryPlan* of the data plane REST API and used by the POST method *runFederatedQuery*.

In the case of aggregate queries, these could be directly carried out by PrestoDB for simple metrics, but it would be necessary to coordinate between 3D image processing services in the case of more complex aggregates, such as the average chest 3D scans of a certain segment of the population.

In this way, both the blockchain network and the data providers can have an exhaustive control over what information has been accessed by which user, thus facilitating monetization of data if necessary.

Finally, to conclude, it is important to note that the data sharing approach designed and implemented for the BodyPass project is accessible through the public endpoint http://145.239.67.20:3001/explorer/#/, through which an end user could access the functionalities behind the data sharing solution described in this chapter.

A Jason Web Token (JWT) will be required for accessing BodyPass endpoint in addition to a valid identity for accessing the private blockchain network that supports BodyPass. The identity will be provided by one of the existing Certification Authorities (CA) in the BodyPass blockchain private network.

5 The Application of BodyPass Results in Healthcare: Obesity

BodyPass generates tools to access huge data sets extracting useful 3D data information for assessment of body shape and its relationship to the amount of fat and body distribution. BodyPass has adapted and created tools for the secure exchange of information of 3D body shape and CT scan fat distribution.

Overweight and obesity are conditions highly prevalent worldwide, and 60% of adults in Europe meet the criteria that define these conditions [20]. Worldwide, obesity is a growing health concern. Endocrinologists treat patients who are obese because of metabolic and hormonal problems and want to provide insights into metabolic and cardiovascular disease risk [21]. The classification of obesity using BMI (Body Mass Index) does not fully encompass the complex biology of excess adiposity [22]. Obesity is closely related to metabolic risk factors and is associated with significant cardiovascular morbidity and mortality. Obese patients with metabolic abnormalities have insulin resistance, atherogenic dyslipidemia, low-grade inflammation and hypertension with a high risk to develop type 2 diabetes, atherosclerosis and cardiovascular diseases. However, not all obese subjects have these cardiometabolic abnormalities, and it is crucial to know which patients are at risk and which are not, since prognosis and therapeutic approach are different in those named Obese Healthy and Obese Unhealthy [23, 24].

The different metabolic statuses are related to fat distribution. In the Obese Healthy, fat accumulates in the subcutaneous tissue of the abdomen, around the hip and in the legs, while in the Obese Unhealthy accumulation is mainly in the mesenterium, liver, mediastinum, muscle and epicardial.

Assessment in the clinic has been the object of multiple research studies trying to properly identify measurements that predict fat distribution. Two kinds of approaches have been used, a combination of anthropometric parameters and different kinds of scans [25].

The most used among the anthropometric measurement were BMI, which indicates the presence or not of overweight and obesity but not fat distribution, abdominal circumference, waist-to-hip ratio and body shape index calculated with [WC (cm) $\times$ BMI$^{0.66}$ $\times$ height (m)$^{0.3}$]. All these parameters have the challenge of low accuracy and variability when measured in the clinic [26].

The scan methods obtain images from CT scan (Computed Tomography), MRI (Magnetic Resonance Imaging) and MRS (Magnetic Resonance Spectroscopy). The challenges of these methods is irradiation of patients, cost of the equipment, maintenance, cost of operators and time consuming. Using the BodyPass ecosystem it is possible to develop an easy, fast, accurate and inexpensive method to assess fat distribution.

Two hospitals in Italy and Spain developed a pilot with patients in order to test BodyPass, the first one includes the recording and integration of 3D Body Surface shape with data of internal fat amount and disposition in subcutaneous and visceral territories in a research environment and the second collect and transfer 3D Body Surface from a clinical environment. A summary or the pilot process is described below.

Patients recruitment was based on inclusion and exclusion criteria. This should be done in the hospital when subjects come to take CT.

Inclusion criteria: Age 19 or above, both sexes, thorax-abdominal CT and signed consent form. *Exclusion criteria:* Limitations for stand up and no signed consent form.

Data collection. In the Radiology Department a researcher recorded in a research protocol several data including:

- Demographic, clinical and anthropometric parameters to phenotype the subjects: age, gender, weight, height, BMI, waist circumference, blood pressure, personal history (diabetes, hypertension, dyslipidemia, atherosclerosis disease, other medical conditions).
- Biochemical parameters: fasting glucose, creatinine, urea, uric acid, total cholesterol, triglycerides, HDL cholesterol, AST/ALT.
- 3D scan and thoracic-abdominal CT are taken.
- Phenotype, biochemical parameters, 3D scan and CT data are gathered and stored by the researcher following the security protocol, Fig. 4.

Data integration: Integration of demographic and biochemical parameters with the results of the 3D Body Scan and CT images obtained the following data:

- Total volume of ectopic fat.
- Anatomical distribution of ectopic fat (localization).
- Total volume of VAT (Visceral Adipose Tissue) and fat in the mesenterium, liver, mediastinum, pericardium and psoas muscle.
- 3D images as an Avatar (IBV) from the subjects to explore correlation with classical clinical, anthropometric and biological parameters that are used for obesity classification and for subject's risk (BMI, waist circumference, definition of metabolic syndrome)
- Correlation of ectopic fat total volume and 3D Avatar images.

The Data Node consists of a number of software components, which, besides the extraction of measurements, also processes data queries for average measurements and body shapes, resisting data de-anonymization and providing graphical tools for exploration of the data. The software components extend the BARISTA platform which is an integrated suite of tools, developed in Philips Research, enabling study-oriented data collection, AI algorithm creation and rapid implementation in user-facing workflows; see https://barista.eu1.phsdp.com. The two components of

Barista, CUSI and SDR, have different functions. SDR is a web-based repository organized in the studies and datasets with a separate access control. SDR is used to store data within the workflow of processing individual raw data and within the data query workflow. Section 5.1 describes in more detail the workflow of processing individual raw data, extracting measurements, such as *organ fat* from data and the anthropometric data. Section 5.2 describes the workflow for processing data queries coming from the BodyPass system. CUSI is a graphical web interface that allows data manipulation, data visualization and annotation tools. Within the Data Node CUSI is used as a GUI to control the workflows, for visualization of the avatars and for annotation of the DICOM images. During the BodyPass project doctors used CUSI to define new volumes inside the parametric body for fat measurements and to get the visualization of the computed avatars.

This pilot has demonstrated that BodyPass ecosystem is a promising tool for developing new, less-invasive methods to measure fat than current ones.

Below, we describe the process for:

(a) CT image processing.
(b) Data query processing.

5.1 CT Image Processing

The primary aim of DICOM processing is to extract internal and external body measurements required by healthcare and consumer goods pilots, see Table 1.

Table 1 Body measurements

Internal measurements	FatFractionHeart, VolumeHeart, FatFractionLiver, VolumeLiver, FatFractionKidneys,[a] VolumeKideneys, FatFractionSubcutaneous,[b] VolumeSubcutaneous, VolumeBody, FatFractionPsoas, VolumePsoas, FatFractionSinus, VolumeSinus, FatFractionViceral, VolumeViceral
External measurements	HeadGirth, NeckGirth, UpperArmGirth, WristGirth, BustGirthContoured, WaistGirth, HipGirth, Mid_ThighGirth, LowerKneeGirth_Calf_, AnkleGirth, UpperArmLength, LowerArmLength, HandLength, UpperLimbLength, InsideLegHeight, HipBreadth, ShouldersLength, TrunkLength, ForearmGirth, ThighLength, CalfLength, stature
Reported in the DICOM header	Age, gender, weight, height
Low-resolution skin avatars	Full-body avatar

[a]The small size of kidney and psoas organs put high demand on the accuracy of the localization of the organs for fat and fat fraction counting, which requires further feasibility and accuracy evaluations
[b]Since raw DICOM scans are limited to thorax-abdomen scans, the measurement model of the subcutaneous volume is also limited to the thorax-abdomen area

Table 2 Different surfaces and volumes in the avatar

Body part	# Points	# Triangles	# Tetrahedrons/q5k
Full body	82, 880	106, 119	384, 868
Subcutaneous	40, 071	20, 270	219, 053
Liver	33, 034	7000	92, 785
Kidneys	30, 994	5000	179, 703
Psoas	31, 095	5000	180, 955
Lungs	36, 064	9670	213, 988
Heart	42, 122	25, 000	217, 730
Sinus	29, 923	3600	174, 238

Fig. 11 Introduction of new fat measurements using BARISTA CUSI

The software relies on proprietary Philips algorithms. Since these algorithms have never been published, we can only disclose a brief outline on the underlying computations.

We start with pre-processing of the raw DICOM scans including automatic segmentation of some key 3D surfaces of the human body's anatomy. At the beginning of the BodyPass project, we had a choice whether to extract fat measurements directly from the CT scans, which would require to find and segment each organ of interest, or to register a 3D body avatar and use it as organ atlas. We have chosen the second approach because it allows to define and add new fat measurement locations after the avatar registration. Thus we register the volumetric 3D body avatar to the DICOM scan volume. The volumetric 3D body avatar is a graph in Euclidean space consisting, in our case, of 356,106 vertices, where neighbouring vertices are connected by 185,259 triangles and 1,937,558 tetrahedrons. Table 2 shows an example of the allocation of the avatar elements to different organs of interest.

Barista CUSI provides a collection of data annotation tools, and, amongst others, a tool for manual segmentation of slices on a reference CT scan, which can be used to define new volumes for fat measurements, see Fig. 11.

The registration process consists of two stages: at the first stage an approximate avatar is computed from the *Age, Gender, Weight* and *Height* parameters of the patient, using statistical shape models trained on about 30,000 3D scans. In the second stage the avatar is consistently refined so that the boundaries of organs on the avatar become aligned to the boundaries of the organs segmented on the DICOM image. In the initial version of the registration algorithm only three organs are automatically segmented—skin, subcutaneous volume and lungs—which limits the accuracy of the registration in smaller organs of the visceral cavity like kidney and psoas. There is ongoing work to improve the registration accuracy by adding more automatically segmented organs. The registered avatar is then used as a sort of 'atlas' to compute the local fat volumes according to the parametrically defined sampling positions. This results in the creation of 'internal' measurements. Please, note that the DICOM images are acquired in the laying pose with hands lifted up, while, according to D3.1, it is required that the measurements are collected in the standing A-pose. Therefore, an experimental algorithm was implemented that, for every registered volumetric avatar, computes another avatar in the standing pose. The external measurements are defined parametrically, as close as possible to ISO 8559-1:2017 according to the anthropometric data dictionary previously defined. This assures data harmonization with the consumer goods sector.

5.2 Data Query Processing

The software implements asynchronous query processing where the BodyPass system has to submit new queries, using forms to the dataset 'MeasurementQueries' in the study called 'BodyPass'; and then retrieve the same forms augmented with the query results. The avatar query forms should have filled the 'TemplateId' field with a web-link and a randomized 3D body template. The form field 'Sql' specifies the cohort selection, followed by an optional 'average' or 'regression' operator. Figure 12 shows an example query form and some possible SQL specifications. The processed form received contains a 'DataLink' pointing to the form with the query results that are stored in the dataset specified by 'UseCaseName' in the Study corresponding to the 'ClientName'.

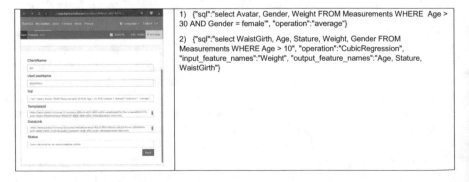

Fig. 12 CUSI Data query form (left), and example specifications of 'Sql' field (right)

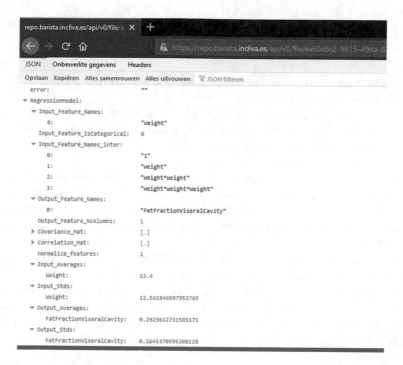

Fig. 13 The polynomial regression model

Table 3 Execution times

Operation	Time
Query transfer from 'edge' to 'internal' Barista + (de)serialization	16 s
SQL parsing (in memory active DB synchronization)	<1 s
De-anonymization control (after many queries, threshold = 10	<6 s in avg
Random template registration	<40 s
Average measurement/avatar computation	<1 s
Query transfer from 'internal' to 'edge' Barista + (de)serialization	16 s

Figure 13 shows an example response of the system for the query to compute on-premise at INCLIVA hospital the polynomial regression of the fat percentage in the visceral cavity as a function of weight.

Observe that the polynomial regression models can be easily aggregated, therefore providing opportunity for federated learning.

Table 3 shows the time required to execute different stages of the query processing, averaged over 100 queries on a data set of 37,000 records.

6   The Application of the BodyPass Ecosystem in the Apparel Industry

We describe in this section the application of the BodyPass ecosystem by two companies in the apparel industry.

P&R Têxteis S.A., founded on May 13, 1982, operates in the sector of the clothing industry, more specifically in the Apparel Sport Technical segment. Since its foundation, P&R's investment policy priorities have focused on the permanent readjustment of the Company's structure to its markets. This strategy was reflected in a constant focus on productive innovation, as well as in research and development of new innovative products that surprise the market, in differentiating factors such as quality, environment, responsiveness, flexibility and customer proximity service. P&R applies BodyPass for improving the design process of sports technical clothing.

ELSE Corp is an Italian startup that offers B2B and B2B2C solutions to brands, retailers, manufacturers and independent designers. ELSE designs and develops a Cloud SaaS platform that puts together the front-end retail processes such as product personalization and virtual 3D commerce of exclusive, personalized, possibly made to measure products. ELSE applies BodyPass for developing better online services for the apparel industry.

6.1   The Use of 3D Data for Designing Sports Technical Clothing

Sports garments require precise fitting, especially in the market segment aimed at high-performance at athletes. The traditional process needs to manufacture several prototypes until the final result is obtained: a perfect fit! The importance of this development is related to the fact that garments worn by athletes influence their performance, achievements and results.

The process of design and engineering of functional clothing design is based on the outcomes of an objective assessment of many requirements of the user, such as physiological, biomechanical, ergonomic and psychological [27]. All these requirements intensify its importance, when we are talking about high-performance athletes, such as Olympics athletes, when all the details, which could influence winning or losing by seconds, matter.

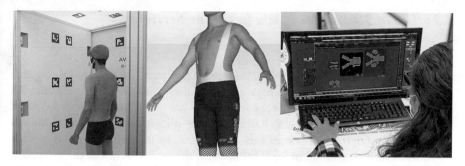

Fig. 14 Design process: scanner (left), 3D avatar (middle), pattern design (right)

For example, badly fitted clothing can cause friction and injury: loose shorts can cause drag on pedalling motions, a tight top can prevent fluid movement, among so many others. In this sense, 3D design is more than a tendency from the market, it's a need to the textile industry, among other criteria like quality and sustainability, to guarantee competitiveness in a worldwide market.

The API-Ecosystem developed by BodyPass is used for processing and exchanging 3D data in a secure manner, respecting privacy protocols. The BodyPass API's are used for:

(a) Processing the 3D data obtained with in-house 3D body scanner in order to retrieve accurate 3D human models and metrics for the development of sports garments. The 3D data obtained from BodyPass and the metrics are used by patternmaking software (Fig. 14). In this way, P&R reduces the number of prototypes needed to achieve a perfect fit for customized products is reduced, improving the efficiency of the process.

(b) BodyPass also allows access to specific 3D information from specific target consumers (e.g. by country or age), in view of the possibility to create, for example, a new collection for the segment of Winter bikewear directed to taller athletes (e.g. Northen Europe). This statistical 3D data could transmit pertinent information that will help in the development of the collection, having in concern the requirements of the market.

Looking to the future, BodyPass represents the next step in the secure transmission of 3D data between sectors, improving the ability of customization. A possibility to have access to a customer scan from the other side of the world (taken in another company) that will be used to produce an in-house garment perfectly fitting this customer, without the need for further fitting travels.

6.2   Use of 3D Personal Data in Online Services for the Apparel Industry

The apparel industry is going digital and every day more brands are joining this inevitable process. From the business and consumer point of view, traditional 2D images are becoming outdated because products are now presented and delivered digitally. Garments should be elaborated in a three-dimensional format that demonstrates the real physical properties such as material, texture, color and the product physical construction. In addition, 3D clothes are shown on realistic bodies and body parts. This makes BodyPass a crucial element of the new digital environment for the apparel industry.

BodyPass ecosystem can be made available and easily used by retail companies and tech providers. We have tested in three different scenarios how the data comes from the consumer, passes through the BodyPass ecosystem and is transformed to make it available via APIs. The processed information can be incorporated by applications such as ELSE Corp's Virtual Retail platform, which is used directly by brands.

- **Manufacturing Scenario**, also called Industrial Made to Measure: To enable companies to produce garments which are almost made to measure but still manufactured in an industrial way. This allows individual fitting by finding the most suitable items for the consumer.
- **Design Process**: To help brands and designers to create new collections by being oriented to the concrete avatars, group of people and target markets.
- **Marketing and Operations**: Reliable 3D data allows better segmentation and understanding of specific markets. The information can be used to take more accurate decisions regarding brand positioning and distribution channels.

6.2.1   Manufacturing

The apparel manufacturing process is a complex and detailed work. In the case of made to measure production, the complexity and operations increase significantly. Made to measure items must fit the persons body, and at the same time be produced together with other orders to maintain efficiency in an industrial level. The use of reliable information of body measurements allow manufacturers to create more flexible production lines. 3D data can be utilized for improved production planning and to achieve better understanding of orders placed.

Nowadays privacy is one of the most valuable assets for customers. Automation of virtual fitting on 3D body avatars in an anonymous way is one of the obligatory tasks, and BodyPass database allows to find similar bodies by measurements and create individual online orders without disclosing private information.

6.2.2 Design

In the design scenario, the BodyPass 3D anthropometric data is key. Currently, one of the most important stages of design collection creation and production is determining the size stage, which is traditionally based on statistical data. This data is often outdated because it is difficult to renew such a huge amount of statistical information in a short period of time. It represents an inconvenience for the apparel manufacturing industry, because the body parameters of the average person change more often than when the statistics are updated. Based on the dimensions of the past, the so-called "bullwhip effect" is created, when a single size range that differs significantly from reality is still used by designers. As a result, the market is overflowing with clothing that doesn't fit the potential buyer.

Based on the value created by BodyPass, it is now becoming a reality for the clothing industry to create products, which covers all parameters of the human body. It is important for producers to classify data according to different criteria (gender, age, geographical area, etc.). Updating BodyPass data regularly will be necessary because these parameters are dynamic.

Traditional body data in tables can often give the designer a blurry view of the body. But with the use of 3D modelling, it is easier for designers to understand the overall body shape and make measurements of any part of the body that is necessary for sewing a separate unit of clothing. The BodyPass technology allows not only to produce a relevant product, but to reduce the production of unsuitable clothes, creating only what will be worn.

BodyPass is not just a database of human body data. Digital analogs of the human figure can help to understand the proportions of the body. The use of BodyPass and designer's creativity together provide the ability to create a unique style based on individual characteristics of the body. The correct selection of clothing models will emphasize the advantages and provide aesthetic enhancements based on precise measurements.

6.2.3 Marketing and Operations

In the Era of Body Positivity, which focuses on challenging body standards, it becomes important for people to be able to wear clothing that will be tailored to their individual standards. This approach can increase brand positioning due to product comfort and tailored services. In addition, companies can support and execute strategies based on personalized made to measure items. This can act as a communication tool to refresh brand positioning and enhance customers' perception.

Creating clothes according to individual requirements is a trend that has been around for years already. Examples there are many, from luxury brands that have offered made to measure services along their history, to more affordable brands that have democratized product customization. In fact, according to Deloitte [28], on average 36% of consumers expressed an interest in purchasing personalized products or services. Nowadays it is possible to save the parameters of body types and measurements for future use, and thanks to 3D modeling brands are able to modify collections according to the figure of a specific person or market, producing products that fit perfectly without compromising any of its features.

By getting statistical data about body measurements, it is possible to reduce costs because knowing the exact size stage, brands can buy the necessary amount of materials. After adjusting some necessary parameters (height, chest, waist, etc.) and using 3D modeling as a tool, the perfect collection for a market can be created and sent to production. Another advantage that 3D data offers is the possibility of segmenting orders by body type and regions, which enables businesses to optimize their distribution channels and to reduce transportation and warehouse costs.

Nowadays e-commerce represents an essential revenue stream for businesses around the world. Even though online sales are already a developed form of retail, there are still disadvantages specially for products that have a measure for fitting. Sizing issue is a reason for shoppers returning online orders. According to a research performed by Global Web Index in the USA and UK, 52% of people had to return an item because the fit was not right and they could not try it on before they bought it [29]. The use of avatars and anthropometric data segmented by range of age and country could solve the fitting problem for online retailers.

The benefits of a reliable database of body measurements do not stop there. The information can be utilized to develop better shipping and return policies. Almost 50% of online shoppers buy multiple sizes of a product in order to ensure the right fit [30]. In addition, 14% of customers buy items they don't need so they could qualify for free shipping, with the intention of returning them after [29]. In the US alone, a more accurate size segmentation and access to reliable data, could help online retailers to reduce return expenses for more than US$107 billion lost yearly [31].

Thus, the usage technology of BodyPass in online services for the apparel industry will contribute to the digital transformation of the fashion industry, optimizing production process, reducing costs and waste.

7 Conclusions

BodyPass has developed an ecosystem of APIs and tools that allow the exchange of 3D anthropometric data that preserves IP rights and personal privacy. This is achieved through the use of:

- Semantic data annotation with a data dictionary compatible with ISO and CEN (European) standards.
- GDPR compliance with the use of tools that create anonymous synthetic data in 3D, the implementation of an architectural solution in hospitals that guarantees additional protection to sensitive data and the use of off-chain storage blockchain.

BodyPass ecosystem offers a novel approach for effective data sharing of 3D human data between data silos. The tools developed in BodyPass contribute specifically to two of the technical priorities defined in the framework of the European Big Data Value Strategic Research and Innovation Agenda: data management and data protection.

Acknowledgements This research was funded by the BodyPass project, which has received funding from the European Union's Horizon 2020 research and innovation programme under grant agreement No 779780.

References

1. Zillner, S., Curry, E., Metzger, A., Auer, S., & Seidl, R. (2017). *European Big Data Value. Strategic Research & Innovation Agenda*. Big Data Value Association. https://bdva.eu/sites/default/files/BDVA_SRIA_v4_Ed1.1.pdf
2. ASTM D5219 Standard Terminology Relating to Body Dimensions for Apparel Sizing, ASTM D5219 (2015).
3. *ISO 7250-1:2017—Basic human body measurements for technological design – Part 1: Body measurement definitions and landmarks*. (2017). https://www.iso.org/standard/65246.html
4. ISO 8559:1989 Garment construction and anthropometric surveys—Body dimensions, ISO, 8559 (1989).
5. *ISO 18825-1:2016—Clothing – Digital fittings – Part 1: Vocabulary and terminology used for the virtual human body*. (2016). https://www.iso.org/standard/61643.html
6. Gordon, C. C., & Bradtmiller, B. (1992). Interobserver error in a large scale anthropometric survey. *American Journal of Human Biology, 4*(2), 253–263.
7. Gordon, C. C., Churchill, T., Clauser, C. E., Bradtmiller, B., & McConville, J. T. (1989). *1988 Anthropometric survey of US army personnel: Methods and summary statistics* (Technical Report Natick/TR-89/044). Anthropology Research Project Inc. Yellow Springs OH.
8. Kouchi, M., & Mochimaru, M. (2008, June 17). *Evaluation of accuracy in traditional and 3D anthropometry*. https://doi.org/10.4271/2008-01-1882
9. Kouchi, M., Mochimaru, M., Tsuzuki, K., & Yokoi, T. (1996). Random errors in anthropometry. *Journal of Human Ergology, 25*, 12,155–12,166.
10. Ballester, A., Parrilla, E., Uriel, J., Pierola, A., Alemany, S., Nacher, B., Gonzalez, J., & Gonzalez, J. C. (2014). 3D-based resources fostering the analysis, use, and exploitation of available body anthropometric data. In *5th International Conference on 3D Body Scanning Technologies*.
11. Ballester, A., Pierola, A., Parrilla, E., Uriel, J., Ruescas, A. V., Perez, C., Dura, J. V., & Alemany, S. (2018). 3D human models from 1D, 2D and 3D inputs: Reliability and compatibility of body measurements. In *Proceedings of 3DBODY.TECH 2018 – 9th International Conference and Exhibition on 3D Body Scanning and Processing Technologies, Lugano, 16–17 Oct. 2018* (pp. 132–141). https://doi.org/10.15221/18.132
12. Durá-Gil, J. V., Remon, A., & Ballester, A. (2020). *bodypass-project/Python_GUI: First Public Version* (v1.0) [Computer software]. Zenodo. https://doi.org/10.5281/ZENODO.4269292
13. Allen, B., Curless, B., & Popović, Z. (2003). The space of human body shapes: Reconstruction and parameterization from range scans. *ACM SIGGRAPH 2003 Papers* (pp. 587–594). https://doi.org/10.1145/1201775.882311
14. Mulligan, C., Scott, J. Z., Warren, S., & Rangaswami, J. P. (2018). Blockchain beyond the hype: A practical framework for business leaders. *White Paper of the World Economic Forum*.
15. IBM. (2018). *Why new off-chain storage is required for blockchains*. https://www.ibm.com/downloads/cas/RXOVXAPM
16. Lyons, T., Courcelas, L., & Timsit, K. (2018). *Blockchain and the GDPR* [Thematic Report]. The European Union Blockchain Observatory & Forum. https://www.eublockchainforum.eu

17. Hyperledger Fabric. (n.d.). Hyperledger. Retrieved 3 November 2020, from https://www.hyperledger.org/use/fabric
18. Hyperledger Composer. (n.d.). Retrieved 3 November 2020, from https://hyperledger.github.io/composer/latest/
19. PrestoDB. (n.d.). Retrieved November 3, 2020, from http://prestodb.github.io/
20. WHO. (2017). *Prevalence of obesity and overweight worldwide 2016* (Bulletin WHO).
21. Stefan, N. (2020). Causes, consequences, and treatment of metabolically unhealthy fat distribution. *The Lancet Diabetes & Endocrinology, 8*(7), 616–627. https://doi.org/10.1016/S2213-8587(20)30110-8
22. Tsatsoulis, A., & Paschou, S. A. (2020). Metabolically healthy obesity: Criteria, epidemiology, controversies, and consequences. *Current Obesity Reports, 9*(2), 109–120. https://doi.org/10.1007/s13679-020-00375-0
23. Brandão, I., Martins, M. J., & Monteiro, R. (2020). Metabolically healthy obesity—Heterogeneity in definitions and unconventional factors. *Metabolites, 10*(2), 48. https://doi.org/10.3390/metabo10020048
24. Fava, S., Fava, M.-C., & Agius, R. (2019). Obesity and cardio-metabolic health. *British Journal of Hospital Medicine, 80*(8), 466–471. https://doi.org/10.12968/hmed.2019.80.8.466
25. Bosy-Westphal, A., Braun, W., Geisler, C., Norman, K., & Müller, M. J. (2018). Body composition and cardiometabolic health: The need for novel concepts. *European Journal of Clinical Nutrition, 72*(5), 638–644. https://doi.org/10.1038/s41430-018-0158-2
26. Antonopoulos, A. S., & Tousoulis, D. (2017). The molecular mechanisms of obesity paradox. *Cardiovascular Research, 113*(9), 1074–1086. https://doi.org/10.1093/cvr/cvx106
27. Gupta, D. (2011). Design and engineering of functional clothing. *Indian Journal of Fibre and Textile Research, 9.*
28. Deloitte. (2019). *The Deloitte Consumer Review Made-to-order: The rise of mass personalisation.* Deloitte. https://www2.deloitte.com/content/dam/Deloitte/ch/Documents/consumer-business/ch-en-consumer-business-made-to-order-consumer-review.pdf
29. Gilsenan, K. (2018, December 3). Online shopping returns: Everything retailers need to know. *GWI.* https://blog.globalwebindex.com/chart-of-the-week/online-shopping-returns/
30. Berthene, A. (2019, September 12). Sizing issue is a top reason shoppers return online orders | Why online shoppers make returns. *Digital Commerce 360.* https://www.digitalcommerce360.com/2019/09/12/sizing-issue-is-a-top-reason-shoppers-return-online-orders/
31. Rakuten. (2018). *Imperfect fit and the $100 billion cost of returns.* Rakuten. https://fits.me/wp-content/uploads/2018/01/whitepaper.pdf

Using a Legal Knowledge Graph for Multilingual Compliance Services in Labor Law, Contract Management, and Geothermal Energy

Martin Kaltenboeck, Pascual Boil, Pieter Verhoeven, Christian Sageder, Elena Montiel-Ponsoda, and Pablo Calleja-Ibáñez

Abstract This chapter provides insights about the work done and the results achieved by the Horizon 2020-funded Innovation Action "Lynx—Building the Legal Knowledge Graph for Smart Compliance Services in Multilingual Europe." The main objective of Lynx is to create an ecosystem of multilingual, smart cloud services to manage compliance based on a Legal Knowledge Graph (LKG), which integrates and links heterogeneous compliance data sources including legislation, case law, regulations, standards, and private contracts. The chapter provides a short introduction in regards to the market needs, gives an overview of the Lynx services available on the Lynx Services Platform (LySP), and provides valuable insights into three real-world compliance solutions developed on top of LySP together with Lynx's industry partners, namely, (1) Labor Law (Cuatrecasas, Spain), (2) Contract Management (Cybly, Austria), and (3) Geothermal Energy (DNV.GL, the Netherlands).

Keywords Compliance · Legal knowledge graph · Multilingualism · NLP · Labor law · Contract management · Geothermal energy

M. Kaltenboeck (✉)
Semantic Web Company, Vienna, Austria
e-mail: martin.kaltenboeck@semantic-web.com

P. Boil
Cuatrecasas, Barcelona, Spain

P. Verhoeven
DNV, Utrecht, The Netherlands

C. Sageder
Cybly, Salzburg, Austria

E. Montiel-Ponsoda · P. Calleja-Ibáñez
Universidad Politécnica de Madrid, Madrid, Spain

© The Author(s) 2022
E. Curry et al. (eds.), *Technologies and Applications for Big Data Value*,
https://doi.org/10.1007/978-3-030-78307-5_12

Insights into the industry solutions include the concrete business cases, the problem statements and requirements, the relevant data identified and used, and the LySP AI services combined to realize powerful multilingual compliance solutions in the respective fields. The chapter closes with findings and learnings from the implementation phase and a future outlook for further developments, specifically for the three vertical solutions and LySP.

The chapter relates to the technical priorities of Data Management and Data Analytics of the European Big Data Value Strategic Research and Innovation Agenda [1]. It addresses all challenges of the horizontal concern Data Management and some of the challenges of the horizontal concern Data Analytics of the BDV Technical Reference Model. It addresses the vertical concerns: (a) Big Data Types and Semantics (with a focus on Text data, including Natural Language Processing data and Graph data, Network/Web data and Metadata) as well as (b) Standards (standardization of Big Data technology areas to facilitate data integration, sharing, and interoperability). The chapter relates to the Reasoning and Decision Making cross-sectorial technology enablers of the AI, Data and Robotics Strategic Research, Innovation and Deployment Agenda [16].

1 Introduction: Building the Legal Knowledge Graph for Smart Compliance Services in Multilingual Europe

Currently, European small and medium-sized enterprises (SMEs) and companies operating internationally or wanting to branch out to other countries and markets, *face multiple difficulties to engage in trade abroad and to localize their products and services to other countries*, owing to legal and also to language barriers in Europe. As reported by the European Commission, only 7% of European SMEs sell across borders. SMEs that sell their products and services internationally exhibit 7% job growth and 26% innovation in their offering, compared to 1% and 8% for SMEs that do not go outside their local markets [2]. A key challenge for businesses in Europe is, thus, how to engage with customers effectively across the legal and language barriers.

One of the main problems is the *management of compliance across different countries*. "Compliance is a term generally used to refer to the conformance to a set of laws, regulations, policies, or best practices" [3]. When companies want to sell a product or offer a service in a new market they must comply with the applicable legislation (European, regional, local), implement different standards (e.g., from ISO, AENOR, or DIN [4]) and possibly follow sector-specific best practices. Dealing with *legal and regulatory compliance data* is a cumbersome task usually delegated to legal and consultancy firms that obtain documents from several data sources, published by various institutions according to different criteria and formats.

While data analytics is trying to address the issue of data heterogeneity from a technical viewpoint, the more human side of the data still remains a greenfield, i.e., the inherent incompatibility of multiple natural languages, which not only involve different words but also different syntax and different semantics. Europe is determined to make the most of the linguistic wealth that characterizes the continent. An increasing number of voices are in favor of a stronger commitment towards a multilingual Digital Single Market (DSM) as the key for becoming the most competitive market in the world [5]. As per the former EC vice president Andrus Ansip: "Overcoming language barriers is vital for building the DSM, which is by definition multilingual. It is now time to reduce and remove the language barriers that are holding back its advance, and turn them into competitive advantages" [6].

With the aim of addressing the challenges posed by the European market, currently fragmented into legal silos and split into more than 20 linguistic islands, constituting a competitive disadvantage for SMEs and large companies in general, and in line with other initiatives in Europe that share the same spirit (e.g., Digital Single Market, ISA, CEF.AT [7]), Lynx has created an ecosystem of smart cloud services that exploit a multilingual Legal Knowledge Graph (LKG) of legislation, regulations, policies, and standards from multiple jurisdictions.

This cloud of services integrated in the Lynx Services Platform (LySP) provides mass-customized regulatory information to European businesses. Additionally, it supports the creation of a common legal ICT infrastructure that will contribute to unlocking the potential of a multilingual and truly single digital market.

In order to achieve these objectives, the Lynx platform managed to: (1) create a novel and unique knowledge base related to compliance, integrating information from heterogeneous data and content sources; (2) provide a set of multilingual and smart core services to extract value from the knowledge base, and (3) translate its value into the market in the form of three business-driven pilots, making use of LySP.

In the first step, the LySP acquired—and continuously maintains—data and documents related to compliance from multiple jurisdictions in different languages, as well as interlinked terminologies and language resources, open standards, and sectorial best practice guidelines. This collection of structured data and unstructured documents, obtained from open sources, was the base of the LKG [8].

In the second step, a set of core domain-agnostic services was put in place to analyze and process the documents and data in order to integrate them into the LKG. Existing multilingual terminologies, semantic tools, and machine learning mechanisms were adapted and customized to the legal domain and used to annotate, structure, and interlink the LKG contents. Iteratively and incrementally, the LKG has been developed and augmented by linking to external databases and corpora, by discovering topics and entities linked implicitly, as well as by using translation services to translate documents not previously available in certain languages.

In the third and final step, these services have been configured in three real-world pilots according to the industry needs represented by the Lynx business cases. These vertical solutions exploit the knowledge available in the LKG and have been driven forward and evaluated by companies with an existing customer base.

The main objective of Lynx is to *facilitate compliance for companies in internationalization processes*, by leveraging the existing European *legal and regulatory open data* seamlessly *interlinked* and offered through a set of *cross-sectorial, cross-lingual smart services in the Lynx Services Platform: LySP*. SMEs and other organizations can benefit from LySP through: (1) companies directly making use of the LySP Services and (2) companies in the portfolio of law firms and consultancy companies making use of LySP, both through using LySP Services either standalone (as a self-service) or integrated into existing IT systems.

2   The Lynx Services Platform: LySP

LySP is a cloud of smart and multilingual services working on top of the LKG and acting as a basis for training and operation of end user services. As illustrated in Fig. 1, the LKG contains law and legal information, directives, regulations, and other

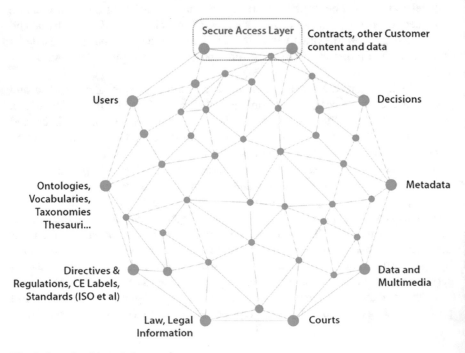

Fig. 1 Lynx legal knowledge graph

relevant data and information harvested from public sources, integrated and enriched by making use of the Lynx data model [15]. Additionally, the LKG has been expanded—in a secure layer—by data and information of the pilot applications.

LySP Services have been developed (1) from scratch by the Lynx partners: Universidad Politécnica de Madrid (Spain), Semantic Web Company (Austria), Cybly GmbH (Austria), Deutsches Forschungszentrum für Künstliche Intelligenz GmbH (Germany), Alpenite (Italy), or (2) through the adaptation of existing software components, namely: Tilde Translator (https://tilde.com/products-and-services/machine-translation) by Tilde, Lexicala (https://www.lexicala.com/) by KDictionaries, and PoolParty Semantic Suite (https://www.poolparty.biz) by Semantic Web Company, all of them docked onto and trained by the LKG to ensure the full value of LySP Services in the field of LegalTech and compliance. Through the orchestration of these services, a broad portfolio of real-world use cases can be created [9]. In the framework of the project, three pilots have been developed, as further explained in Sect. 3.

The key principles according to which LySP is being built are summarized in the following [10]:

- Token-based OAuth2 protocol for authorization together with the centralized access control and authorization rules management based on Keycloak.
- An established LynxDocument schema according to the LKG ontology.
- Containerized deployment in an orchestrated application platform making use of Red Hat OpenShift.
- Workflow Manager based on Camunda.
- LinkedDataPlatform-inspired Document Manager.
- Common rules for the development of web APIs: REST + API gateway patterns, including OpenAPI 3 description.

In its current status, LySP provides 16 services that accomplish different purposes, as listed below:

LySP Enrichment Services

LySP Annotation Services

1. Temporal Expression Recognition (TimEx): finds temporal expressions in documents.
2. Named Entity Recognition (NER): finds named entities using state-of-the-art methods.
3. Geographical NER (Geo): finds geographical entities in documents.
4. Relation Extraction (RelEx): extracts relations between entities.
5. Entity Linking (EL): identifies and links entities, provides annotations, including word sense disambiguation.

LySP Conversion Services

1. Machine Translation: translates documents.
2. Summarization: summarizes the content of a document.

LySP Search and Information Retrieval Services
1. Question Answering (QADoc): retrieves the most relevant answer for a given question.
2. Cross Lingual Search (Sear): searches a text string in documents across different languages.
3. Semantic Similarity (SeSim): calculates similarity between any two documents.
4. Terminology Query (TermQ): obtains information about a certain term with examples and notes of use.

LySP Vocabulary Services
1. Dictionary Services (DA): queries domain-independent dictionaries from SPARQL endpoint.
2. Terminology Extraction (TermEx): extracts terminology from document corpus.

LySP Platform Services
1. Workflow Manager (WM): manages workflows defined in BPMN.
2. Document Manager (DCM): manages documents and annotations in the LKG.
3. Authentication and Identity Management (APIM): provides Lynx identity, OAuth2 flows, and social login.

At the moment, most of the services are available for the languages English, German, Spanish, and Dutch [11]. The current version of the LySP Architecture can be seen in Fig. 2, where arrows and colors illustrate the principal workflows. Broadly speaking, a collection of documents (corpus) is ingested into the platform where TermEx performs a terminology extraction process (red arrow). Next, documents are annotated by means of LySP Enrichment Services. Some services depend on the annotations produced by specific services, whereas others can run in parallel. To efficiently orchestrate the different services, a dedicated Workflow Manager based on Camunda is used. The result of this process is what we call an Enriched Lynx Document. The service in charge of efficiently storing, updating, and retrieving documents is the Document Manager. Enriched documents are then stored for subsequent retrieval by LySP Search and Information Retrieval Services (Storage and Information Retrieval box in Fig. 2). For more details on how the services are orchestrated in LySP, we refer the interested reader to [9]. All three Lynx compliance solutions described in detail in the next section are built on top of LySP and the workflows explained above.

3 Lynx Compliance Solutions

The purpose of this section is to describe the three real-world compliance solutions that have been developed on top of LySP together with Lynx's industry partners, namely: (1) Labor Law (Cuatrecasas, Spain), (2) Contract Management (Cybly, Austria), and (3) Geothermal Energy (DNV.GL, the Netherlands).

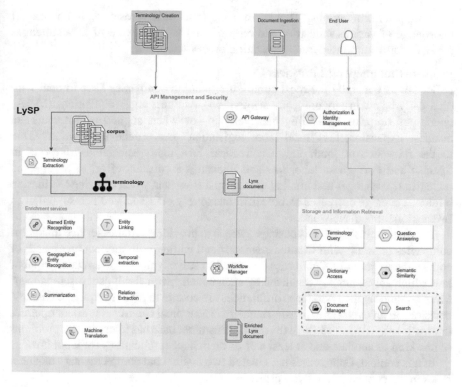

Fig. 2 Architecture of LySP

Each solution is described according to the same structure. First, the industry partner involved in the solution is introduced, to provide the context for the needs and requirements of each business case in what we have called "Problem Statement and Business Case." Next, the solution is spelled out and, finally, details of the Lynx services involved are provided.

3.1 Compliance Services in Labor Law (Cuatrecasas, Spain)

About Cuatrecasas

Cuatrecasas (www.cuatrecasas.com) is an international law firm with headquarters in Barcelona, Madrid, and Lisbon. The firm is specialized in all areas of business law, applying a sectoral approach and covering all types of business. It represents several of the largest international companies, advising them on their investments in the major markets in which they operate. Cuatrecasas is present in the main financial centers of Europe, America, Africa, and Asia through international offices; European Network teams in Germany, France, and Italy; and international desks covering over 20 regions. Thanks to its international presence, it has expert knowledge of various industries and regions, and is well aware of the challenges posed to companies in internationalization processes.

Problem Statement and Business Case

Labor law is generally the most common regulation companies have to deal with on a daily basis. In any company acquisition—Mergers and Acquisitions (M&A) operations or prior Due Diligence analysis—or when supporting international business expansion, the "local labor legislation" has crucial implications. Due to the relevance of labor law, Cuatrecasas has highly specialized lawyers in Spanish and Portuguese labor law and dedicates a constant effort to be updated on legislative changes and binding precedents. However, when crossing the Iberian borders, coverage decreases and the firm has to rely on associated firms (similar to International Legal Networks).

The main objective of this business case is to provide a reliable service that helps companies (starting with Cuatrecasas itself and ending with any company and, of course, the majority of Cuatrecasas' clients) to solve typical issues related to labor law, which are commonly regulated by each country with significant discrepancies. More commonly than not, these differences are crucial to making strategic business decisions in an overseas expansion strategy. The typical Cuatrecasas client operates internationally. Around 30% of the current customer base has international problems of one kind or another, and at least half of them have implications for labor law.

In this context, Cuatrecasas has created two distinct but complementary business cases in the context of the Lynx project:

Internal Usage. A tool for Cuatrecasas' lawyers

The first business case is based on an internal approach that would result in time and cost savings for Cuatrecasas' legal assessment related to country-specific labor law. The firm, typically as part of an M&A (Mergers and Acquisitions) operation or Due Diligence, normally sends out questionnaires on labor law with frequently asked questions (one questionnaire has between 10 and 50 questions) to an average of six to ten partner firms from different countries (jurisdictions). Usually, the completion of these questionnaires is subcontracted to local partner law firms. The partial cost savings estimation gives a basis for the ROI justification. The pilot developed during the Lynx project lifetime covers the four countries/languages of the Lynx project (ES, IT, DE, EN) due to available resources. However, the numbers are also interesting when other countries (languages and jurisdictions) of interest for Cuatrecasas clients (e.g., Russia, China, Mexico, Brazil) are included. In the second scenario (EU and non-EU countries), the expected cumulative cost reduction benefits would be close to the 3 MM € in 5 years.

External Usage. A tool for Cuatrecasas customers

The second business case is based on a SaaS (Software as a Service) approach (new line of business income), putting the solution directly in the hands of Cuatrecasas' big customers with a high level of internationalization. This scenario is not a very aggressive one regarding pricing, since it could be considered rather a loyalty system than a software product itself. The cumulative figures of this second use case are quite similar to the first one. Estimating 3MM € in a 5-year plan is a conservative projection, given that a minimum of 25 existing Cuatrecasas' clients could use the Lynx platform.

The Solution

In Lynx we focused on the internal use case in the Lynx project. The idea was to test, improve, and evaluate accuracy, and show internal value before presenting the solution to customers. As a support tool for Cuatrecasas' lawyers this application should be executed internally (inside the corporate Cuatrecasas' network).

Cuatrecasas provides a range of services to clients, including (1) specific operations, which typically involve a project with a limited scope and period; and (2) general legal advice, which is usually categorized by practice area (e.g., labor, tax, and corporate) due to the different legal specializations required. Moreover, lawyers at Cuatrecasas may work on more than one matter and with multiple clients at a given time. For this reason, the system must provide lawyers with the tools they need to organize and optimize their tasks, enabling them to configure and save their favorite options (more common/default): personal or client/company.

Although Cuatrecasas has offices in multiple countries, the firm's official languages are Spanish, English, and Portuguese. Despite being specialized in Spanish and Portuguese jurisdictions, the firm offers global international coverage to its clients, with a focus on Latin America. Typical clients of Cuatrecasas include large (Spanish and Portuguese) companies with businesses around the world. The main problem that nonlocal lawyers usually face is accessing and understanding foreign local laws and regulations that are often unavailable in other languages.

For this internal use case, users are assumed to be legal experts. Often, they are junior lawyers who are tasked to investigate external regulations. Currently, these lawyers have to contact the internal Knowledge and Innovation Team to find out about (1) the legal particularities of a specific country/jurisdiction, (2) the legal sources available and (3) whether they can count on local lawyers from partnering institutions that can be contacted, if necessary. These lawyers are accustomed to use legal databases and other information resources (e.g., the ones provided by LexisNexis, Thomson Reuters or vLex). Moreover, they usually have a good command of the legal terminology in their own language and in English, but limited knowledge of the legal terminology in other languages.

To fulfill the requirements of this business case, an application has been developed in which the user formulates a complete query in natural language (Spanish, English, German, Dutch) about labor law and workers' regulations, specifying one or more jurisdictions (Spain, Austria, Netherlands). Then, the system returns the most relevant results based on the direct texts of the law, and translated to the language previously selected by the user, including the following:

- The most precise answer possible (when the question is specific, asking for a value, data and name).
- The paragraph(s) related with the topic/question, where the possible answer appears as part of the text (ideally highlighted).

- The context by showing the article (and section) from which the paragraph(s) is extracted, showing the number and title, and allowing the user to view the full text of the article and law, which the user should be able to access and download.

Complex legal questions are almost impossible to answer by only highlighting parts of the law. Context and additional information are often needed. This additional information is sometimes difficult to incorporate into a question and these context words are not always easy to find directly mentioned in laws. To palliate this issue, the system is designed to be used as an intelligent search tool, providing legal guidance to lawyers, to help substitute or minimize some of their less-valued work.

The Use of LySP Services
The Cuatrecasas Lynx Pilot is comprised of four main parts/components (see Fig. 3): a Front-end Application, with the presentation layer responsible for the user experience; a Back-end Application and business logic layer, to encapsulate the defined modules (login, configuration, and Q&A modules) and provide all the required application functionalities; an Application database, to ensure data persistency; and, finally, the Cuatrecasas-Lynx API, a middleware component to encapsulate and centralize interaction with Lynx Services.

The LySP Services used in the Cuatrecasas Lynx Pilot are described below:

- *WM, DCM, and LKG*

    The Cuatrecasas Lynx Pilot makes use of the LySP Services, as defined in Sect. 2, namely, the Workflow Manager (WM), responsible for the effective orchestration of the LySP Services, and the Document Manager (DCM) service, where documents are stored and maintained. As already mentioned, the basic

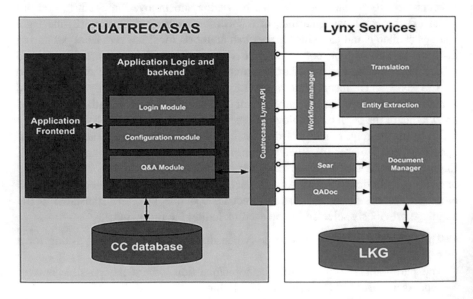

Fig. 3 Pilot modules and components schema of the Cuatrecasas Lynx Pilot

functionality of the DCM includes storing documents and their annotations, particularly in regards to the support of their synchronization, providing read and write access, as well as updates of documents and annotations. The DCM can be queried in terms of annotations (e.g., "which documents mention this entity?"), as well as in terms of documents (e.g., "what are the contents/annotations of document X?"). The interface includes a set of APIs to manage the following resources within LySP: collections, documents, and annotations. DCM is responsible for storing the LKG (Legal Knowledge Graph) and the documents once they have been processed through the different workflows.

Additionally to the LySP Platform services, this application relies on some LySP Annotation and LySP Search and Information Retrieval Services, as specified below:

- *SEAR Service*

  The Cross-lingual search service is used to retrieve documents from collections previously defined by end users. Documents are retrieved from the Document Manager based on metadata and content filters. The service generates a first list of ranked candidate answers (previously broken down into paragraphs), and highlights the text segment that is responsible for the selection. The SEAR service relies on the document enrichment processes performed by the LySP Enrichment Services to allow filtering out searches and to score the results based on the query. Additionally, this service uses Query Expansion (QE) mechanisms to improve search precision and cover the main use case requirements.

- *QADoc Service*

  The QADoc service receives a query posed in natural language and a source text to find a precise answer within it. Only when the service returns a result with a high level of confidence, the application will show this result to the user.

- *TimEx Service*

  Temporal expressions are very relevant in any legal document. For example, expressions for deadlines or regulated procedures are common in the labor context, such as "something has to be done 10 days after the contract is signed," "the probationary period does not exceed six months," or "the cost of dismissing an employee is 20 days per worked year." The pilot makes use of this service to identify time expressions that may contain the answer to a question.

- *Machine Translation Service*

  The translation service provides automated machine translation by using the Tilde MT cloud platform. Currently, the translation service provides support for a runtime scenario and an endpoint for the Lynx platform's asynchronous process in the background. Neural Machine Translation (NMT) systems were trained for the project languages. In the domain of labor law, specific legal and business data was gathered and processed before training the NMT systems on a mix of broad-domain and in-domain data that is able to translate both in-domain and out-of-domain texts.

3.2   Smart Contract Management (Cybly, Austria)

About Cybly

Cybly (www.cybly.tech) is a legal tech company based in Salzburg and Vienna, Austria. It combines two brands or product lines under one roof—"LawThek" and "Legalnetics." "LawThek" is a legal database with content from EUR-Lex (directive, regulation, and decisions), Austria (federal and state laws, decisions), and Germany (federal laws), offering cross-platform access to standardized and interlinked sources of law. In addition to the desktop version, the RIS:App (Right Information System) is distributed. This enables mobile access and is available for free download in the Apple and Google app stores. "LawThek" is complemented by the high-end products and services of "Legalnetics." The range of services offered by "Legalnetics" includes process-oriented, integrated IT solutions in the areas of law, finance, and compliance, as well as all other areas with a legal or legal information background.

Problem Statement and Business Case

Contracting is a common activity in companies, but managing contracts systematically, which means keeping track of changes or updates, is a cumbersome activity only few companies are effective at. Many SMEs (small and medium enterprises) do not have a database with all the information of their contracts, which prevents them from easily finding information or applying changes.

Let us imagine the following situations in the context of a company:

1. There is a change in law, and you need to know which contracts are affected.
2. An overview on all obligations with a certain company is needed.
3. A contract is needed urgently and no one knows where to find the latest version because the responsible employee left the company. Moreover, the opposing party confronts you with a signed amendment or a subsidiary agreement you've never seen before.

Countless organizations are confronted with similar scenarios, although we are all significantly shaping our legal reality by concluding various contracts. Abstractly, the problem can be summarized as follows: Contracts and contract-relevant documents are physically and electronically distributed across the entire organization and tools, e.g., file server, emails, physical documents. As a result, there is often no overview, which leads to inconsistent applications, breaches of contracts, and (financial) disadvantages.

The implementation of a comprehensive cross-organizational contract management process appears to be the solution. Flitsch [14] defines contract management as the creation of ideal structures for: contract planning, contract design, contract negotiations, implementation of contracts, contract administration, and contract archiving. In many cases, organizations are lacking these structures.

When it comes to contracts, there are very few tools being used: a word processor to create the contracts, email to communicate with the client or the other party, a file storage in a defined directory structure, and/or a legal software to store the documents. To keep history recorded, they often add the different versions of the document and individual mail communication to this software or put them on the file system. With this in mind, we are focusing on automated contract administration and archiving.

The Solution

The aim of our solution is not to change the existing workflow, which is well-established in most companies and law firms, but to provide an integrated solution to the existing toolset and workflows.

The starting point and, at the same time, the simplest use case is the analysis of a single contract/document. However, the reality is much more complex: Regularly, a large number of contracts of diverse nature and purposes need to be analyzed and kept track of, taking into account various regulatory frameworks. In order to achieve this, we have two approaches. On the one hand, we have a pure back-end solution, and, on the other hand, we provide a visualization of the created data space for end users.

In the course of the Contracts Management Lynx Pilot, our goal is to develop reasonable strategies for automated contract analysis and contract archiving. Contract administration and management are crucial when it comes to defining the application:

To harvest documents, a command line tool has been implemented to provide the following two main functionalities: Recursively send all documents of a directory and its subdirectory to the Document Service for processing them. Monitoring a given directory and its subdirectory and send notifications when the contents of the specified files or directories are modified. With this tool it is possible to ingest a large set of documents to the system and also to monitor this set for any subsequent changes. Other external systems can use the REST interface of the Document Service to ingest contract-related documents too.

To make use of the harvested document, the Conversion Service converts documents in different data formats into a Lynx-Document that includes metadata and document structure where possible. The following main document formats are currently supported: Microsoft Office document formats, Open Document Format, iWorks, HTML, PDF, Images, Outlook Messages (*.msg), MIME Messages. The newly created Lynx Document is then annotated by the Annotation Service which orchestrates the calls to the different LySP Annotations Services and also to others. The document and its extracted information are stored within the LawThek Document Store. To do so, LawThek has been extended with the possibility to store Lynx Documents with the annotations beside the original document.

Through the front-end solution a single user has the possibility of managing (add, delete, update, group, search, etc.) contracts/documents. The user can view a single contract and related annotations or get a broader view of the corresponding data space, e.g., legislation, similar contracts, other contracts with the same partner, etc.

The search builds on top of the Lynx SEAR service. It is possible to search for documents by full text, document type, annotations, metadata, e.g., document date (facets) and any combination. It is therefore possible to search for newly added/modified documents or for certain document names, etc.

The Use of LySP Services
To ingest documents into the system, the following tasks are performed: (1) convert files into Lynx Documents with the Converter Service; (2) store, update, delete the file in Customers' local LawThek Document Store, which is a neo4j database for metadata and relationships in combination with file storage to persist the original file; and (3) store, update, delete the document in the search index.

The Lynx Services used in this pilot (at the time of writing or in a near future) are specified in the following:

- *TimEx*—Temporal Expression Recognition—used to detect temporal expressions in documents, e.g., the date when the offer was made.
- *NER*—Named Entity Recognition—used to identify named entities such as persons and organizations (companies).
- *RelEx*—Relation Extraction between entities within a single document—used to find, e.g., cause-effect relationships, such as "The agreement ends by 20.1.2020."
- *EntEx + WSID*—Entity Extraction and Word Sense Disambiguation Service— used to enrich documents with entities from a previously defined vocabulary.
- *Geo*—Geographical NER—used to find geographical expressions in documents, mainly addresses.
- *SEAR*—Cross Lingual Search—provides the ability to search for documents by full text, document type, annotations, metadata, e.g., document date (facets) and any combinations. It is therefore possible to search for newly added/modified documents, or for certain document names, for example.
- *APIM*—Authentication and Identity Management—exposes RESTful API to first-party clients, end users, and administrators. It will represent the main entry point to the Lynx Services in the future.

3.3 Compliance Solution for Geothermal Energy (DNV GL, the Netherlands)

About DNV GL
DNV GL (www.dnvgl.com) is the independent expert in risk management and assurance, operating in more than 100 countries. Through its broad experience and deep expertise, DNV GL advances safety and sustainable performance, sets industry benchmarks, and inspires and invents solutions.

Whether assessing a new ship design, optimizing the performance of a wind farm, analyzing sensor data from a gas pipeline, or certifying a food company's supply chain, DNV GL enables its customers and their stakeholders to make critical decisions with confidence.

Driven by its purpose, to safeguard life, property, and the environment, DNV GL helps tackle the challenges and global transformations facing its customers and the world today and is a trusted voice for many of the world's most successful and forward-thinking companies.

Problem Statement and Business Case

Geothermal Energy is an emerging source of sustainable energy. Its application is expected to show accelerated growth driven by the need for the global energy transition. To achieve sustainable and controlled growth, modernization of legislation and regulations, as well as industry standards and best practices, is required. Stakeholders—such as project developers, regulators, and engineers—typically struggle to find this information, resulting in delayed application and imposing additional risks. The pilot developed during the Lynx project aims to demonstrate how the structuring of documents in the Legal Knowledge Graphs can help users to find and select relevant regulatory documents (e.g., permits) and recommended reading on safety and environmental risks. For this pilot, it was essential to find and correctly link entities from the regulations to taxonomies (in the enrichment phase) and to quickly and reliably estimate the semantic similarity between the user's document and the previously collected documents. Moreover, to improve the accessibility and discoverability of data, it was essential to translate the documents automatically.

Governments play a crucial role in legislating and assuring compliance to mitigate safety and environmental risks, in all sectors, and in the energy industry in particular, due to the transition which is currently undergoing. With the expected growth in sustainable energy alternatives, continuous standardization of technology to bring down costs and risks can be expected. Most countries will develop policies and laws individually or together with other countries. Governments will seek balance in the use of subsidy schemes to accelerate growth and develop regulation or legislation to mitigate safety and environmental risks to guide the sustainable growth of technologies and markets. Companies active in these supply chains are likely to seek cross-border growth in order to develop economies of scale and bring costs down. If cross-border growth is envisioned, keeping up with the latest legal and regulatory rules is likely to become a challenge as country-specific clauses and local languages complicate when trying to gain an overview.

In the DNV GL Lynx Pilot, this specific context and challenge is explored for the geothermal energy domain as a proxy of the wider renewable energy domain. Geothermal energy is heat generated in the sub-surface of the earth. A geothermal fluid or steam carries the geothermal energy to the earth's surface. Geothermal energy operators drill a production and an injection well (also known as a doublet) to a certain depth (between 100 m and 4000 m) to circulate fluid to produce "heat." Depending on the temperatures, this fluid can be used to produce clean electricity, or as a baseload for municipal district or industry heating or cooling. Geothermal energy is seen as a promising sustainable energy alternative, and the industry (supply) and its users (demand) are at the dawn of accelerated growth.

To prove the value of LySP, two business cases were designed to explore the typical problems and challenges in this domain:

1. National actors in the geothermal energy supply chain facing regulatory risks, missing potential opportunities, are taking poor decisions due to compliance information being fragmented over multiple information sources. The first geothermal energy challenge is better expressed by the following question: "Can value be generated by connecting machine-readable regulatory information resources for geothermal energy?"
2. International actors in the geothermal energy supply chain struggle with a lack of understanding of country-specific regulatory frameworks (which is a competitive disadvantage), thus limiting international competition and the potential benefits of economies of scale as well as standardization. The second geothermal energy challenge is: "Can internationalization be stimulated by providing the same level of access to relevant compliance information for, and from, different EU countries?"

The Solution: The Use of LySP Services

To address these two challenges, a web application "Recommender" (see Fig. 4) was developed on top of LySP. It facilitates searching for relevant documents in multilingual corpora. The Recommender accepts plain text and PDF documents. The documents are preprocessed and plain text is extracted. The plain text is then annotated by the *Entity Linking* (EL) service. The annotated documents are processed by the *Semantic Similarity* (SeSim) service (see Fig. 4). On the left the original document title and content are displayed (A) with highlighted entities from the LKG, identified through the EL service. The SeSim service not only creates similarity scores, but it is also the reasoning behind these scores visualized as a

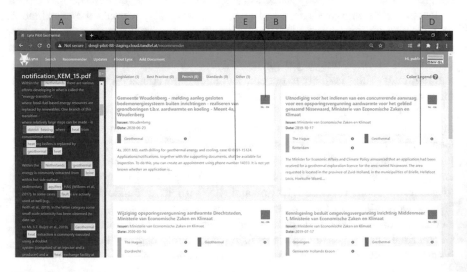

Fig. 4 Geothermal use case: recommender results page

table (D) and relevant metadata (C). The documents are translated (B) using the *MachineTranslation* (MT) service and presented to the user in the user's language (E).

4 Key Findings, Challenges, and Outlook: LySP—The Lynx Services Platform

To summarize, the Lynx Services Platform (LySP) provides a total of 16 smart services that can be used either standalone or orchestrated in specific combinations to provide powerful solutions for multilingual compliance-related applications. The services have been developed and trained on top of the Legal Knowledge Graph (LKG) to ensure domain-specific solutions with high precision, whereby the LKG consists of both: open data from public sources as well as solution-specific data and information inside a secure layer that can only be used by the respective vertical solution.

LySP Services include: 7 Enrichment Services (5 Annotation Services, 2 Conversion Services), 4 Search and Information Retrieval Services, 2 Vocabulary Services, 3 Platform Services, and are available in 4 languages—English, German, Spanish, and Dutch at the moment—and trained for legal, regulatory, and compliance use cases. However, it is worth noting that LySP Services are developed for a generic use, meaning that LySP Services can be trained for other domains (e.g., health information of financial industry) and for other languages (e.g., French and Portuguese) to allow for future scalability and the exploitation of LySP.

As a result, LySP Services can become an integral part of the European Digital Single Market [12] to be used and provided via the continuously growing number of European Data Markets and Data Spaces. Services can be trained on specific domains and languages and can thereby be used either as a core service of a Data Market and/or offered as a service for customers of Data Markets and Data Spaces. A first exercise into this direction is in progress with the European Language Grid, ELG [13] and ongoing discussions in regards to industrial Data Spaces and Markets are currently taking place.

The biggest challenges in the realization of LySP and the three industry business cases can be summarized as below:

1. The specification and the development of the LGK in regards to the harvesting of available law and regulations and other relevant information as well as the regular update of these, as such legal information is only partly available as open data, and is available in different formats and through different access paths.
2. The training of LySP Services for the three different industry use cases, as of the requirement to make use of additional data and information and as of the training effort required for specific areas of law. Again in regards to identification, specification and harvesting of data, but also the training of some services by domain experts.

3. The development of a LySP pricing model that takes into account dynamic infrastructure costs, as well as continuous maintenance costs of services and the Legal Knowledge Graph, to establish a stable and sustainable pricing model in a complex market of services available on the internet.

At the time of writing, the Lynx consortium is working on the exploitation strategy of LySP to bring LySP to the market in 2021. Besides the commercial offering of LySP, the pilot partners are going to use LySP Services internally and for their businesses, and the Lynx technology partners are integrating LySP Services into their own product and professional service offerings.

References

1. Zillner, S., Curry, E., Metzger, A., Auer, S., & Seidl, R. (Eds.). (2017). *European big data value strategic research & innovation agenda*. Big Data Value Association.
2. Press Release, Brussels 06 July 2010: Internationally active SMEs yield better results. https://ec.europa.eu/commission/presscorner/detail/en/IP_10_895, link opened 01/2021.
3. Silveira, P., Rodríguez, C., Casati, F., Daniel, F., D'Andrea, V., Worledge, C., & Taheri, Z.. (2010). On the design of compliance governance dashboards for effective compliance and audit management. In *Service-oriented computing*. ICSOC/ServiceWave 2009 (pp. 208–217). Berlin: Springer.
4. ISO is the International Standards Office, AENOR is Asociación Española de Normalización y Certificación, DIN is Deutsches Institut für Normung.
5. See the "META-NET Strategic Research Agenda for Multilingual Europe 2020", and the more recent "Strategic Research and Innovation Agenda for the Multilingual Digital Single Market" reports.
6. https://medium.com/@JochenHummel/how-multilingual-is-europes-digital-single-market-9b8d908fce6c, link opened 12/2020.
7. Connecting Europe Facility, Automated Translation: machine translation for selected Europe's public digital services.
8. Even if the name LKG refers to 'legal', it must be understood in a broad sense, comprising also regulatory data and the terminologies to facilitate cross-lingual analysis.
9. Orchestrating legal NLP services for a portfolio of use cases, medium, 19/10/2020, Artem Revenko. https://medium.com/semantic-tech-hotspot/lynx-service-platform-architecture-ac8d88c754f6
10. Details of the used and listed technologies: Keycloak: https://www.keycloak.org/, LKG ontology: https://www.lynx-project.eu/doc/lkg/, Red Hat OpenShift: https://www.openshift.com/, Camunda: https://camunda.com/, and OpenAPI3: https://swagger.io/specification/
11. Overview of LySP Services plus links to documentation, videos, services websites and the respective service API: https://lynx-project.eu/doc/api/
12. https://ec.europa.eu/digital-single-market/en/shaping-digital-single-market, link opened 12/2020.
13. The ELG, the European Language Grid: https://www.european-language-grid.eu/, link opened 12/2020.
14. Flitsch, M. (2010). *Verträge und Vertragsmanagement in Unternehmen*. Linde Verlag.
15. LKG Ontology. Accessed 01/2021, from https://www.lynx-project.eu/doc/lkg/
16. Zillner, S., Bisset, D., Milano, M., Curry, E., García Robles, A., Hahn, T., Irgens, M., Lafrenz, R., Liepert, B., O'Sullivan, B., & Smeulders, A. (Eds.). (2020). *Strategic research, innovation and deployment agenda – AI, data and robotics partnership*. Third Release. September 2020, Brussels. BDVA, euRobotics, ELLIS, EurAI and CLAIRE.

Big Data Analytics in the Banking Sector: Guidelines and Lessons Learned from the CaixaBank Case

Andreas Alexopoulos, Yolanda Becerra, Omer Boehm, George Bravos,
Vasilis Chatzigiannakis, Cesare Cugnasco, Giorgos Demetriou,
Iliada Eleftheriou, Lidija Fodor, Spiros Fotis, Sotiris Ioannidis,
Dusan Jakovetic, Leonidas Kallipolitis, Vlatka Katusic, Evangelia Kavakli,
Despina Kopanaki, Christoforos Leventis, Mario Maawad Marcos,
Ramon Martin de Pozuelo, Miquel Martínez, Nemanja Milosevic,
Enric Pere Pages Montanera, Gerald Ristow, Hernan Ruiz-Ocampo,
Rizos Sakellariou, Raül Sirvent, Srdjan Skrbic, Ilias Spais, Giorgos Vasiliadis,
and Michael Vinov

Abstract A large number of EU organisations already leverage Big Data pools
to drive value and investments. This trend also applies to the banking sector. As
a specific example, CaixaBank currently manages more than 300 different data
sources (more than 4 PetaBytes of data and increasing), and more than 700 internal

A. Alexopoulos · S. Fotis · L. Kallipolitis · I. Spais
Aegis IT Research LTD, London, UK

Y. Becerra · C. Cugnasco · M. Martínez · R. Sirvent
Barcelona Supercomputing Center, Barcelona, Spain

O. Boehm · M. Vinov
IBM, Haifa, Israel

G. Bravos · V. Chatzigiannakis
Information Technology for Market Leadership, Athens, Greece

G. Demetriou · V. Katusic · H. Ruiz-Ocampo
Circular Economy Research Center, Ecole des Ponts, Business School, Marne-la-Vallée, France

I. Eleftheriou · E. Kavakli · R. Sakellariou
University of Manchester, Manchester, UK

L. Fodor · D. Jakovetic · N. Milosevic · S. Skrbic
Faculty of Sciences, University of Novi Sad, Novi Sad, Serbia

S. Ioannidis
Technical University of Crete – School of Electrical and Computer Engineering, Crete, Greece

Foundation for Research and Technology, Hellas – Institute of Computer Science, Crete, Greece

D. Kopanaki · C. Leventis · G. Vasiliadis
Foundation for Research and Technology, Hellas – Institute of Computer Science, Crete, Greece

273
E. Curry et al. (eds.), *Technologies and Applications for Big Data Values*,
https://doi.org/10.1007/978-3-030-78307-5_13

and external active users and services are processing them every day. In order to harness value from such high-volume and high-variety of data, banks need to resolve several challenges, such as finding efficient ways to perform Big Data analytics and to provide solutions that help to increase the involvement of bank employees, the true decision-makers. In this chapter, we describe how these challenges are resolved by the self-service solution developed within the I-BiDaaS project. We present three CaixaBank use cases in more detail, namely, (1) *analysis of relationships through IP addresses*, (2) *advanced analysis of bank transfer payment in financial terminals* and (3) *Enhanced control of customers in online banking*, and describe how the corresponding requirements are mapped to specific technical and business KPIs. For each use case, we present the architecture, data analysis and visualisation provided by the I-BiDaaS solution, reporting on the achieved results, domain-specific impact and lessons learned.

Keywords Self-service solution · Banking · Security applications · Big data analytics · Advanced analytics · visualisations

1 Introduction

Collection, analysis and monetisation of Big Data is rapidly changing the financial services industry, upending the longstanding business practices of traditional financial institutions. By leveraging vast data repositories, companies can make better investment decisions, reach new customers, improve institutional risk control and capitalise on trends before their competitors. But given the sensitivity of financial information, Big Data also spawns a variety of legal and other challenges for financial services companies.[1]

Following this digitalisation trend, CaixaBank has been developing its own Big Data infrastructure since years and has been awarded several times (e.g. '2016 Best Digital Retail Bank in Spain and Western Europe' by Global Finance). With almost 14 million clients across Spain (and Portugal under their subsidiary brand BPI), CaixaBank has a network of more than 5000 branches with over 40,000 employees

---

[1] https://www.wilmerhale.com/uploadedFiles/Shared_Content/PDFs/Services/WilmerHale-BigData-FinancialServices.pdf

M. M. Marcos · R. M. de Pozuelo (✉)
CaixaBank, Barcelona, Spain
e-mail: rmartindepozuelo@caixabank.com

E. P. P. Montanera
ATOS, Madrid, Spain

G. Ristow
Software AG, Darmstadt, Germany

and manages an infrastructure with more than 9500 ATMs, 13,000 servers and 30,000 handhelds. All those figures represent a massive amount of data collected every day by all the bank systems and channels, gathering relevant information of the bank's operation from the clients, employees, third-party providers and autonomous machines. In total, CaixaBank has more than 300 different data sources used by their consolidated Big Data models and more than 700 internal and external active users enriching their data every day, which is translated into a Data Warehouse with more than 4 PetaBytes (PBs), which increases by 1 PB per year.

Much of this information is already used in CaixaBank by means of Big Data analytics techniques, for example, to generate security alerts and prevent potential frauds—CaixaBank faces around 2000 attacks per month. However, CaixaBank is one of the banking leaders in the European and national collaborative research, taking part in pre-competitive research projects. Within the EU I-BiDaaS project (funded by the Horizon 2020 Programme under Grant Agreement 780787), CaixaBank identified three concrete use cases, namely (1) *analysis of relationships through IP addresses,* (2) *advanced analysis of bank transfer payment in financial terminals* and (3) *Enhanced control of customers in online banking* to study the potential of a Big Data self-service solution that will empower its employees, who are the true decision-makers, giving them the insights and the tools they need to make the right decisions in a much more agile way.

In the rest of this chapter, Sect. 2 discusses the requirements and challenges for Big Data in the banking sector. Section 3 details the different use cases considered, together with their technical and business KPIs. In Sect. 4, for each use case, we present the architecture, data analysis and visualisation of the I-BiDaaS solution, reporting on the achieved results and domain-specific impact. It also relates the described solutions with the BDV reference model and priorities of the BDV Strategic Research and Innovation Agenda (SRIA) [1]. Section 5 summarises the lessons learned through all the experiments deployed by CaixaBank and the rest of I-BiDaaS partners, especially on how to handle data privacy and how to iteratively extend data usage scenarios. Finally, Sect. 6 presents some conclusions.

2 Challenges and Requirements for Big Data in the Banking Sector

The vast majority of banking and financial firms globally believe that the use of insight and analytics creates a competitive advantage. The industry also realises that it is sitting on a vast reservoir of data, and insights can be leveraged for product development, personalised marketing and advisory benefits. Moreover, regulatory reforms are mainly leading to this change. Ailing business and customer settlements, continuous economic crisis in other industry verticals, high cost of new technology and business models, and high degree of industry consolidation and automation are some of the other growth drivers. Many financial services currently focus on improving their traditional data infrastructure as they have been addressing issues such as customer data management, risk, workforce mobility and multichannel effectiveness. These daily problems led the financial organisation to deploy Big Data as a long-term strategy and it has turned out to be the fastest growing technology adopted by financial institutions over the past 5 years.[2]

Focusing on the customer is increasingly important and the critical path towards this direction is to move the data analytics tools and services down to the employees with direct interaction with the customers, utilising Big-Data-as-a Self-Service solutions[3] [2].

Another critical requirement for financial organisations is to use data and advanced analytics for fraud and risk mitigation and achieving regulatory and compliance objectives. With cyber security more important than ever, falling behind in the use of data for security purposes is not an option. Real-time view and analysis are critical towards competitive advantage in the financial/banking sector.

The usage of Big Data analytics is gradually being integrated in many departments of the CaixaBank (security, risks, innovation, etc.). Therefore, there is a heterogeneous group of experts with different skills but the bank also relies on several Big Data analytics experts that provide consultancy services. However, the people working with the huge amount of data collected from the different sources and channels of CaixaBank can be grouped into the following categories (which indeed could be fairly generalised to other financial entities):

- IT and Big Data expert users: employees and third-party consultants with excellent programming skills and Big Data analytics knowledge.
- Intermediate users: People with some notion on data analytics that are used to work with some Big Data tools, especially for visualisation and Big Data visual analysis (such as QlikSense/QlikView[4]). They are not skilled programmers, although they are capable of programming simple algorithms or functions with Python or R.
- Non-IT users: People with an excellent knowledge of the field and the sector; they could interpret the data, but they lack programming skills or Big Data analytics knowledge.

Although 'IT and Big Data expert users' are getting more involved and being a relevant part of the day-by-day business operations of the entity, there are few compared to the 'Intermediate' and 'Non-IT users'. Reducing the barriers and the knowledge required by those user categories in exploiting efficiently the collected data represents one of the most relevant challenges for CaixaBank.

With all this, the I-BiDaaS methodology for eliciting CaixaBank requirements (see Table 1) took into consideration the specific challenges faced by CaixaBank, as

---

[2] https://www.mordorintelligence.com/industry-reports/big-data-in-banking-industry

[3] http://www.gartner.com/it-glossary/self-service-analytics

[4] https://www.qlik.com/

Table 1 CaixaBank consolidated requirements

Business requirements	
R1	To speed up the implementation of new big data analytics applications (business goal)
R2	To be able to test new data analytics tools and algorithms outside CaixaBank premises whilst assuring maximum level of security/privacy (business goal)
R3	To enable third parties to efficiently implement and test new tools and algorithms without accessing real data (business goal)
R4	To ensure accuracy and reliability of analytics process (quality business goal)
R5	To improve efficiency of the analytics process (quality business goal)
R6	Time efficiency
R7	Cost reduction
User requirements	
R8	Data is collected by several different sources (ATMs, online banking services, employees' workstations, external providers' activity, network devices, etc.) (data provider requirement)
R9	Data are owned by CaixaBank and are not publicly available. They can be shared with third parties only once the data is anonymised (data provider requirement)
R10	Support the use of techniques related to log analysis, such as process mining algorithms or similar (big data analytics provider requirement)
R11	Users will be able to download results (in several formats such as .csv, .xls, etc.) in order to analyse them by their own or send them to other employees of the Security Operation Centre (data consumer requirement)
R12	Intermediate users will be able to modify parameters of the algorithms and refine the initial results (data consumer requirement)
System requirements	
R13	The system should enable the generation of anonymised and synthetic data to enable safe experimentation and testing (functional requirement)
R14	The system should support diversified, analytic processing, machine learning and decision support techniques to support multiple stages of analysis (functional requirement)
R15	The system should ensure security of sensitive data (non-functional requirement)

well as the literature on Requirements Engineering (RE) approaches specifically for Big Data applications [3].

In particular, the I-BiDaaS methodology followed a goal-oriented approach to requirements engineering [4], whereby elicitation of requirements was seen as the systematic transformation of high-level business goals that reflect the company vision with respect to the Big Data analytics activity or project, the user requirements of the groups of stakeholders involved (e.g., data providers, Big Data capability providers, data consumers) and finally the specific system functional and non-functional requirements, which describe the behaviour that a Big Data system (or a system component) should expose, or the capabilities it should own in order to realise the intentions of its users.

The requirements elicitation process was carried out in collaboration with both CaixaBank stakeholders and Big Data technology providers. It involved two steps: the first step was to extract specific requirements based on the characteristics of each

CaixaBank use case; the second step involved the consolidation of all requirements in a comprehensive list. Appropriate questionnaires were used to assist participants in expressing their requirements. Requirements consolidation was guided by generic requirements categories identified through the review of RE works for big data applications [5].

Although described in a linear fashion, the above activities were carried out in an iterative manner resulting in a stepwise refinement of the results being produced. The complete list of all requirements elicited is described in detail in [6].

3 Use Cases Description and Experiments' Definition: Technical and Business KPIs

The CaixaBank experiments aim at evaluating and validating the self-service Big Data platform [7] proposed in the framework of the I-BiDaaS project, and its implementation in the specific CaixaBank use cases. More precisely, the experiments aim to test the efficiency of the I-BiDaaS platform for reducing the costs and the time of analysing large datasets whilst preserving data privacy and security.

The definition of the experiments follows a goal-oriented approach, whereby for each experiment: the experiment's goal(s) towards which the measurement will be performed are first defined. Then a number of questions are formed aiming to characterise the achievement of each goal and, finally, a set of Key Performance Indicators (KPIs) and the related metrics are associated with every question in order to answer it in a measurable way.

Such KPIs have been defined at the business level during the user requirements elicitation phase (see Sect. 2). However, they need to be further elaborated and refined so that they can be mapped onto specific indicators at the Big Data application and platform level. This ensures that (a) both business and technical requirements and (b) the traceability among business and application performance are taken into consideration. In addition, for each KPI, the baseline (current) value and the desired improvement should also be defined, whose measurement relates to the achievement (or not) of the specific indicator.

The definition of each experiment also included the definition of the experiment's workflow in terms of the type and order of activities (workflow) involved in each experiment, as well as the definition of the experimental subjects that will be involved in the experiment.

Taking all the aforementioned into account, CaixaBank proposed three different use cases and evaluated the I-BiDaaS tools from the perspective of potential usage by those different groups of employees:

- Analysis of relationships through IP addresses.
- Advanced analysis of bank transfer payment in financial terminals.
- Enhanced control of customers in online banking.

The rest of the section includes the final use cases definitions in chronological order as developed and deployed in the project. We also refer the reader to Sect. 4 for further details of the use cases corresponding solutions and Sect. 5, which provides a complementary description, collecting the lessons learned during the respective processes.

3.1 Analysis of Relationships Through IP Addresses

Analysis of relationships through IP addresses was the first use case selected to test the I-BiDaaS Minimum Viable Product (MVP). In this use case, CaixaBank aims to validate the usage of synthetic data and the usage of external Big Data analytics platforms. It is deployed in the context of identifying relationships between customers that use the same IP address when connecting to online banking. CaixaBank stores information about their customers and the operations they perform (bank transfer, check their accounts, etc.) using channels such as mobile apps or online banking, and they afterwards use this data for security and fraud-prevention processes. One of the processes is to identify relationships between customers and use them to verify posterior bank transfers between linked customers. Such operations are considered with lower possibility to be fraudulent transactions. It allows CaixaBank's Security Operation Centre (SOC) to directly discard those bank transfers during the revision processes. The goal of this experiment is to validate the use of synthetic data for analysis (i.e. one of the I-BiDaaS platform features), evaluate the quality of the synthetic data (i.e. if the algorithm can find the same amount and patterns of connections using the real and the synthetic datasets) and to test the time efficiency of the I-BiDaaS solution.

3.2 Advanced Analysis of Bank Transfer Payment in Financial Terminals

The second CaixaBank use case that was studied in I-BiDaaS is *advanced analysis of bank transfer payment in financial terminals*. This use case aims to detect the differences between reliable transfers and possible fraudulent cases. The goal of this experiment is to test the efficiency of the I-BiDaaS solution in the context of anomaly detection in bank transfers from employees' workstations (*financial terminals*).

For that reason, the first step was to identify all the contextual information from the bank transfer (i.e. time execution, transferred amount, etc.), the sender and receiver (e.g. name, surname, nationality, physical address, etc.), the employee (i.e. employee id, authorisation level, etc.) and the bank office (e.g. office id, type of bank office, etc.). All this information is coming from several relational database tables stored in the CaixaBank Big Data infrastructure (called 'datapool'). The meaningful information was extracted and flattened in a single table. This task is particularly challenging because it is needed to identify events and instances from the log file corresponding to the money transfer operations carried out by an employee from a bank centre and to connect those related to the same bank transfer. The heterogeneous nature of the log files, as saved in the CaixaBank datapool, makes this task even more difficult. There is a total of 969,351,155 events in the log data just for April 2019. These events are heterogeneous in nature and arise from mixing of disparate operations associated with services provided by the different types of bank offices. After a laborious table flattering and composition process, a table of 32 fields was obtained and then tokenised.

3.3 Enhanced Control of Customers in Online Banking

In this use case, we focused on analysing the mobile-to-mobile bank transfers executed via online banking (web and application). It focuses on the assessment that the controls applied to user authentication are adequately implemented (e.g. Strong Customer Authentication (SCA) by means of second-factor authentication) according to PSD2 regulation and depending on the context of the bank transfer. With that aim, we wanted to cluster a dataset collected from mobile-to-mobile transfers. Most of the information of this dataset does not need encryption because only a few fields were sensitive. The main objective of the use case is to identify useful patterns of mobile-to-mobile bank transfers and enhance current cybersecurity mechanisms. A set of mobile-to-mobile bank transfers and the authentication mechanisms applied by CaixaBank to authenticate the senders of those transfers is analysed in order to identify if different mechanisms should be applied to specific subsets of transfers. The use case tries to identify small subsets of transfers that present similar patterns to larger sets of transfers with other types of authentication mechanisms, to re-assess if the correct authentication mechanism is being applied to this subset of transfers or if the level of security in the authentication process should be increased.

4 I-BiDaaS Solutions for the Defined Use Cases

4.1 Analysis of Relationships Through IP Addresses

4.1.1 Architecture

The architecture uses a traditional component-based architecture where the components communicate via a message queue (Universal Messaging component). This approach is important for a scalable and flexible hardware resource organisation. The architecture includes a batch and a stream processing subcase, complemented

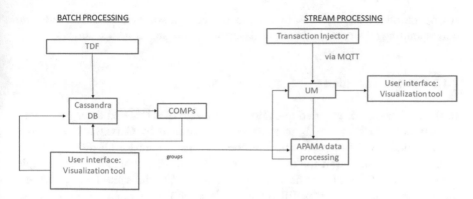

Fig. 1 Updated data flow for the components of the architecture

with the Universal Messaging component (mentioned earlier) for easing communication between components, an Orchestration layer that coordinates their interaction and a Visualisation layer providing an extensible visualisation framework that helps with data inspection and user interaction with the system. The Universal Messaging component uses a message queue system that allows easy, robust and concurrent communication between components. The Orchestration layer uses Docker for managing other components, which are all running individually as Docker containers. Figure 1 depicts the components of the architecture, as well as their interactions.

The batch processing subcase starts with the creation of a file of realistic synthetic data in SQLite format (TDF component[5]), which is then imported in a Cassandra Database, which is specifically used for its distributed properties, and COMPSs [8] with Hecuba[6] are used to run the analysis. In the streaming subcase, transactions of users are created and published via the Message Queuing Telemetry Transport (MQTT) protocol (Universal Messaging), and later an APAMA[7] GPU-enabled data processing application loads the data analysis created in the batch subcase, and compares any data coming from the stream to it, generating a new message if there is a match. 3D data analysis through visualisation is also available via the Qbeast tool [9].

The rationale for using realistic synthetic data (TDF component in Fig. 1) is that technology development and testing processes can be simplified and accelerated, before or in parallel with carrying out the processes of making real data available (e.g. a tedious data tokenisation process). The incorporation of realistic synthetic data is done with care and is subject to data quality assessment (see Sect. 4.1.2)

---

[5] TDF: https://www.ibm.com/il-en/marketplace/infosphere-optim-test-data-fabrication

[6] Hecuba: https://github.com/bsc-dd/hecuba

[7] APAMA: https://www.softwareag.com/corporate/products/apama_webmethods/analytics/overview/default.asp

An operational-ready solution then replaces the realistic synthetic data with the corresponding real, tokenised data, as described in the subsequent sections.

4.1.2 Data Generation

In this first use case, we tried to evaluate the usage of fabricated data, which was created using TDF according to a set of rules defined by CaixaBank. The rules were refined several times in order to create realistic data for all different fields considering the format of the real data. It is difficult to distinguish a data sample from a field in the synthetic dataset and a sample from the same field in the real dataset. Some properties were difficult to model as constraint rules, e.g. the concrete time connectivity patterns that the real data follows, and thus they were not included in the specification of the synthetic dataset. Constraints for parameters which were not critical for the relationship analysis that was performed in the use case were sometimes relaxed as long as they allowed the synthetic dataset to remain valid for assessing that there exist the same percentage of relationships as in the real dataset.

4.1.3 Data Analytics

The goal of the case is to find relations between people, given a set of connections to IP addresses, maximising the detection of close relations between users. This application has been implemented using the COMPSs programming model and Hecuba as the data interface with the Cassandra database.

We have defined several parallel tasks, not only to exploit parallelism but also to benefit from the automatic detection of the dependencies from COMPSs. Using Cassandra to store the data allows us to delegate on the database the management of the global view of the data. This approach frees programmers from implementing an explicit synchronisation between those parallel tasks that modify the data structure. This way, removing the synchronisation points, we are able to maximise the parallelism degree of the application and thus the utilisation of the hardware resources. Notice that the interface for inserting data in Cassandra is asynchronous with the execution of the application, this way overlapping data storage with computation.

The approach to solve this implementation has been to define a clustering-based analysis of CaixaBank's IP address connections using a synthetic dataset. The purpose of the analysis is to provide additional modelling possibilities to this CaixaBank's use case. The obtained results should be understood relative to the fact that the data set utilised is synthetic, even though the initial feedback from CaixaBank about the usefulness of the developed process is positive, and the approach is promising. The data set contains 72,810 instances, with each instance containing the following attributes:

- *User ID*—representing a unique identification number for each user.
- *IP address*—representing the IP address of the connection of the user.
- *Date*—representing the date and the time of the connection made by the user.
- *Operation*—representing the code of the business operation made by the user.
- *Status*—representing the code of the status of the operation made by the user.

Initially, the dataset is transformed as follows: each user represents a sample, while each IP address represents a feature. In such a data matrix, the value in position (i, j) represents the number of times user i connected via IP address j. Such a dataset turns out to be extremely sparse. In order to tackle this problem and retain only meaningful data, the next pre-processing step is to drop all the IP addresses that were used by only one user (intuitively, such IP addresses represent home network, etc. and thus cannot be used to infer relationships between users). After dropping all such IP addresses, 1075 distinct IP addresses remain from the initial 22,992 contained in the original dataset. Subsequently, we filter out the users that are not connected to any of the remaining IP addresses.

To infer relationships between users, we applied clustering algorithms. In particular, we used K-means [10] and DBSCAN [11], which are both available in the dislib library [12]. Additionally, we used the t-distributed Stochastic Neighbour Embedding (t-SNE) method [13] to visualise the reduced dataset in 2D. The visualisation is presented in Fig. 2.

Both K-means and DBSCAN offer some interesting hyperparameters. In particular, K-means allowed us the flexibility of setting the desired number of clusters. A preset number of clusters could be a limitation, especially in an exploration phase. However, the possibility to set up the number of clusters allowed us to define them according to the number of authentication mechanisms, for example, which was useful in the analysis of the use case. On the other hand, DBSCAN decides

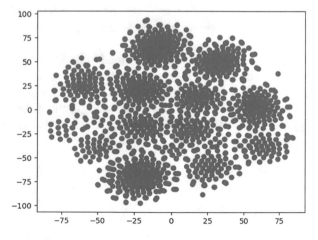

Fig. 2 t-SNE 2D visualisation

on the number of clusters internally while providing us with the parameters that represent the minimum number of samples in a neighbourhood for a point to be considered a core point, and the maximum distance between two samples for them to be considered as in the same neighbourhood. These parameters are to be set by an end-user based on experimentation and domain knowledge and are tuneable through the I-BiDaaS user interface.

Moreover, the evaluation of this use case was especially focused on analysing the validation of fabricated data for identifying patterns and number of connections. Therefore, a more advanced analysis with K-means and DBSCAN was done using both, the synthetic dataset and a tokenised version of a real dataset. The data tokenisation process included the encryption of all the fields of the dataset. The analysis performed over this dataset allowed the inference of conclusions and relationships in the real non-encrypted data.

4.1.4 Visualisations

The visualisation of the use case includes several graphic types. First, a graph shows the distribution of relationships detected based on their IP addresses (Fig. 3a).

Using these relationships, visualisation of real-time bank transfers in the form of a continuous stream of sender-receiver records is used to emulate real-time detection of possibly fraudulent transactions (Fig. 3b). The visualisation utilises the previously detected relationships to display a graph of connected users so as to aid operators in determining possible relationships between users and decide whether further actions should be taken.

4.1.5 Results

Results obtained from both real tokenised data and the synthetic data using those algorithms showed that the majority of the clusters found were 2-point clusters, indicating a good similarity for this use case.

An additional evaluation process was performed to determine a specific utility score, i.e. the similarity of results of analyses from the synthetic data and the original data. The propensity mean-squared-error (pMSE) was used as a general measure of data utility to the specific case of synthetic data. As specific utility measures we used various types of data analyses, confidence intervals overlap and standardised difference in summary statistics, which were combined with the general utility results (Fig. 4).

By randomly sampling 5000 datapoints from real and synthetic datasets, and using logistic regression to provide the probability for the label classification, we were able to show that the measured mean pMSE score for the 'analysis of relationships through IP addresses' dataset is 0.234 with a standard deviation of 0.0008.

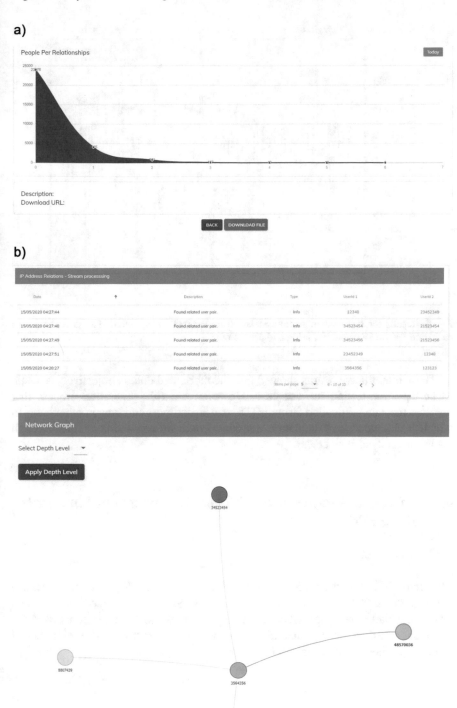

Fig. 3 (**a**) User groups per IP relationships. (**b**) Real-time relationship detection

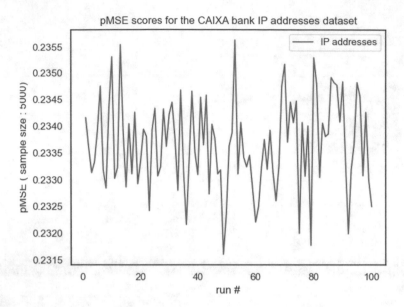

Fig. 4 Results for 100 random sampling taken from the real and synthetic data (5 K datapoints each) and the pMSE calculated using a logistic model

Those quantitative results showed that the fabricated data is objectively realistic to be used for testing the use case. However, the rule-generation process that involves the data fabrication through TDF can be complex and long in other cases in which the knowledge of the data is not complete or the extraction of rules through statistical analysis is not clear.

4.2 Advanced Analysis of Bank Transfer Payment in Financial Terminals

4.2.1 Architecture

The architecture (i.e. the specific components of the I-BiDaaS general architecture) in this use case is the same as the one described in Sect. 4.1.1, focused on the batch processing part. Therefore, what essentially changes are the algorithms used for processing the data (i.e. the bank transfers conducted by employees on their financial terminals). These algorithms will be described in the next sections.

4.2.2 Data Analytics

This CaixaBank use case is focused on advanced analysis of bank transfers executed by employees on financial terminals to detect possible fraud, or any other potential anomalies that differ from the standard working procedure. The used dataset is composed of different attributes which record the different steps that the employee performs and other important data such as the account, client or amount of money transferred. All the data is encrypted using the Dice Coefficient [14], which codifies the data without losing important information.

All data processing techniques, like the K-means, PCA (Principal Component Analysis) [15] and DBSCAN have been performed using the dislib library. Also, the data structure used by dislib has been modified to be stored on a Cassandra Database using the Hecuba library.

The received dataset must be pre-processed before using the data transformation techniques from dislib. First, the attributes which contained the same value for all the registers have been deleted, as they do not give any relevant information. Also, all nulls and blank registers have been transformed into 0 values. Finally, for those categorical attributes, we transform the variable categories into columns (1, 0), a transformation known as one-hot encoding [16].

Due to the encoding transformations, the number of attributes has increased considerably from 89 to 501. This large amount of attributes made it difficult to perform K-means, and for this reason, it was decided to apply a PCA transformation to reduce the number of dimensions to 3, to also be able to represent it graphically. Before applying the PCA transformation, and due to the differences in the magnitude of the attributes, we have standardised the data using the *scikit-learn* method *StandardScaler*.

Finally, we have executed two different clustering algorithms: DBSCAN and K-means. As K-means requires the desired number of clusters as an input parameter, we have executed first DBSCAN and we have used the obtained number of clusters as the input parameter of K-means.

4.2.3 Visualisations

For this use case, a 3D graph of the data and detected anomalies has been developed. Users can select parts of the graph to focus on and can also extract the specific data samples that are included in the selection (Fig. 5).

4.2.4 Results

Figure 6a shows the graphical representation of the clusters generated by DBSCAN in a three-dimensional space, where the third dimension of the PCA is shown as the Z-axis. The result of K-means can be examined in Fig. 6b, we can observe that

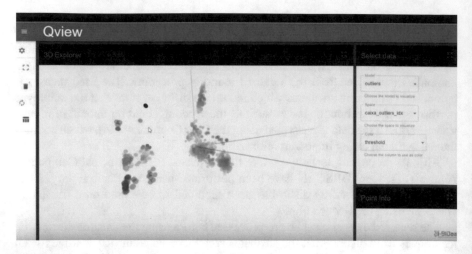

Fig. 5 Visualisation of detected anomalies in 3D graph

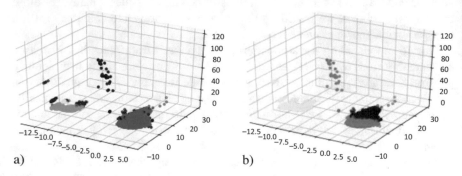

Fig. 6 (**a**) DBSCAN representation. (**b**) K-means representation

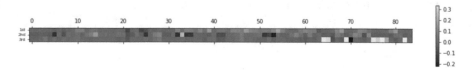

Fig. 7 Heat-plot of the 84 most relevant attributes from the 501 original attributes

some values in the Z-axis are far away from the main cluster and, thus, are potential anomalies in the data.

The PCA reduced the attributes from 501 to 3, thus it is difficult to understand which is the correlation between the resultant three dimensions and the 501 original attributes. In Fig. 7, we have printed the mentioned correlation. We only show the first 84 because they are the most interesting with respect to the third dimension of the Z-axis. We can appreciate that this third dimension is heavily influenced by attributes from 64 to 82.

4.3 Enhanced Control of Customers in Online Banking

4.3.1 Architecture

As in Sect. 4.2.1, the set of components used to analyse bank transfers which were executed using online banking options (both web and mobile app) are the same as the ones described in Sect. 4.1.1, also with a main focus on the batch processing part, and selecting a different set of algorithms to analyse the data, as will be described in the following sections.

4.3.2 Data Analytics

Following the objective of the use case in Sect. 3.3, this use case tried to identify useful patterns of mobile-to-mobile bank transfers and enhance current cybersecurity mechanisms by identifying if there is a set of transactions in which the level of security in the authentication process should be increased. For that reason, we decided to analyse a dataset collecting the information of all mobile-to-mobile bank transfers from clients for a month and work on non-supervised methods such as clustering. That cluster was done on a categorical database so that most known algorithms lost efficacy. The first attempt was to apply a K-means. However, since the vast majority of available variables were not numerical, calculating the distances for grouping in K-means algorithm was no longer so simple (e.g. if there are three types of enhanced authentication, should the distance between them be the same? Should it be greater since some of them are more restrictive than the others?). This type of question affects the result of the model; therefore, a transformation was made to the data. We applied one-hot encoding [16]. This transformation allowed to eliminate the problems of calculating the distance between categories. Even so, the results were not satisfactory. Given the situation, a search/investigation process was carried out for an appropriate model for this case series. We find the k-modes library that includes algorithms to apply clustering on categorical data.

The k-modes algorithm [17] is basically the already known K-means, but with some modification that allows us to work with categorical variables. The k-modes algorithm uses a simple matching dissimilarity measure to deal with categorical objects, replaces the means of clusters with modes, and uses a frequency-based method to update modes in the clustering process to minimise the clustering cost function.

Once the algorithm has been decided, we must calculate the optimal number of clusters for our use case. For this, the method known as the *elbow* method is applied, which allows us to locate the optimal cluster as follows. We first define:

- *Distortion*: It is calculated as the average of the squared distances from the cluster centres of the respective clusters.
- *Inertia*: It is the sum of squared distances of samples to their closest cluster centre.

Fig. 8 Number of clusters
selection for 'Enhanced
control of customers in online
banking'

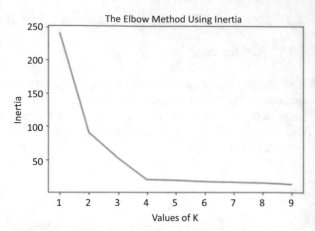

Fig. 8 Number of clusters selection for 'Enhanced control of customers in online banking'

Then we iterated the values of k from 1 to 10 and calculated the values of distortion for each value of k and calculated the distortion and inertia for each value of k in the given range. The idea is to select the number of clusters that minimise inertia (separation between the components of the same cluster) (Fig. 8).

To determine the optimal number of clusters, we had to select the value of k at the 'elbow' in the point after which the distortion/inertia starts decreasing in a linear fashion. Thus, for the given data, we conclude that the optimal number of clusters for the data is 4. Once we know the optimal number of clusters, we apply k-modes with $k = 4$ and analyse the results obtained.

4.3.3 Visualisations

A dynamically updated chart depicting the clusters in which the monitored transactions fall into was used for this use case. The number of clusters is automatically updated to reflect new ones being detected by the processing pipeline (Fig. 9).

4.3.4 Results

With this use case, I-BiDaaS allowed CaixaBank's 'Intermediate users' and 'Non-IT users' to modify the number of clusters and run the algorithm over a selected dataset of transactions in a very fast and easy way. It was used for exploring clients' mobile-to-mobile transaction patterns, identifying anomalies in the authentication methods and potential frauds, allowing fast and visual analysis of the results in the platform (Fig. 10).

The results were checked with the Digital Security and Security Operation Centre (SOC) employees from CAIXA in order to correctly understand if the clustering algorithm applied allowed to identify potential errors in our automated authentication mechanisms in mobile-to-mobile bank transfers. The obtained clusters

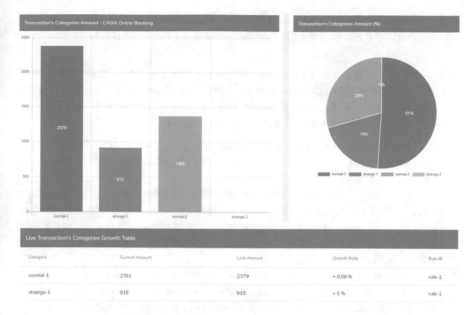

Fig. 9 Sample of the I-BiDaaS graphical interface showing the identified clusters of incoming mobile-to-mobile bank transfers

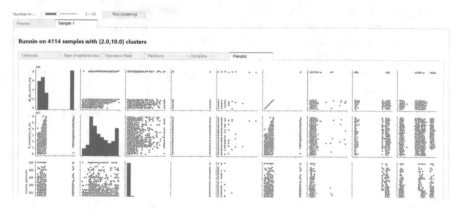

Fig. 10 Sample of the 'Enhanced control of customers in online banking' use case clustering results in the I-BiDaaS platform

of entries were useful to identify the different mobile-to-mobile bank transfers patterns and reconsider the way we are selecting the authentication method to proceed with the transfer. I-BiDaaS tools made it easier for the SOC employees and Digital Security department to analyse and identify bank transfer patterns in which a higher level of security would be beneficial. The rules on the authentication mechanism application for those patterns were redefined, applying a more restrictive authentication mechanism in around 10% of the mobile-to-mobile bank transfers (in a first iteration).

4.4   Relation to the BDV Reference Model and the BDV Strategic Research and Innovation Agenda (SRIA)

The described solution can be contextualised within the BDV Reference Model defined in the BDV Strategic Research and Innovation Agenda (BDV SRIA) [1] and contributes to the model in the following ways. Specifically, the work is relevant to the following BDV Reference Model horizontal concerns:

- *Data visualisation and user interaction*: We develop several advanced and interactive visualisation solutions applicable in the banking sector, as illustrated in Figs. 3, 5, and 9.
- *Data analytics*: We develop data analytics solutions for the three industrial use cases in the banking sector, as described in Sects. 4.1–4.3. While the solutions may not correspond to state-of-the-art advances in algorithm development, they clearly contribute to revealing novel insights into how Big Data analytics can improve banking operations.
- *Data processing architectures*: We develop an architecture as shown in Fig. 1 that is well suited for banking applications where both batch analytics (e.g., analysing historical data) and streaming analytics (e.g., online processing of new transactions) are required. A novelty of the architecture is the incorporation of realistic synthetic data fabrication and the definition of scenarios of usefulness and quality assurance of the corresponding synthetic data.
- *Data protection*: We describe in Sect. 5 how data tokenisation and realistic synthetic data fabrication can be used in baking applications to allow for more agile development of Big Data analytics solutions.
- *Data management*: We present innovative ways for data management utilising efficient multidimensional indexing, as described in Sect. 4.3.

Regarding the BDV Reference Model vertical concerns, the work is relevant to the following:

- *Big Data Types and Semantics*: The work is mostly concerned with structured data, meta-data and graph data. The work contributes to a generation of realistic synthetic data from the corresponding domain-defined meta-data.
- *Cybersecurity*: The presented solutions that include data tokenisation correspond to novel best practice examples for securely sharing sensitive banking data outside bank premises.

Therefore, in relation to BDV SRIA, we contribute to the following technical priorities: data protection, data processing architectures, data analytics, data visualisation and user interaction.

Finally, the chapter relates to the following cross-sectorial technology enablers of the AI, Data and Robotics Strategic Research, Innovation and Deployment Agenda [18], namely: Knowledge and Learning, Reasoning and Decision Making, and Systems, Methodologies, Hardware and Tools.

5   Lessons Learned, Guidelines and Recommendations

CaixaBank, as many entities in critical sectors, was initially very reluctant to use any Big Data storage or tool outside its premises. To overcome that barrier, the main goal of CaixaBank when enrolling in the I-BiDaaS project was to find an efficient way to perform Big Data analytics outside its premises, which would speed up the process of granting new external providers to access CaixaBank data (which usually encompasses a bureaucratic process that takes weeks or even a month). Additionally, CaixaBank wanted to be much more flexible in the generation of proof-of-concept (PoC) developments (i.e. to test the performance of new data analytics technologies to be integrated into its infrastructure). Usually, for any new technology testing, even a small test, if any hardware is needed to be arranged, it should be done through the infrastructure management subsidiary who will finally deploy it. Due to the size and level of complexity of the whole CaixaBank infrastructure and rigid security assessment processes, its deployment can take months.

For those reasons, CaixaBank wanted to find ways to bypass these processes without compromising the security of the entity and the privacy of its clients. General Data Protection Regulation (GDPR)[8] really limits the usage of the bank customers' data, even if it is used for potential fraud detection and prevention and for enhancing the security of its customers' accounts. It can be used internally to apply certain security policies, but how to share this data with other stakeholders is still an issue. Furthermore, bank sector is strictly regulated, and National and European regulators are supervising all the security measures taken by the bank in order to provide a good level of security for the entity and, at the same time, maintain the privacy of the customers at all times. The current trend of externalising many services to the cloud also implies establishing a strict control of the location of the data and who has access to it for each migrated service.

The I-BiDaaS CaixaBank roadmap (Fig. 11) had a turning point, in which the entity completely changed its approach from a non-sharing real data position to looking for the best way possible to share real data and perform Big Data analytics outside its facilities. I-BiDaaS helped to push for internal changes in policies and processes and evaluate tokenisation processes as an enterprise standard to extract data outside their premises, breaking both internal and external data silos.

Results obtained from the first use case validated the usage of rule-based synthetically generated data and indicated that it can be very useful in accelerating the onboarding process of new data analytics providers (consultancy companies and tools). CaixaBank validated that it could be used as high-quality testing data outside CaixaBank premises for testing new technologies and PoC developments, streamlining grant accesses of new, external providers to these developments, and thus reducing the time of accessing data from an average of 6 days to 1.5 days. This analysis was beneficial for CaixaBank purposes, but it was also concluded that the

---

[8] https://eur-lex.europa.eu/eli/reg/2016/679/oj

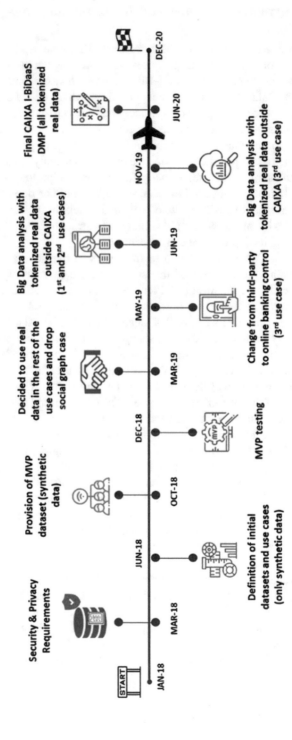

Fig. 11 CaixaBank roadmap in I-BiDaaS project

Table 2 Summary of the impact of the CaixaBank use cases studied in I-BiDaaS

Benefits	KPIs
To increase the efficiency and competitiveness in the management of its vast and complex amounts of data	75% time reduction in data access by external stakeholders using synthetic data (from 6 to 1.5 days)
To break data silos not only internally, but also fostering and triggering internal procedures to open data to external stakeholders	Real data accessed by at least six different external entities skipping long-time data access procedures
To evaluate big data analytics tools with real-life use cases of CaixaBank in a much more agile way	I-BiDaaS overall solution and tools experimentation with three different industrial use cases with real data

analysis of rule-based fabricated data did not enable the extraction of new insights from the generated dataset, simply the models and rules used to generate the data.

The other two use cases focused on how extremely sensitive data can be tokenised to extract real data for its usage outside CaixaBank premises. By tokenising, we mean encrypting the data and keeping the encryption keys in a secure data store that will always reside in CaixaBank facilities. This approach implied that the data analysis will always be done with the encrypted data, and it can still limit the results of the analysis. One of the challenges of this approach is to find ways to encrypt the data in a way that it loses as little relevant information as possible. Use case 2 and use case 3 experimentation was performed with tokenised datasets built by means of three different data encryption algorithms: (1) format-preserving encryption for categorical fields, (2) order-preserving encryption for numerical fields and (3) a bloom-filtering encryption process for free text fields. This enabled CaixaBank to extract the dataset, upload it to I-BiDaaS self-service Big Data analytics platform and analyse it with the help of external entities without being limited by the corporate tools available inside CaixaBank facilities. I-BiDaaS Beneficiaries proceeded with an unsupervised anomaly detection in those use cases, identifying a set of pattern anomalies that were further checked by CaixaBank's Security Operation Center (SOC), helping to increase the level of financial security of CaixaBank. However, beyond that, we consider this experimentation very beneficial, and should be replicated in other commercial Big Data analytics tools, prior to their acquisition.

The main benefits obtained by CaixaBank due its participation in I-BiDaaS (highlighted in Table 2) directly relate to the evaluation of the different requirements presented in Sect. 2 (Table 1).

We were able to speed up the implementation of Big Data analytics applications (R1), test algorithms outside CaixaBank premises (R2) and test new tools and algorithms without data privacy concerns by exploring and validating the usage of synthetic data and tokenised data (R3) in three different use cases, improving the efficiency in time and cost (R5, R6, R7) by means of skipping some data access procedures and being able to use new tools and algorithms in a much more agile way. User requirements regarding the availability of 'Intermediate and Non-IT users' to analyse and process the data of the use cases were also validated through several internal and external workshops[9] in which the attendees from several departments of CaixaBank and other external entities (data scientists, business consultants, IT and Big Data managers) provided very positive feedback about the platform usability. Moreover, use cases 2 and 3, as mentioned previously, were also validated by the corresponding business processes employees, being able to extract the results by themselves.

Last but not least, it is important to highlight that those results should be applicable to any other financial entity that faces the same challenges and tries to overcome the limitations of data privacy regulation, the common lack of agility of large-scale on-premise Big Data infrastructures and very rigid but necessary security assessment procedures.

6   Conclusion

The digitalisation of the financial sector and the exploitation of the incredible amount of sensitive data collected and generated by the financial entities day by day makes their Big Data infrastructure very difficult to manage and to be agile in integrating innovative solutions. I-BiDaaS integrated platform provided a solution to manage it in a much more friendly manner and makes Big Data analytics much more accesible to the bank employees with less technical and data science knowledge. It also explored ways to reduce the friction between data privacy regulation and the exploitation of sensitive data for other purposes, showcasing it in several use cases on enhancing an entity's cybersecurity and preventing fraud toward their clients.

Acknowledgements The work presented in this chapter is supported by the I-BiDaaS project, funded by the European Union's Horizon 2020 research and innovation programme under Grant Agreement 780787. This publication reflects the views only of the authors, and the Commission cannot be held responsible for any use which may be made of the information contained therein.

References

1. Zillner, S., Curry, E., Metzger, A., Auer, S., & Seidl, R. (2017). *European big data value strategic research & innovation agenda.*
2. Passlick, J., Lebek, B., & Breitner, M. H. (2017). A self-service supporting business intelligence and big data analytics architecture. In *13th International Conference on Wirtschaftsinformatik*, St. Gallen, February 12–15, 2017.
3. Arruda, B. D. (2018). Requirements engineering in the context of big data applications. *SIGSOFT Software Engineering Notes, 43*(1), 1–6.
4. Horkoff, J. (2019). Goal-oriented requirements engineering: An extended systematic mapping study. *Requirements Engineering, 24*, 133–160.
5. NIST Big Data Public Working Group: Use Cases Requirements Subgroup: National Institute of Standards and Technology (NIST). Big Data Interoperability Framework: Volume 3,

---

[9] https://www.ibidaas.eu/blog/%e2%80%9cI-BiDaaS-Application-to-the-Financial-Sector%e2%80%9d-Workshop/

Use Cases and General Requirements. Technical report, National Institute of Standards and Technology, Special Publication 1500-3. 2015.

6. I-BiDaaS Consortium. (2018). D1.3: Positioning of I-BiDaaS. https://doi.org/10.5281/zenodo.4088297

7. Arapakis, I., et al. (2019). Towards specification of a software architecture for cross-sectoral big data applications. *2019 IEEE World Congress on Services (SERVICES)* (Vol. 2642). IEEE.

8. Badia, R. M., Conejero, J., Diaz, C., Ejarque, J., Lezzi, D., Lordan, F., & Sirvent, R. (2015). Comp superscalar, an interoperable programming framework. *SoftwareX, 3*, 32–36.

9. Cugnasco, C., Calmet, H., Santamaria, P., Sirvent, R., Eguzkitza, A. B., Houzeaux, G., Becerra, Y., Torres, J., & Labarta, J. (2019). The OTree: Multidimensional indexing with efficient data sampling for HPC. In *2019 IEEE International Conference on Big Data* (pp. 433–440). IEEE. 2019.

10. Bock, H.-H. (2017). *Clustering methods: A history of k-means algorithms. Selected contributions in data analysis and classification* (pp. 161–172). Berlin: Springer.

11. Ester, M., Kriegel, H. P., Sander, J., & Xu, X. (1996). A density-based algorithm for discovering clusters in large spatial databases with noise. In *Kdd* (Vol. 96, No. 34, pp. 226–231).

12. Álvarez Cid-Fuentes, J., Solà, S., Álvarez, P., Castro-Ginard, A., & Badia, R. M. (2019). dislib: Large scale high performance machine learning in python. In *Proceedings of the 15th International Conference on eScience, 2019* (pp. 96–105).

13. Hinton, G. E., & Roweis, S. (2002). Stochastic neighbour embedding. *Advances in Neural Information Processing Systems, 15*, 857–864.

14. Jimenez, S., Becerra, C., & Gelbukh, A. (2012). Soft cardinality: A parameterized similarity function for text comparison". In *SEM 2012: The First Joint Conference on Lexical and Computational Semantics–Volume 1: Proceedings of the main conference and the shared task, and Volume 2: Proceedings of the Sixth International Workshop on Semantic Evaluation (SemEval 2012)* (pp. 449–453).

15. Pearson, K. (1991). On lines and planes of closest fit to systems of points in space. *The London, Edinburgh, and Dublin Philosophical Magazine and Journal of Science, 2*(11), 559–572.

16. Okada, S., Ohzeki, M., & Taguchi, S. (2019). Efficient partition of integer optimization problems with one-hot encoding. *Scientific Reports, 9*(1), 1–12.

17. Huang, Z. (1998). Extensions to the k-modes algorithm for clustering large data sets with categorical values. *Data Mining and Knowledge Discovery, 2*(3), 283–304.

18. Zillner, S., Bisset, D., Milano, M., Curry, E., García Robles, A., Hahn, T., Irgens, M., Lafrenz, R., Liepert, B., O'Sullivan, B., & Smeulders, A. (Eds.). (2020). *Strategic research, innovation and deployment agenda – AI, data and robotics partnership*. Third Release. September 2020.

Data-Driven Artificial Intelligence and Predictive Analytics for the Maintenance of Industrial Machinery with Hybrid and Cognitive Digital Twins

Perin Unal, Özlem Albayrak, Moez Jomâa, and Arne J. Berre

Abstract This chapter presents a Digital Twin Pipeline Framework of the COG-NITWIN project that supports Hybrid and Cognitive Digital Twins, through four Big Data and AI pipeline steps adapted for Digital Twins. The pipeline steps are Data Acquisition, Data Representation, AI/Machine learning, and Visualisation and Control. Big Data and AI Technology selections of the Digital Twin system are related to the different technology areas in the BDV Reference Model. A Hybrid Digital Twin is defined as a combination of a data-driven Digital Twin with First-order Physical models. The chapter illustrates the use of a Hybrid Digital Twin approach by describing an application example of Spiral Welded Steel Industrial Machinery maintenance, with a focus on the Digital Twin support for Predictive Maintenance. A further extension is in progress to support Cognitive Digital Twins includes support for learning, understanding, and planning, including the use of domain and human knowledge. By using digital, hybrid, and cognitive twins, the project's presented pilot aims to reduce energy consumption and average duration of machine downtimes. Data-driven artificial intelligence methods and predictive analytics models that are deployed in the Digital Twin pipeline have been detailed with a focus on decreasing the machinery's unplanned downtime. We conclude that the presented pipeline can be used for similar cases in the process industry.

Keywords Digital Twin · Data-driven Artificial Intelligence · Big Data pipeline · AI pipeline · Hybrid Digital Twin · Cognitive Digital Twin · Predictive Analytics · Predictive Maintenance

P. Unal (✉) · Ö. Albayrak
Teknopar Industrial Automation, Ankara, Turkey
e-mail: punal@teknopar.com.tr

M. Jomâa
SINTEF Industry, Oslo, Norway

A. J. Berre
SINTEF Digital, Oslo, Norway

1   Introduction

The COGNITWIN project,[1] "Cognitive Plants Through Proactive Self-Learning Hybrid Digital Twins", is a 3-year project in the Horizon 2020 program for the Process Industry, from 2019 to 2022. The project focuses on using Big Data and AI for the European Process Industry through the development of a framework with a toolbox for Hybrid and Cognitive Digital Twins.

This chapter describes the Digital Twin approach and demonstrates its application to one of the project's six use case pilots: Spiral Welded Machinery (SWP) in the steel pipe industry. The chapter relates to the data spaces, platforms, and "Knowledge and Learning" cross-sectorial technology enablers of the AI, Data and Robotics Strategic Research, Innovation & Deployment Agenda [32]. The chapter is organised as follows:

Section 2 presents the *Digital Twin Pipeline Framework* from the COGNITWIN project supporting Digital, Hybrid and Cognitive Digital Twins, through the four pipeline steps. Section 3 briefly introduces the state of the art and state of the practice in maintenance of industrial machinery. Section 4 presents the pilot Application of *Maintenance of Industrial Machinery for Spiral Welded Steel*, focusing on the Digital Twin support for Predictive Maintenance for Machines in the Welded steel plant/factory. Section 5 describes the *Big Data and AI Technology* selections of the Digital Twin system applied for the steel industry case related to the different technology areas in the BDV Reference Model. Section 6 details COGNITWIN Digital Twin Pipeline Architecture and the Platform Developed and presents the four pipeline steps of the COGNITWIN Digital Twin Pipeline realised in the context of the Spiral Welded Steel pilot case. Section 6.1 explains how *Digital Twin Data Acquisition and Collection* is taking place from factory machinery and assets with connected devices, and controllers through protocols and interfaces like OPC UA and MQTT. Section 6.2 exemplifies *Digital Twin Data Representation* in various forms, based on the sensor and data sources connections involving event processing with Kafka and storage in relevant SQL and NoSQL databases combined with Digital Twin API access opportunities being experimented with, such as the Asset Administration Shell (AAS). Section 6.3 presents *Digital Twin Hybrid (Cognitive) Analytics* with AI/Machine learning models based on applying and evaluating different AI/machine learning algorithms. This is further extended with first-principles physical models—to form a Hybrid Digital Twin with examples of data and electrical and mechanical models for a DC motor to support predictive maintenance. Section 6.4 describes the pipeline step for Digital Twin Visualisation and Control, including the use of 3D models and dashboards suitable for interacting with Digital Twin data and further data access and system control through control feedback to the plant/factory. Finally, the conclusion in Sect. 7 presents a summary

---

[1] http://cognitwin.eu/

of this chapter's contributions and the plans for future improvement of technologies and use case pilots in the COGNITWIN project.

2   Digital Twin Pipeline and COGNITWIN Toolbox

Many of the advancements and requirements related to Industry 4.0 are being fulfilled by the use of Digital Twins (DT). We have in earlier papers introduced our definitions for Digital, Hybrid, and Cognitive Twins [1–3]—which also aligns with definitions of others [4, 5]: "A DT is a digital replica of a physical system that captures the attributes and behaviour of that system" [6]. The purpose of a DT is to enable measurements, simulations, and experimentations with the digital replica to gain an understanding of its physical counterpart. A DT is typically materialised as a set of multiple isolated models that are either empirical or first-principles based. Recent developments in artificial intelligence (AI) and DT bring more abilities to the DT applications for smart manufacturing.

A hybrid twin (HT) is a set of interconnected DTs, and being an extension of a HT, a cognitive twin (CT) is a self-learning and proactive system [6]. The concepts of HT and CT are introduced as elements of the next level of process control and automation in the process and manufacturing industry. In the COGNITWIN project, we define an HT as a DT that integrates data from various sources (e.g., sensors, databases, simulations, etc.) with the DT models, and applies AI analytics techniques to achieve higher predictive capabilities, while optimising, monitoring, and controlling the behaviour of the physical system. A Cognitive Twin (CT) is defined as an extension of HT incorporating cognitive features to enable sensing complex and unpredicted behaviour and reason about dynamic strategies for process optimisation. A CT will combine expert knowledge with the power of HT.

We have adopted the Big Data and AI Pipeline that have been described in [7] and specialised this for the context of Digital Twins as shown in Fig. 1.

The proposed pipeline architecture starts with data acquisition and collection to be used by the DT. This step includes acquiring and collecting data from various sources, including streaming data from the sensors and data at rest.

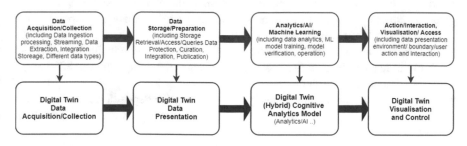

Fig. 1   Big Data and AI Pipeline architecture—applied for Digital Twins

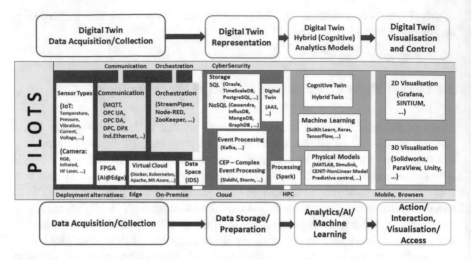

Fig. 2 Cognitive Twin Toolbox with identified components for the various pipeline steps

Following the data acquisition and collection, the next step is the DT data representation in which the acquired data is stored and pre-processed. The DT (Hybrid) Cognitive Analytics Models step of the pipeline enables integration of multiple models and the addition of cognitive elements to the DT through data-analytics. Finally, the DT Visualisation and Control step of the pipeline provides a visual interface for the DT, and it provides interaction between the twin and the physical system.

Figure 2 shows various components in the COGNITWIN Toolbox that can be selected in order to create Digital Twin pipelines in different application settings.

The COGNITWIN Toolbox is being used to create operational Digital Twin pipelines in a set of use cases as follows:

- Operational optimisation of gas treatment centre (GTC) in aluminium production, with support for the recommendation of optimal operating parameters for adsorption based on real-time data gathered about conditions such as the pressure, temperature, humidity, etc., from sensors.
- Minimise health and safety risks and maximise the metallic yield in Silicon (Si) production to provide best estimates of when the furnace can be emptied to the ladle for further operations.
- Real-time monitoring of finished steel products for operational efficiency with an ability to react on its own to situations requiring intervention, thus stabilising the production process further.
- Improving heat exchanger efficiency by predicting the deposition of unburnt fuel mixtures, ash, and other particles on the heat-exchanger tubes based on both historical practices and real-time process.

In the following, we illustrate the approach for creating an operational Digital Twin pipeline for the use case on predictive maintenance for Steel Pipe Welding. We plan to apply it to improve the Steel Pipe Welding process further as follows:

- Life cycle optimisation of Spiral Welded Machine (SWP) in steel pipe production, where CT of the SWP monitors the condition and health of the machinery, offers early warnings, and suggests optimised predictive maintenance plans for the machinery based on real-time data gathered from sensors such as the pressure, temperature, vibration, etc., and alarm and status information.
- Improving operational performance of the production process by predicting and identifying the optimal operating parameters based on both historical practices and real-time process and thus improving the overall productivity of the plant.
- Improving energy consumption efficiency by monitoring and predicting the energy analyser and operational parameters based on both historical practices and real-time process.
- Enhanced utilisation of computing infrastructure with virtual machines and containerisation technologies to achieve optimised RAM and CPU usage.
- Minimise health and safety risks and maximise the human operator performance by early warning of machine and system problems.
- Real-time monitoring of parameters like pipe diameter, pitch angle, belt width, production speed, pipe diameter and wall thickness for semi-finished and finished steel products for ensuring operational efficiency and stabilising the production process.

3   Maintenance of Industrial Machinery and Related Work

Maintenance in the production industry has always been an important building block providing essential requirements, such as cost minimisation, prolonged machine life, and increased safety and reliability. On the other hand, Predictive Maintenance (PM) has been a popular topic of research for decades with hundreds of papers having been published in the area. Since machine learning techniques came into prominence in the field with the emergence of industry 4.0, PM has become an even more important area of interest [8].

There are three maintenance strategies. The first is Reactive Maintenance (RM) in which little or no maintenance is undertaken until the machine is broken. The second type is Preventive Maintenance, which is based on the repair or replacement of equipment on a fixed calendar schedule regardless of their condition. This approach has benefits over RM, but it can lead to the unnecessary replacement of components that may still be in good working condition, resulting in increased downtime and waste. The last and most recent method is Predictive Maintenance, in which the main goal is to precisely estimate the Remaining Useful Life (RUL) of machinery based on various readings of sensor data (heat, vibration, etc.).

Unplanned downtime of machinery often results in economic losses for companies. Thus, predicting timely machine needs for maintenance can result in financial gain [9] and reduce unnecessary maintenance operation due to the preventive approach [10]. Another advantage of PM is cost minimisation, including minimising fatal breakdowns and reducing certain components' replacement, which is closely related to the other benefits [11].

There are many uses for PM in industrial application. A large amount of event data related to errors and faults in the internet of things (IoT) and digital platforms are continuously collected. An event records the behaviour of an asset, and by nature it comes in the form of data-streams. Event Processing is a method for processing streams of events to extract useful information, and complex event processing (CEP) is used for detecting anomalies and for predictive analytics [12]. Although event processing has been a paradigm for data processing for nearly three decades, there have been recent advancements in the last decade due to novel applications by the IoT and machine learning technologies. The presence of large amounts of streaming sensor data that can be widely generated is another reason for the growing interest in event processing. Predictive analytics uses the streams of data to make predictions of future events.

Making use of Predictive Analytics and CEP together provides synergy in PM performance [13]. Lately, there has been increased usage of event processing platforms that use open source technologies for Big Data and stream processing. Sahal, Breslin and Ali used an open-source computation pipeline and showed that the aggregated event-driven data, such as errors and warnings, are associated with machine downtime and can be qualified as predictive markers [14]. Calabrese et. al. performed equipment failure prediction from event log data based on the SOPHIA architecture, which is an event-based IoT and machine learning architecture for PM in Industry 4.0 [15]. Aivaliotis, Georgoulias and Chryssolouris investigated PM for manufacturing resources by utilising physics-based simulation models and the DT concept [16].

This research aimed to approach PM based on an open-source event processing platform and allow for the accurate prediction of time to failure to increase machine availability. Data-driven models are accompanied by a machine learning library and the DT concept to analyse the components of a machine's health status. The DT enables the platform under study to be a PM system, rather than a predictive analytics system. The DT concept used together with the open-source event-driven platform is detailed in [6], and the DT developed together with the platform is presented in [2].

Maintenance approaches that can monitor equipment conditions for diagnostic and prognostic purposes can be grouped into three main categories: statistical, artificial intelligence, and model based [17]. The model-based approach requires mechanical and theoretical knowledge of the equipment to be monitored. The statistical approach requires a mathematical background, whereas, in artificial intelligence, data are sufficient; thus, despite the challenges in the data science pipeline (data understanding, integration, cleaning, etc.), the last approach has been increasingly applied in PM applications.

In this study [18], the authors defined three types of PM approaches: (1) The data-driven approach, also known as the machine learning approach, uses data; (2) the model-based approach relies on the analytical model of the system; and (3) the hybrid approach, combining both methods. Dalzochio et. al. evaluated the application of machine learning techniques and ontologies in the context of PM and reported the application areas as fault diagnosis, fault prediction, anomaly detection, time to failure, and remaining useful life estimation, which refer to the several stages of PM [19].

According to [20], artificial neural networks (ANNs) are widely used for PM purposes. Carvalho et. al. noted that random forest (RF) was the most used method in PM applications [21]. Compared to the other machine learning methods in PM applications, RF was found to be more complex and it took more computational time [21]. In their review, the authors also noted that ANNs were one of the most common and applied machine learning algorithms in industrial applications since they were only based on previous data, requiring minimum expert knowledge, and the need for coping with challenges of data science pipeline.

Another study [22] evaluated the performance of k-Nearest neighbour (kNN), back-propagation feed-forward neural network (FFNN), DecisionTree, RF, support vector machine (SVM), and naïve Bayesian and assessed the results for time-series prediction. Rivas et. al. used the long short-term memory (LSTM) recurrent neural network (RNN) model, another type of ANN capable of using memory, for failure prediction [23]. Kanawaday and Sane discussed the use of autoregressive integrated moving average to predict possible failures and quality defects [24]. They basically used this method to predict future failure points in the data series for the diagnosis of machine downtimes.

Another research [15] presented an architecture that used tree-based algorithms to predict the probability of a failure where the gradient boosting machine generated the models that obtained the best results for classification when compared to the distributed RF models and extreme gradient boosting models. The use of deep learning algorithms is another promising area in PM. In this context, [25] used convolutional neural networks in performing the task of extracting features, and [26] used deep neural networks to reduce the dimensionality of data.

4 Maintenance of Spiral Welded Pipe Machinery

Figure 3 presents the Spiral Welded Steel (SWP) pipe production process. The SWP pilot of the COGNITWIN project is one of the six pilots that aims at enabling predictive maintenance (PM) for the SWP machinery presented in Fig. 4. NOKSEL is one of the use case partners of the COGNITWIN project [27]. The figure shows the machinery used in NOKSEL's facilities, and the process followed by this machinery to produce steel pipes. With the Digital Twin-supported condition monitoring platform, an infrastructure that aims to analyse the operational and

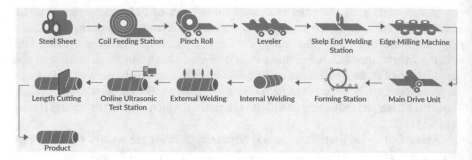

Fig. 3 Spiral welded steel pipe process

Fig. 4 SWP Machinery and process for which PM is developed

automation data received from sensors and PLC/SCADA will be used for PM, which will help increase the overall equipment performance.

In the steel pipe sector, operations run on a 24/7 basis. Due to the multi-step and interdependent nature of the production process, a single malfunction in one of the work stations can bring the whole production process to a halt. Thus, the cost of machine breakdown is very high. Under the COGNITWIN project scope, the Digital Twin is developed for the production process of Spiral Welded Steel Pipe machinery (SWP). The goal is to make use of developed models and analyse multiple sensors' data streams in real-time and enable predictive maintenance to reduce downtimes by Digital Twins. The main targets to be achieved are:

- 10% reduction in energy consumption
- 10% reduction in the shifted average duration of downtimes

A DT on NOKSEL's production process of Spiral Welded Steel Pipes (SWP) collects, integrates, and analyses multiple sensors' data. CT will be built upon DT with an aim to autonomously detect changes in the process and to know how to respond in real time to the constantly changing scenario with minimal human intervention. The first iteration of DT and CT systems will be an HT where human input will be required to take action based on the CT system's feedback. The CT will have cognitive capabilities by using real-time operational data to enable understanding, self-learning, reasoning, and decisions.

All the parameters in the first-order model are set as same as the real ones and expert knowledge is integrated into the model for HT. The data collected from the plant and the results of the simulation model are compared to ensure consistency in an iterative process. The model is modified, and simulation is repeated iteratively until the difference between the simulation results and the data retrieved is negligible. The collected data are used to train machine learning models and make predictions. In the next step, the predictions from the data-driven model will be taken as the observation value of the hybrid algorithms to adjust the theoretical values. The theoretical values coming from first-order models and data coming from the digital platform will be fused by hybrid algorithms like Kalman filtering and particle filtering algorithms.

In the NOKSEL pilot case, CT will introduce improved decision-making by integrating human knowledge into decision-making process. The anomalies, alarms, and early warnings of machine and system problems will be tackled by CT, and the decision-making process will emulate the experienced human operator with an embedded knowledge base. CT will augment expert knowledge for unforeseeable cases on HT and DT. The human operator's experience is added to process knowledge model and physics-based models with parametric values as well as thresholds and causality relations. Expert knowledge on the causes of breakdowns is collected with the series of problematic operations and the initial causes that trigger the successive reactions.

Data Analytics is deployed to extract knowledge from data. Machine learning techniques are used for this purpose, and a machine learning library is built. In the top layer, Data Visualisation and User Interaction, an advanced visualisation approach is provided for improved user experience with low latency. In vertical layers Cybersecurity and Trust is supported by IDS Security component and the Communication and Connectivity layer is composed of REST API, Kafka [28], MQTT [29], and OPC [30] components.

In Fig. 5, relevant components from the COGNITWIN toolbox have been used in the creation of a DT pipeline for the Steel Pipe Welding. The platform is aimed to be used for use cases where continuous availability, high performance, flexibility, robustness, and scalability are critical.

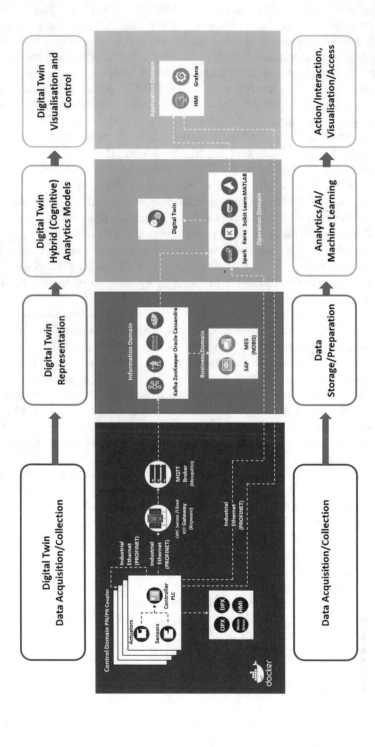

Fig. 5 Digital Twin pipeline for Steel Pipe Welding

5 Components and Digital Twin Pipeline for Steel Pipe Welding

The Digital Twin pipeline components are in the following mapped to the BDV Reference Model of the Big Data Value Association [31], which serves as a common reference framework to locate Big Data technologies on the overall IT stack. It has been presented in Fig. 6 and detailed in BDVA SRIA [32].

Things, assets, sensors, and actuator (IoT, CPS, edge computing) layer contain PLC as the main source of data, as well as the control units containing sensor data, alarms, and states of assets. In the Cloud and High-Performance Computing layer, the Big Data processing platform and data management operation are supported by the effective use of a private cloud system and computing infrastructure. Docker [33] is used here as a packaging and deployment methodology to easily manage the variety of the underlying hardware resources efficiently.

To meet system needs, Data Management is handled to collect and store raw data and manage the transformation of these data into the required form. The protection of these data is handled via privacy and anonymisation mechanisms like encryption, tokenisation and access control in the Data Protection layer. In the Data Processing Architecture layer, an optimised and scalable architecture is developed for the analytics of both batch and stream processing via SIMATIC and Spark [34] stream processing. Data clean-up and pre-processing are also handled in this layer.

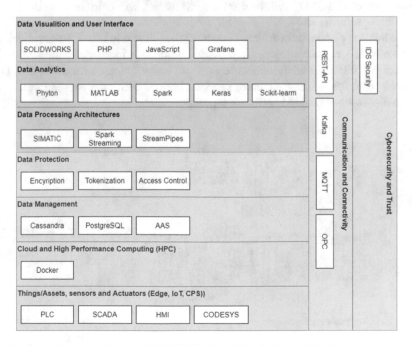

Fig. 6 The architecture aligned with BDVA Big Data Value Reference Model

6   COGNITWIN Digital Twin Pipeline Architecture

The four-step Digital Twin pipeline also has a mapping to the technical areas in the SRIDA for AI, Data and Robotics Partnership [45] as follows: Digital Twin Data Acquisition and Collection relate to enablers from *Sensing and Perception* technologies. Digital Twin Data Representation relates to enablers from *Knowledge and Learning* technologies, and also to enablers for *Data for AI*. Hybrid and Cognitive Digital Twins relate to enablers from *Reasoning and Decision*. Digital Twin Visualisation and Control relate to enablers from *Action and Interaction*. These four steps are described in more detail in the following.

6.1   Digital Twin Data Acquisition and Collection

IIoT refers to IoT technologies used in industry, and it has been the primary building block of the systems facilitating the convergence and integration of operational technology (OT) and information technology (IT) for gathering data from sites [35]. This section details the IIoT system used for data acquisition and collection.

Figure 7 shows the hardware topology of the previously existing system. With the Digital Twin-supported condition monitoring platform to be developed, and infrastructure that aims to analyse the operational and automation data received from sensors and PLC/SCADA will be used for PM, which will help increase the overall equipment performance.

The topology of the infrastructure established is shown in Fig. 7. Communication between these two topologies is provided with the industrial communication protocol PROFINET, and the two structures will communicate with each other. Data required from the existing structure can be obtained using the existing controller. Figure 8 presents the added hardware topology.

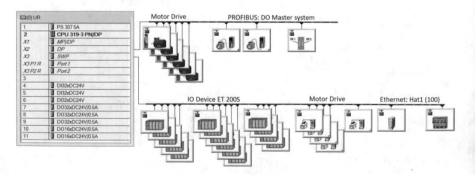

Fig. 7 Existing hardware topology

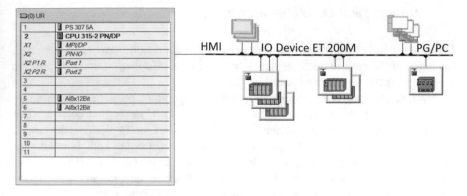

Fig. 8 Added hardware topology

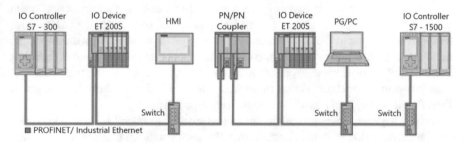

Fig. 9 Coupling of the PROFINET subnets with the PN/PN Coupler

The current PLC model used for process control is S7 300. The operation details of the components; status information; process information, such as speed and power; production details; and system alarms are kept on this PLC while the newly added sensor data and alarms will be located in the S7 1500 PLC. The existing PLC data will be transferred to the S7 1500 PLC through the PN/PN Coupler module, allowing all data tracking to be carried out over the new PLC.

The PN/PN Coupler module provides the simple connection of two separate PROFINET networks. The PN/PN Coupler enables data transmission between two PROFINET controllers. The output data from one network becomes the input data of the other. For data transfer, additional function blocks are not required and the transfer is realised without latency. For the new sensors added to the system not to affect the existing process, a new PLC is employed and the controls are implemented over it. The communication structure between the PLCs is designed using the PN/PN Coupler module as shown in Fig. 9.

PLC transmits the data it receives from the sensors to OPC, which then transfers the data to the platform via MQTT. The received data is transmitted to Kafka, which passes it on to the Cassandra [36] database to be stored for further processing or later access.

6.2   Digital Twin Data Representation

DT representation step follows the data acquisition/collection step. In this step, data collected in the data acquisition step is stored in the information domain, and the stored data is used by the business domain.

The data obtained from the sensors, such as temperature, pressure and vibration, voltage, and current, are transmitted to MQTT over OPC in the first tier, and then to Kafka in the JSON format. Apache Kafka is a data streaming platform developed specifically to transmit real-time data with a low error margin and short latency. Kafka achieves superior success in systems with multiple data sources, such as sensor data, and reduces the inter-system load. It has an integration that can also process Big Data coming from sensors operating at high frequencies.

Instant data received by Kafka is transmitted to the Python [37]-based server, where the attribute extraction process begins. Incremental principal component analysis (PCA), which is the most well-known method used in the Big Data flow, applies PCA stages to the instantaneous data using data in a certain window range, and thus large data that cannot fit into the memory can also be processed effectively. PCA performs dimensional reduction by making the incoming high-dimensional data low-dimensional, providing more accurate results for machine learning, and therefore it is frequently used for categorisation problems.

A fully asynchronous communication structure with the event-bus method is used for the transmission of data collected from the source with OPC. Data transmission is provided in the JSON format. In the architecture managed based on Microservices, Cassandra is used as the NoSQL database with a database presented as a log file to users. Cassandra is a database that provides continuous availability, high performance, and scalability. PostgreSQL [38], a relational database (RDBMS), is used by the interface program that provides user interaction to display time-series data in real time.

A total of 120 sensor values are monitored on the SWP machine to capture data on temperature, vibration, pressure, current, oil temperature, and viscosity. A value is taken every 10 ms from the vibration sensors, once every 100 ms from the current sensors, and every 1000 ms from the temperature and pressure sensors plus alarm and status fields. The SWP machinery has a total of 120 sensor values, 122 alarms, and 175 status data which create 11 GB of incoming data in 1 day (24 h).

To organise the collected data in a Digital Twin structure, we have analysed several emerging Digital Twin open source-initiatives and standards, as reported in the COGNITWIN project survey paper "Digital Twin and Internet of Things-Current Standards Landscape" [3]. Based on this, we have selected to use the Asset Administration Shell (AAS) for further Digital Twin API development. AAS was developed by Platform Industry 4.0, and similar to DT, AAS is a digital representation of a resource [3]. Descriptions of AAS can be serialised using JSON, XML, RDF, AutoML and OPC UA [39]. To realise COGNITWIN vision, the AAS model and APIs are not sufficient. For COGNITWIN, the models and the components using different technologies should be reusable. For this purpose,

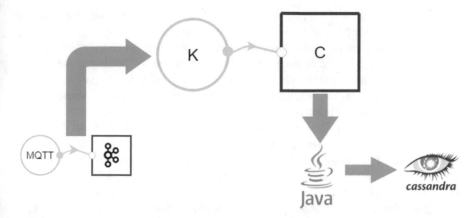

Fig. 10 An example pipeline created in Apache StreamPipes

it is decided to utilise Apache StreamPipes [40] for the IIoT. StreamPipes is expected to provide an environment to host different models, components, and services, and to orchestrate them. Figure 10 presents a sample pipeline generated in Apache StreamPipes. As shown, MQTT data is retrieved by Kafka and stored in the Cassandra database in a pipeline. The main purpose of the developed pipeline is to create a path from sensory data to the neural network output, depicting the state of the tool, through several pipeline elements. Besides, this pipeline triggers notification when the value of a particular property goes above a certain threshold and shows results. The below sample shows the pipeline elements regarding data collection on data storage.

6.3 Hybrid and Cognitive Digital Twins

In the Machine Learning Library (MLL) module, different machine learning algorithms are applied through the incremental PCA stage to detect anomalies. Prediction results are produced using various machine learning libraries. First, is Spark MLlib produced entirely by Spark, which uses Spark's engine optimised for large-scale data processing. Keras library utilises TensorFlow, and is used for deep learning. The LSTM algorithm of this library is utilised. This open-source neural network library makes it simpler to work with artificial neural networks through its user interface facilities and modular structure. The Scikit-Learn [41] library is another open-source machine learning library that contains several algorithms for regression, classification, clustering. We used algorithms like RF, GBT, LSTM, SVM, KNN and multi-layer perceptron (MLP) from Scikit-Learn library for data modeling and prediction.

The MLL module is used for comparing the different machine learning models. When setting up a machine learning model, it is difficult to predict which model architecture will provide the best results. The parameters that can affect the model architecture are called hyper-parameters. For each machine learning algorithm used, hyper-parameters tuning is performed by comparing the previously determined success criteria and selecting the best result combination by looking at the results obtained by testing possible combinations of the values of the hyper-parameters in a certain range. For each ML algorithms used, in addition to different parameters, such as precision, recall, F1 score, error detection rate, total training time, total test time, average training time, Type 1 error, and Type II error were calculated and displayed to the user. Besides, the user is offered a voting option for deciding the algorithm to use. Selected graphical user interfaces of the application are provided in Fig. 6. The application enables users to select the machine learning model for a given set of data and then compares the output using graphical elements. For developing and testing purposes, AML Workshop dataset from Microsoft [42] is used in the MLL module.

Hybrid Digital Twin

The above-described platform is enhanced by DT, which contains two related models: Data-Driven and Physics-Driven (first-order principal models); thus a Hybrid DT is generated.

In the context of hybrid Digital Twins, Physics-Driven models can be beneficial over the Data-Driven ones in many aspects, including but not limited to:

- **Generating synthetic data in case of "data-poor" cases**: A typical example is training a ML pipeline for predictive maintenance. Meanwhile, very often, when a machine is new, it does not have historical sensor data that can be used to train a data-driven approach. When carefully designed, the virtual physics-based twin can generate the needed supervised training dataset.
- **Quality control of data-driven Digital Twin**: When operating a critical infrastructure or asset, it is seen as a risky approach to fully rely on data-driven approaches in taking real-time decisions. To mitigate such risks, it is possible to build a control pipeline in which the physics-based model will be a controller to the data-driven predictor. A broker needs to be designed to integrate the two approaches in a seamless way.

On the other hand, data-driven models can be used to continuously calibrate physics-based models. In fact, machine degradation, wearing of parts, environment, and other factors impact the overall process performance over time. The state of the practice is that an operator will manually recalibrate the control system when a deviation is identified. Such manual operation can be replaced by setting a data-driven model that will identify and calibrate critical process variables that will be fed into a physics-based model that will optimise the process's control system.

An example of first-order model tools we have is a DC motor, which is a commonly used component. The motor is made of two coupled electrical and mechanical models based on the governing equations from Newton's second law ($v = R{\cdot}i + L{\cdot}di/dt + ve$) and Kirchhoff's voltage law ($Te = TL + B{\cdot}\omega + JL{\cdot}d\omega/dt$) where v is the voltage [V], i is the current [A], ve is the back electromotive force

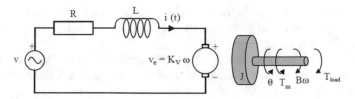

Fig. 11 Physics-based schematic of a DC motor (http://pubs.sciepub.com/ajme/4/7/27/)

```
Input: Description of the List of Failures
1)Develop Detailed Physics-Based Model of the Process.
2)Develop and Implement modelling Strategies of the Failures.
3)Realistic Range of Variables Responsible of the Failures.


Repeat
Randomly vary the Variables Responsible of the Failures.
Run (1)
Until Enough Data

Output: Supervised Dataset for Predictive Maintenance
```

Fig. 12 Synthetic data generation for Predictive Maintenance Pipeline

[V], ω is the angular velocity [rad/s], TL is the Load torque [Nm], JL is Load inertia + Rotor inertia [kg·m^2], ve = Kv·ω, Te = Kt·i, Kt is Torque constant [Nm/A], Kv is the back EMF constant [V/(rad/s)], R is the Phase Resistance [Ohm], L is the Phase Inductance [H], J is the Rotor Inertia [kg·m^2], and B is the Rotor Friction [Nm/(rad/s)] (Fig. 11).

A hybrid Digital Twin for predictive maintenance of a DC motor is built using the principles given above and the architecture given in Fig. 12. For the physics-based model, the MATLAB environment has been used to model the physical elements of the electric DC motor and related electric and mechanical components. Real data has been obtained by means of real sensor data collected, but they are not with enough failure cases to train an ML model. On the DC motor, the following sensors have been installed (current, voltage, temperature, and vibration). The physics-based model is being calibrated by comparing the measured data with the MATLAB predicted one. Results show a qualitative similarity between the synthetic and the sensors data. The models will later be added to the knowledge base by integrating the cognitive elements into the model.

Cognitive Digital Twin

In the Steel Pipe Welding pilot case, CT will introduce improved decision-making by integrating human knowledge into the decision-making process. The anomalies, alarms, and early warnings of machine and system problems will be tackled by CT, and the decision-making process will emulate the experienced human operator with the embedded knowledge base. CT will augment expert knowledge for unforeseeable cases on HT and DT. The human operator's knowledge is reflected to process knowledge and physics-based models with parametric values as well as thresholds and causality relations. Expert knowledge on the causes of breakdowns is collected with the series of problematic operations and the initial causes that trigger the successive reactions.

Cognition will be integrated to support Cognitive Digital Twins for learning, understanding, and planning, including the use of domain and human knowledge by making use of the ML algorithms, ontologies, and Knowledge Graphs (KG) to capture background knowledge, entities, and their relationships. CT will bring life cycle optimisation by reacting to early warnings and suggesting optimised predictive actions to improve operational performance by optimising operational parameters and enhance the utilisation of computing infrastructure and energy usage.

6.4   Digital Twin Visualisation and Control

This component contains dashboards suitable for sensor data, error detection, and transfer of regular information obtained from data processing to the real-time status monitoring system, and development of end-user applications.

Three.js, an open-source JavaScript library, was used to develop animated or non-animated 3D applications that can be opened in the web browser using WebGL. Three.js is supported by all WebGL-supported web browsers. In addition to Three.js for visualisation of the Digital Twin elements, Solidworks [43] is used for 3D visualisation.

For the web interface, the JSON data received with JavaScript have been parsed and then transferred to PHP pages. In this communication, the post method has been used in the requests sent with JavaScript. With the help of PHP, the information is placed in HTML objects. Grafana [44] is used in the process of placing graphics within the card object. Dynamic graphics created on Grafana are placed on cards in iframe tags.

Sensor data on temperature, vibration, pressure, current, and oil temperature are given in dashboards. Alarm and status information are provided as they occur. Besides sensor, alarm, and status values, real-time monitoring of parameters like pipe diameter, pitch angle, belt width, production speed, motor speed (RPM), energy consumption, instant output power, pipe diameter, and wall thickness are visualised in the dashboards.

7   Conclusions

This chapter has presented the Digital Twin Pipeline Framework of the COGNITWIN project that supports Hybrid and Cognitive Digital Twins, through four Big Data and AI pipeline steps. The approach has been demonstrated with a Digital Twin pipeline example for Spiral Welded Steel Pipe Machinery maintenance. The components used in this pipeline have also been mapped into the different technical areas of the BDV Reference Model.

The COGNITWIN project has further five use case pilots which include also other types of sensors, in particular for image and video analytics with RGB and infrared cameras, and support for image analytics through deep learning, including AI@Edge support through FPGA hardware. This includes also high temperature processes for aluminium, silicon, and steel production and Digital Twin pipeline support including analytics for nonlinear model predictive control combined with machine learning.

Synergies between and combinations of different elements for hybrid Digital Twin are now being further enhanced through the use of orchestration technologies like StreamPipes and Digital Twin access APIs like AAS.

The COGNITWIN project is now proceeding with further Hybrid Digital Twins in all of the six pilot cases, extending this also with cognitive element for self-learning and control, for the establishment of a Toolbox with relevant and reusable tools within each of the four pipeline steps. This will be further applied and evaluated for the six pilot use cases of the project, including the steel welding pilot described in this chapter.

Acknowledgements The COGNITWIN project has received funding from the European Union's Horizon 2020 research and innovation programme under GA No.870130. The platform developed by TEKNOPAR [45] is partially funded by the COGNITWIN project and has been validated by the NOKSEL pilot of the project. We thank the COGNITWIN consortium partners for fruitful discussions related to the Digital Twin pipeline architecture and the use case presented in this chapter.

References

1. Abburu, S., Berre, J. A., Jacoby, M., Roman, D., Stojanovic, L., & Stojanovic, N. (2020, November). Cognitive digital twins for the process industry. In *Cognitive technologies, the twelfth international conference on advanced cognitive technologies and applications (COGNITIVE 2020)*.
2. Albayrak, Ö., & Unal, P. (2020). Smart Steel pipe production plant via cognitive digital twins: a case study on digitalization of spiral welded pipe machinery. In *The proceedings of the ESTEP workshop on impact and opportunities of artificial intelligence in the steel industry*.
3. Jacoby, M., & Usländer, T. (2020). Digital twin and internet of things—Current standards landscape. *Applied Sciences, 10*, 6519. Retrieved from https://www.mdpi.com/2076-3417/10/18/6519
4. Eirinakis P., Kalaboukas, K., Lounis, S., Mortos, I. Rozanec, J. M., Stojanovic, N., & Zois, G. (2020). Enhancing cognition for digital twins. In *IEEE International conference on engineering, technology and innovation (ICE/ITMC), Cardiff, United Kingdom, 2020* (pp. 1–7). https://doi.org/10.1109/ICE/ITMC49519.2020.9198492.
5. Fuller, A., Fan, Z., Day, C., & Barlow, C. (2020). Digital twin: enabling technologies, challenges and open research. *IEEE Access, 8*, 108952–108971. https://doi.org/10.1109/ACCESS.2020.2998358
6. Abburu, S., Berre, J. A., Jacoby, M., Roman, D., Stojanovic, L., & Stojanovic, N. (2020, June 15–17). COGNITWIN—hybrid and cognitive digital twins for the process industry.

In *Proceedings of the 2020 IEEE international conference on engineering, technology and innovation (ICE/ITMC), Cardiff, UK* (pp. 1–8).

7. Berre, A. J., Tsalgatidou, A., Francalanci, C., Ivanov, T., Lobo, T. P., Saiz, R. R., Novalija, I., & Grobelnik, M. (2021). Big data and AI pipeline framework—Technology analysis from a benchmarking perspective. In *TABDV book*. Springer Open Access.

8. Krupitzer, C., Wagenhals, T., Züfle, M., Lesch, V., Schäfer, D., Mozaffarin, A., Edinger, J., Becker, C., & Kounev, S. (2020). A survey on predictive maintenance for Industry 4.0. *arXiv*, preprint arXiv:2002.08224.

9. Wu, S. J., Gebraeel, N., Lawley, M. A., & Yih, Y. (2007). A neural network integrated decision support system for condition-based optimal predictive maintenance policy. *IEEE Transactions on Systems, Man, and Cybernetics-Part A: Systems and Humans, 37*(2), 226–236. https://doi.org/10.1109/TSMCA.2006.886368

10. Baidya, R., & Ghosh, S. K. (2015). Model for a predictive maintenance system effectiveness using the analytical hierarchy process as analytical tool. *IFAC-PapersOnLine, 48*(3), 1463–1468.

11. Deloux, E., Castanier, B., & Bérenguer, C. (2009). Predictive maintenance policy for a gradually deteriorating system subject to stress. *Reliability Engineering & System Safety, 94*(2), 418–431.

12. Dayarathna, M., & Perera, S. (2018). Recent advancements in event processing. *ACM Computing Surveys (CSUR), 51*(2), 1–36.

13. Fülöp, L. J., Beszédes, Á., Tóth, G., Demeter, H., Vidács, L., & Farkas, L. (2012, September). Predictive complex event processing: a conceptual framework for combining complex event processing and predictive analytics. In *Proceedings of the fifth Balkan conference in informatics* (pp. 26–31).

14. Sahal, R., Breslin, J. G., & Ali, M. I. (2020). Big data and stream processing platforms for Industry 4.0 requirements mapping for a predictive maintenance use case. *Journal of Manufacturing Systems, 54*, 138–151.

15. Calabrese, M., Cimmino, M., Fiume, F., Manfrin, M., Romeo, L., Ceccacci, S., Paolanti, M., Toscano, G., Ciandrini, G., Carrotta, A., Frontoni, E., Kapetis, D., & Mengoni, M. (2020). SOPHIA: An event-based IoT and machine learning architecture for predictive maintenance in Industry 4.0. *Information, 11*(4), 202.

16. Aivaliotis, P., Georgoulias, K., & Chryssolouris, G. (2019). The use of digital twin for predictive maintenance in manufacturing. *International Journal of Computer Integrated Manufacturing, 32*(11), 1067–1080.

17. Jardine, A. K., Lin, D., & Banjevic, D. (2006). A review on machinery diagnostics and prognostics implementing condition-based maintenance. *Mechanical Systems and Signal Processing, 20*(7), 1483–1510.

18. Paolanti, M., Romeo, L., Felicetti, A., Mancini, A., Frontoni, E., & Loncarski, J. (2018, July). Machine learning approach for predictive maintenance in industry 4.0. In *2018 14th IEEE/ASME international conference on mechatronic and embedded systems and applications (MESA)* (pp. 1–6). IEEE.

19. Dalzochio, J., Kunst, R., Pignaton, E., Binotto, A., Sanyal, S., Favilla, J., & Barbosa, J. (2020). Machine learning and reasoning for predictive maintenance in Industry 4.0: Current status and challenges. *Computers in Industry, 123*, 103298.

20. Li, Z., Wang, Y., & Wang, K. S. (2017). Intelligent predictive maintenance for fault diagnosis and prognosis in machine centers: Industry 4.0 scenario. *Advances in Manufacturing, 5*(4), 377–387.

21. Carvalho, T. P., Soares, F. A., Vita, R., Francisco, R. D. P., Basto, J. P., & Alcalá, S. G. (2019). A systematic literature review of machine learning methods applied to predictive maintenance. *Computers & Industrial Engineering, 137*, 106024.

22. Schmidt, B., Wang, L., & Galar, D. (2017). Semantic framework for predictive maintenance in a cloud environment. *Procedia CIRP, 62*, 583–588.

23. Rivas, A., Fraile, J. M., Chamoso, P., González-Briones, A., Sittón, I., & Corchado, J. M. (2019, May). A predictive maintenance model using recurrent neural networks. In *International*

workshop on soft computing models in industrial and environmental applications (pp. 261–270). Springer.

24. Kanawaday, A., & Sane, A. (2017, November). Machine learning for predictive maintenance of industrial machines using IoT sensor data. In 2017 8th IEEE international conference on software engineering and service science (ICSESS) (pp. 87–90). IEEE.

25. Yuan, Y., Ma, G., Cheng, C., Zhou, B., Zhao, H., Zhang, H. T., & Ding, H. (2018). Artificial intelligent diagnosis and monitoring in manufacturing. arXiv, preprint arXiv:1901.02057. https://doi.org/10.1093/nsr/nwz190

26. De Vita, F., Bruneo, D., & Das, S. K. (2020, April). A novel data collection framework for telemetry and anomaly detection in industrial IoT systems. In 2020 IEEE/ACM fifth international conference on internet-of-things design and implementation (IoTDI) (pp. 245–251). IEEE.

27. NOKSEL. http://www.noksel.com.tr/index.php/en/

28. Apache Kafka. https://kafka.apache.org/

29. MQTT Broker. https://mqtt.org/

30. Open Platform Communications, OPC. https://opcfoundation.org/about/what-is-opc/

31. BDVA, Big Data Value Association. https://www.bdva.eu

32. Zillner, S., Bisset, D., Milano, M., Curry, E., García Robles, A., Hahn, T., Irgens, M., Lafrenz, R., Liepert, B., O'Sullivan, B., & Smeulders, A. (Eds.) (2020, September). Strategic research, innovation and deployment agenda—AI, data and robotics partnership. Third release. Brussels. BDVA, euRobotics, ELLIS, EurAI and CLAIRE.

33. Docker. https://www.dockcr.com/

34. Apache Spark. https://spark.apache.org/

35. Unal, P. (2019). Reference architectures and standards for the internet of things and big data in smart manufacturing. In 2019 7th international conference on future internet of things and cloud (FiCloud) (pp. 243–250). IEEE.

36. Apache Cassandra. https://cassandra.apache.org/

37. Python. https://www.python.org/

38. PostGreSQL. https://www.postgresql.org/

39. OPC UA. https://opcfoundation.org/about/opc-technologies/opc-ua.

40. Apache StreamPipes. https://streampipes.apache.org/

41. Scikit-Learn. https://scikit-learn.org/

42. AML Data Set, https://github.com/microsoft/AMLWorkshop/tree/master/Data

43. Solidworks. https://www.solidworks.com/tr

44. Grafana. https://grafana.com/

45. TEKNOPAR. https://TEKNOPAR.com.tr

Big Data Analytics in the Manufacturing Sector: Guidelines and Lessons Learned Through the Centro Ricerche FIAT (CRF) Case

Andreas Alexopoulos, Yolanda Becerra, Omer Boehm, George Bravos, Vassilis Chatzigiannakis, Cesare Cugnasco, Giorgos Demetriou, Iliada Eleftheriou, Spiros Fotis, Gianmarco Genchi, Sotiris Ioannidis, Dusan Jakovetic, Leonidas Kallipolitis, Vlatka Katusic, Evangelia Kavakli, Despina Kopanaki, Christoforos Leventis, Miquel Martínez, Julien Mascolo, Nemanja Milosevic, Enric Pere Pages Montanera, Gerald Ristow, Hernan Ruiz-Ocampo, Rizos Sakellariou, Raül Sirvent, Srdjan Skrbic, Ilias Spais, Giuseppe Danilo Spennacchio, Dusan Stamenkovic, Giorgos Vasiliadis, and Michael Vinov

Abstract Manufacturing processes are highly complex. Production lines have several robots and digital tools, generating massive amounts of data. Unstructured, noisy and incomplete data have to be collected, aggregated, pre-processed and transformed into structured messages of a common, unified format in order to be

A. Alexopoulos · S. Fotis · L. Kallipolitis · I. Spais
Aegis IT Research LTD, London, UK

Y. Becerra · C. Cugnasco · M. Martínez · R. Sirvent
Barcelona Supercomputing Center, Barcelona, Spain

O. Boehm · M. Vinov
IBM, Haifa, Israel

G. Bravos · V. Chatzigiannakis
Information Technology for Market Leadership, Athens, Greece

G. Demetriou · V. Katusic · H. Ruiz-Ocampo
Circular Economy Research Center, Ecole des Ponts, Business School, Marne-la-Vallée, France

I. Eleftheriou · E. Kavakli · R. Sakellariou
University of Manchester, Manchester, UK

G. Genchi · J. Mascolo · G. D. Spennacchio (✉)
Centro Ricerche FIAT, Orbassano, Italy
e-mail: giuseppedanilo.spennacchio@crf.it

S. Ioannidis
Technical University of Crete - School of Electrical and Computer Engineering, Crete, Greece

Foundation for Research and Technology, Hellas - Institute of Computer Science, Crete, Greece

321
E. Curry et al. (eds.), *Technologies and Applications for Big Data Value*,
https://doi.org/10.1007/978-3-030-78307-5_15

analysed not only for the monitoring of the processes but also for increasing their robustness and efficiency. This chapter describes the solution, best practices, lessons learned and guidelines for Big Data analytics in two manufacturing scenarios defined by CRF, within the I-BiDaaS project, namely 'Production process of aluminium die-casting', and 'Maintenance and monitoring of production assets'. First, it reports on the retrieval of useful data from real processes taking into consideration the privacy policies of industrial data and on the definition of the corresponding technical and business KPIs. It then describes the solution in terms of architecture, data analytics and visualizations and assesses its impact with respect to the quality of the processes and products.

Keywords Big Data · Self-service solution · Manufacturing · Die-casting · Maintenance and Monitoring · Advanced analytics and visualizations

1 Introduction

The manufacturing industry transforms material or assembles components to produce finished goods that are ready to be sold in the marketplace. The organizational structure of manufacturing companies is very complex and involves many business and operative functions with different roles and responsibilities in order to guarantee efficiency at every level [1]. The fourth industrial revolution [2, 3] has initiated many changes in the industrial value chain, transforming the shop floor, which is the production part of the manufacturing industries. Companies are introducing process equipment provided with several robots and digital tools. In this way, it is possible to set and control processes in an automated manner that speeds up production with a high level of accuracy [4]. Furthermore, large volumes of data are generated every day that may be collected and analysed for increasing process robustness and efficiency and building a technical cycle that reduces the consumption of energy and material. However, despite the potential benefits offered by the exploitation of Big Data, its usage is still at an early stage in many manufacturing companies [5].

Centro Ricerche FIAT (CRF) is one of the main private research centres in Italy and represents Fiat Chrysler Automobiles (FCA) in European and national collaborative research projects. In the context of the European Horizon 2020 I-

D. Kopanaki · C. Leventis · G. Vasiliadis
Foundation for Research and Technology, Hellas - Institute of Computer Science, Crete, Greece

D. Jakovetic · N. Milosevic · S. Skrbic · D. Stamenkovic
University of Novi Sad - Faculty of Sciences, Novi Sad, Serbia

E. P. P. Montanera
ATOS, Madrid, Spain

G. Ristow
Software AG, Darmstadt, Germany

BiDaaS project,[1] CRF identified two use cases, in which complex datasets are retrieved from real processes. By exploiting Big Data analytics in these two cases, CRF aims to improve the process and product quality in a much more agile way through the collaborative effort of self-organizing and cross-functional teams, reducing costs due to further processing and predicting faults and unnecessary actions. This requires solutions that will allow manufacturing experts to interact with Big Data [6] in order to understand how to easily utilize important information often hidden in raw data. In other words, *the first best practice (1)*[2] *is the correlation between the value of Big Data technology and the skills of people involved in the data management process.* The I-BiDaaS approach follows this best practice and develops a self-service [7] Big Data analytics platform that enables different CRF end-users to exploit Big Data in order to gain new insights assisting them to make the right decisions in a much more agile way.

The aim of this chapter is to demonstrate how advanced analytic tools can empower end-users [8] in the manufacturing domain (see Sect. 5) to create a tangible value from the process data that they are producing, and to identify a number of best practices, guidelines and lessons learned. For future reference, we list here the main best practices with the identified guidelines and lessons learned, while they will be discussed in detail throughout the chapter:

- The correlation between the value of Big Data technology and the skills of people involved in the data management process with the involvement of different departments belonging to the same or different organizations in order to extract the value of all data collected from several sources and levels (breaking data silos).
- The alignment of the Big Data requirements with the business needs and the definition of appropriate experiments with the identification of Big Data technologies most suitable for the specific identified business requirements.
- The management of the type of data generated with the identification of the types of data useful for the analysis, their anonymization and generation of synthetic data in parallel with the process of data anonymization.
- The development of a solution that satisfies Big Data requirements of specific use cases by mapping the identified functional and non-functional concerns into a concrete software architecture with the development of Advanced Visualization tools for showing high-value Big Data analytics solutions for domain experts and operators.

The remainder of this chapter is organized as follows. Section 2 describes the process followed for the identification of the Big Data requirements in the manufacturing sector and demonstrates how it was applied to elicit the requirements of the CRF use cases, which are imposed the design of the I-BiDaaS Big Data

---

[1] http://www.ibidaas.eu/

[2] As explained below, we identify throughout the chapter several best practices for the application of Big Data analytics in manufacturing.

solution. Furthermore, CRF requirements guide the definition of the experiments for assessing the developed system, described in Sect. 3. The architecture of the I-BiDaaS solution is described in Sect. 4. Finally, Sect. 5 reports on the lessons learned, challenges and guidelines reflecting the experience of the I-BiDaaS project. Section 5 also provides the connection of the described work with the Big Data Value (BDV) reference model and its Strategic Research and Innovation Agenda (SRIA) [9]. Finally, Sect. 6 concludes the chapter.

2   Requirements for Big Data in the Manufacturing Sector

Alignment between business strategy and Big Data solutions is a critical factor for achieving value through Big Data [10]. Manufacturers must understand how the adoption of Big Data technologies [11] is related to their business objectives in order to identify the right datasets and increase the value of the analytics results. *Therefore tailoring Big Data requirements to the business needs is the second best practice (2) reported in this chapter.*

In more detail, the I-BiDaaS methodology for eliciting CRF requirements draws on work in the area of early Requirements Engineering (RE), which considers the interplay between business intentions and system functionality [12, 13]. In particular, the requirements elicitation followed a (mostly) top-down approach whereby business goals reflecting the company's vision were progressively refined in order to identify the user requirements of specific stakeholder groups (i.e. data providers, Big Data capability providers and data consumers). Their analysis resulted in the definition of system functional and non-functional requirements, which describe the behaviour that a Big Data system (or a system component) should expose in order to realize the intentions of its users. This process was facilitated by the use of appropriate questionnaires. In the cases that information on the requirements was available (either collected in the context of the project setup phase, or identified through a review of related literature [10, 14]), this was used to partly pre-fill the questionnaires and minimize end-users' effort. Evidently, users were asked to check pre-filled fields and ensure that documented information was valid and accurate.

Table 1 gives a summary of the CRF requirements. Although it provides only an excerpt of the elicited CRF requirements, it demonstrates the application of the I-BiDaaS way-of-working in the CRF use cases.

In more detail, the strategic CRF business goal (R1) was refined into a number of more operational business goals that need to be satisfied through Big Data analytics (R3). In addition, a number of relevant KPIs (R6) were defined that can be used to assess the proposed solution (see Sect. 3). Continuing, at the user requirements level, requirements were described in terms of the characteristics of different data sources that are planned to be used (requirements R7 and R8), the analytics capability of the proposed solution envisaged (R9) and the different interface requirements of the end-users that will consume the analytics results (R10–R12). Finally, analysis of

Table 1 CRF Big Data requirements

Business requirements	
R1	Improve and optimize business processes and operations (business goal)
R2	Improve monitoring and maintenance of production assets (business goal)
R3	Improve decisions about production line re-planning based on the analysis of maintenance data (business goal)
R4	Maintain efficiency (quality business goal)
R5	Cost reduction (KPI)
R6	Product/service quality (KPI)
User requirements	
R7	Data is stored locally in the Manufacturing Execution System (data provider requirement)
R8	Real-time data on the operating status of the machines is obtained from SCADA sensors in real time (data provider requirement)
R9	MES and SCADA sensor data information will be combined and proceed to real-time re-planning (Big Data analytics provider requirement)
R10	Line operators will only visualize the results (data consumer requirement)
R11	Data scientist will customize and then analyse data (data consumer requirement)
R12	Process manager will collaborate with the data scientist to decide on the action to actuate as a consequence of the analysis (data consumer requirement)
System requirements	
R13	The system should enable aggregation of both attribute level and transaction level data coming from a variety of internal data sources and in multiple formats (FR)
R14	The system should support multilevel access control at resource and application level (NFR)
R15	The system should enable near real-time re-planning (NFR)

the above user requirements resulted in the generation of the system requirements, both functional (R13) and non-functional (R14 and R15). Although described in a linear fashion, the above activities were carried out in an iterative manner, resulting in a stepwise refinement of the results being produced. The complete list of CRF requirements elicited is described in detail in [15].

Further to forming the baseline of the I-BiDaaS solution (see Sect. 4), these requirements also assist the definition of experiments as described in Sect. 3.

3 Use Cases Description and Experiments' Definition: Technical and Business KPIs

The aim of experimentation is to assist stakeholders' acceptance of any new Big Data solution. The definition of appropriate experiments is thus another best practice (3) reported in this chapter. In particular, the definition of CRF experiments aims at evaluating and validating the I-BiDaaS solution and its implementation in the context of CRF use cases. It follows a goal-oriented approach, whereby the experiment's goal(s) towards which the measurement will be performed are defined, then a number of questions are formed aiming to characterize the achievement of each goal and, finally, a set of Key Performance Indicators (KPIs) and associated metrics is associated with every question in order to answer it in a measurable way.

The definition of each experiment also involved the specification of the experiment's workload in terms of the use case datasets and type of analysis envisaged, as well as the definition of the experimental subjects that will be involved in the experiment, as reported in the following Sects. 3.1 and 3.2 that discuss, respectively, the 'Production process of aluminium die-casting' and 'Maintenance and monitoring of production assets' use cases.

3.1 Production Process of Aluminium Die-Casting

The 'Production process of aluminium die-casting' use case generates complex datasets from the production process of the engine blocks. During the die-casting process [16, 17], molten aluminium is injected into a die cavity, mounted in a machine, in which it solidifies quickly. In this case, we have a large number of interconnected process parameters that influence the flow behaviour of molten metal inside the die cavity, and, consequently, the productivity and the quality [18–20]. *Henceforth, the fourth best practice (4) is to identify the type of data generated.* Data collected from several sources can be disorganized and in different formats and data may not be exploited.

In this use case, the data provided for the analyses consist of a collection of casting process parameters, such as piston speed in the first and second phase, intensification pressures and others. In addition to the process data, CRF also provided a large dataset of thermal images of the engine block casting process, under a hypothesis that there is a correlation among process data, thermal data and the outcome of the process.

For the mentioned complexity of the process, it is important to not only carefully design parameters and temperatures but also to control them because they have a direct impact on the quality of the casting.

Analysis of the datasets aims to predict whether an engine block will be produced correctly during the casting process in order to avoid further processing and scraps, which would lead to financial savings for the manufacturers.

To test the efficiency of the I-BiDaaS solution in this context, an experiment has been defined, as shown in Table 2. As seen in Table 2, the Business KPI 'Product/service quality' identified during requirements elicitation (see Sect. 2) was further elaborated in order to define appropriate metrics (quality control levels related to good and defective products) and to map it to appropriate indicators at the I-BiDaaS solution level (execution time, data quality, cost).

For each KPI, a baseline value for evaluating the performance of the I-BiDaaS solution has also been defined. For example, an increase of 2–6% of the quality control level related to good products and a decrease of 1–4% and 0.05–2% of the

Table 2 Overview of the 'Production process of aluminium die-casting' experiment

Experiment's goals	To test the efficiency of I-BiDaaS solution in the context of correlating defects with the production process parameters.	
Experiment's questions	*Q1. What is the quality of the analytics results?* Q1.1 What is the accuracy of new models with respect to internal CRF aluminium die-casting models? *Q2. How efficient is the process of data analytics?* Q2.1 How efficient is the performance of the analytics application (algorithm)? Q2.2 How efficient is the visualization of the analytics solution to allow a quick intervention with specific actions?	
KPIs	Indicator	Metric
Business level	Product quality	Quality control 1; Quality control 2; Quality control 3.
Application level	Execution time	Time to produce automated decisions
Platform level	Data quality	Accuracy of new models with respect to internal CRF aluminium die-casting models
	Cost	Cost regarding personnel time spent on using the system (for analysis process), e.g. time spent for data anonymization
Experimental subjects	Quality assurance and control managers, data analysts, financial administrators, infrastructure engineers, IT security personnel	

two quality control levels related to defective products is sought in order to satisfy manufacturers' requests in terms of product quality.

3.2 Maintenance and Monitoring of Production Assets

In this use case, data have been retrieved from sensors mounted on several machines (e.g. linear stages, robots, elevators) along the production line of vehicles. Many related works are conducted in this field concerning, e.g., sensor applications in tool condition monitoring in machining [21], predictive maintenance of industrial robots [22] and assessing the health of sensors using data historians [23].

We focused on welding lines in which robots are used to assemble vehicle components, and flexibility is required for the continual changes of the types of components and vehicles. A data server gathers sensor data, which is categorized into two different datasets, namely SCADA and MES. The SCADA dataset contains production, process and control parameters of daily vehicle production and is structured as in Table 3.

There are over 100 sensors and each one is identified by a specific number (id). The other columns report on the value of the specific sensor, the unit of measurement and the timestamp.

Table 3 Structure of the dataset for the SCADA data

	Id	Value	Unit	Timestamp
Example	667	49.75	Mg	23/04/2018

Table 4 Structure of the dataset for the MES data

	Date	Time	OP020.Passo20	modello_op_020
Format	Date	Hour	Boolean	Number
Example	06/10/2018	09:44:22	0	11

The MES dataset contains specific data associated with the type of vehicle being produced and is structured as in Table 4.

When OP020.Passo20 changes from 0 to 1, a new vehicle enters into the area provided with sensors and modello_op_020 indicates the model of the vehicle being processed.

Analysis of this data aims at predicting unnecessary actions and the improvement of the efficiency of manufacturing plants by reducing production losses. Once again, an experiment has been defined in order to test the efficiency of the I-BiDaaS solution in this context. The key points of the 'Maintenance and monitoring of production assets' experiment are shown in Table 5. In particular, data was analysed to obtain thresholds for anomalous measurements for all sensors. *The fifth best practice (5) is the building of a foundational database with the history of anomalies that may help end-users to plan maintenance through prevision of asset failures only when it is necessary.*

As shown in Table 5, the business KPIs reported during requirements elicitation were further elaborated to identify related metrics (Overall Equipment Effectiveness (OEE) [24, 25] and maintenance costs [26]) and to map them on specific indicators at the Big Data solution level (execution time, data quality and cost).

For each KPI, a baseline value for evaluating the performance of the I-BiDaaS solution has been defined. For example, the prediction of unnecessary actions and the improvement of the efficiency should reduce production losses and achieve greater competitiveness of the company by an increase of 0.05% of the current Overall Equipment Effectiveness (OEE) and a decrease of 50% in maintenance costs.

4   I-BiDaaS Solutions for the Defined Use Cases

The final best practice (6) reported in the following sections relates to the development of a solution that satisfies Big Data requirements of specific use cases by mapping the identified functional and non-functional concerns into a concrete software architecture [27]. In particular, the general requirements reported in Sect. 2 were further clarified, taking into consideration the specific context of each use

Table 5 Overview of the 'Maintenance and monitoring of production assets' experiment

Experiment goals	To test efficiency of I-BiDaaS solution in the context of anticipation of maintenance events (alarm).	
Experiment questions	*Q1. What is the quality of the analytics results?* Q1.1 What is the accuracy of new models with respect to internal CRF models in use (geographical representation of the process)? *Q2. How efficient is the process of data analytics?* Q2.1 How efficient is the performance of the analytics application (algorithm)? Q2.2 How efficient is the visualization of the analytics solution to allow the workers a quick intervention with specific actions?	
KPIs	Indicator	Metric
Business level	Product/ service quality	Overall Equipment Effectiveness (OEE) Job per Hour (JpH)
	Cost reduction	Maintenance cost
Application level	Execution time	Time to produce automated decisions
	Data quality	Accuracy of new models with respect to internal CRF models
Platform level	Cost	Cost regarding personnel time spent on using the system (for analysis process), e.g. time spent for data anonymization
Experimental subjects	Quality assurance and control managers, data analysts, financial administrators, infrastructure engineers, IT security personnel	

case (described in Sect. 3), resulting in customized solutions per use case described in Sects. 4.1 and 4.2.

For both use cases, data gathered from the production lines are sent to CRF, where they are manipulated and masked. After the anonymization, data are sent to the I-BiDaaS Platform, hosted in a Virtual Machine. This represents a bridge between the I-BiDaaS infrastructure and CRF internal server, created by the I-BiDaaS technical partners. The same bridge is used to send the analytics results to the production plant end-users, as seen in Fig. 1.

Fig. 1 Flow of data and results

4.1   Production Process of Aluminium Die-Casting

In this section, the architecture, data analytics, visualization and results for the 'Production process of aluminium die-casting' use case are described.

4.1.1   Architecture

Figure 2 shows the architecture of this use case, which consists of several well-defined components. The Universal Messaging component is used for communication with most of the other components. To start with describing the data flow for this use case, we first consider the dataset. Data is transferred from CRF's internal server to the I-BiDaaS platform server. Therein, the data is pre-processed and cleaned— this step is important as the data needs to be prepared for model training and inference tasks. Then, the data is given to the Machine Learning algorithm from the I-BiDaaS pool of ML algorithms. In this use case, the model is a complex neural network implemented in PyTorch[3] and trained jointly from thermal images and sensor datasets. The Machine Learning component outputs two results: training metrics/results for visualization purposes—used in the Advanced Data Visualization component—and the trained model used for inference. Both these results are transferred through Universal Messaging. In the end, for inference purposes, the Model Serving (Inference) Service component is used. In the initial phases of development, before the real data is fully prepared (e.g. retrieved, anonymized, etc.), the architecture uses realistic synthetic data for initial components development. The use of synthetic data can make the development significantly more agile, but is utilized with care and under a quality assurance process. For example, a final trained ML model has to be delivered on real data. We refer to Sect. 4.1.5 for details on realistic synthetic data generation and quality assessment.

4.1.2   Data Analytics

In this section, we describe in more detail the data analytics solution that corresponds to the four respective modules in Fig. 2 (Data pre-processing, PyTorch neural network model, Trained model and Training results) and that analyses the thermal images and the sensors datasets.

   Under the hypothesis that there is a correlation among sensor data, thermal data and the outcome of the process, a further task is to classify combined image and sensor data inputs to see whether the cast engine blocks are without any production faults. Formally, data analytics here corresponds to an M-ary supervised

---

[3] http://pytorch.org

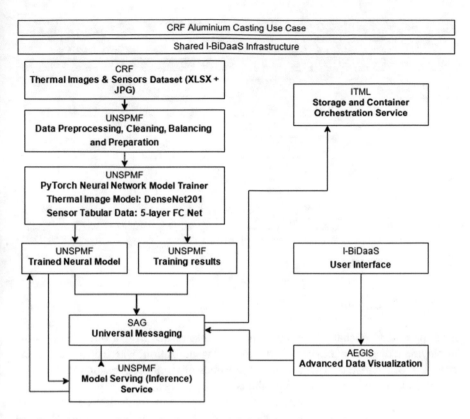

Fig. 2 Architecture of the 'production process of aluminium die-casting' use case

classification task [28]. As the dataset involves image classification, for this task we utilize Deep Convolutional Neural Networks [29].

We tried three approaches during this use-case analytics development regarding the input data: unmodified thermal images, grayscale thermal images and raw sensor data. For raw sensor data the thermal camera provides a matrix of values which is the same dimension as the image, which when normalized provides very similar (almost the same, depends on the normalization process) input to the grayscale image from the computing standpoint. While the grayscale image and the raw sensor data did have faster training times (one channel for convolutions versus three for thermal images) from our experiments the thermal images gave best accuracy/precision/recall metrics so we decided to keep using them. We suspect that this is the case because modern neural network architectures we are using (e.g. DenseNet [29, 30]) are optimized to work with coloured images (e.g. ImageNet dataset [31]). The corresponding results are reported in Sect. 4.1.4.

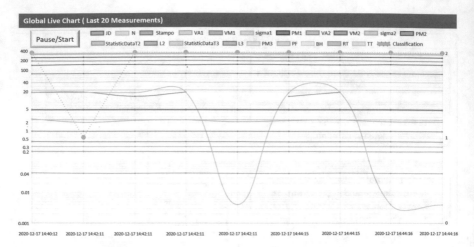

Fig. 3 Real-time aggregated results

4.1.3 Visualizations

The approach to visualize the die-casting process results in real time involves the deployment of a number of constantly updated visualizations which offer a complete overview of the results. These include the values of monitored sensor variables and the final classification of the end products of the process.

We report here, as example, the Global Live Chart that allows end-users to timely visualize the trend of the main parameters (e.g. velocity, pressure, standard deviation, etc.) and to check the classification levels (Fig. 3).

4.1.4 Results

The models described in Sect. 4.1.2 were trained on both the original and the newly balanced datasets. We favour the model trained on the balanced dataset as it learns to recognize faulty engine blocks much better than the model trained on the imbalanced dataset, even though the overall accuracy is lower—simply because we have less faultless engines. In Fig. 4, we see the accuracies of both models on the training and testing datasets (standard 80/20 split). The orange (top) line is the model trained on the full dataset and the pink line is the model trained on the balanced dataset[4] [32].

---

[4] Visualization with TensorBoard: https://www.tensorflow.org/tensorboard

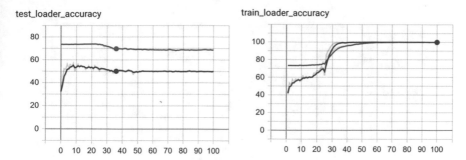

Fig. 4 Training and testing accuracy for the two joint neural network models: when trained on full imbalanced data (orange line) and when trained on sub-sampled balanced data (pink line)

4.1.5 Synthetic Data Generation and Quality Assessment

An initial development of the use case solution was carried out with realistic synthetic data. In parallel with the process of data anonymization, making data structured, etc., it was useful to carry out a synthetic data generation for early development stages with particular caution when extracting insights from synthetic data.

The fabrication of synthetic data that exhibits similar characteristics and similar distribution as the real data is a challenging task. The IBM Test Data Fabrication technology (TDF) was used for that purpose. TDF requires constraint rules that model the relationships and dependencies between the data and leverages a Constraint Satisfaction Problems (CSP) solver to fabricate data that satisfies these constraints. The rules for the production of synthetic data were set by CRF with the help of IBM. The correlation between the real parameters and the synthesized parameters was further refined after reiteration of the data analysis.

For the initial evaluation of the synthetic data, we performed empirical and analytical validations. The empirical technique consisted of delivering these data to the expert production technicians, which were not able to indicate any difference with the actual production data, as there was no distinguishing factor for them. The second analytical technique was carried out by the CRF research team. They used the K-Means algorithm [33] as their desired technique. Further evaluation was carried out by IBM while striving to perform a qualitative generic evaluation process for the real data compared with the fabricated data. This evaluation was concerned with methods to judge whether the distributions of the fabricated data and the original data were comparable, what is commonly referred to in the literature as the general utility of the datasets. In addition to the general utility, IBM also considered the specific utility, i.e. the similarity between the synthetic data and the original data.

The propensity mean-squared-error (pMSE) [34] was used as a general measure of data utility to the specific case of synthetic data. Propensity scores represent probabilities of group memberships. If the propensity scores are well modelled, this

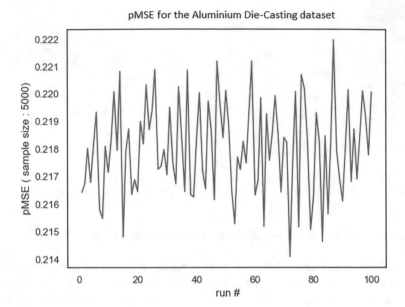

Fig. 5 Results for 100 random sampling taken from the real and the synthetic data (5K datapoints each) and the pMSE calculated using a logistic model

general measure should capture relationships among the data that methods such as the empirical Cumulative Distribution Function (CDF) may miss.

The method is a classification problem where the desired result is poor classification (50% error rate), giving better utility for low values of the pMSE.

Randomly sampling 5000 data points from the real and synthetic datasets, and using a logistic regression to provide the probability for the label classification, we were able to show that the measured mean pMSE score for the 'Production process of aluminium die-casting' dataset is 0.218 with a standard deviation of 0.0017, as shown in Fig. 5.

4.2 Maintenance and Monitoring of Production Assets

In this section, the architecture, data analytics, visualization and results for the 'Maintenance and monitoring of production assets' use case are described.

4.2.1 Architecture

Figure 6 shows the architecture, which consists of several well-defined components. The Universal Messaging component is used for communication in most of the

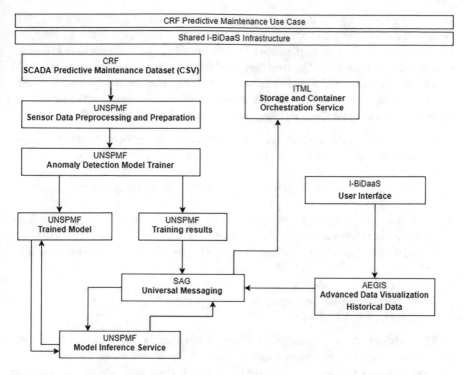

Fig. 6 Architecture of the 'maintenance and monitoring of production assets' use case

components. To start to describe the data flow, we start with the dataset. Data are sent from CRF to the I-BiDaaS platform. There, the data is pre-processed and prepared for model training with an outlier detection model. The outlier detection model outputs two results: training results for visualization purposes—used in the Advanced Data Visualization component, and the trained model used for inference. Training results are transferred through Universal Messaging. In the end, for inference purposes, the Model Inference Serving component is used. It is also important to say that all the components use containerized (i.e. Docker[5]) backbone from the Storage and Container Orchestration Service. Data is visualized and the jobs are scheduled through the I-BiDaaS User Interface component.

4.2.2 Data Analytics

Data, described in Sect. 3, has been transformed into separate time series—one per sensor so that each sensor can be monitored separately. Since the measurements were not labelled (anomalous/non-anomalous), outlier detection algorithms arose

---

[5] https://www.docker.com/

as natural candidates for this use case [35]. We constructed an outlier detection model for each of the time series. While more advanced algorithms can be used, we adopted a simple, easy-to-implement and computationally cheap, yet here effective, solution based on the Inter-Quartile Range (IQR) test. Results of these models could be used for suggesting if a measurement is an outlier and for discovering the pairs of sensors that have anomalous measurements at similar timestamps. Preparation of these models was done using Python, and it consisted of the following steps:

1. For each sensor, obtain thresholds for anomalous measurements using a modified interquartile range (IQR) test. Three different variants of IQR-like tests were performed:
   $(Q_1, Q_3) \in \{(5\text{th}, \ 95\text{th}), (10\text{th}, \ 90\text{th}), (25\text{th}, \ 75\text{th})\}$ where Q_1 and Q_3 are the corresponding percentiles.
2. With obtained thresholds, filter the time series such that only anomalous measurements were kept, as shown in Fig. 8.
3. Calculate the Dynamic Time Warping (DTW) [36] distance between outlier time series.
4. Rescale distances to [0, 1].
5. Group pairs of sensors by the distance into groups:
   $[0, 0.1), [0.1, 0.2) \ldots [0.9, 1]$.

Time series with anomalous measurements obtained in step 2 enabled us to see the outlier trends for each sensor and to compare their behaviour. Comparison of anomalous trends was made using steps 3, 4 and 5. If the distance obtained in step 5 is small, it means that two sensors output anomalous measurements in a similar fashion. Therefore, if one of them fails, then the other sensor in the pair should also be inspected. We present the distribution of sensors' similarity in Fig. 9.

4.2.3 Visualizations

Data stemming from the aforementioned analysis are presented using a multi-step approach that allows operators drill down to sensory data and detected anomalies in an intuitive and easy-to-use way. Starting from a given month, operators then select the category of sensors they wish to see and immediately have an overview of the ones having anomalies detected, as shown in Fig. 7. Upon selection of a sensor, operators see the anomalies detected during the selected month and can furthermore select a specific day to see the actual values and therefore review the actual anomaly that was detected, as shown in Fig. 8.

4.2.4 Results

The obtained boundaries (from step 2 in Sect. 4.2.2) could be used for daily analysis of sensors and various visualization tasks, such as showing the number of anomalous

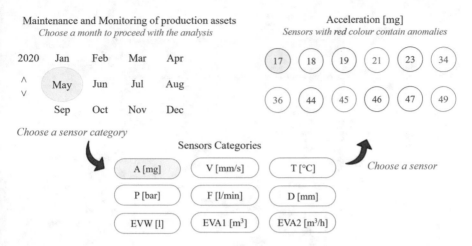

Fig. 7 Sensor selection

measurements for the current day, as seen in Fig. 8, comparing the number of outliers between two sensors for the given time window, etc., as shown in Fig. 9.

5   Discussion

Reflecting on CRF's experience and all the work done within the I-BiDaaS Project, this section develops several recommendations addressed to any manufacturing company willing to undertake Big Data projects. This section also positions the I-BiDaaS solution within Big Data Value (BDV) reference model and Strategic Research and Innovation Agenda (SRIA).

5.1   Lessons Learned, Challenges and Guidelines

The I-BiDaaS project developed an integrated platform for processing and extracting actionable knowledge from Big Data in the manufacturing sector. Based on the challenges experienced and lessons learned through our involvement in I-BiDaaS, we propose a set of guidelines for the implementation of Big Data analytics in the manufacturing sector, with respect to the following concerns:

1. **Data storage and ingestion from various data sources and its preparation:** In a production line deploying digital instruments, there are many devices which setup operating values and adjust and control parameters during the production processes. Depending on whether we want to act on the quality of the production process or on the maintenance of the equipment, the first

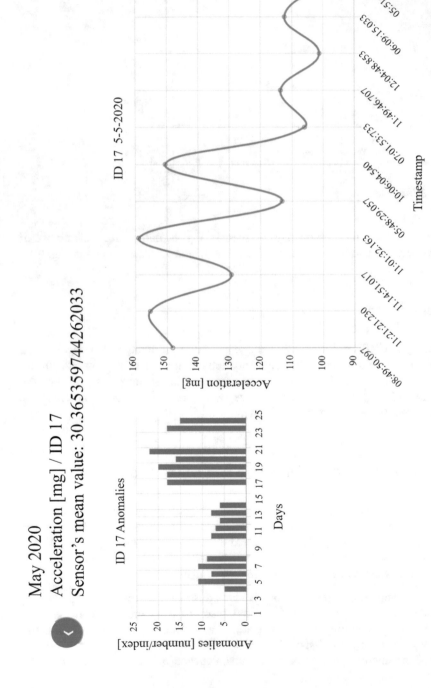

Fig. 8 Sensor history and details

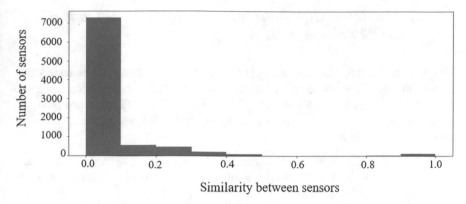

Fig. 9 Number of outliers between sensors

challenge is to understand how data will be ingested and managed from data sources over time and who will be able to access them. Furthermore, this aspect highlights the importance of breaking data silos by extracting the value of all data collected from several sources and levels and may be necessary to involve different departments belonging to the same or different organizations.

2. **Data cleaning:** A second important aspect is to understand which types of data can be useful for analysis. This implies the importance of data cleaning in order to identify incomplete, inaccurate and irrelevant parts of the generated dataset.

3. **Fabrication of realistic synthetic data for experimentation and testing:** Data are strictly confidential, so another challenge is to decide how data will be shared if external analysis is required. In this case, manufacturers need to evaluate the possibility of fabrication of realistic synthetic data for experimentation of the analytical models that will be developed and then to test the same models with anonymized real data.

4. **Batch and stream analytics for increasing the speed of data analysis:** After collecting and analysing data, it is necessary to understand which Big Data technologies are most suitable for the specific identified business requirements. Batch and stream analytics cover all aspects, which may occur in real-world environments, including cases that require a deeper analysis of large amounts of data collected over a period of time (batch) or those that require velocity and agility for the events that we need to monitor in real or near-real-time (streaming).

5. **Simple, intuitive and effective visualization of results and interaction capabilities for the end-users:** Advanced visualization tools which provide the insights, value and operational knowledge extracted from available data need to consider both expert and non-expert end-users (e.g. manufacturers, engineers and operators)

5.2   Connection to BDV Reference Model, BDV SRIA, and AI, Data and Robotics SRIDA

The described solution for the defined manufacturing use cases can be contextualized within the BDV reference model defined in the BDV Strategic Research and Innovation Agenda (BDV SRIA). They contribute to the BDV reference model in the following ways. Specifically, regarding the BDV reference model horizontal concerns, we address:

- **Data visualization and user interaction**: By developing several advanced and interactive visualization solutions applicable in the manufacturing sector, as detailed in Sects. 4.1.3 and 4.2.3.
- **Data analytics**: By developing data analytics solutions for the two industrial use cases in the manufacturing sector, as described in Sects. 4.1.2 and 4.2.2. While the solutions may not correspond to state-of-the-art advances in AI/machine learning algorithms development, they clearly contribute to revealing novel insights and best practices on how Big Data analytics can improve manufacturing operations.
- **Data processing architectures**: We develop architectures as shown in Figs. 2 and 6 that are well suited for manufacturing applications wherein both batch analytics (e.g. analysing historical data) and streaming analytics (e.g. online processing of the data that correspond to a newly manufactured engine) are required.
- **Data protection and data management**: Real data were anonymized by CRF that manipulated and masked them after they were retrieved from an internal proprietary server.

Regarding the BDV reference model vertical concerns, we address the following:

- **Big data types and semantics**: Our work here is mostly concerned with structured sensory data, meta-data and thermal images data (which corresponds to the Media, Image, Video and Audio data types according to the BDV nomenclature). The work also contributes to best practices in the generation of realistic synthetic data from the corresponding domain-defined meta-data, as well as a systematic way to assess the quality and usefulness of the generated synthetic data.
- **Communication and connectivity**: the work describes innovative ways to communicate with and retrieve data from an internal manufacturing company proprietary server, as described in Sect. 4 and outlined in Fig. 1.

Therefore, in relation with BDV SRIA, the I-BiDaaS solution contributes to the following technical priorities: Data protection; Data Processing Architectures; Data Analytics; and Data Visualization and User Interaction.

Furthermore, in relation to the BDVA SRIA priority areas in connection with Factories of the Future with EFFRA, we address the following dimensions:

(a) Excellence in manufacturing: advanced manufacturing processes and services for zero-defect and innovative processes and products
(b) Sustainable value networks: manufacturing driving the circular economy
(c) Inter-operable digital manufacturing platforms: supporting an ecosystem of manufacturing services

In more detail, CRF use cases have been selected in order to develop innovative tools and solutions that may ensure better product quality towards zero-defect manufacturing. In particular, the existing production lines may be improved to maximize the quality of their product through the integration of solutions that exploit Big Data technologies. A better process efficiency can result in energy saving and cost reduction in the context of circular economy and allow manufacturers to reach a high level of competitiveness and sustainability.

Finally, the chapter relates to the following cross-sectorial technology enablers of the AI, Data and Robotics Strategic Research, Innovation & Deployment Agenda [37], namely: Knowledge and Learning, Reasoning and Decision Making, and Systems, Methodologies, Hardware and Tools.

6 Conclusion

The increasing levels of digitalization in the manufacturing sector contribute to generate a large amount of data that often contain a high value of hidden information. This is due to the complexity of real processes that require several interconnected stages to obtain finished goods. Variables and parameters are set for the operation of each digital machine and just like we assemble components, we need to pull together data generated from different sources and levels if we want to improve the quality of processes and products. I-BiDaaS developed an integrated platform, taking into consideration how complex data can be managed and how to help manufacturers who are not sufficiently enabled to analyse complex datasets, by empowering them to easily utilize and interact with Big Data technologies.

Acknowledgements The work presented in this chapter is supported by the I-BiDaaS project, funded by the European Union's Horizon 2020 research and innovation programme under Grant Agreement 780787. This publication reflects the views only of the authors, and the Commission cannot be held responsible for any use which may be made of the information contained therein.

References

1. Nahm, A., Vonderembse, M., & Koufteros, X. (2003). The impact of organizational structure on time-based manufacturing and plant performance. *Journal of Operations Management, 21*, 281–306.
2. Schwab, K. (2016). *The fourth industrial revolution*. Franco Angeli.

3. Chen, B., Wan, J., Shu, L., Li, P., Mukherjee, M., & Yin, B. (2017). Smart factory of Industry 4.0: Key technologies, application case, and challenges. *IEEE Access, 6*, 6505–6519.
4. Groover, M. P. (2018). *Automation, production systems, and computer-integrated manufacturing*. Pearson.
5. Yadegaridehkordi, E., Hourmand, M., Nilashi, M., Shuib, L., Ahani, A., & Ibrahim, O. (2018). Influence of big data adoption on manufacturing companies' performance: an integrated DEMATEL-ANFIS approach. *Technological Forecasting and Social Change, 137*, 199–210.
6. O'Donovan, P., Leahy, K., Bruton, K., & O'Sullivan, D. T. J. (2015). Big data in manufacturing: A systematic mapping study. *Journal of Big Data, 2*, 20.
7. Passlick, J., Lebek, B., & Breitner, M. H. (2017). A self-service supporting business intelligence and big data analytics architecture. In *13th international conference on Wirtschaftsinformatik, St. Gallen, Switzerland.*
8. Bornschlegl, M. X., Berwind, K., & Hemmje, M. (2017). Modeling end user empowerment in big data applications. In *26th International conference on software engineering and data engineering at: San Diego, CA, USA.*
9. Zillner, S., Curry, E., Metzger, A., Auer, S., & Seidl, R., (Eds.). (2017). *European big data value. Strategic research & innovation agenda*. Springer.
10. Arruda, D. (2018). Requirements engineering in the context of big data applications. *SIGSOFT Software Engineering Notes, 43*(1), 1–6.
11. Raguseo, E. (2018). Big data technologies: An empirical investigation on their adoption, benefits and risks for companies. *International Journal of Information Management, 38*(1), 187–195.
12. Nuseibeh, B., & Easterbrook, S. (2000). Requirements engineering: A roadmap. In *Proceedings of the conference on the future of software engineering, ICSE'00* (pp. 35–46).
13. Paech, B., Dutoit, A. H., Kerkow, D., & Von Knethen, A. (2002). Functional requirements, non-functional requirements, and architecture should not be separated—A position paper. In *Proceedings of the 8th international working conference on requirements engineering.*
14. Horkoff, J., Aydemir, F. B., Cardoso, E., Li, T., Maté, A., Paja, E., Salnitri, M., Piras, L., Mylopoulos, J., & Giorgini, P. (2019). Goal-oriented requirements engineering: An extended systematic mapping study. *Requirements Engineering, 24*, 133–160.
15. I-BiDaaS Consortium. (2018). D1.3: Positioning of I-BiDaaS. Available at: https://doi.org/10.5281/zenodo.4088297.
16. Murray, M. T., & Murray, M. (2011). High pressure die casting of aluminium and its alloys. In *Fundamentals of aluminium metallurgy production, processing and applications*. Woodhead Publishing series in metals and surface engineering (pp. 217–261).
17. Lumley, R. N. (2011). Progress on the heat treatment of high pressure die castings. In *Fundamentals of aluminium metallurgy production, processing and applications*. Woodhead Publishing series in metals and surface engineering (pp. 262–303).
18. Winkler, M., Kallien, L., & Feyertag, T. (2015). Correlation between process parameters and quality characteristics in aluminum high pressure die casting. In *Conference: NADCA.*
19. Fiorese, E., & Bonollo, F. (2016). *Process parameters affecting quality of high-pressure diecast Al-Si alloy*. Doctoral Thesis, University of Padova.
20. Chandrasekaran, R., Campilho, R. D. S. G., & Silva, F. J. G. (2019). Reduction of scrap percentage of cast parts by optimizing the process parameters. *Procedia Manufacturing, 38*, 1050–1057.
21. Bhuiyan, M. S. H., & Choudhury, I. A. (2014). Review of sensor applications in tool condition monitoring in machining. *Reference Module in Materials Science and Materials Engineering, Comprehensive Materials Processing, 13*, 539–569.
22. Borgi, T., Hidri, A., Neef, B., & Nauceur, M. S. (2017). Data analytics for predictive maintenance of industrial robots. In *International conference on advanced systems and electric technologies (IC_ASET).*
23. Eren, H. (2012). Assessing the health of sensors using data historians. In *IEEE sensors applications symposium proceedings.*

24. Dal, B., Tugwell, P., & Greatbanks, R. (2000). Overall equipment effectiveness as a measure of operational improvement—A practical analysis. *International Journal of Operations & Production Management, 20*(12), 1488–1502.
25. Ljungberg, Ö. (1998). Measurement of overall equipment effectiveness as a basis for TPM activities. *International Journal of Operations & Production Management, 18*(5), 495–507(13).
26. Galar, D., Sandborn, P., & Kumar, U. (2017). *Maintenance costs and life cycle cost analysis.* CRC Press.
27. Arapakis, I., Becerra, Y., Boehm, O., Bravos, G., Chatzigiannakis, V., Cugnasco, C., Demetriou, G., Eleftheriou, I., Mascolo, J. E., Fodor, L., Ioannidis, S., Jakovetic, D., Kallipolitis, L., Kavakli, E., Kopanaki, D., Kourtellis, N., Marcos, M. M., de Pozuelo, R. M., Milosevic, N., Morandi, G., Montanera, E. P., & Ristow, G. H. (2019). Towards specification of a software architecture for cross-sectoral big data applications. In *IEEE world congress on services (SERVICES)* (Vol. 2642). IEEE.
28. Bishop, C. M. (2006). *Pattern recognition and machine learning.* Springer.
29. Gu, J., Wang, Z., Kuen, J., Ma, L., Shahroudy, A., Shuai, B., Liu, T., Wang, X., Wang, G., Cai, J., & Chen, T. (2018). Recent advances in convolutional neural networks. *Pattern Recognition, 77*, 354–377.
30. Huang, G., Liu, Z., Van der Maaten, L., & Weinberger, K. Q. (2017). Densely connected convolutional networks. In *Proceedings of the IEEE conference on computer vision and pattern recognition (CVPR)* (pp. 4700–4708).
31. He, K., Zhang, X., Ren, S., & Sun, J. (2016). Deep residual learning for image recognition. In *Proceedings of the IEEE conference on computer vision and pattern recognition (CVPR)* (pp. 770–778).
32. I-BiDaaS Consortium. (2020). D3.3: Batch Processing Analytics module implementation final report. Available at: https://doi.org/10.5281/zenodo.4608346
33. Bock, H. H. (2017). Clustering methods: a history of k-means algorithms. In *Selected contributions in data analysis and classification.* Springer (pp. 161–172).
34. Snoke, J., Raab, G. M., Nowok, B., Dibben, C., & Slavkovic, A. (2016). General and specific utility measures for synthetic data. *Journal of the Royal Statistical Society Series A (Statistics in Society), 181*(3).
35. Gupta, M., Gao, J., Aggarwal, C. C., & Han, J. (2014). Outlier detection for temporal data: a survey. *IEEE Transactions on Knowledge and Data Engineering, 26*(9), 2250–2267.
36. Petitjean, F., Forestier, G., Webb, G. I., Nicholson, A. E., Chen, Y., & Keogh, E. (2014). Dynamic time warping averaging of time series allows faster and more accurate classification. In *IEEE international conference on data mining.*
37. Zillner, S., Bisset, D., Milano, M., Curry, E., García Robles, A., Hahn, T., Irgens, M., Lafrenz, R., Liepert, B., O'Sullivan, B., & Smeulders, A. (Eds.) (2020, September). *Strategic research, innovation and deployment agenda—AI, data and robotics partnership.* Third release. Brussels. BDVA, euRobotics, ELLIS, EurAI and CLAIRE.

Next-Generation Big Data-Driven Factory 4.0 Operations and Optimization: The Boost 4.0 Experience

Oscar Lázaro, Jesús Alonso, Philip Ohlsson, Bas Tijsma, Dominika Lekse, Bruno Volckaert, Sarah Kerkhove, Joachim Nielandt, Davide Masera, Gaetano Patrimia, Pietro Pittaro, Giuseppe Mulè, Edoardo Pellegrini, Daniel Köchling, Thanasis Naskos, Ifigeneia Metaxa, Salome Leßmann, and Sebastian von Enzberg

Abstract This chapter presents the advanced manufacturing processes and big data-driven algorithms and platforms leveraged by the Boost 4.0 big data lighthouse project that allows improved digital operations within increasingly automated and intelligent shopfloors. The chapter illustrates how three different companies have been able to implement three distinct, open, yet sovereign cross-factory data spaces

O. Lázaro (✉) · J. Alonso
Asociación de Empresas Tecnológicas Innovalia, Bilbao, Spain
e-mail: olazaro@innovalia.org; jalonso@innovalia.org

P. Ohlsson · B. Tijsma
Philips Consumer Lifestyle, Amsterdam, The Netherlands
e-mail: philip.ohlsson@philips.com; bas.tijsma@philips.com

D. Lekse
Philips Electronics Netherland, Eindhoven, The Netherlands
e-mail: dominika.lekse@philips.com

B. Volckaert · S. Kerkhove · J. Nielandt
Interuniversitair Microelectronica Centrum, Leuven, Belgium
e-mail: bruno.volckaert@ugent.be; sarah.kerkhove@ugent.be; joachim.nielandt@ugent.be

D. Masera
Centro Ricerche Fiat, Orbassano, Italy
e-mail: davide.masera@stellantis.com

G. Patrimia · P. Pittaro
Prima Industrie, Collegno, Italy
e-mail: gaetano.patrimia@primapower.com; pietro.pittaro@primapower.com

G. Mulè · E. Pellegrini
Siemens, Milan, Italy
e-mail: giuseppe.gm.mule@siemens.com; edoardo.pellegrini@siemens.com

© The Author(s) 2022
E. Curry et al. (eds.), *Technologies and Applications for Big Data Value*,
https://doi.org/10.1007/978-3-030-78307-5_16

under a unified framework: the Boost 4.0 big data reference architecture and Digital Factory Alliance (DFA) service development framework.

Keywords Data Visualization · Data Quality · Data Lake · Big Data Stream Analytics · Brownfield · Trial · Predictive Maintenance 4.0 · Injection Moulding 4.0 · SUMA 4.0 · Intra-logistics

1 Introduction

The rapidly growing number of sensors, embedded systems and connected devices as well as the increasing horizontal and vertical networking of value chains result in a huge continuous data flow. In fact, *the manufacturing sector generates more data annually than any other sector in the EU or US economy, and the manufacturing industry (83%) expects data to have a big impact on decision-making in 5 years.* As highlighted by the European data strategy [1], by 2025, we will experience a 530% increase in global data volume from 33 zettabytes in 2018 to 175 zettabytes, and data will represent an economic value of 829 million € in the EU27 economy compared to the €301 million that it represented in 2018 (2.4% of the EU GDP).

Big Data will have a profound economic and societal impact in the Industry 4.0 sector, which is one of the most active industries in the world, contributing to approximately 15% of EU GDP. According to the World Economic Forum report on Digital Transformation of Industry [2], *Big Data is expected to take off in the consumer market to a value at stake of over $600 billion for industry and $2.8 trillion for society in improved customer service and retailing experience.* Moreover, the total value that companies can create in five key areas of data sharing is estimated to be more than $100 billion. In fact, 72% of the factories consider that sharing data with other manufacturers can improve operations and 47% find enhanced asset optimization to be the most relevant application area. Big Data as part of European Industrial Digitization could see manufacturing industry *add gross value worth 1.25 T€*—or more importantly *suffer the loss of 605 BEUR* in foregone

D. Köchling
BENTELER Automobiltechnik, Paderborn, Germany
e-mail: daniel.koechling@benteler.com

T. Naskos · I. Metaxa
ATLANTIS Engineering, Athens, Greece
e-mail: naskos@abe.gr; metaxa@abe.gr

S. Leßmann
it's OWL Clustermanagement, Paderborn, Germany
e-mail: s.lessmann@its-owl.de

S. von Enzberg
Fraunhofer Institute for Mechatronic Systems Design, Paderborn, Germany
e-mail: sebastian.von.enzberg@iem.fraunhofer.de

value added if it fails to incorporate new data, connectivity, automation and digital customer interface enablers and getting their digital manufacturing processes ready, i.e. cognitive and predictive, in automotive engineering and logistics. The European Commission foresees that *advanced analytics in predictive maintenance systems only could reduce equipment downtime by 50% and increase production by 20%.* Overall, *only the top 100 European manufacturers could save around 160 BEUR* thanks to improved error-correcting systems and the ability to adjust production in real time. Additionally, *10% production efficiency improvement* can be realized in top 100 EU manufacturers with an associated *265 BEUR* gain for the industry.

Despite the big data promises, interestingly (1) only 3% of useful manufacturing data is tagged and even less is analysed, (2) manufacturing industry are currently losing up to 99% of the data value they capture since evidence cannot be presented at the speed decisions are made and (3) only half of industry is currently using any data to drive decisions with a much lower 15% of EU industry employing Big Data solutions as part of value creation and business processes.

Boost 4.0 [3] is a European lighthouse initiative for the large-scale trial of big data-driven factories. The Boost 4.0-enabled Connected Smart Factory 4.0 vision is one where digital design technologies enable short times to market, resources are optimally planned, downtime is predicted and prevented, waste and defects are eliminated, surplus production is minimized, machine behaviour is optimized as conditions change and systems can make context-based 'next best' actions. Connected devices in the factory report their status, giving operations personnel and decision-makers access to real-time, actionable information. Wearable technology tracks employee location and status in case of emergency. A global ecosystem of partners ensures that specific parts are replenished based on automated, real-time needs analysis. *Data is at the heart of Industry 4.0, the experience economy and the manufacturing digital transformation towards 'servitised' product service systems and outcome-based digital business models; as opposed to traditional product ownership business models.* But the massively growing information flow brings little value without the right analytics techniques.

Full adoption of a data-driven Factory 4.0 has been largely hampered by: unclear ownership and access right definition in the data value chain; need to harmonise cross-border heterogeneous flows of data; limited availability of open datasets to feed industrial ecosystems; insufficient diffusion of advanced technologies to preserve data confidentiality and privacy. All these issues, which broadly relate to data sovereignty, remain largely un-addressed challenges by current Digital Manufacturing Platforms solutions. The lack of such reference framework has the following drawbacks:

- Manufacturing big data sets are highly heterogeneous in nature and are spread across the product and factory lifecycles.
- Manufacturing big data is highly unstructured, hard to analyse and distributed across various sectors and different stakeholders involved in the product and factory lifecycles.

- Data is usually duplicated across many digital manufacturing platforms and systems (data multi-homing), thereby making it difficult to maintain 'quality' and updated data to base decisions upon.
- Data analysis necessarily implies a loose of control over the use of data from the data owner since the transfer of data across digital platforms and enterprises is mandatory for data consolidation and processing.
- Often valuable data is measured for real-time use in specialized systems, but not stored for later processing or recorded in a way suitable for data collation across individual systems.
- Data-driven decision support is slow and contextualization of information is cumbersome and involves intensive manual operation on data sources.
- Data transactions (grant of data access rights, data transfer) are slow, mediated and cumbersome.
- Machine- and shopfloor-generated data is usually not ready for sharing with external stakeholders.
- Engineering, production, IT and IoT data remain as isolated silos that make costly and complex the development of smart services on top of smart products.

In addition, Big Data will be exponentially created, processed and stored in the coming years—see EU Data strategy projections above—but no single infrastructure, let alone a single stakeholder, can do the job on its own. Boost 4.0 has addressed the lack of European and global standards and an Industry 4.0 big data reference framework that ensures data sovereignty, while enabling the agile and value-driven creation of ad hoc trusted data networks across currently isolated consumer experience data, usage-context data, production and engineering data 'clouds' (Fig. 1).

1.1 Big Data-Centric Factory 4.0 Operations

Data-centric operations is one of the fundamental cornerstones of modern industrial automation technologies and also one of the bases for decision-making and control operation. While the use of statistical data analysis for control is well established, recently the diffusion of big data methodologies added a new dimension, by providing additional approaches to data-centric automation. A big data-centric approach to Factory 4.0 operations opens the door to migration from an asset-centric decision process towards truly real-time, predictive and coordinated multi-level process centric decision processes. Such *process-centric, holistic and integrated data space* for Factory 4.0 operations calls for significant improvements in speed, flexibility, quality and efficiency (Fig. 2).

Within the Boost 4.0 project, pilots have leveraged Industrial Internet of Things (IIoT), big data space technologies, advanced visualization, predictive analytics and collaborative AI engineering and decision support systems for the benefit of significant improved operations. As shown in Fig. 3, Boost 4.0 is consider-

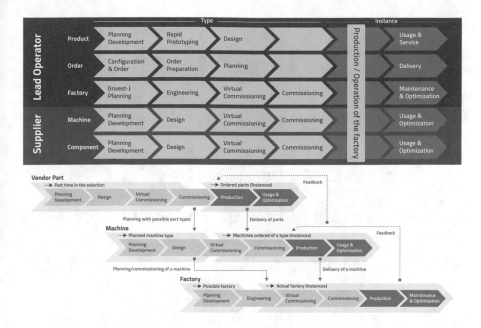

Fig. 1 Boost 4.0 'whole' lifecycle digital thread synchronization big data challenge

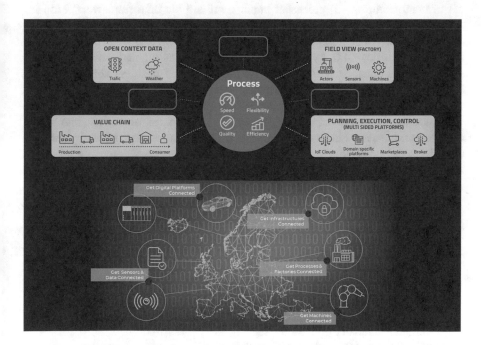

Fig. 2 Boost 4.0 'process-centric' data space connecting industrial things and platforms

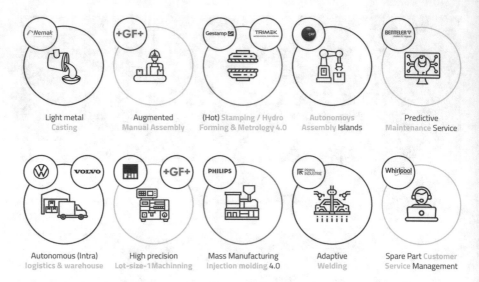

Fig. 3 Boost 4.0 manufacturing 4.0 processes supported

ing the development and evaluation of process-centric, data-centric, AI-powered advanced manufacturing 4.0 processes. Boost 4.0 is considering a highly diverse set of manufacturing 4.0 processes under a unified big data framework, ensuring high portability and replicability. The manufacturing 4.0 processes supported by Boost 4.0 range from light metal casting to augmented manual assembly, hot stamping, metrology 4.0, hydroforming, autonomous automated assembly islands, predictive maintenance, autonomous intra logistics and business network tracing, high-precision lot-size machining, mass manufacturing injection moulding 4.0, adaptive welding and spare part management customer services. Moreover, these processes are implemented across a number of sectors (automotive, white goods, high-end textiles, machine tool industry, ceramics, elevation, aero), thereby ensuring that highly varied sectors are amenable to big data transformations. The interested reader is referred to the additional chapters in this book for more details on the manufacturing processes implemented.

These Boost 4.0 data-driven manufacturing processes are supported by advanced big data technologies—e.g. data streaming, batch and predictive analytics, Machine Learning (ML) and Artificial Intelligence (AI)—which are applied seamlessly across the full product and process lifecycle (Smart Digital Engineering, Smart Digital Planning & Commissioning, Smart Digital Workplace & Operations, Smart Connected Production, Smart Service & Maintenance). Boost 4.0 has thereby leveraged a number of high-performance big data algorithms and platform features that, as illustrated by the implemented trials, can deliver high impact and performance improvements in factory operations; see Fig. 4.

The three Boost 4.0 lighthouse pilots that are presented in this chapter have introduced into their processes new big data methodologies to optimize different aspects

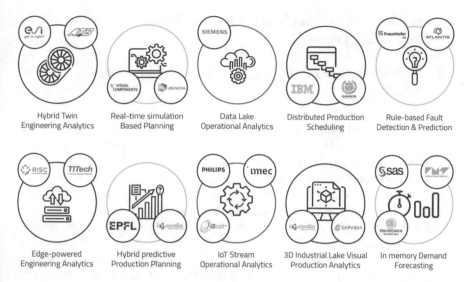

| Hybrid Twin Engineering Analytics | Real-time simulation Based Planning | Data Lake Operational Analytics | Distributed Production Scheduling | Rule-based Fault Detection & Prediction |
| Edge-powered Engineering Analytics | Hybrid predictive Production Planning | IoT Stream Operational Analytics | 3D Industrial Lake Visual Production Analytics | In memory Demand Forecasting |

Fig. 4 Boost 4.0 big data algorithms and platforms

of the product lifecycle, from the production itself to the distribution of spares for the after-sales services. In the next section, Philips present their *Injection Moulding Smart Operations & Digital Workspace*, in which the Drachten premises pave the way for a generic platform usable for the full fleet of injection-moulding machines across Philips' factories. Afterwards, the BENTELER Automotive lighthouse trial is discussed, which deployed a big data platform for smart maintenance of industrial assets, focusing on the example of a hydraulic press. A novel predicted structured and effective approach toward assets' failure management and synchronization with higher level plant management system has been provided, where predicted failures and estimated RUL are dynamically assessed in real time for their severity and potential impact on the plant, evaluating their criticality in order to provide the right recommendation for remedy actions. Next, the FCA trial is introduced. In this use case, the focus is on the smart collaboration between mobile robots, more specifically AGVs and laser machine. Finally, some conclusions are drawn.

This chapter relates mainly to the technical priorities Data Analytics and Advanced Visualization and User Experience of the European Big Data Value Strategic Research & Innovation Agenda [4]. It addresses the horizontal concerns of analytics frameworks and processing, predictive and prescriptive analytics and interactive visual analytics of multiple-scale data of the BDV Technical Reference Model. It addresses the vertical concerns regarding the use of standards to facilitate integration of data end-points from legacy and heterogenous systems and development of trusted and sovereign data spaces across production sites and development of third-party applications and services. The work in this chapter relates mainly, but not only, to the Reasoning and Decision Making cross-sectorial technology enablers of the AI, Data and Robotics Strategic Research, Innovation & Deployment Agenda [5].

2   Mass Injection Moulding 4.0 Smart Digital Operations: The Philips Trial

2.1   Data-Driven Digital Shopfloor Automation Process Challenges

Philips Drachten encompasses a large suite of highly automated processes used during the manufacturing of electric shavers. Of these manufacturing processes, injection moulding is of particular importance, as it is used during the fabrication of plastic components for electric shavers. Injection moulding is a competitive market, which makes it essential for Philips Drachten to continuously improve on quality, production performance, and costs where this process is concerned.

All the plastic parts are manufactured on-site at Drachten, requiring approximately 80–90 moulding machines of multiple vendors, models and generations. For large manufacturing sites, generalization is key to deploy data-driven solutions. It is simply not feasible to develop a specialized solution for each machine in the machine park. Thus, in this pilot our main challenge is to develop scalable solutions.

Furthermore, specialized custom solutions do not yield a positive business case in the case of moulding; plastic is relatively cheap, meaning that fall-off is not that expensive. Building a solution per machine type would simply be too costly: the time investment required to build these custom solutions is too high compared to the potential annual savings. However, focusing on the fall-off rate of the entire plastic-part-making departments and all such departments, the financial gains are significant. By lowering the amount of time required to enable analytic capabilities for each machine, we can transform it into a positive business case. This is why in this pilot the focus has also been on developing general predictive maintenance and process control solutions that are cloud-enabled and thus easily scalable.

Another challenge that has been tackled is the interaction of data-driven digital processes with the current manufacturing processes and how data-driven decision should be translated into actionable insights within production.

From a strategic standpoint, it is expected that the technologies developed using data-driven processes can be developed into new autonomous modes of manufacturing. Production customization has been made possible, implying frequent product changeover and smaller batch sizes, so-called Innovative Big Data Cognitive Manufacturing Processes. This pilot has deployed a series of technologies that facilitate increased quality and productivity, while also investigating generalization and scalability of these technologies in an industrial setting.

2.2 Big Data-Driven Shopfloor Automation Value for Injection Moulding 4.0

Philips Drachten wants to remain a pilot location for Industry 4.0-related activities. The business experiences need to become more data driven, while the effort of achieving this should reduce over time. Over the years, there have been data-driven solutions demonstrated and implemented within production; however, they have never been successful in scaling up nor maintaining those data-driven solutions, as they have focused on special solutions for unique cases.

The data business process value lies in developing generic automated solutions capable of scaling up across multiple injection moulding machines. A failure prediction model is one example of a generic automated solution, which can be applied for multiple machines and it results in reduction of fall-off rate and reduction of machines' downtime.

The application of Big Data and fact-based decision-making, along with seamless connectivity in the manufacturing process, results in efficient ramp-up times between different moulds, along with full traceability along the process chain all the way to the customer. A new data collection and storage infrastructure has been deployed to effectively integrate various types of data into a single common repository. This includes state-of-the-art technologies like streaming, edge computing and cloud computing in order to provide our operators with actionable insights. The results of the data monitoring and machine learning must be made available to process engineers, assembly line operators and data scientists.

2.3 Implementation of Big Data-Driven Quality Automation Solutions for Injection Moulding 4.0

To successfully implement Big Data solutions within the production process, an architecture map was made for the pilot phase during Boost 4.0. From this pilot setup, we identified the basic components and tested the concepts of connecting machines to a 'data collection' platform. In addition, several technical elements were identified that needed to be taken care of in order to build a fully scalable platform for Philips' injection moulding machines.

Although the eventual end goal is to prepare for a generic platform usable for the full fleet of injection moulding machines across Philips' factories, the final architecture of this trial has been instantiated for the Drachten site only. This also allows building an on-premise platform to manage all local data and consider offloading and/or management connection to external platforms, e.g., the cloud and/or the Industrial Data Spaces framework (IDS). The Drachten facility is a so-called brown-field factory, which means we need to comply with the local architecture as implemented. Philips has teamed up with two technology providers,

Philips Research (PEN) and IMEC, to support the implementation. PEN has a long heritage of pioneering innovation (inventions related to x-ray, optical recording, CD, DVD, etc.), currently focusing on data-driven research and service orientation, and IMEC is a world-leading research and innovation hub in nanoelectronics and digital technologies, combining widely acclaimed leadership in microchip technology and profound software and ICT expertise. Both technical partners (PEN & IMEC) provided input on where to put their proposed solutions:

- **Cloud connectivity** using a custom-build gateway service (based on the Microsoft Edge framework).
- **Data broker** to allow easy data acquisition for (historical) data, used for data analysis and machine learning models.
- **Machine learning models** that use real-time and historical data for predicting failures of machines.
- **Dashboard** (real-time) visualization of machine data, including pre-processing and machine learning models deployed as services.

Boost 4.0 big data platforms and techniques (Fig. 6) comply with RAMI 4.0 Digital Factory Alliance (DFA) service development reference architecture and ISO 20547 Big Data Reference Architecture (see chapter 'Big Data Driven Industry 4.0 Service Engineering Large Scale Trials: The Boost 4.0 Experience' in this book). Thus, it is possible to map into the Boost 4.0 Big Data Reference Architecture (RA) [6] the Philips predictive quality architecture (Fig. 5).

The Boost 4.0 architecture is based on multiple (Big Data) IT solutions being integrated via open APIs (Fig. 6). Some of which are essential and are part of the backbone, while others optional and extend functionality beyond the core functionalities of the platform.

- **Machine connector**: Allow to acquire (time-series) data from the machine controller. This is highly dependent on the machine interfaces available on the equipment itself. Typical machine connectors include OPC(-UA), CodeSys, Serial, Modbus, Canbus, EtherCAT, etc.
- **Protocol translation:** Translate an industrial protocol to another (open) standard. In this case, this is done by KEPWARE [7] and transforms data to OPC-UA-formatted data.
- **Semantics injection:** Make data understandable by adding semantic information (standard names, units, location, source, etc.)
- **Streaming data ingestion:** Transform OPC-UA to JSON formatted data and put them on a Kafka bus.
- **Streaming data bus:** A publish/subscribe enabled pipeline for real-time data transport. In this case, the data is JSON, based on the JSON-Header-Body (JHB) standard for industrial applications.
- **Micro-service architecture** for deployment of (Docker) containers (connected to the data bus), usable for data (pre-)processing, data analysis and data visualization.

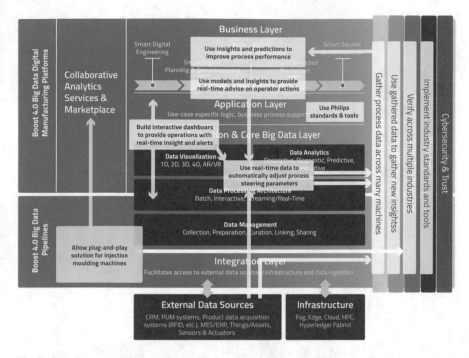

Fig. 5 Shopfloor automation trial mapping in Boost 4.0 big data reference architecture

- **A data historian** for long-term storage of time-series data. In the current version this is handled by Inmation [8] (based on MongoDB) or by Azure Time Series Insights (cloud storage, based on compressed JSON files in Parquet format).
- **Data broker** for providing different users with data from different sources in a standardized format, used for analysis. It supports real-time connections and is custom built.
- **Data analysis** is mainly taken care of by Python code (deployed in a container). Depending on the solution, multiple packages are used (like Pandas, SciKit, Keras, TensorFlow, etc.).
- **Rancher** solution to manage all micro-service containers from an easy-to-use web interface.
- **Open-source tools** for visualization of (live) data, based on Web technology, including Vue.JS, Quasar Framework, HTML, etc.).

2.4 Big Data Shopfloor Quality Automation Large-Scale Trial Performance Results

The instantiation and deployment of the Boost 4.0 reference model and implementation of big data pipelines into the Philips Drachten factory has translated in

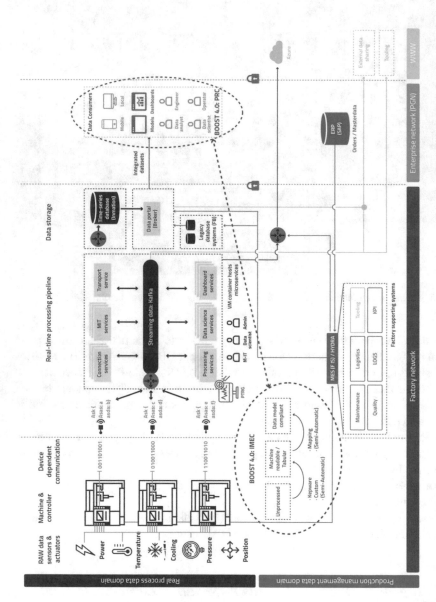

Fig. 6 Philips reference architecture and pipelines for predictive big data-driven quality automation solutions for injection moulding 4.0

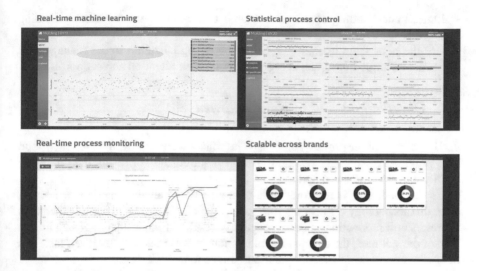

Fig. 7 Shopfloor dashboards experience and advanced data processing tools

significant shopfloor performance improvements in terms of flexibility, efficiency, quality and time to market. This is directly related to the ability to implement advanced decision support dashboards that reduce decision-making time and allow anticipating unplanned events, leveraging close to 10% production performance improvements (Fig. 7).

The main quantitative efficiency achievements are summarized as follows:

- A 10% reduction in fall-off rate and a 9% reduction in downtime
- Increased availability of process parameters data available from every 20 minutes to real-time information and increased number of parameters from 10 to ~80 per machine per cycle
- Collected over 400k of individual shots in 5 months of injection moulding data, which can be used to build more advanced models
- By automating the process of machine data end-point, increased and more homogeneous data quality and decreased time needed for connecting a machine to the 'real-time' platform from 2 weeks to around 4 h
- Reduction of 70% in the amount of (non)valuable and unnecessary control actions by operators

The adoption of the Boost 4.0 universal big data pipeline has also translated into increased quality of work in the shopfloor in the following way:

- Providing technicians with a more efficient tool for solving production issues and becoming part of a *better method for troubleshooting*.
- Take *wiser and more informed decisions* based on facts, for instance avoiding acquiring new machines by using the current machine park more efficiently.
- Better understanding of the current state of the art regarding IT, semantics and machine Learning.

- Better understanding of the time-related behaviour of the injection moulding process.
- Use the pilot setup to showcase the value of digitalization to the management board of Philips Corporate, in order to obtain their attention and support.

2.5 Observations and Lessons Learned

In the first year of the project, some basic interfaces were deployed on the shopfloor. This provided valuable lessons on the exact requirements for deploying predictive quality processes on the shopfloor.

As the main goal is to provide the operators with valuable insights, it became clear that technology (IT) is only one part of the challenge. Working together with operators and productions engineers quickly results in other challenges. With the help of our partners, the technical implementation was built and deployed fast. It is of crucial importance to keep 'operations' in the loop at all times.

Bringing IT solutions to the shopfloor (and essentially making them part of the production system) also implies requirements that were not as visible at the start of the project. These requirements must make sure operation can rely on the performance as well as the availability of solutions, and include actions such as training, coaching, providing support, but also continuously monitor solutions, preferably 24/7. When these measurements are taken into consideration, the results of the experimentations will already have a significant impact on the quality control process.

3 Production Data Platform Trials for Intelligent Maintenance at BENTELER Automotive

BENTELER is a global, family-owned company serving customers in automotive technology, the energy sector and mechanical engineering. As an innovative partner, it designs, produces and distributes safety-relevant products, systems and services. In the 2019 financial year, Group revenues were €7.713 billion. Under the management of the strategic holding BENTELER International AG, headquartered in Salzburg, Austria, the Group is organized into the divisions BENTELER Automotive and BENTELER Steel/Tube. Around 30,000 employees at 100 locations in 28 countries offer first-class manufacturing and distribution competence—all dedicated to delivering a first-class service wherever their customers need it. BENTELER Automotive is the development partner for the world's leading automobile manufacturers. Around 26,000 employees and more than 70 plants in about 25 countries develop tailored solutions for their customers. BENTELER Automotive's products include components and modules in the areas of chassis, body, engine and exhaust systems, as well as solutions for electric vehicles.

3.1 Data-Driven Digital Shopfloor Maintenance Process Challenges

Intelligent, self-regulated maintenance is a key element in Industry 4.0. The networking of machines and plants and the availability of machine data allows continuous monitoring and evaluation of the health status of a production system in real time. Failures and malfunctions can be detected or even foreseen at an early stage, and measures to protect the functionality and performance of the production system can be derived from them. The aim of Smart Maintenance is to increase the performance of production technology, for example through increased plant availability, optimized process quality and improved planning.

The basic technologies for Smart Maintenance solutions are already available. Seventy per cent of machine and plant manufacturers are developing or piloting Smart Maintenance offerings or already offer them [9]. Market-ready solutions are offered in particular by component suppliers from the automation and drive technology sector, as they can be transferred to a large quantity of systems. Nevertheless, the application of Smart Maintenance in manufacturing is below expectations, even though solutions for individual components are available: On the side of machine operators, maintenance knowledge is required for a large number of different machine types and systems. This know-how is hardly ever bundled, documented or made available by means of standardized processes. According to Acatech [10], 47% of German manufacturing companies record information on malfunctions and failures only manually. Fifty-seven per cent of companies still initiate measures without any data at all. Only 4% make decisions based on real-time data.

The biggest challenge in developing a fault detection system is the availability of fault data. Compared to the total amount of data available, failures and errors occur only rarely. Many machine learning methods (especially so-called supervised learning methods) are therefore not or only partially applicable. Hence, mainly methods of anomaly detection were used during the development. Thereby characteristics for normal behaviour are derived from the signal courses in regular production use. During operation, deviations from this normal behaviour are detected and reported.

3.2 Implementation of a Big Data Production Platform for Intelligent Maintenance in Automotive

The goal of the trial is the implementation of a global (cross-factory) system for automated detection of failures and recommendations for actions in the context of machine health monitors with notification and planning of actions (Fig. 8). The availability of a platform for the storage and processing of production data is a prerequisite for the implementation of such centralized intelligent maintenance in

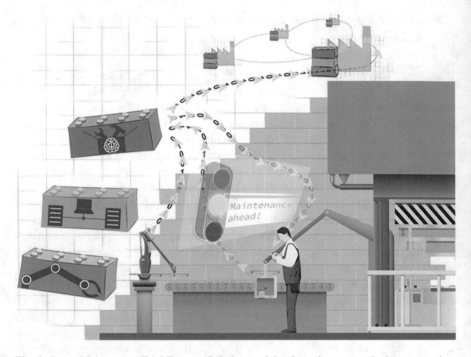

Fig. 8 Smart Maintenance Trial Factory: Solution modules from the systematic data connection, data infrastructure and intelligent data processing are the basis for the successful implementation of use cases (source: IEM)

production. As part of the Boost 4.0 project, BENTELER Automotive, the Fraunhofer Institute for Mechatronic Systems Design IEM and ATLANTIS Engineering have built a Smart Maintenance pilot factory within the Leading-Edge Cluster it's OWL—Intelligent Technical Systems OstWestfalenLippe.

Solutions for different problem dimensions have been developed. In addition to the technical infrastructure for industrial data analysis, the development of data evaluation, process integration on the application level and the methodical procedure for the implementation of Smart Maintenance have been analysed. The maintenance of a hydraulic press, as well as a material handling system have been considered as examples. The functional core of the pilot factory is the Smart Production Data Platform. The platform operated by BENTELER IT fulfils three central tasks:

- The central provision of current and historical production data
- The execution of data analysis such as error detection
- The visualization and return of results to the user

For data provision, the machine controls were connected by means of standard interfaces (e.g. OPC-UA) and well over a thousand data sources have already been tapped in the plant. Signal changes in the range of less than 1 s are recorded, so that several million data points are recorded and made available every hour. In addition to real-time data, several years of historical data recordings can be accessed. These are necessary for the development and testing of data analysis methods, such as machine learning methods.

Dashboard usability and data interpretability are of prominent importance to ensure effective data visualization and decision support experience. Standard solutions like Grafana enable employees on the shopfloor to develop dashboards and individual displays independently. Individual machine data as well as the results of an anomaly detection are both available as data sources. The capability of the workforce to easily create alarms has been introduced in the decision workflow. The result is a significant time reduction in the response to unexpected events or even anticipation to failures. These new features also allow that out-of-range critical values or the frequent occurrence of anomalies can be reported immediately and addressed effectively to allow reduction of unplanned breakdowns. For further improvement of fault detection, employee feedback on the store floor by means of a decision support system is installed.

The production data platform, see Fig. 9, complies with the modular approach to Boost 4.0 big data pipeline development and open digital factory reference framework. It deploys modern technologies for container management, which allows the utilization of reusable software modules, for example for data provision, error detection, reporting or visualization. The individual modules can be flexibly combined to form new services, and a service can be transferred to another plant in just a few steps. The error detection for material handling systems developed in the Paderborn plant has already been tested in other BENTELER plants. The so-called micro-service architecture allows the fast and flexible development, adaptation and

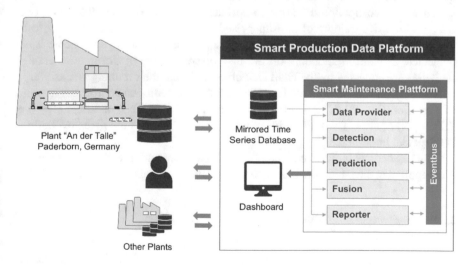

Fig. 9 Production data platform: A Smart Maintenance service accesses the provided data, uses modules for data analysis and visualizes the results in a dashboard (source: IEM/ATLANTIS Engineering)

testing of smart maintenance solutions. At the same time, it provides a future-proof architecture for other applications in production, such as process optimization or Smart Quality. The platform is already being used in other OWL research projects, for example ML4Pro [11]—Machine Learning for Production and its products.

3.3 Big Data-Driven Intelligent Maintenance Large-Scale Trial Performance Results

The implementation of the production data platform enables the deployment of software solutions that take advantage of Industry 4.0 technologies. One example is the Smart Maintenance Platform (SMP), which is able to monitor the machinery equipment of potentially all the BENTELER plants that are connected to the platform from a central remote location. Based on virtualization technologies, like Docker, and utilizing a micro-service architecture, SMP is able to *scale* its resources both vertically (e.g. adapt the system resources like CPU cores) and horizontally (e.g. deploy more instances in parallel of the Anomaly Detection micro-service), in order to cope with demanding data streaming scenarios.

Apart from the *scalability challenge* of the Big Data processing, SMP should also address the *transferability challenge*, in order to enable its application in different scenarios and use cases among the connected BENTELER plants. As already stated, the supervised learning-based monitoring approaches require the existence of fault data (i.e. machinery failures and errors), which in most of the crucial cases are rare due to preventive maintenance. However, SMP offers a set of Fault Detection tools, which utilize unsupervised learning approaches. Hence, only configuration over the data-intensive training of the supervised approaches is required in order to be applied. Of course, the great potential of the supervised predictive approaches is not neglected, as Fault Prediction tools are also offered by the platform, once enough fault data are collected by the Fault Detection tools and their training is feasible.

The performance of both the Fault Detection and Prediction approaches in terms of *Precision* (i.e. TP / (TP + FP)), *Recall* ((TP) / (TP + FN)) and *Accuracy* (i.e. (TP + TN) / (TP + TN + FP + FN)), where TP, TN, FP, FN are given by Table 1, is of special importance.

Table 2, depicts indicative results for both Fault Detection and Fault Prediction tools applied in the Paderborn plant analysing data of 1.5 years. The Fault Prediction

Table 1 The four outcomes of the data analysis tools

		Actual case	
		Fault	Normal
Predicted case	Normal	(False Negative) FN	(True Negative) TN
	Fault	(True Negative) TP	(False Negative) FP

Table 2 Fault Prediction and Detection results of the Hydraulic Press and the Material Handling use case in the Paderborn plant

		Precision	Recall	Accuracy
Hydraulic Press use case	Fault Prediction	0.5	1	0.93
	Fault Detection	0.99	1	0.99
Material Handling system	Fault Detection	0.93	0.95	0.99

was applied only in the Hydraulic Press use case as the behaviour of the Material Handling System was unpredictable. The results of the prediction approach should not be compared with the results from the detection approach as they are computed differently; however, the low precision of the prediction shows the difficulty of the approach to be trained properly as only three incidents occurred in 1.5 years.

Applying the Fault Prediction and Fault Detection tools to the Paderborn production line has already shown promising results that have the potential to remain at the same or even better levels, once the tools are adopted at a larger scale. The key performance indicators of interest for BENTELER from the business point of view are:

- Reduction in maintenance cost
- Reduction in MTTR (Mean Time To Repair)
- Increase in MTBF (Mean Time Between Failures)
- Increase in OEE (Overall Equipment Efficiency)

It should be mentioned that the application of the tools for certain equipment has already indicated the possibility of reducing the MTTR by 30% and of at least doubling the MTBF for certain types of failures.

3.4 Observations and Lessons Learned

The implementation of smart maintenance use cases posed not only technical challenges, but also challenges in project organization, e.g. communication with stakeholders within the company, knowledge management and its transfer between stakeholders and acceptance of developed solutions. In terms of domain and data understanding, using semi-formal models was a key to successful knowledge transfer. Constructing easy-to-understand, interdisciplinary models in joint work-shops also increases acceptance and awareness for the involved stakeholders. The development of user-friendly and easily understandable dashboards allowed the demonstration of benefit of the smart production platform at shopfloor level. The utilization of reusable software modules facilitated the quick construction of a solution and transfer to other plants.

The implementation of a production data platform has proven to play a central role in the digitalization of BENTELER plants. It provides the basis for all data-driven use cases and data-driven decision-making: transparency about individual production machines as well as extensive production processes, monitoring and alerting, and advanced data analytics not only in smart maintenance, but also smart quality and process optimization. The decision to invest in the implementation of a production data platform thus is a complex matter, since it involves a comprehensive benefit analysis that is difficult to quantify. It is a mostly strategic decision, setting the roadmap for further approaches to factory operation and optimization.

4 Predictive Maintenance and Quality Control on Autonomous and Flexible Production Lines: The FCA Trial

4.1 Data-Driven Digital Process Challenges

The main challenges related to the pilot regard firstly the data management, starting from their collection, which can be difficult because of the different sensing systems implemented on the shopfloor production actors (e.g. accelerometers on AGVs and power meter on the laser cells). The presence of heterogeneous devices means the need to deal with specific communication protocols and different data acquisition speeds.

Another aspect, linked to the previous one, concerns the possible speed mismatch between production process, with the related data generation, and the information flow. As a consequence, one of the main challenges consists in the reduction of this time that has to be approximately equal to zero.

Then, an important challenge source is data protection, since security of the data is a crucial element of the pilot as we are dealing with the industrial field and especially with the production sector (e.g. production levels) and the quality sector (e.g. level of default). In particular, the management of data exposition to external providers on the cloud platform becomes very important, which necessitates careful data subdivision.

An additional field of action is represented by the communication between the industrial field and the cloud platform, because of the presence of security policies regarding the data flow which have to be respected, and that could represent a strong constraint for the pilot development.

Moving then to data utilization, the understanding, organization and use of the expansive datasets made available in new and better ways pursuing data uniformity and standardization across the entire product development lifecycle bring incredible challenges for data exploitation.

Then, regarding the data processing and analytics, we have that the entire process from the data acquisition, which sometimes could comprehend some edge pre-processing in order to reduce the volume of stored data, to data transfer and cloud analysis, in several cases has to be fast enough in order to enable near-real-time process feedback.

Finally, different challenges may come from the fields of data visualization and user interaction, since several end-users are considered in the pilot, from operators at the shopfloor level to maintenance operators at the information/operational level.

4.2 Big Data Manufacturing Process Value in Autonomous Assembly Lines in Automotive

The initial business scenario is about the implementation of the concept of autonomous production, where the traditional linear process is removed and mobile robots, such as Automated Guided Vehicles (AGVs); collaborative robots with vision capabilities; and fixed production cells collaborate together. In the traditional production processes, mobile robots have only duties related to logistics (e.g. replenishment, preparation of components, etc.) or manufacturing (e.g. carrying work in progress), and the control of fleets of such AGVs and their availability and reliability to respect cycle time and lead-time is crucial to ensure the stability and throughput of the production systems.

Planning, control, monitoring and maintenance of the mobile robots are required due to the fact that currently there is no specific approach to store and analyse data related to the missions of the vehicles, their wear-out and availability, taking into account the lead time for delivery and the uncertainty related to the interaction with the presence of human operators.

One of the main objectives is to ensure that the new technology is robust enough to avoid business interruption (e.g. stock-out, unwanted waiting or idle time for the machine), delays and reduction of throughput to transfer the autonomous production to the rest of the plants.

The autonomous assembly line aims to provide the maximum flexibility to potential changes in the demand or to issues/delays/changes in the logistics or productive systems by means of using available and new datasets (such as flows of components in the plants and their precise localization) ensuring business continuity. At the same time, the over-dimensioned fleet of robots is reduced and the (big) data are shared among the whole value chain (providers, maintenance services, etc.) (Fig. 10).

AGVs are used to replenish and handle material or work-in-process between the different production islands, in particular the assembly and welding cells, and to/from the warehousing areas. Production actors are connected to the different production management platforms.

In the new scenario, production data coming from AGVs and laser cells are collected and enriched using FIWARE technology. Then they are sent through MindConnect technology and stored in a data lake on cloud provided by SIEMENS MindSphere Platform.

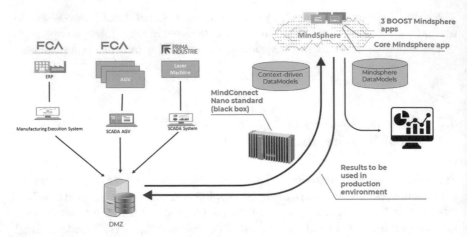

Fig. 10 FCA trial big data pipeline for autonomous assembly lines (AGV and Laser cells)

Different algorithms elaborate the production data in order to monitor the quality of the produced components, detect malfunctions enabling the definition of a maintenance schedule and optimizing the allocation of production missions. Besides, the different data are made accessible to external service providers, in order to enable the development of innovative applications based on proprietary data. To this end, and to ensure data privacy and security, open data models have been developed and IDS technology (e.g. IDS connector) has been implemented.

4.3   Implementation of Big Data Solutions

The pilot development began with a prototype application, which gathered sensors data from an assembly cell, replenished by an AGV, located in the Campus Melfi shopfloor. Data were collected from the machines to a central database, and they were visualized by the prototype application through a dashboard.

Successively, it has evolved into an industrial experimentation site, which started from the results provided by the prototype application in order to progressively develop and test the complete pilot architecture. It began with a first phase, in which data were gathered from different sources into MindSphere [12], which implemented the IDS architecture [13] and has been structured on the basis of the data sovereignty principles. The data sources were represented by an AGV owned by FCA and located in the Melfi Campus, and a laser machine owned by PRIMA and located within the Prima's labs. MindSphere hosted on the cloud and data from the shopfloor were exposed to external service providers. Specific APIs, called Mindlib [14], were used to send data from the source systems (data provider) to MindSphere Services Platform (data consumer). A datamodel has been created and used to set

Fig. 11 FCA Boost 4.0 Melfi Campus Experimental Site and several pilot production actors

up the platform, to be able to collect the data. A visualization App (MindApp) displayed the data (Fig. 11).

Then, in the second phase a scenario of interaction between the robots (AGVs) and the production cells (fixed machines from PRIMA) was implemented and tested. Within this scenario, the manufacturing capability w granted by the correct functioning of both the robots and the fixed machines. Three specific apps were defined in the MindSphere environment and mostly the first two, the PRIMA Fleet Management App which monitors the main parameters trends analysing and correlating different types of data and the Smart Scheduling App which optimizes the mission allocation to the different production actors, were developed. The number of data sources from AGVs and Laser machines were widened and a DMZ (demilitarized zone) was set up in CRF in order to permit the interface between the industrial environment and MindSphere.

Lastly, during the third phase, the pilot was extended to the full industrial scale and so the architecture was adapted to the final number of data sources and data amount, the connector to the Fiware Orion Context Broker was developed and inserted in the data flow system and an app for anomaly detection using data coming the AGVs was finalized.

4.4 Large-Scale Trial Performance Results

The implementation of an Industrial IoT Data Space based on the MindSphere platform has allowed FCA to develop a number of MindApps with the collaboration of external service providers. This approach has delivered three big-data driven innovative services that are currently being operated at the Melfi Campus:

- MindApp for production optimization
- MindApp for welding quality control
- MindApp for AGV anomaly detection (Fig. 12)

These three apps are significantly improving the performance and flexibility of the autonomous assembly lines, adapting the production scheduling to real-time sensitivity to production quality variation and asset maintenance needs to allow zero unplanned breakdown.

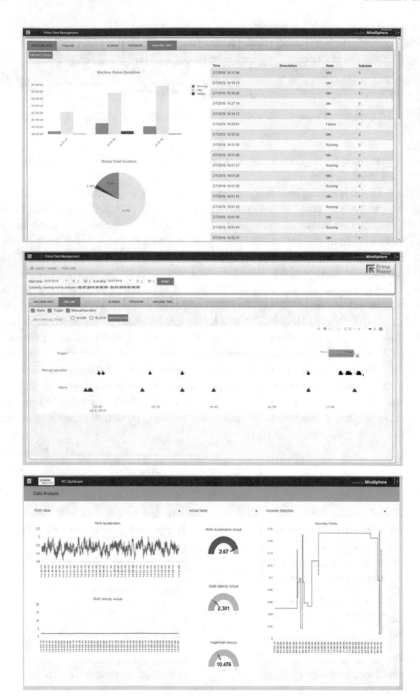

Fig. 12 MindApps dashboards developed by PRIMA and third-party developers at FCA Melfi Campus Experimental Site based on the IIoT MindSphere industrial data space

4.5 Observations and Lessons Learned

The pilot development and implementation presented some barriers that had to be overcome. Regarding the software development, the main effort has been represented by the connectivity aspects. In order to make the applications work properly, it was necessary to have data in the correct aggregation and format as required by MindSphere. Therefore, the development of connectors and format converters was the most expensive part, along with the coding and application deployment, in terms of resources and effort. Moving then to the data flow architecture, the main issue here was the choice and the development of a solution which allowed the exposition of the industrial data to external partners avoiding to put in danger the security of the entire company's internal network. The identification and the development of the solution, which consists in a Demilitarized Zone, required a huge effort in collaboration with the IT and security departments in order to, on one side, respect all the company policies and, on the other, meet all the project requirements that would have led then to the development of the data transfer infrastructure.

5 Conclusions

This chapter has presented the advanced manufacturing processes and big data-driven algorithms and platforms leveraged by the Boost 4.0 big data lighthouse project that allow improved digital operations within increasingly automated and intelligent shopfloors. It has demonstrated how three different companies have been able to implement three distinct, open, yet sovereign cross-factory data spaces under a unified framework, i.e. Boost 4.0 big data reference architecture and Digital Factory Alliance (DFA) [15] service development framework. Philips has provided evidence of the significant benefits that data spaces and integrated data pipelines can bring to their Drachten brownfield production lines in terms of implemented increasingly predictive quality control and fact-based automated decision support processes. BENTELER Automotive, has equally demonstrated the benefits of a modular and data-space approaches to deliver high cross-factory transferability of smart maintenance 4.0 services from their factory in Padeborn, all based on the use of advanced software containerization and virtualization as well as open source technology for the implementation of data spaces and data pipelines. Finally, FCA has demonstrated the benefits and challenges that the operation of Industrial IoT data spaces supported by MindSphere entail to support the implementation of flexible, modular autonomous assembly cells. FCA has demonstrated how the implementation of such data spaces in their Melfi Campus facilities based on open APIs allows not only a better integration of the shopfloor assets but also opens up the opportunity for the development of high-value customized services and data-driven apps that positively impact the performance of the digital shopfloor and allow a more resilient and adaptive scheduling of production. The chapter has shown that maintenance performance improvements (main time between failures) can be improved by 600%, overall equipment efficiency (OEE) by 14% and production efficiency by 10%. These figures are close to those estimated by literature studies and can be achieved by means of adopting a unified big data approach provided by the Boost 4.0 reference model, the implementation of industrial data spaces and the realization of advanced decision support dashboards that reduce the time to decision and action data. Boost 4.0 has demonstrated that industry can cost-effectively implement effective means for data integration, even in brownfield production lines with significant legacy equipment.

Acknowledgements This research work has been performed in the framework of the BOOST 4.0 Big Data Lighthouse initiative, a project that has received funding from the European Union's Horizon 2020 research and innovation programme under grant agreement No. 780732. This data-driven digital transformation research is also endorsed by the Digital Factory Alliance (DFA) www.digitalfactoryalliance.eu

References

1. European Economic and Social Committee and the Committee of the Regions. (2020). *A European strategy for data*. European Commission. Available at: https://ec.europa.eu/info/sites/info/files/communication-european-strategy-data-19feb2020_en.pdf. Accessed January 2021
2. World Economic Forum. (2016). Digital transformation of industries demystifying digital and securing $100 trillion for society and industry by 2025.
3. Boost 4.0. https://www.boost40.eu. Accessed January 2021.
4. Zillner, S., Curry, E., Metzger, A., Auer, S., & Seidl, R. (Eds.). (2017). *European big data value strategic research & innovation agenda*. Big Data Value Association.
5. Zillner, S., Bisset, D., Milano, M., Curry, E., García Robles, A., Hahn, T., Irgens, M., Lafrenz, R., Liepert, B., O'Sullivan, B., & Smeulders, A. (Eds.). (2020, September). *Strategic research, innovation and deployment agenda—AI, data and robotics partnership*. Third release. Brussels. BDVA, euRobotics, ELLIS, EurAI and CLAIRE.
6. D2.5 Boost 4.0 reference architecture specification. Available at: https://boost40.eu/wp-content/uploads/2020/11/D2.5.pdf. Accessed January 2021.
7. Kepware. https://www.kepware.com. Accessed January 2021.
8. Inmation. https://www.inmation.com. Accessed January 2021.
9. Roland Berger GmbH. (2017). Predictive maintenance. Service der Zukunft—und wo er wirklich steht. Studie im Auftrag des VDMA.
10. Henke, M., Heller, T., & Stich V. (Eds.). (2019). *Smart maintenance—Der Weg vom Satus quo zur Zielvision (acatech STUDIE)*. utzverlag GmbH.
11. ML4Pro. https://www.iem.fraunhofer.de/de/referenzen/forschungsprojekte/benteler-ml4pro2.html. Accessed January 2021.
12. Mindshpere. https://siemens.mindsphere.io. Accessed January 2021.
13. International Data Spaces Association. (2019). Reference architecture model. Online. Available at: https://www.internationaldataspaces.org/wp-content/uploads/2019/03/IDS-Reference-Architecture-Model-3.0.pdf. Accessed January 2021.
14. MindLib. https://community.sw.siemens.com/s/article/data-structure-and-data-upload-to-mindsphere-via-mindlib. Accessed January 2021.
15. Digital Factory Alliance (DFA). https://digitalfactoryalliance.eu. Accessed January 2021.

Big Data-Driven Industry 4.0 Service Engineering Large-Scale Trials: The Boost 4.0 Experience

Oscar Lázaro, Jesús Alonso, Paulo Figueiras, Ruben Costa, Diogo Graça, Gisela Garcia, Alessandro Canepa, Caterina Calefato, Marco Vallini, Fabiana Fournier, Nathan Hazout, Inna Skarbovsky, Athanasios Poulakidas, and Konstantinos Sipsas

Abstract In the last few years, the potential impact of big data on the manufacturing industry has received enormous attention. This chapter details two large-scale trials that have been implemented in the context of the lighthouse project Boost 4.0. The chapter introduces the Boost 4.0 Reference Model, which adapts the more generic BDVA big data reference architectures to the needs of Industry 4.0. The Boost 4.0 reference model includes a reference architecture for the design and implementation of advanced big data pipelines and the digital factory service development reference architecture. The engineering and management of business network track and trace processes in high-end textile supply are explored with a

O. Lázaro (✉) · J. Alonso
Asociación de Empresas Tecnológicas Innovalia, Derio, Spain
e-mail: olazaro@innovalia.org; jalonso@innovalia.org

P. Figueiras · R. Costa
UNINOVA- Instituto de desenvolvimento de novas tecnologías – Associacao, Caparica, Portugal
e-mail: paf@uninova.pt; rddc@uninova.pt

D. Graça · G. Garcia
Volkswagen Autoeuropa, Anjo, Portugal
e-mail: diogo.graca@volkswagen.pt; gisela.garcia@volkswagen.pt

A. Canepa
Fratelli Piacenza, Pollone, Italy
e-mail: alessandro.canepa@piacenza1733.it

C. Calefato · M. Vallini
Domina Srl, Ragusa, Italy
e-mail: caterina.calefato@domina-biella.it; marco.vallini@domina-biella.it

F. Fournier · N. Hazout · I. Skarbovsky
IBM Israel, Haifa, Israel
e-mail: fabiana@il.ibm.com; nathanh@il.ibm.com; inna@il.ibm.com

A. Poulakidas · K. Sipsas
Intrasoft International S.A., Luxembourg City, Luxembourg
e-mail: Athanasios.Poulakidas@intrasoft-intl.com; Konstantinos.SIPSAS@intrasoft-intl.com

373

E. Curry et al. (eds.), *Technologies and Applications for Big Data Value*,
https://doi.org/10.1007/978-3-030-78307-5_17

focus on the assurance of Preferential Certification of Origin (PCO). Finally, the main findings from these two large-scale piloting activities in the area of service engineering are discussed.

Keywords Reference architecture · ISO 20547 · ISO/IEC/IEEE 42010 · DIN 27070 · Sovereignty · Data spaces · Track & Trace · Blockchain · FIWARE · Virtual commissioning · Testbed · Trial · Business networks 4.0 · SUMA 4.0 · Intralogistics

1 Introduction

Over the last few years, the potential impact of big data for the manufacturing industry has received enormous attention. However, although big data has become a trend in the context of manufacturing evolution, there is not yet sufficient evidence on how and if big data will leverage such impact in practical terms. New concepts in the area of Industry 4.0 such as digital twins, digital threads, augmented decision support dashboards and systems, and simulation-based commissioning systems rely significantly on advanced engineering and operation of big data techniques and technical enablers. The emergence of data-driven techniques to increase data visibility, analytics, prediction and autonomy has been immense. However, those techniques have been developed in many cases as individual efforts, without the availability of an overarching framework making the transfer of such applications to other industries at scale cumbersome. Moreover, the development of such big data applications is not necessarily realized in context with reference architectures such as the European Reference Architectural Model Industry 4.0 (RAMI 4.0), which serves as reference in the sector for Industry 4.0 digital transformation. Big data promises to impact Industry 4.0 processes at all stages of the product life-cycle.

The aim of this chapter is to present the advances made in the area of service engineering and commissioning in the context of H2020 EU large-scale piloting project Boost 4.0 [1]. It gathers the first set of experiences, best practices and lessons learned during the deployment of the two lighthouse trials in the scope of the Boost 4.0 project: the most ambitious European initiative in big data for Industry 4.0. It presents the experiences of two European manufacturing leaders (large industry and SME) in the engineering and management at large scale of data-driven and traceable intra-logistics and supply chain processes. Intra-logistic processes will be addressed by the Volkswagen Autoeuropa (Portugal) plant in the automotive sector, whereas supply chain business network engineering and management will be addressed by the Italian SME Piacenza in the high-end textile sector. This chapter addresses initial data value innovation elicitation and presents and assesses how a common RA can be used to leverage advanced service engineering practices at large scale, as well as lessons learned and impact evaluation.

This chapter relates mainly to the technical priorities Data Management Engineering and optimized architectures for analytics of data-at-rest and data-in-motion of the European Big Data Value Strategic Research & Innovation Agenda [2]. It addresses the horizontal concerns of heterogeneity, scalability and processing of data-in-motion and data-at-rest of the BDV Technical Reference Model. It addresses the vertical concerns of communication and connectivity, engineering and DevOps for building big data value systems areas to facilitate timely access and processing of big data and evolving digital twin models. The work in this chapter relates mainly but not only to the Systems, Methodologies, Hardware and Tools cross-sectorial technology enablers of the AI, Data and Robotics Strategic Research, Innovation & Deployment Agenda [3].

The chapter is organized as follows: First the Boost 4.0 initiative is introduced with a focus on the instantiation of the Boost 4.0 common big data-driven Reference Architecture (RA). This RA is aligned with the big data RA proposed by Big Data Value Association (BDVA) and harmonized with the Digital Factory Alliance (DFA) overall digital factory open reference model. Next, the big data intra-logistic process engineering trial and lessons learned at Volkswagen Autoeuropa are introduced. Next, the engineering and management of business network track and trace processes in high-end textile supply are presented with a focus on assurance of Preferential Certification of Origin (PCO). Finally, the main findings extracted from these two large-scale piloting activities in the area of service engineering are discussed.

2 Boost 4.0 Universal Big Data Reference Model

Boost 4.0 (Big Data Value Spaces for Competitiveness of European Connected Smart Factories 4.0) is the biggest European initiative in big data for Industry 4.0. With a 20 M€ budget and leveraging 100 M€ of private investment, Boost 4.0 has led the construction of the European Industrial Data Space to improve the competitiveness of Industry 4.0. Since January 2018, it has guided the European manufacturing industry in the introduction of big data in the factory, providing the industrial sector with the necessary tools to obtain the maximum benefit of big data.

Since the beginning of the project, Boost 4.0 has demonstrated in a realistic, measurable and replicable way an open, certifiable and highly standardized shared data-driven Factory 4.0 model through 11 lighthouse factories, and has also demonstrated how European industry can build unique strategies and competitive advantages through big data across all the phases of product and process lifecycle.

2.1 Boost 4.0 Objectives

Boost 4.0's overall mission is to accelerate the adoption of Industry 4.0 big data-intensive smart manufacturing services through highly replicable lighthouse activities that are intimately connected to current and future Industry 4.0 investments, resolving the smart connected product and process data fragmentation and leveraging the Factory 4.0 data value chain.

To accomplish this mission, Boost 4.0 has defined the following objectives:

- **Global Standards:** Contribution to the International Data Space data models and open interfaces aligned with the European Reference Architectural Model Industry 4.0 (RAMI 4.0).
- **Secure Digital Infrastructure:** Adaptation and extension of cloud and edge digital infrastructures to ensure high-performance operation of the European Industrial Data Spaces, i.e. support of high-speed processing and analysis of huge and very heterogeneous industrial data sources.
- **Trusted Big Data Middleware:** Integration of the four main open-source European initiatives (International Data Space, FIWARE, Hyperledger, Big Data Europe) to support the development of open connectors and big data middleware.
- **Digital Manufacturing Platforms:** Opening of interfaces for the development of big data pipelines for advanced analysis services and data visualization supported by the main digital engineering, simulation, operations and industrial quality control platforms.
- **Certification:** Development of a European certification programme for equipment, infrastructures, platforms and big data services for operation in the European Industrial Data Space.

2.2   Boost 4.0 Lighthouse Factories and Large-Scale Trials

In Boost 4.0, some of the most competitive factories, from three strategic economic sectors that drive not only European manufacturing economy but also the IoT/smart connected market development (i.e. automotive, manufacturing automation and smart home appliance sectors) join forces to set up 11 lighthouse factories and 2 replication factories (Fig. 1) that are a coherent, complementary and coordinated big data response to the 5 EFFRA Factory 4.0 Challenges, i.e. (1) lot size one distributed manufacturing, (2) operation of sustainable zero-defect processes and products, (3) zero break down operations, (4) agile customer-driven manufacturing value network management and (5) human-centred manufacturing.

Boost 4.0 leverages five widely applicable big data transformations: (1) networked commissioning and engineering, (2) cognitive production planning, (3) autonomous production automation, (4) collaborative manufacturing networks and (5) full equipment and product availability—across each of the five key product and process lifecycle domains considered: (1) Smart Digital Engineering, (2) Smart Production Planning and Management, (3) Smart Operations and Digital Workplace, (4) Smart Connected Production and (5) Smart Maintenance and Service.

+GF+

PHILIPS

United Technologies
Research Center

BENTELER ▽
makes it happen

∧Nemak

TRIMEK
METROLOGICAL ENGINEERING

FBI YOUR FUTURE

RI4 STONE

Ⓦ

CENTRO
RICERCHE
FIAT

Gestamp

Whirlpool

⦿ Lighthouse Factory 4.0
⦿ Replication Factory 4.0

Fig. 1 BOOST 4.0 big data driven lighthouse and replication factories 4.0

2.3 *Boost 4.0 Universal Big Data Reference Architecture*

One of the main ambitions of Boost 4.0 is to define and develop highly replicable big data solutions to ensure the impact of the project beyond the project lifetime. One of the main challenges Industry 4.0 faces when designing their big data solutions is first of all to effectively address the design and development of high-performance big data pipelines for advanced data visualization, analytics, prediction or prescription. Then, the challenge lies in how to successfully integrate such big data pipelines in the digital factory engineering and production frameworks. In this sense, to facilitate the replicability of the lighthouse trials and big data solutions implemented, Boost 4.0 has relied on two reference models. On one hand, the BDVA Big Data Reference Model (BD-RM) [4] to drive Industry 4.0 big data pipelines and process engineering and operation. The goal of this RM is to ensure universality and transferability of trial results and big data technologies as well as economies of scales for big data platform and technology providers across sectors.

On the other hand, Boost 4.0 has developed and applied a RAMI 4.0 [5] compliant Service Development Reference Architecture (SD-RA) for big data-driven factory 4.0 digital transformation. This model is now maintained by the Digital Factory Alliance (DFA) [6]. The goal is to ensure a perfect alignment between big data processes, platforms and technologies with overall digital transformation and intelligent automation efforts in manufacturing factories and connected manufacturing networks.

As illustrated in Fig. 2, the Boost 4.0 BD-RA [7] is composed of four main layers: Integration Layer, Information and Core Big Data Layer, Application and Business Layers. This approach is aligned with the ISO 20547 Big Data Reference Architecture—Big Data Application Provider layer—from Data Acquisition/Collection through Data Storage/Preparation (and sharing) further to any Analytics/AI/Machine Learning and also environmental Action/Interaction including Visualization.

These four layers allow the implementation of a big data pipeline and the integration of such pipelines in specific business processes supporting the Factory 4.0 product, process and service lifecycle, i.e. smart digital engineering, smart digital planning and commissioning, smart digital workplace and operations, smart connected production and smart servicing and maintenance. These four Boost 4.0 layers are supported by a set of transversal services, in particular data sharing platforms, engineering and DevOps, Communications and Networking, Standards and Cybersecurity and Trust. These layers enact the manufacturing 4.0 entities and leverage a data 4.0 value chain that transforms raw data sources into quality data that can be interpreted and visualized, providing mining and context for decision support. This value chain is developed as data is aggregated, integrated, processed, analysed and visualized across the Factory 4.0 layers (product, device, station, workcentre, enterprise and connected world). The Boost 4.0 BD-RA adopts the BDVA RM and adapts it to the specific needs of Industry 4.0.

However, the generic Boost 4.0 BD-RA needs to be articulated and instantiated with the support of specific platforms, solutions and infrastructures so that the big data-driven manufacturing processes can actually be realized. So, even if, as shown in Fig. 2, the BDVA big data reference model can in fact be adapted to Industry 4.0 needs and aligned with the RAMI 4.0 model, a more formal harmonization and integration of the BDVA RM is required to facilitate development of big data services in the context of a digital factory exhibiting high transferability and replication capabilities for big data-driven manufacturing processes. This is further facilitated with the application of the DFA Digital Factory Service Development RA (SD-RA), which ensures a broad industrial applicability of digital enablers, mapping the technologies to different areas and to guide technology interoperability, federation and standard adoption. The DFA SD-RA design complies with ISO/IEC/IEEE 42010 [7] architectural design principles and provides an integrated yet manageable view of digital factory services. In fact, DFA SD-RA integrates functional, information, networking and system deployment views under one unified framework. The DFA SD-RA address the need for an integrated approach to how (autonomous) services can be engineered, deployed and operated/optimized in the context of the digital factory. With this aim, the DFA SD-RA is composed of three main pillars, as depicted in Fig. 3:

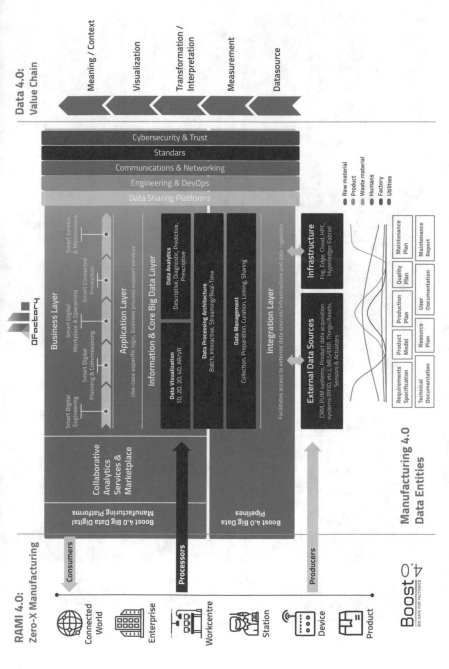

Fig. 2 Boost 4.0 Universal Big Data Reference Architecture 4.0

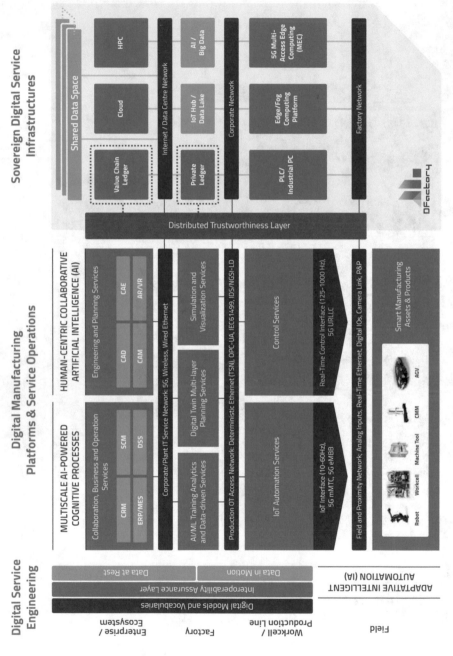

Fig. 3 Digital Factory Alliance Reference Architecture for Industry 4.0

1. **Digital Service Engineering**. This pillar provides the capability in the architecture to support *collaborative model-based service enterprise approaches* to digital service engineering of (autonomous) data-driven processes with a focus on supporting smart digital engineering and smart digital planning and commissioning solutions to the digital factory. The pillar is mainly concerned with the harmonization of digital models and vocabularies. It is this pillar that should develop interoperability assurance layer capabilities with a focus on mature digital factory standards adoption and evolution towards an "industry commons" approach for acceleration of big data integration, processing and management. It is this pillar where "security by design" can be applied both at the big data, manufacturing process and shared data space levels.
2. **Digital Manufacturing Platforms and Service Operations**. This pillar supports the deployment of services and DMPs across the different layers of the digital factory to enact data-driven smart digital workplaces, smart connected production and smart service and maintenance manufacturing processes. The pillar is fundamental in the development of three enabling capabilities central to the gradual evolution of autonomy in advanced manufacturing processes, i.e. multi-scale AI-powered cognitive processes, human-centric collaborative intelligence and adaptive Intelligent Automation (IA). The enablement of both knowledge-based (multi-scale artificial intelligence) and data-driven approaches (collaborative intelligence) to digital factory intelligence is facilitated by the support of service-oriented and event-driven architectures (interconnected OT and IT interworking event and data buses) embracing international and common standard data models and open APIs, thereby enabling enhanced automated context development and management for advanced data-driven decision support.
3. **Sovereign Digital Service Infrastructures**. The operation of advanced digital engineering and digital manufacturing platforms relies on the availability of suitable digital infrastructures and the ability to effectively develop a *digital thread* within and across the digital factory value chain. DFA SD-RA relies on infrastructure federation and sovereignty as the main design principles for the development of the data-driven architecture. This pillar is responsible for capturing the different digital computing infrastructures that need to be resiliently networked and orchestrated to support the development of different levels and types of intelligence across the digital factory. In particular, the DFA SD-RA considers three main networking domains for big data service operation; i.e. factory, corporate and internet domain. Each of these domains needs to be equipped with a suitable security and safety level so that a seamless and cross-domain distributed and trustworthy computing continuum can be realized. The pilar considers from factory-level digital infrastructure deployment such as PLC, industrial PC or Fog/Edge to the deployment of telecom-managed infrastructure such as 5G multi-access edge computing platforms (MEP). At the corporate level, the reference architecture addresses the need for the development of IoT Hubs that are able to process continuous data streams as well as dedicated big data lake infrastructures, where batch processing and advanced analytic/learning services can be implemented. It is at this corporate level that private ledger

infrastructures are unveiled. Finally, at the internet or data centre level, the digital factory deploys advanced computing infrastructures exploiting HPC, Cloud or value chain ledger infrastructures that interact with the federated and shared data spaces.

The DFA RA is aligned with ISO 20547 Big Data Reference Architecture. The DFA Sovereign Digital Service Infrastructures pillar allows Boost 4.0 reference model to additionally address the ISO 20547 Big Data Framework Provider layer. The DFA RA is composed of four layers that address the implementation of the 6 big data "C" (Connection, Cloud/edge, Cyber, Context, Community, Customization), enables four different types of intelligence (smart asset functioning, reactive reasoning, deliberative reasoning and collaborative decision support) to be orchestrated and maps to the 6 layers of the RAMI 4.0 (product, devices, station, workcentre, enterprise and connected world), which target all relevant layers required for the implementation of AI-powered data-driven digital manufacturing processes:

1. The lower layer of the DFA RA contains the *field devices in the shopfloor*: machines, robots, conveyer belts as well as controllers, sensors and actuators are positioned. Also in this layer the smart product would be placed. This layer is responsible for supporting the development of different levels of autonomy and *smart product and device (asset) services* leveraging on intelligent automation and self-adaptive manufacturing asset capabilities.
2. The *workcell/production line* layer represents the individual production line or cell within a factory, which includes individual machines, robots, etc. It covers both the services, that can be grouped in two those that provide information about the process and the conditions (*IoT automation services*), and the actuation and control services (*automation control services*); and the infrastructure, typically represented in the form of PLC, industrial PCs, edge and fog computing systems or managed telecom infrastructures such as MEC. This layer is responsible for developing reactive (fast) reasoning capabilities (automated decision) in the SD-RA and leveraging augmented distributed intelligence capacities based on enhanced management of context and cyber-physical production collaboration.
3. At the *factory level*, a single factory is depicted, including all the work cells or production lines available for the complete production, as well as the factory-specific infrastructure. Three kinds of services are typically mapped in this layer: (1) *AI/ML training, analytics and data-driven services*; (2) *digital twin multi-layer planning services*; and (3) *simulation and visualization services*. The infrastructure that corresponds to this layer is the IoT Hubs, data lakes and AI and big data infrastructure. This layer is responsible for supporting the implementation of deliberative reasoning approaches in the digital factory with planning (analytical, predictive and prescriptive capabilities) and orchestration capabilities, which combine and optimize the use of analytical models (knowledge and physics based), machine learning (data-driven), high-fidelity simulation (complex physical model) and hybrid analytics (combining data-driven and model-based methods) under a unified computing framework. This

leverages in the architecture collaborative assisted intelligence for explainable AI-driven decision processes in the manufacturing environment.

4. The higher layer refers to the enterprise/ecosystem level, that encompasses all *enterprise and ecosystem* (connected world) services, platforms and infrastructures as well as interaction with third parties (value chains) and other factories. The global software systems that are common to all the factories (*collaboration business and operation services* as well as *engineering and planning services*) are supported usually by Cloud or HPC infrastructures. It is this layer that supports the implementation of shared data spaces and value-chain-level distributed ledger infrastructures for implementation of trusted information exchange and federated processing across *shared digital twins* and asset administration shells (AAS). This layer leverages a human-centric augmented visualization and interaction capability in the context of data-driven advanced decision support or generative manufacturing process engineering.

2.4 Mapping Boost 4.0 Large-Scale Trials to the Digital Factory Alliance (DFA) Service Development Reference Architecture (SD-RA)

This chapter aims to present two Boost 4.0 lighthouse trials that focus on the engineering and process planification services, using big data technologies and exploiting the digital twin capabilities to improve the overall production process (Fig. 4). Each section corresponds to one trial:

Section 2 describes the trial that was deployed in Volkswagen Autoeuropa Plant in Palmela (Portugal). This lighthouse factory has deployed a big-data-based solution to plan intra-logistic processes, which fully integrates the material flow from the unloading docks to the point of fit.

Section 3 introduces the Piacenza lighthouse trial, discussing how a business network can be developed in the high-end textile sector with the support of blockchain technology to guarantee traceability and visibility through the supply chain.

3 Big Data-Driven Intra-Logistics 4.0 Process Planning Powered by Simulation in Automotive: Volkswagen Autoeuropa Trial

Volkswagen Autoeuropa (VWAE) belongs to an automotive manufacturing industry located in Portugal (Palmela) since 1995 and is a production plant of Volkswagen Group. VWAE plays a strategic role in the Portuguese automotive industry, as it is the largest automotive manufacturing facility in the country and is responsible for

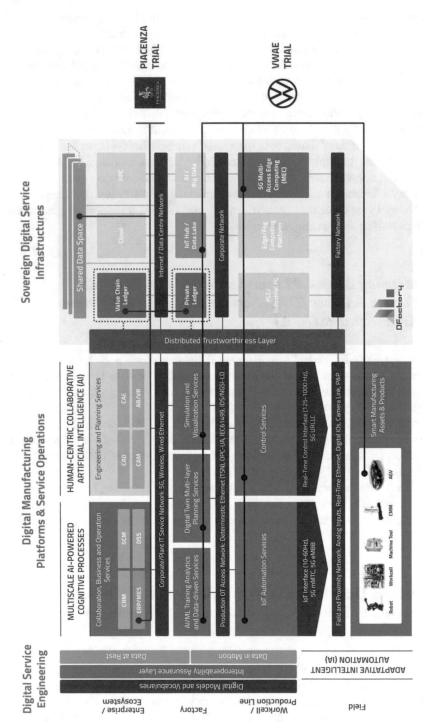

Fig. 4 Mapping of the two service engineering trials to the DFA SD-RA

around 10% of all Portuguese exportations. The plant employs around 6000 workers and, indirectly, it employs close to 8000 people through the more than 800 suppliers that provide materials, components and parts to the facility.

The goal of VWAE, within the Boost 4.0 project, is to take advantage of the latest big data technology developments and apply them to an industry environment with non-stop cycles and with high up-times. In the end, the desirable target is to transform an environment overwhelmed with manual complex processes with one that brings modular flexibility and automation.

The expected benefits with the implementation of a data-driven autonomous warehouse would translate into financial benefits for the Volkswagen Group, increase in flexibility (which is key specially during the introducing of a new model), minimization of human dependency for manual operations and, thus, an increase in the process efficiency. The automation and control of the process through a big data architecture enables a business intelligence approach to the warehouse system. Tools such as reporting, Digital Twin simulation, monitoring and optimization-support offer the opportunity to analyse and improve the system with real-world big data.

3.1 Big Data-Driven Intra-Logistic Planning and Commissioning 4.0 Process Challenges

Currently the logistics process is heavily reliable on manual processes and in addition to that, the operation is performed inside the factory, where space is limited. On the receiving area, trucks are traditionally unloaded by a manual forklift operation, and then the unit loads are transported to the warehouse where they will be stored either in shelves or block storage concept. System wise there is one database to control the parts coming from each truck and then a separate database which registers the unloading, transportation and storing of the material in the warehouse.

Figure 5 represents the data silos used throughout the process to collect the necessary logistics information. Besides the labour-intensive tasks within the logistics at VWAE, the data silo-based architecture does not allow the monitorization and optimization of the overall logistics process. Apart from the data silos for receiving, unloading, warehousing and sequencing, there is a lack of information about the transport operations between these phases in the process. Furthermore, data is captured and collected manually, which contributes to loss of time in the process and potentiates the existence of errors in the collected data.

Hence, the main challenge is to transform the siloed nature of data storage within the logistics process to support a true big data architecture, from which valuable insights can be extracted so as to optimize the whole logistics process and to aid in the optimization and automation efforts within the logistics process at VWAE. To achieve the transformation to a big data context, the integration of data present in the various silos is of the utmost importance. Such data integration efforts will enable the application of big data processing and analytics methods that will support

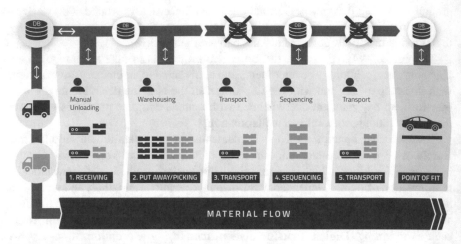

Fig. 5 Intra-logistic silo-based system flow of current process

the capitalization on valuable insights within the process. Moreover, the envisaged big data architecture will also form a basis for the development of a digital twin of the logistics area, which will enable real-world simulation, testing and validation of new automated solutions without the need for actual application in real-world, ready-for-production scenarios.

The planning and commissioning of advanced intra-logistics 4.0 processes therefore presents clear big data challenges in the velocity (real-time warehouse data streaming), veracity (accuracy of digital twin simulations), variety (breaking intralogistics information silos) and volume (data deluge) dimensions.

3.2 Big Data Intra-Logistic Planning and Commissioning Process Value

The expected future scenario aims at achieving a full integration of the material flow, from receiving up to the point of fit. Figure 6 shows the system flow integration as it is foreseen in VWAE. The main objective of the VWAE trial is to eliminate human intervention or at least reduce to a minimum at all phases from receiving up to the point of fit.

In order to test and validate the future scenario, a recurrent issue was chosen as a proof-of-concept: the issue of optimum stock in the logistics area. Due to the lack of data-supported, informed decisions in the process of supply ordering, the logistics area is often in a situation of overstock, meaning that there is always a surplus of parts that goes beyond the envisaged safety stock. The safety stock exists to tackle problems of parts' delivery, due to transportation issues or other obstacles, such as supplier shortfalls due to demand instability. Overstock has several consequences, from overspending and time-in-shelf issues to more concrete problems, such as part rejection due to its temporal validity.

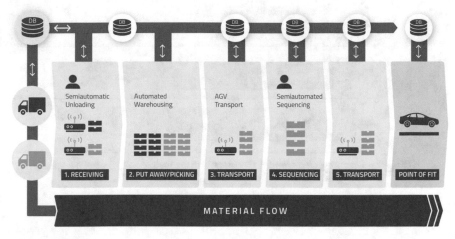

Fig. 6 System flow of data-driven intra-logistic 4.0 process

Hence, the chosen proof-of-concept was the overstock of batteries, since batteries are perishable parts (they have temporal validity) and the overstock situations for this type of part is a known problem and mitigating it represents real business value due to the unit price involved.

3.3 Big-Data Pipelines for Intra-Logistic Planning and Commissioning Solutions in Automotive

Figure 7 shows the general big data architecture and core open source big data technologies that support most of data ingestion, processing and management work, namely to efficiently gather, harmonize, store and apply analytic techniques to data generated within the intra-logistics process. The use of big data technologies with parallel and distributed capabilities is essential to address the processing of large batch/stream data with different levels of velocity, variety and veracity. Therefore, the architecture must meet requirements such as scalability, reliability and adaptability.

The architecture is mainly split into four layers: Data ingestion layer, Data Storage layer, Data Processing layer, and Data Querying/Analytics/Visualization layer. For data processing and collection, Apache Spark [8] is used in conjunction with the IDSA Connectors [9], enabling direct linkage with the IDSA Ecosystem, while for big data storage, the chosen technologies were PostgreSQL [10] and MongoDB [11]. Finally, for data querying and access, data analytics and data visualization, the chosen tools were, respectively, Apache Hive [12], Spark Machine Learning Library and Grafana [13].

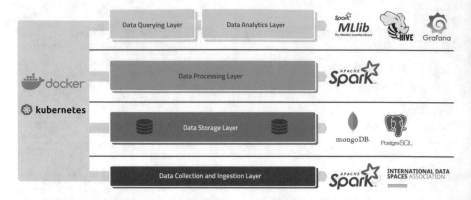

Fig. 7 Big data architecture for the VWAE trial

3.4 Large-Scale Trial of Big Data-Driven Intra-Logistic Planning and Commissioning Solutions for Automotive

The large-scale trial connects the Visual Components simulation environment with the suite of big data Analytics and Machine Learning tools, provided by UNINOVA, in a bidirectional way, as shown in Fig. 8: First, big data and machine learning technologies are used to aggregate real-time logistics operations' data, perform prediction over the data if needed, and send the results to the Visual Components simulation environment. Then, after the simulation ends, analytics and machine learning techniques are used in order to analyse key performance indicator data returned by the simulation environment, in order to find patterns, anomalies or possible points of optimization for future reference.

The 3D simulation environment replicates the trial scenario within the virtual world, i.e. a digital twin. The real model provides the logistics process data, which, after simulation, are validated with the current production outcomes. Once the simulation scenario is validated, simulation data can be analysed to be reused in the simulation to improve process performance and building the digital twin.

When the first version of the simulation, or digital twin, was developed, there was a need to propose actual key performance indicators (KPIs) extracted from the real logistics processes, focusing on the arrival and storage of batteries, with the simulation itself. Several KPIs were selected, such as the number of batteries, per battery type, in the warehouse and in the sequencing area at any given time, the occupation percentage of workers in the several logistics steps and the overall execution times of the different processes.

The first KPI to be validated in this phase was the reduction of truck arrivals, and consequent decrease of battery palettes in stock. The reduction in KPI corresponds to a decrease of 5% of the stock for the so-called high runners: the types of batteries that are most used in the production line. So, the test was performed as follows:

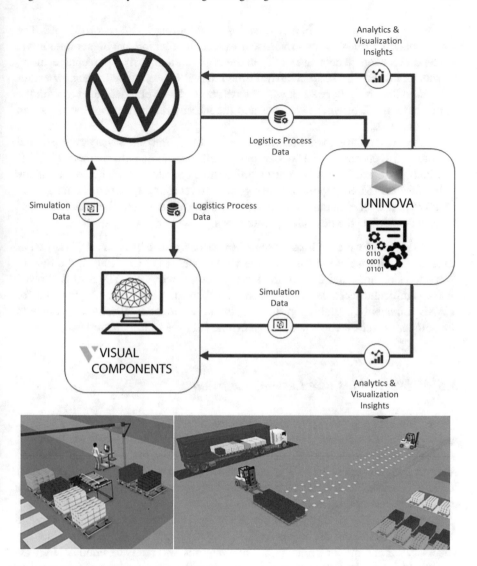

Fig. 8 VWAE Digital Twin Analytics trial data flow and planning platform

1. Real data corresponding to the truck arrivals, and to the usage of batteries in production was injected into the simulation, via Orion Context Broker. From this data injection, the Digital Twin produced a benchmark for the battery palettes' arrival percentages and truck arrival times.
2. The selected KPI was to decrease the arrival of high-runner palettes by 5%, while increasing the time interval between trucks, also saving in CO_2 emissions and direct costs for transport and stock, but maintaining the current production rates. The percentages of arriving palettes were arranged so that there would be a cut of

5% in the high runners while maintaining a total throughput/arrival of 100%. The time between truck arrivals was also arranged in order to have bigger intervals.

3. Hence, an average decrease of 5% in the high runners' arrival percentage, along with an increase in the truck arrival time interval was simulated. The new values showed that the battery stock was always above the level required, even with the decrease of 5% arrival of batteries and the increase in the time interval between truck arrivals.

4. Finally, a prediction model for the battery stock optimization was developed and tested. The chosen model was an optimized long short-term memory (LSTM) which is an artificial recurrent neural network model. This choice was made because LSTM are reportedly very good at forecasting time series data and do not require a lot of parameterization for multivariate datasets. Historical data was used to estimate the possible optimizations.

In 2018, there were multiple cases of overstock of car batteries at VWAE. For instance, in the case of the batteries, the warehouse was at least half of the time in overstock situations and 25% of the time in severe overstock. The results showed that a significant decrease in stock can be achieved, along with real benefits for VWAE, financially, by cutting in stock costs, and environmentally, by reducing both the number of truck arrivals and the occurrence of past-validity batteries.

3.5   Observations and Lessons Learned

The fusion between the big data architecture, developed in the Boost 4.0 project, and the Visual Components simulation environment, in order to create a true Digital Twin, was proven to be a crucial decision-support system, in the sense that it helped relevant stakeholders at VWAE to better understand the limitations in the current logistics process, but also to optimize critical aspects of this process, such as in the case of the overstock situation.

Furthermore, the overall system is ready for full scale-up, since it is capable of ingesting data from the whole logistics process, and for all the parts that are necessary for automotive production. The system is also ready to simulate, in near-real-world conditions, all phases of the logistics process, apart from the arrival of trucks. Hence, it will be a powerful aid in achieving the future automation requisites of VWAE, by enabling the simulation of new, automated and optimized scenarios for the logistics processes.

4 From Sheep to Shop Supply Chain Track and Trace in High-End Textile Sector: Piacenza Business Network Trial

Piacenza company is based in the Italian textile district of Biella, where all its production is carried out, and is one of the oldest textile industries in the world, founded in 1733 and from then on owned by the Piacenza family. Piacenza is one of the few undisputed worldwide leaders in high fashion fabrics and accessories production, with a competitive strategy focused on the maximum differentiation of the product, in terms of raw material choice, style, and colour. Fabric production includes more than 70 production passages or steps, which starts in the countries of origin of the natural fibres used for fashion fabrics (cashmere, vicuna, alpaca, mohair, silk, wool, linen, etc.) and can be summarized into three main changes of material status: raw material → yarn → fabric.

High textile fabric production is characterized by an extremely high number of product variables, deep customization, hardly predictable demand, length of production cycle (60–90 days from raw materials to receipt), physical prototyping and sampling, fragmented distribution and very small batches due to high customization. The combination of these aspects leads to a very complex production, which must properly balance the request of a very fast and demanding market with the length and rigidity of a fragmented and long value chain.

4.1 Data-Driven Textile Business Network Tracking and Tracing Challenges

The garment and footwear industry has one of the highest environmental footprints and risks for human health and society. The complexity and opacity of the value chain makes it difficult to identify where such impacts occur and to devise necessary targeted actions. Key actors in the industry have identified interoperable and scalable **traceability** and **transparency** of the value chain, as crucial enablers of more responsible production and consumption patterns, in support of Sustainable Development.

—United Nations Economic and Social Council [14].

Textile and clothing play a significant role in climate change with 1.7 million tons/year of CO_2 emissions [15], 10% of substances of potential concern to human health, 87% of the workforce (mainly women) gets below living wages. Permitted by lowered cost, a garment is worn an average of 3 times in its lifecycle, with 400 billion euros lost a year due to discarding clothes which can still be worn, 92 million tons of fashion waste every year, 87% of clothes ending up in landfills. In addition, the market for counterfeit clothing, textiles, footwear, handbags, cosmetics, amounted to a whopping $450 billion per year—and growing. The producers of these counterfeit goods, usually located in developing countries, do not adopt sustainable, circular and ethical models, and cause great harm to European companies that are seriously committed to implementing them.

On the contrary, fashion and luxury consumers are becoming more and more demanding with regard to sustainability of the products they are buying; 66% of consumers are ready to pay more for products or services from companies committed to sustainability [16]. But sustainability is only possible when supported by production traceability, which demonstrates how and where the manufacturing process is carried out. In addition, in recent years, duties have been increased as the most evident aspect of international commercial turbulences. Since they are calculated on the basis of the Preferential Certification of Origin (PCO), a proper traceability of production is becoming mandatory to simplify the export procedures and to address the increasing requirements of custom agencies.

4.2 Supply Chain Track and Trace Process Value

Traceability by blockchain technology provides all the information to support informed purchase decisions of consumers, favouring real sustainable products. We apply blockchain technology to build a shared tamper-proof ledger that tracks the fabric manufacturing from source to sales. Our *sheep to shop track and trace blockchain*-based solution records the transformation of raw materials into fabrics and enables verification of EU PCO.

The expected impact is providing a complete and controlled set of information to support the efforts of the Piacenza company in the field of sustainability, environmental protection and ethical respect. The proposed solution leverages the competitive positioning of Piacenza and its customers, by providing final consumers with full provenance of items and documents. In addition, blockchain enables the full visibility of textile manufacturing by a safe and not modifiable process, which prevents the market from being affected by counterfeiting and unfair competition.

4.3 Distributed Ledger Implementation for Supply Chain Visibility

The blockchain solution implemented in the Piacenza trial records all steps and documents in the production process in a general way, storing documents hash on the ledger (on-chain) and a reference to their physical location while assuring their authenticity.

Figure 9 illustrates the main components of the supply chain visibility solution. Real data flows from Piacenza's ERP system through a wrapper so data can be written to the blockchain ledger through a RESTful API. The wrapper extracts the data from the ERP system in JSON format that matches the blockchain data model.

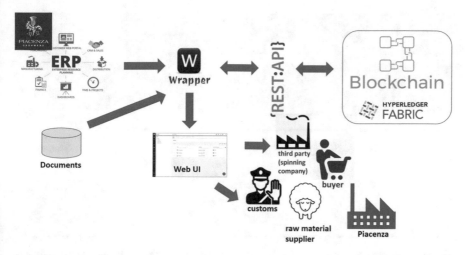

Fig. 9 High-level overview of the solution

The wrapper also feeds a dedicated Web UI whose role is to show the provenance of a specific selected item along with the corresponding documents as follows: The UI feeds the data from the API wrapper. The wrapper has a recursive function to retrieve every element in the chain. The recursive function calls the blockchain API to retrieve the information. The same information stored in the blockchain is then displayed in the UI. The PCO and other tracked documents information is displayed in the UI with a link to download the document. In other words, for a selected tracked item it graphically depicts its provenance and the (validated) documents stored on the ledger. This web UI can serve all participants in the network to trace a specific item and to check for specific documents (e.g., customs asking for a specific PCO).

Figure 10 shows the main modules for our sheep to shop blockchain application:

- **Blockchain infrastructure**: We apply Hyperledger Fabric [17] components. Our data model consists of two primary entities: *trackedItem* and *document*.
- **Smart contracts layer**: Smart contracts (chaincodes in Fabric) embed the business logic of the solution. Smart contracts functions are accessed through the Hyperledger Fabric Client (HFC) Software Development Kit (SDK) in Node.js.

  - Query functions enable accessing and fetching information stored in the ledger, including trackedItems and documents.
  - Invoke functions include the possibility of creating trackedItems and documents, and connecting a new document to an existing trackedItem.
  - Administration functions enable the management of the channels implemented as well as basic functions such as enrolment and registration.

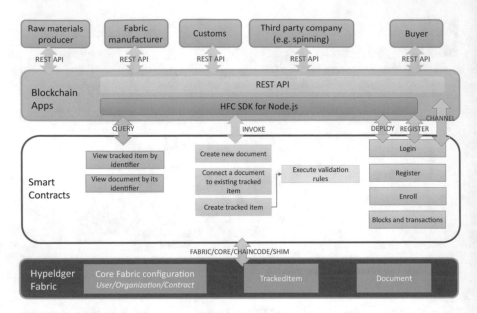

Fig. 10 Blockchain solution high-level design

- **Blockchain apps**: HFC SDK allows developing a blockchain client application which can use the SDK to invoke smart contract functions. This client can serve as a middle layer between frontend applications and the backend blockchain platform by providing RESTful APIs to be used by frontend applications.

4.4 Observations and Lessons Learned

Our achievements include the blockchain backend (released to open source under Apache v2 license [18]) and a (private) repository containing the dedicated code developed for extracting data from Piacenza's ERP system and enabling the display of the provenance of items in the chain along with scripts, data and documents. The trial has emulated a complete blockchain business network. Obviously, the most natural way of extending and exploiting the results of the trial is by gradually incorporating Piacenza partners to the business network (e.g. customs and buyers) through the APIs provided. The provided blockchain backend is generic so this can be done in a straightforward manner.

The more challenging part is, therefore, not the technical but the business one, by defining a business model of onboarding, how to manage the network and how to monetize the savings and costs of participating and managing such a network. There are compelling evidences that show a great potential for this first trial solution to be extended to a full production environment for a full transparent and trackable solution towards a sustainable textile supply chain.

5 Conclusions

This chapter has introduced two large-scale trials that have been implemented in the context of the lighthouse project Boost 4.0. The Chapter has introduced the Boost 4.0 Reference Model, which adapts the more generic BDVA big data reference architectures to the needs of Industry 4.0. The Boost 4.0 reference model includes, on one hand, a reference architecture for design and implementation of advanced big data pipelines and, on the other hand, the digital factory service development reference architecture. Thus, Boost 4.0 can fully address ISO 20547 for Industry 4.0.

This chapter has demonstrated that the BDVA big data reference architecture can indeed be adapted to the needs of the Industry 4.0 and aligned with an overall digital factory reference architecture, where big data-driven processes will have to extend advanced manufacturing processes such as smart engineering, smart planning and commissioning, smart workplaces and operations, smart connected production and smart maintenance and customer services. Such digital factory service development architecture can indeed host and accommodate the needs of advanced big data-driven engineering services.

The chapter has demonstrated that both intra-logistic process planning and connected supply chain track and tracing can achieve significant gains and extract significant value from the deployment of big data-driven technologies. The evolution from traditional data analytic architecture into big data architectures will enable increased automation in simulation and process optimization. The combination of Industry 4.0 data models such as OPC-UA, AML and IoT open APIs such as FIWARE NGSI allows for dynamic and real-time optimization of intra-logistic processes compared to off-the-shelve commercial solutions. Moreover, big data architectures allow a higher granularity and larger simulation scenario assessment for a high-fidelity intra-logistic process commissioning.

The use of open-source big data technology suffices to meet the challenge of very demanding big data processes in terms of variety, velocity and volume as the VWAE trial has demonstrated. This trial has also shown that digital twin operations can be greatly improved if supported by advanced big data streaming technologies, and the use of shared data spaces demonstrates the suitability of such technologies to break information silos and increase efficiency and scale up intralogistics processes. This chapter has also shown that distributed ledger technology can be seamlessly integrated with distributed data spaces and support business network traceability and visibility in the high-end textile sector (Piacenza trial). The chapter has also provided evidence on how the extensive use of open technologies, APIs and international standards can greatly support the large-scale adoption and uptake of big data technologies across large ecosystems. The chapter has provided compelling evidences that big data can greatly improve performance of Industry 4.0 engineering services, particularly when development and exploitation of digital threads and digital twins come into operation. The interested reader is also referred and invited to browse the content in chapters "Next Generation Big Data

Driven Factory 4.0 Operations and Optimisation: The Boost 4.0 Experience" and "Model Based Engineering and Semantic Interoperability for Trusted Digital Twins Big Data Connection Across the Product Life Cycle" focused on the Boost 4.0 lighthouse project; these chapters discuss how further trials have incorporated big data technologies as part of the business processes for increased competitiveness.

This research is opening the ground for implementation of more intelligent, i.e. cognitive and autonomous, intra-logistic processes. As the diversity of parts considered and the autonomy in decision process increase, further research is needed in terms of development of sovereign and large-scale distributed data spaces that can provide access to the necessary data for AI model training beyond pure data analytics and digital twin simulation. The Boost 4.0 big data framework calls for further research on the development of federated learning models that can combine highly tailored models matching and optimized to the specificities of the factory layout with more general models that can be shared and work at higher levels of abstractions; thus, speed and long-term planning can be combined in new forms of autonomous shopfloor and supply chain operations.

Acknowledgements This research work has been performed in the framework of the BOOST 4.0 Big Data Lighthouse initiative, a project that has received funding from the European Union's Horizon 2020 research and innovation programme under grant agreement No. 780732. This data-driven digital transformation research is also endorsed by the Digital Factory Alliance (DFA) www.digitalfactoryalliance.eu

References

1. Boost 4.0 https://www.boost40.eu
2. Zillner, S., Curry, E., Metzger, A., Auer, S., & Seidl, R. (Eds.). (2017). *European Big Data Value Strategic Research & Innovation Agenda*. Big Data Value Association.
3. Zillner, S., Bisset, D., Milano, M., Curry, E., García Robles, A., Hahn, T., Irgens, M., Lafrenz, R., Liepert, B., O'Sullivan, B., & Smeulders, A. (Eds.) (2020) *Strategic Research, Innovation and Deployment Agenda - AI, Data and Robotics Partnership. Third Release*. September 2020, Brussels. BDVA, euRobotics, ELLIS, EurAI and CLAIRE.
4. Big Data Value Association. (2017). *Strategic Research and Innovation Agenda v.4.0*. Available at: https://www.bdva.eu/sites/default/files/BDVA_SRIA_v4_Ed1.1.pdf
5. VDI/VDE-Gesellschaft Mess- und Automatisierungstechnik. (2015). *Status Report: Reference Architecture Model Industrie 4.0 (RAMI4.0) (Vol. 0)*. https://www.zvei.org/Downloads/Automation/5305. Publikation GMA Status Report ZVEI Reference Architecture Model.pdf.
6. Digital Factory Alliance (DFA). https://digitalfactoryalliance.eu/
7. *D2.5 boost 4.0 reference architecture specification*. Available at: https://boost40.eu/wp-content/uploads/2020/11/D2.5.pdf
8. International Organization for Standardization. (2011). *ISO/IEC/IEEE 42010. System and Software Engineering – Architecture Description*. Available at: https://www.iso.org/standard/50508.html
9. *Apache Spark*. https://spark.apache.org/
10. International Data Spaces Association. (2019). *Reference architecture model*. Online. Available at: https://www.internationaldataspaces.org/wp-content/uploads/2019/03/IDS-Reference-Architecture-Model-3.0.pdf
11. *PostgreSQL*. https://www.postgresql.org/

12. *MongoDB*. https://www.mongodb.com/
13. *Apache Hive*. https://hive.apache.org/
14. *Grafana*. https://grafana.com/
15. Center for Trade Facilitation and Electronic Business, Economic Commission for Europe, UN. (2019). *Briefing note on sustainable textile value chains in the garment and footwear domain For SDG12*. [online] Geneva. Available at: https://undocs.org/pdf?symbol=en/ECE/TRADE/C/CEFACT/2019/26
16. Center for Trade Facilitation and Electronic Business, Economic Commission for Europe, UN. (2019). *Report of the Centre for Trade Facilitation and Electronic Business on its twenty-fifth session*. [online] Geneva. Available at: https://www.unece.org/fileadmin/DAM/cefact/cf_plenary/2019_plenary/ECE_TRADE_C_CEFACT_2019_02E_Report.pdf
17. NIELSEN. (2015). *The sustainability imperative. New insights on consumer expectations*. Online. Available at: https://www.nielsen.com/wp-content/uploads/sites/3/2019/04/global-sustainability-report-oct-2015.pdf
18. *Hyperledger Fabric*. https://www.hyperledger.org/use/fabric
19. Available at: https://gitlab.com/boost4-piacenza/public-artifacts

Model-Based Engineering and Semantic Interoperability for Trusted Digital Twins Big Data Connection Across the Product Lifecycle

Oscar Lázaro, Jesús Alonso, Roxana-Maria Holom, Katharina Rafetseder, Stefanie Kritzinger, Fernando Ubis, Gerald Fritz, Alois Wiesinger, Harald Sehrschön, Jimmy Nguyen, Tomasz Luniewski, Wojciech Zietak, Jerome Clavel, Roberto Perez, Marlene Hildebrand, Dimitris Kiritsis, Hugues-Arthur Garious, Silvia de la Maza, Antonio Ventura-Traveset, Juanjo Hierro, Gernot Boege, and Ulrich Ahle

Abstract With the rising complexity of modern products and a trend from single products to Systems of Systems (SoS) where the produced system consists of multiple subsystems and the integration of multiple domains is a mandatory step, new approaches for development are demanded. This chapter explores how Model-Based Systems Engineering (MBSE) can benefit from big data technologies to

O. Lázaro (✉) · J. Alonso
Asociación de Empresas Tecnológicas Innovalia, Derio, Spain
e-mail: olazaro@innovalia.org; jalonso@innovalia.org

R.-M. Holom · K. Rafetseder · S. Kritzinger
RISC Software GmbH, Hagenberg im Mühlkreis, Austria
e-mail: roxana.holom@risc-software.at; katharina.rafetseder@risc-software.at;
stefanie.kritzinger@risc-software.at

F. Ubis
Visual Components Ov, Espoo, Finland
e-mail: fernando.ubis@visualcomponents.com

G. Fritz
TTTech Industrial Automation AG, Wien, Austria
e-mail: gerald.fritz@tttech-industrial.com

A. Wiesinger · H. Sehrschön
Fill Gesellschaft m.b.H., Gurten, Oberösterreich, Austria
e-mail: alois.wiesinger@fill.co.at; harald.sehrschoen@fill.co.at

J. Nguyen · T. Luniewski · W. Zietak
CAPVIDIA, Houston, TX, USA
e-mail: jmmy@capvidia.com; tl@capvidia.com; wz@capvidia.com

J. Clavel · R. Perez
Agie Charmilles New Technologies, Meyrin, Switzerland
e-mail: Jerome.clavel@georgfischer.com; roberto.perez@georgfischer.com

implement smarter engineering processes. The chapter presents the Boost 4.0 Testbed that demonstrates how digital twin continuity and digital thread can be realized from service engineering, production, product performance, to behavior monitoring. The Boost 4.0 testbed demonstrates the technical feasibility of an interconnected operation of digital twin design, ZDM subtractive manufacturing, IoT product monitoring, and spare part 3D printing services. It shows how the IDSA reference model for data sovereignty, blockchain technologies, and FIWARE open-source technology can be jointly used for breaking silos, providing a seamless and controlled exchange of data across digital twins based on open international standards (ProStep, QIF), allowing companies to dramatically improve cost, quality, timeliness, and business results.

Keywords Interoperability · Semantic data model chains · Industry commons · Model based design · IDSA · QIF · FIWARE · Pro-STEP · Digital twin · Digital thread · Testbed · Trial · Maintenance 4.0 · Metrology 4.0 · ZDM

1 Introduction

With a rising complexity of modern products and a trend from single products to Systems of Systems (SoS) where the produced system consists of multiple subsystems and the integration of multiple domains is a mandatory step, new approaches for development are demanded. One of these approaches is Systems Engineering (SE).

Systems Engineering is a transdisciplinary and integrative approach to enable the successful realization, use, and retirement of engineered systems, using systems principles and concepts, and scientific, technological, and management methods. *(INCOSE)* [1]

M. Hildebrand · D. Kiritsis
Ecole Polytechnique Federale de Lausanne, Lausanne, Switzerland
e-mail: marlene.hildebrand@epfl.ch; Dimitris.kiritsis@epfl.ch

H.-A. Garious
ESI Group, Paris, France
e-mail: hugues-arthur.garious@esi-group.com

S. de la Maza
Trimek S.A, Araba, Spain
e-mail: smaza@trimek.com

A. Ventura-Traveset
Innovalia Metrology,Álava, Spain
e-mail: toni.ventura@datapixel.com

J. Hierro · G. Boege · U. Ahle
FIWARE Foundation EV, Berlin, Germany
e-mail: juanjose.hierro@fiware.org; gernot.boege@fiware.org; ulrich.ahle@fiware.org

To tackle this challenge Model-Based Systems Engineering (MBSE) has been introduced.

> Model-based systems engineering (MBSE) is the formalized application of modeling to support system requirements, design, analysis, verification and validation activities beginning in the conceptual design phase and continuing throughout development and later life cycle phases. *(INCOSE Technical Operations 2007)* [2]

With the advent of Model-Based Definition (MBD) and Model-Based Engineering, the 3D CAD model carries details for both human and machine interpretation, taking legacy 2D drawings and practices and updating them to twenty-first century evolutions leading to automation, AI, and improved products and cost savings. In general, parametric modelling and optimization techniques supported by MBSE methods contribute significantly to the process of building CAD simulations. Several design parameters and probably density factors are taken into consideration for simulation sequence. These simulations are very important in analyzing different factors such as sensitivity, optimization, and correlation of the design or structure. Moreover, the rapid growth of the Internet and wireless technology has led to an influx of raw, unlimited data. Companies that are able to collect, analyze, and execute upon internal data have become cultural and business revolutions, such as Google, Facebook, and Amazon. As more industries refine their data, breakthroughs in artificial intelligence, automation, Internet of Things, and predictive analytics begin to showcase the influence of big data—especially the ability to connect different data sets and provide actionables that can impact the bottom line and society. Nearly a quarter of the way into the twenty-first century, while many new industries are creating digital transformation and older industries embracing it, today's manufacturing enterprise is still stuck with last century's practices and mindset, especially when it comes to data that is disconnected and disorganized. Though terms like Industry 4.0, Industrial Internet of Things, and Model-Based Enterprise outline a fundamental need for digital transformation and a basic understanding of its importance—for manufacturing, it is still more theory than practice. And nothing highlights this better than the different data file formats used from design to manufacturing. Additionally, many practical problems usually have several conflicting objectives that need optimization.

Boost 4.0 [3] is the European lighthouse project that has trialed at large scale over 13 industrial leading factories, 40 business processes, and 7 manufacturing sectors a unified standardized big data reference architecture, highly replicable advanced big data solutions, and sovereign industrial data spaces for Industry 4.0. As the implementation of Boost 4.0 large-scale trials have evidenced, the *real big data challenge for Factory 4.0* does not lie just in the actual "storage of data or exchange of assets across digital platforms" but primarily on the speed, transparency, and trustfulness in which highly heterogeneous and multi-domain interoperable data networks can be established and accessed, as well as the real-time synchronization of such data networks across the many cross-sectorial big data lifecycles. In other words, the ability to effectively support the implementation of cross-sectorial data value chains along the product lifecycle and across connected factories; i.e., *seamless "digital*

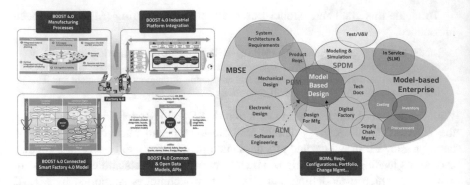

Fig. 1 Model-Based Systems Engineering (MBSE) BOOST 4.0 pillars for digital thread and connected digital factories 4.0

threads" among connected designers, connected suppliers, connected machines, connected boardrooms, connected products, and connected customers. To ensure a unified approach and high replicability, within Boost 4.0 a number of pilots have applied the model-based engineering paradigm, thereby enhancing the capability for multi-objective optimization introducing machine learning and lightweight deep learning architectures to address this issue taking into account significant production features.

BOOST 4.0 is not just a project dealing with factories implementing isolated big data processes. In fact, over the last few years, many Factories of the Future (FoF) projects have already shown that Industrial Internet and big data can bring clear business value to isolated factory operations. However, the connected smart Factory 4.0 is a paradigm shift towards optimizing how data and information are leveraged across new value chains (becoming more integrated and more complex) with interoperable digital manufacturing platforms as central to its vision. The connected smart factory 4.0 pillars (see Fig. 1) *integrate* digital platforms and industrial things and foster *collaboration* across factories and workforce. Factories 4.0 industrial platform horizontal and vertical integration leads to E2E real-time business planning with support of extended data availability and big data analytics for *real-time production scheduling*, dynamic *real-time inventory management* based on demand sensing, and production *quality and maintenance automation and optimization*. Such competitive advantages for European factories can only be made possible through *industrial data model convergence* at many levels, i.e., OT, IT, ET, and IoT and a *core capability in Industry 4.0 frameworks for big data and data analytics*.

This chapter relates mainly to the technical priority Data Management Engineering of the European Big Data Value Strategic Research & Innovation Agenda [4]. It addresses the horizontal concerns of semantic annotation, semantic interoperability, and data lifecycle management of the BDV Technical Reference Model. It addresses the vertical concerns of standardization of big data technology areas to facilitate data integration, sharing, and interoperability. The work in this chapter relates mainly but not only to the Knowledge and Learning cross-sectorial technology enablers of the AI, Data and Robotics Strategic Research, Innovation & Deployment Agenda [5].

This chapter will initially present in Sect. 2 the Boost 4.0 approach to big data-driven smart digital model-based engineering. This section will also address how such an approach has been realized in the Internet of Things Solutions World Congress (IoTWC) [6] testbed. This testbed has been built to highlight the feasibility of integrating model-based engineering methods with big data technologies and Boost 4.0 European industrial data space technology to leverage digital twin and digital thread continuity. Thus, the benefits of model-based engineering to improve the interoperability and data sharing capabilities for trusted digital twin's big data connection across the product lifecycle can be materialized in a concrete workflow and product, in this case a Mars rover.

This demonstrator has then inspired and motivated a number of large-scale pilots that capitalize on the digital thread continuity technologies at scale. The trials showcase the impact of enhanced engineering practices in the machine tool sector. Section 3, presents the large-scale big data trial conducted by FILL GmbH machine tool builder, addressing next-generation machine-tool engineering and digital service provisioning. Section 4, presents an overview of the large-scale pilot conducted by George Fischer's Smart Zero-Defect Factory, focusing on data-driven production improvement for milling machine spindle component manufacturing. Finally, Section 5, presents Trimek large-scale big data pilot for zero defect manufacturing powered by massive metrology that showcase how new big data technologies can significantly increase the capacity for 3D quality control of very large pieces and components in automotive through model-based design and intensive use of QIF (Quality Information Framework), as an open standard developed to enrich CAD models with additional process-related information.

2 Boost 4.0 Testbed for Digital Twin Data Continuity Across the Product Lifecycle

The aim of this section is to introduce the Boost 4.0 approach to big data-driven model-based engineering and the testbed built to demonstrate the feasibility of digital twin and digital thread implementation across design, production, and product operation lifecycle. There is a special bond between the digital twin and the physical world it represents. The digital twin has largely been a PLM concept for design and performance simulation of discrete products. Now, new kinds of digital twins are available to support and improve specific manufacturing plant production processes through Cyber Physical Systems (CPS) and obtain a better understanding of the product performance in operation through IoT. Each of these various kinds of digital twins have been developed as siloed solutions, each dealing with different manufacturing processes across the product lifecycle. The data exchange among digital twins breaking these silos opens manufacturers

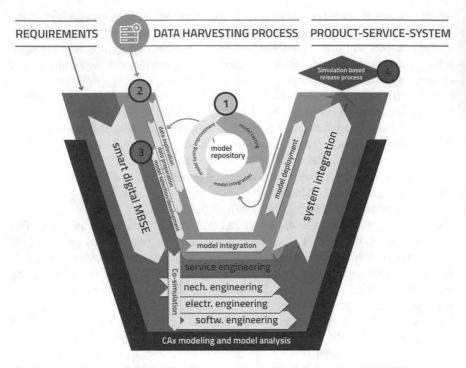

Fig. 2 Boost 4.0 smart digital engineering process using big data (based on VDI 2206)

the door to unprecedented insights, visibility, and automation opportunities leading to efficiency improvements in product design, product performance, behavior and manufacturing process operations like never before.

The approach implemented by Boost 4.0 to leverage data continuity across product lifecycle has been twofold. On one hand, the adoption of an agile V development model (based on VDI 2206) enhanced with big data (Fig. 2). A metadata representation approach has been integrated to define the structure and the relations (i.e., the connections) between the various data sources across the full process lifecycle. The Boost 4.0 smart digital engineering process is interacting with the model-based engineering process, a model repository (1) for trusted digital twins using big data, (2) for better service design, (3) and a simulation-based release process (4) to create product-service-systems (PSS) across the lifecycle.

On the other hand, to support the interconnection of metadata representation across the full lifecycle, the Boost 4.0 approach has been the extension, adoption, and demonstration of ProSTEP chain of Industry 4.0 standards with QIF capabilities (Fig. 3). This would ensure the semantic interoperability across not only product and process design/engineering but also quality control and production system commissioning and optimization.

As illustrated in Fig. 4, the Boost 4.0 testbed demonstrates how ProStep [7] model-based engineering approach, the QIF semantic framework, IDSA data space

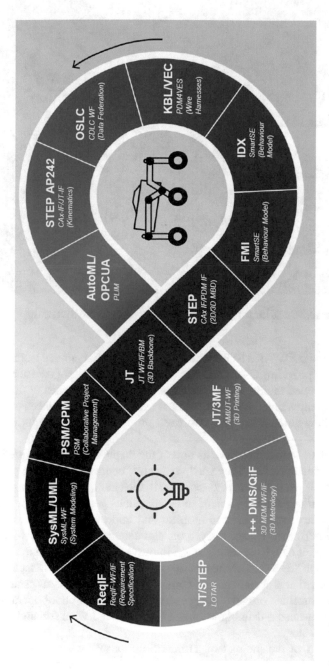

Fig. 3 Boost 4.0 ProSTEP interoperable semantic data model standards chain

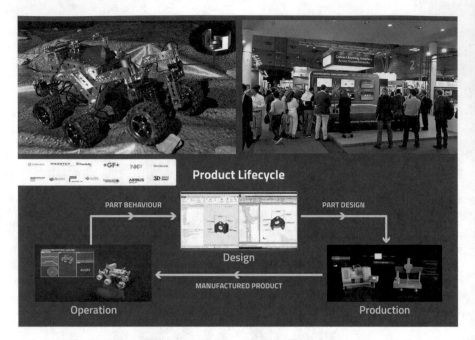

Fig. 4 A Mars rover vehicle trusted digital twin continuity testbed. IoTSWC 2019

[8] technology, and FIWARE NGSI Context Broker [9] open-source technology can be integrated for providing a seamless and controlled exchange of data across digital twins based on open international standards, allowing companies to dramatically improve cost, quality, timeliness, and business results through enhanced traceability, process workflow automation, and improved product and manufacturing process knowledge.

Machines chained to a shop floor as part of the manufacturing setup are typically working as information silos. They are "physically" connected since the part treated by one machine is passed to the next machine in the chain, which in turns treats this part and passes it to the next. Each of those machines generates a large amount of data, which so far has been used to monitor and improve the processes and tasks each machine performs. However, systems associated with each machine are not designed to exploit data from others when improvements can be gained if the data from one machine "feeds" the systems connected to the other and if such exchange is made in a way that is secure: access control terms and conditions established by each individual machine provider are preserved and the shop floor operator is also the final decision-maker, defining what is exchanged and what for, and whether it goes out of the factory.

In the context of the Internet of Things Solutions World Congress that took place in Barcelona in 2019, a group of companies that partner in Boost 4.0 (Capvidia, EPFL, FIWARE Foundation, +GF+, IDSA, Trimek, and Innovalia), in collaboration with key ProStep IVIP partners, presented a testbed that demonstrated

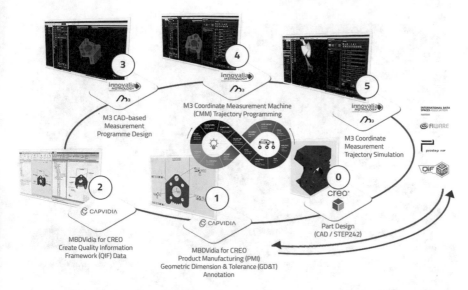

Fig. 5 The Mars rover big data-driven model-based engineering process and federated digital manufacturing platforms chain (PTC, CAPVIDIA, INNOVALIA METROLOGY)

how factories can benefit from IDS concepts and FIWARE open-source technology by bringing enhanced functionalities for the improvement of processes or the support of smart decisions, through management of context data shared across the product lifecycle.

This testbed demonstrator uses the example of a specific component of the Mars rover exploration vehicle to visualize the benefits of how companies can collaborate throughout the product lifecycle. As illustrated in Fig. 5, thanks to the combined exploitation of model-based engineering (MBE supporting standardized open PLM STEP standards [10] and Quality Information Framework (QIF) semantics [11]), *digital threads* and *digital twin* technologies based on *European industrial data space trusted connector technology* that allows product and process information sharing in an environment of trust.

The testbed is focused on the production of a specific component that is a very sensitive piece in the suspension of the Mars rover. In this case, it is manufactured by a +GF+ milling machine. As shown in Fig. 6, this milling machine is connected, through an IDS connector, to a predictive maintenance system deployed at +GF+ cloud systems, so information about the status of the milling spindle is constantly sent to the predictive maintenance system to be analyzed, and maintenance tasks are programmed to avoid breakdowns that force to stop the production.

After the piece is produced, an Innovalia Metrology coordinate measuring machine (CMM) takes care of the dimensional quality control. This machine measures millions of points in a very short period of time, producing what is called a point cloud (a high-resolution, high-fidelity, micron-resolution digital replica of

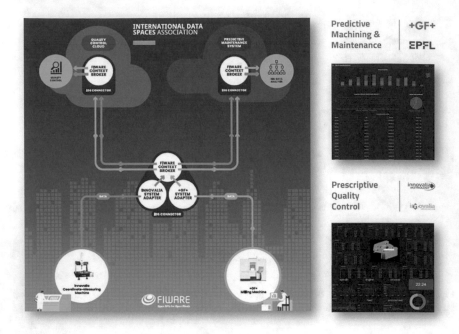

Fig. 6 The Mars rover digital manufacturing excellence threads and integrated decision support dashboards

the physical part). This information is sent to Innovalia's quality control cloud (M3 Workspace) through trusted IDS connector technology, and there it is compared with the 3D CAD model to identify deviations. The result of this analysis is a 3D model with color mapping, which allows the operator to easily find out if these deviations are within the allowed range.

The IDS connector in this setup is the same for both machines. Prior to use the IDS connector, each machine adapter had to be deployed across all machines that have to engaged in secure data sharing to enable trusted communication and data exchange among the machines. Thanks to the privacy-by-design defined by the IDS Reference Architecture, the communication between the machines and their respective cloud systems is isolated and independent.

As part of the demonstration, and to showcase the benefits of using the IDS connector to share information in a trusted and sovereign way, Innovalia can configure the IDS connector to allow +GF+ predictive maintenance systems to gather data from Innovalia's CMM. This would allow the system to enrich their algorithms to include the deviation's information in the analysis, so the system can cross this information with the machine status, improving the predictive maintenance system. In a similar way, +GF+ can allow Innovalia to request specific information from the milling machine to add new functions to the quality management system. The overall schema of the use case is depicted in Fig. 4.

3   FILL GmbH Model-Based Machine Tool Engineering and Big Data-Driven Cybernetics Large-Scale Trial

The expansion of competitiveness and sustainability are fundamental goals of companies. To achieve these, every business needs to deploy digital technologies in a variety of areas, including customer relationships and services, productivity, business model, IT security, and privacy. Digitization and networking are playing an increasingly important role, as the digital data volume will increase significantly.

The growth speed of the data volume, the diversity of data, and the various data sources pose many challenges, such as a collection of sensor data and the mapping of model data underlying the machine, and their integration and interpretation into a structured database system.

The FILL trial builds on the Boost 4.0 testbed concepts and technologies described in Sect. 2, i.e., advanced model-based engineering coupled with product digital thread and digital twin implementations applied to machine tool products. The main achievement of the trial is to demonstrate that such an approach (big data-driven V-model engineering and semantic data model chains) can be applied under real production scenarios, with clear benefits and that the approach can scale to effectively deal with the complexity of machine tool production lines. Semantic information is integrated across the full lifecycle from machine operation to machine design and manufacturing.

Within the Fill trial, it is possible to record the data of their machines standardized by OPC UA data model [12], including semantic information and metadata to be used in analyses and optimizations. Utilizing standardized communication technology (e.g., MQTT, HTTP, REST) the existing specific solution Machine-Work-Flow-Framework is generalized and used for further customer requests. In doing so, Fill, and the pilot partners took a big step forward in the digitization of the data flow on the shopfloor with the expanded machine state model. The Fill pilot pursues the following goals:

1. Cost reduction expected by reducing the time spent on future development and customer projects.
2. Development of data-driven business models in service and support.
3. Identification of optimization potentials in the engineering process for long-term reduction in the development times of machines.

The Fill pilot primarily serves the engineering process of the machine builder. It allows for a better understanding of machinery by detecting cause-and-effect relationships due to anomalies and patterns (semantic interoperability). In addition, maintenance intervals and cycles are optimized and, as a result, quality improvements of the production and the product are achieved.

3.1 Big Data-Driven Model-Based Machine Tool Engineering Business Value

Within the CAx systems, the optimization of products, e.g., machine tools, are done by engineer's expertise and their empirical knowledge. Several loops are performed to optimize products and production processes to fit customer needs. Thus, customization is resource and time-consuming. During the sales and project-planning phase in many cases, no simulations of the process or valid process data are available for frontloading to minimize the risk for the machine builder.

During the engineering process, the start of the project is when the order is placed. Several simulation models in different software tools are generated to avoid failures in the early project phases. The models exist mostly independently and have to be modified by hand if there is a change in the requirements.

In general, parametric modelling and optimization techniques contribute significantly to the process of building CAx simulations. Several design parameters and probability density factors are taken into consideration for simulation sequences. The simulations are most important for analyzing different factors such as sensitivity, optimization, and correlation of the design or structure. Many practical problems usually have several conflicting objectives that need optimization. In these multi-criteria optimization problems, a solution is found iteratively and systematically.

The production concept and the machines engineering solve the multi-criteria equipment effectiveness optimisation function. The machine physical behavior sets the quality of the part and the production time. The material flow concept implemented solves the interaction of production steps and provides the overall logistic concept. These concepts are most important for overall equipment effectiveness.

3.2 Implementation of Big Data-Driven Machine Tool Cybernetics

The V-model (Fig. 2; compare VDI 2206 [13]) describes the development of mechatronic systems based on a systematic analysis of the requirements and a distribution of the requirements and loads among the individual disciplines. The detailed development then takes place in parallel and independently of each other in the individual disciplines. The results are then integrated into subsystems and systems and validated regarding compliance with the requirements. The new proposal for an integrated business process for smart digital engineering using big data extends the V-model by (Fig. 7):

1. Agile model management and development process (model repository)
2. Data analytics process (involving data analysis and machine learning methods)
3. Service development process
4. Simulation-based release process for product service systems

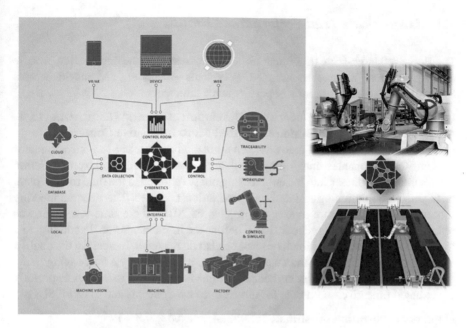

Fig. 7 The Fill big data cybernetics trial and semantic data integration based on Industry 4.0 standards. Digital twin and real line agile production engineering setups

Within the Fill pilot, TTTech provides the edge computing platform that provides real-time data harvesting services and enables the data-driven model approach. The resulting Product Service System (PSS) applications enhance shopfloor functionality utilizing machine data with significantly decreased latency and increased interoperability. To exploit the full potential of big data (and Industry 4.0), companies in the value-added network are willing to cooperate and share data. Openness and trust are crucial factors since the long-term strategy aims to create partner ecosystems, where different productions of different companies in the value-added network are connected, and individual processes across the companies are coordinated. The availability of process data across companies opens up great potential for agile production systems.

The main tasks of RISC in the Fill trial are focusing on the selection of the appropriate machine learning and data analysis methods that are suitable for very large data and that have the potential for parallel implementation (step (2) from above). An architectural concept [14] that combines big data technologies (such as Apache Spark [15] on Hadoop) with semantic approaches (Apache Avro [16] and SALAD [17]) has been defined to facilitate the exploration and analysis of large volumes of data from heterogenous sources, adhering to the Boost 4.0 unified big data pipeline and service mode proposed in the chapter *"Big Data Driven Industry 4.0 Service Engineering Large Scale Trials: The Boost 4.0 Experience"* of this book.

3.3 Large-Scale Trial Performance Results

The Boost 4.0 large-scale trial impact has been assessed in the engineering process on Fill products, accurate robotic NDT systems Accubot®, and machine tools Syncromill ®, addressing four key performance indicators.

Long-term Reduction of Machine Development Times (Estimated: 15%, Achieved: 26.4%) or the Option for Better, More Innovative Concepts
Reducing time-to-market of innovative customized products is a key success factor for industrial companies. Integrating big data feedback information from operation and maintenance phases into the engineering phases shortened the time for real plant or factory commissioning in lot-size-1 production facilities. The new engineering process is shorter in time: in several projects of 2019/20 the time-to-market was reduced by 26.4% compared to 2015/16 before the implementation. In addition to a reduction in time, optimization loops can be also dedicated to increasing the quality, efficiency and flexibility of the machines. Thus, it is definitely a cost-saving issue, but also a key to attract customers because of fast ramp-up of production. Especially the reduced time gives extra time for innovative new concepts and to fulfill SDGs.

Unplanned Downtimes (Estimated: 20%, Achieved: 20.8%)
To use this new model-based and big data-driven process engineering methodology it is essential to establish a pattern and anomaly detection framework. With this framework, different behavior models as well as artificial intelligence and machine learning algorithms are developed and provided in a model repository. The models are used in the engineering process to get better insights of the physical, logistics, or other behaviors. This accelerates the reduction of unplanned downtimes and therefore improvements in the sense of time, quality, and costs. The models are stored in a model repository and can be further used by the customer, e.g., as virtual sensors. This leads to new business models, such as "Model as a Service." Since the period of Boost 4.0, the amount and duration of downtimes due to maintenance actions at the customer's site was reduced by 20.8%.

Service Cost Reduction (Estimated: 15%, Achieved: 19%)
The pilot focuses on three states: as-engineered (state after the engineering was finished) as-manufactured (state after manufacturing and in-house commissioning), and as-operated (how the customer operates the production system, simulate historical and real-time data). In order to extend the field of application from pure simulation and monitoring usage, the requirements of service design and service engineering has been integrated into the digital twin. This enables requirements analysis for new projects, failure analysis, and designing a service process with focus on service cost reduction. The measured projects in Boost 4.0 achieved a reduction of 19%; this is mainly due to designed remote service actions and failure prevention concepts in the engineering phase.

Simulation-Based Release Process Reducing Commissioning Time (Estimated: 10%, Achieved: 14.7%)

With the efficient use of CAD models, interfaces between Visual Components simulation tools and the engineering management tools of Fill have been established; thus a better integration of the simulation process into the proposed engineering process has been achieved. This led to a simulation management in which saving, loading, and version control are well integrated. Moreover, design changes are updated faster in the simulation. The use of virtual commissioning was measured in a reduction of commissioning time by 14.7%. In future, the target from Visual Components is to create a generic PLM interface which can be tailored with add-ons for the PLM solutions in the market.

> "Beside the Fill Pilot that we have been involved in, by realizing customized solutions in simulation, our customers benefit in reductions of commissioning time in the range of 15–25%." (Fernando Ubis (Visual Components).

3.4   Observations and Lessons Learned

The development of innovative, highly customized production systems is generally based on a customer-centric approach. Furthermore, for the machine builder, the operations process knowledge is a key success factor. This is mostly expertise and empirical knowledge and is usually built during the operating phase and used in follow-up projects. The intended feedback loop should accelerate the buildup of knowledge and make it possible to secure customer needs and even wishes earlier and increases customer satisfaction. This leads to fewer delays in the business process, like customer-dependent approval processes, e.g., design approval, delivery approval, etc.

> "Fill is a grown family-owned company and focused on its customers success. With the new approach of closing the gap of digitalization between customers and the Fill engineering we also can feel a higher customer satisfaction, which is brought to me by feedback of our customers. Beside the facts and figures of measured KPIs, it is about the people working and their motivation. I think the digitalization and its approach is mandatory to benefit but it is about the mindset to see digitalization as a chance and not a threat. With the transparency on the process and the trust we have built up with our partners and customers we are best prepared for the factory of the future!" Alois Wiesinger (CTO of Fill).

4   +GF+ Trial for Big Data-Driven Zero Defect Factory 4.0

This section presents the +GF+ trial and the Boost 4.0 concepts adopted for model-based engineering, in particular the semantic information framework to deal with product quality information (metrology data) and digital factory/process information. The trial is intended to demonstrate how the implementation of semantic information representation supported by cutting-edge digital platforms and

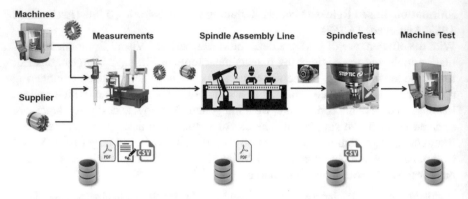

Fig. 8 GF milling machine spindles assembly process

based on agreed standards can break the silos, implement federated workflow and pipelines, and lead to a cost-effective implementation of advanced decision support systems and production cost reduction.

The trial is focused on the manufacturing of the most critical component on +GF+ machine tools, i.e., +GF+ Milling Spindles. These are critical components conveying the highest value in production processes. Their assembly processes currently generate data in isolation, which can be identified in the diagram showed in Fig. 8.

1. Factory machines produce some critical parts, which are measured with different systems producing data in different formats (from manual devices to CMM machines). One key issue is to link all those measurements to one part and avoid dealing with multiple reports.
2. The data acquired on the assembly line is among the most critical ones. It is currently collected by hand, what makes it, it difficult to aggregate and analyze.
3. The quality testing of each spindle also produces data, which today is stored as a .csv file and not correlated to any other data in the process.
4. The data related to the assembly of the spindle into the machine itself is stored into a PostgreSQL database, again not aggregated for analysis.
5. Finally, there is also data coming from service records and sensors (spindles which are returned by the assembly of the machine or by the end customer), with unexploited key information on the component quality.

While the quantity of data created is not a problem nowadays, the major challenge is the complexity to structure and aggregate it so as to have a comprehensive understanding of their meaning, in order to improve the manufacturing process and even machine operation through predictive maintenance applications.

4.1   Value of a Quality Information Framework for Dimensional Control and Semantic Interoperability

QIF (quality information framework) [18] is an open-standard CAD format made specifically for introducing twenty-first-century concepts such as digital transformation, digital thread, and IoT (Internet of Things) to computer-aided technology and engineering applications. The two main points of QIF are interoperability and traceability throughout the entire product lifecycle. From design to planning to manufacturing to analysis, full metadata can be mapped back to the "single source of truth" (native CAD).

QIF is built on the XML framework for easy integration and interoperability with other systems, web/internet applications, and other formal standards – a unified and universal approach.

- **Structured Data**: featured-based, characteristic-centric ontology of manufacturing quality metadata.
- **Modern Approach:** XML technology—simple implementation and built-in code validation.
- **Connected Data**: information semantically linked to model for full information traceability to MBD.
- **Standard Data:** approved ISO and ANSI interoperability standard.

It also contains holistic, semantic PMI (product manufacturing information)/3D product definition and other metadata that is both human-readable and computer-readable for MBD (model-based definition) implementation. QIF is an ANSI and ISO 23952:2020 standard managed by the Digital Metrology Standards Consortium (DMSC), an international leader in the field of metrology. QIF supports Design, Metrology, and Manufacturing as it enters the Industry 4.0 initiative: data that is semantic, machine-readable, standard, and interoperable to enable the smart factory. QIF is a key conversation starter for companies beginning the MBD/MBE (model-based enterprise) process, especially for metrology-related information in PLM (produce lifecycle management) and PDM (product data management).

As the Boost 4.0 + GF+ trial has evidenced, a semantic approach to data integration across the product and process lifecycle has very clear benefits:

1. **Automation**: Defined business process and software compatibility leads to the possibility of automation.
2. **Interoperability**: Enables authority CAD file to be reused on different software by different departments and companies.
3. **Single Source of Truth**: Derivative models for robust, semantic PMI, metrology features, and mapping back to any native CAD model.
4. **Big Data:** Manufacturing data is moved upstream for analytics and design improvements.
5. **Faster Time to Market**: Automation and decreased manual translation and validation begets shorter production cycles.

6. **Cost Savings**: Up to 80% of total hours saved for annotation, control planning, and inspection processes together, meaning less resources needed for a particular task, reducing overhead.
7. **Work Efficiency**: Automation is repeatability, relying less on human involvement (and possible error), and freeing the engineer to focus on other value-adding work.
8. **Process Over Personnel**: Avoiding the "human-in-the-loop" method provides documented process-driven strategy.
9. **Better Product**: Faster time to market leads to more iterations and breakthroughs in product, process, or pricing.
10. **Better Bottom Line**: Automated work processes, less bottlenecks, and faster iteration and feedback for ideation all lead to savings in time and money.

On the other hand, many of today's manufacturing practices depend on disparate data sets and manual transcription and validation that impede the ability for automation. Not all data is created equal. Different data file formats (e.g., PDF, TXT, TIF, CSV, XLS, STEP, JT, IGES, PRT, QIF, XML, etc.) from different software are either proprietary or lacking robust data capabilities to produce true MBD. The incompatibility and inaccessibility prevents connecting data throughout the whole product lifecycle—traceability and automation in the digital thread. With multiple stakeholders throughout the supply chain with their own CAD, CAM, and CMM software and custom-made data, exchanging inoperable data with a disjointed approach results in multiple disconnections in the digital thread. As illustrated in Fig. 9, QIF is an MBD-ready, XML-based, CAD-neutral, and open standard that includes the following features: (1) PMI (Product Manufacturing Information), (2) GD&T (Geometric dimensioning and tolerancing), (3) Measurement plans, (4) Geometry, (5) Bill of Characteristics, (5) Inspection Plans, and (6) Other semantic data.

All these features allow seamless handoff of data downstream, enabling automation to quality control and production with full traceability to the single source of truth: the CAD model. It empowers businesses for a better product, faster process, and bigger bottom line. This is the purpose of MBD and MBE in manufacturing! And this all begins with quality information.

According to the Data Information Knowledge Wisdom (DIKW) Pyramid [19] & QIF stands in the information layer, data is a number while information is data with context. It is well known that decisions made from data require a "human-in-the-loop," while decisions made from information lead to automation. Anyone can collect data, but data in action is wisdom (Table 1).

4.2 Value of Semantically Driven Big Data Zero Defect Factory

The realization of the concept of zero defect factory that underpins the +GF+ trial relies on mastering four critical business processes: (1) spindle component

Fig. 9 QIF structure

Table 1 Decisions from data vs. information

Decisions from data	Decisions from information
Interpretation	Increases speed of task completion
Tedious process	Lowers cost due to decreased labor requirements
High cognitive load	Frees up valuable personnel for other tasks more suited for the human mind
More opportunities for error	Repeatable solutions
Costly solutions	Lower risks
Inconsistent solutions	

manufacturing, (2) spindle assembly and delivery to machine factory, (3) after-sales guarantee management, and (4) service contracts. For each of these processes, specific KPIs have been defined and are now constantly monitored. This will allow delivering well-defined benefits reaching nearly 1 M€/y only for the manufacturing and after-sales guarantee stages (Fig. 10).

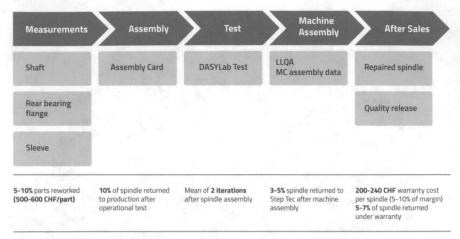

Fig. 10   Spindle component lifecycle and KPIs

4.3   Semantic Big Data Pipeline Implementation for Zero Defect Factories

The +GF+ big data pipeline has been built on the bases of three main types of data. Metrology, machine sensor data and assembly data.

Spindle Metrology Data QIF workflow was implemented on a typical +GF+ part—the main spindle. The part was designed with PTC CREO software and was used as input to the workflow, creating the QIF MBD representation with the help of Capvidia's MBDVidia for CREO software. The workflow was divided into steps. Step 1: the native CAD model is converted into the standard Model Based Definition (QIF). Step 2: the QIF file with the semantically linked Product Manufacturing Information (PMI) is verified for correctness of the Model-Based Design (MBD) definition (this check validates and corrects all semantics in the PMI definition). As a result we get MBD Ready data being 100% machine-readable for the next application (maintain the digital thread). Step 3: This MBD model is used for automatic generation of the First Article Inspection (FAI) document specifying what entities have to be measured in the quality control process. This is an inspection document to verify the quality of the manufacturing process confirming that the physical part complies with the design intent. As explained in the first part, today for one specific part, information about the measurement is spread between three documents most of the time (from different machines, operator measuring manually, etc.). The implementation of the QIF workflow reduces it to one document, making it easier for the quality team to check (Figs. 11 and 12).

In addition, the data, being fully digital, is now traceable and organized in a standard (QIF) data format, guarantying open access and transparency. They are

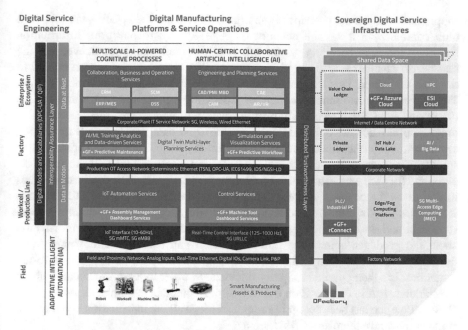

Fig. 11 +GF+ service development architecture

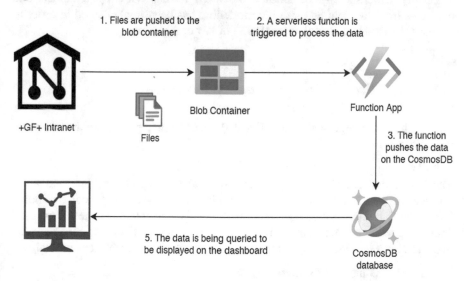

Fig. 12 +GF+ service development architecture and Azure Data Factory Pipeline

also uniquely linked with the physical part, which significantly simplifies the repair and maintenance activities.

Spindle Assembly Data Assembly cards, which contain all the needed information about the produced spindles, have been digitalized, and manual filling process suppressed. An Az-copy script copies these files from the +GF+ intranet to a blob container on a storage account on +GF+'s Azure platform. Information regarding the different parts of the spindles (the shaft, the front sleeve, and the rear bearing flange) is also saved in different CSV files, which are also copied to the blob container. This triggers a piece of logic called an Azure Function App. This app contains two functions: one for processing the assembly cards and one for processing the CSV files containing information to update the spindle parts. The first function reads the assembly card, retrieves the data, transforms it to add the semantics, and pushes the transformed data to the database in order to create a new spindle. The second one read the files containing information about the spindle's parts, and for each part, retrieves the corresponding entity from the database, updates it, and saves the changes to the database. The database is a CosmosDB database which uses the Gremlin API, so as to operate with graph data, which allows us to store data with its semantic metadata. Thanks to this platform, all the information about the produced spindles is now digitalized and centralized into one single database, and can be easily understood and retrieved by all parties thanks to the semantics. It's then possible to query this data, in order to learn information about the spindles, and to display it on a dashboard. As another example, an Azure Data Factory Pipeline has been set up in order to retrieve all the information about all the spindles, and save them as a CSV file.

Finally, **Machine Sensor Data** is retrieved through OPC-UA channels and protocols from the field and specific modelling has been implemented in order to estimate the Residual Useful Time (RUT) of key components.

4.4 Large-Scale Trial Performance Results

The results of the large-scale trials (Fig. 13) are extremely positive in the following two dimensions.

Data Workflow and Visualization
The implementation of the digital data flow enabled keeping a self-service warehouse always up to date. This can be shown in a dashboard with the corresponding KPIs. The Capvidia solution (QIF model) shows also the potential benefit from easy data aggregation around a single model. It appears clearly that this provides value in a short term (time savings, transparency) and in the subsequent steps of analysis.

Machine Learning and Predictive Quality
Thanks to this program, it is possible to apply machine learning in order to detect 30% of manufacturing quality problems. Additionally, the benefit of having an interpretable model will enable us in the future to improve our design and our tolerances by understanding the root causes (Fig. 14).

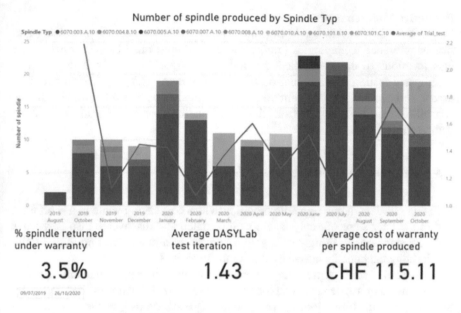

Fig. 13 Example of report showing the potential benefit

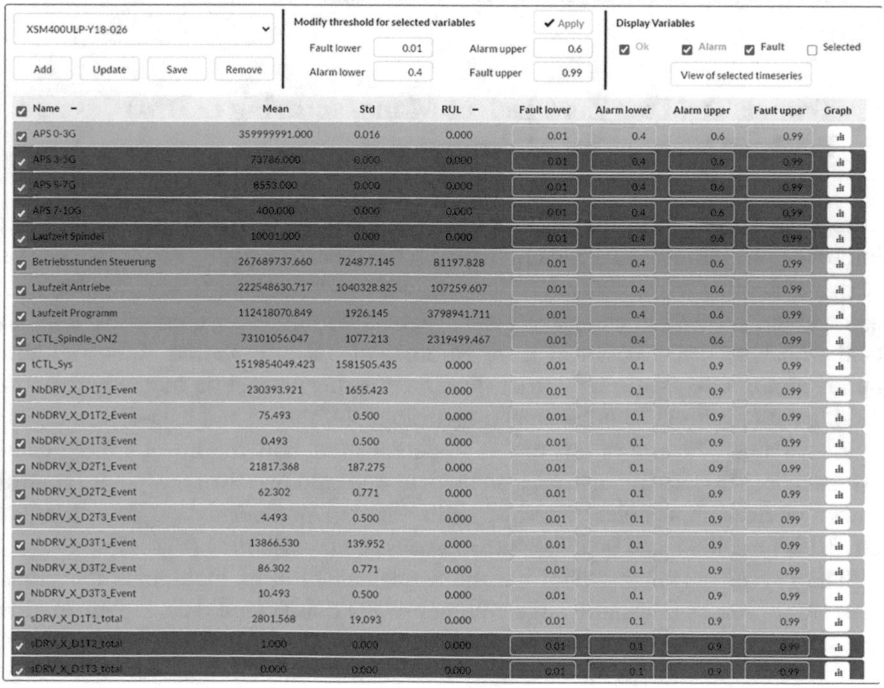

Fig. 14 Predictive maintenance testing environment on Scilab Cloud

Predictive Maintenance

Deployment is a key factor in big data knowledge inference. Preprocessing and further inference technologies (e.g., machine learning) must be performed at the data location. To do so, predictive maintenance algorithm has been deployed in a dedicated testing environment on a Cloud Solution. This application could be used to test and validate different predictive maintenance algorithms, and predictive capabilities can be accessed from the client platform, at the data location, via a dedicated API.

4.5   Observations and Lessons Learned

Bringing our data in one semantic model provides cost saving value in a short term and in the later steps of analysis. Boost 4.0 trialed and discovered also new ways and new standards to bring data together and set up a manufacturing digital twin, highlighting the benefits internally during the project.

As a result, some critical improvements are now possible: machine learning algorithms help to achieve unexpected accuracy (even with limited data) by detecting manufacturing issues; semantic and QIF standards give the possibility to link data together in an efficient way, and the cloud solution gives the possibility to deploy easily AI in testing environments and integrate them through APIs in any platform, making the full deployment swift and valuable.

5   Trimek Trial for Zero Defect Manufacturing (ZDM) Powered by Massive Metrology 4.0

Zero Defect Manufacturing (ZDM) powered by massive metrology is aimed at improving the performance and efficiency of the essential quality control processes in manufacturing lines, which have to deal with very large parts, e.g., automotive, aeronautics, renewable energy systems, and railways, and consequently have to deal with very heavy data and large volumes of 3D information. This trial also builds on the QIF (see +GF+ trial discussed earlier) semantic model and Boost 4.0 big data pipelines to implement a metrology 4.0 thread and data-driven digital twin analytics. The aim of this trial is to demonstrate how the Boost 4.0 big data-driven model-based approach can lead to increased automation of the quality control workflow through seamless interoperability among product design data, product GD&T information, and quality control process commissioning (Fig. 15).

The key pillars of this pilot are the implementation of high-definition inline metrology process to capture large volumes of 3D point cloud data, the integration and analysis of heterogeneous data, both quality and product data from different sources, incorporating to the metrology flow data coming from the product design steps, and finally the development of an advanced human-centric collaborative and

Fig. 15 Massive metrology 4.0 trial scenario

visual analytic process based on advanced color mapping. As a result, a connected and secure quality control process covering the whole metrology workflow and based on QIF standard has been implemented, with innovative and efficient visualization, processing, storage, and analysis capabilities, which definitely improves the decision-making process and reduces the number of defective parts. The ZDM Massive Metrology trial is based on the M3 Big Data Platform [20] for design and production data sharing and QIF standard as dimensional metrology data exchange. The M3 platform is poised to provide a structured solution for Metrology 4.0, an edge-powered quality control analytics, monitoring, and simulation system (Fig. 16).

5.1 Massive 3D Point Cloud Analytics and Metrology 4.0 Challenges

Up to now, the metrology results are usually only visualized in reports based on the Geometric Dimensioning and Tolerancing (GD&T) analysis, which is aimed only for metrology use and only a few control points are considered (10–100 points). Industry 4.0 in general and zero defect manufacturing in particular demand that metrology 4.0 processes scale up data acquisition, processing time, and visualization speed at various orders of magnitude. This is in accordance with the demand for more holistic, flexible, and fast metrology solutions—see requirements and trends from the VDI 2020 Manufacturing Metrology Roadmap.

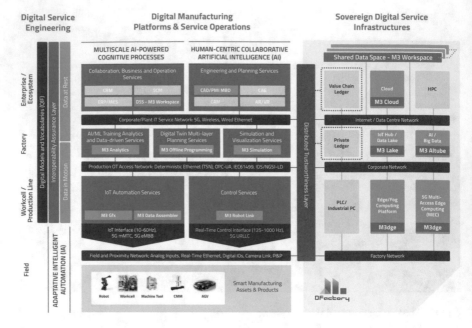

Fig. 16 Boost 4.0 reference model for massive metrology 4.0 based on the M3 platform pipeline

The Trimek trial exploits model-based engineering of metrology solutions to implement massive metrology 4.0 workflows that will address the need for holistic measurement systems capable of dealing with increasing information density, diversity, integration, and data processing automation in reduced measurement times.

The objective of the Trimek trial is to implement a rapid big data pipeline for processing and advanced high-speed high-resolution texturized colormaps for high fidelity visual analysis of massive 3D point clouds (from 10 to 100 million points) in less than 30 s. So the challenge is to demonstrate that future metrology 4.0 platforms can be used to assess and guarantee the fit, performance, and functionality of every part (irrespective of its size and tolerancing requirements) and support the targets of zero defect, zero waste, and carbon neutrality (Fig. 17).

5.2   Implementation of Massive Metrology 4.0 Big Data Workflow

The Trimek ZDM Massive Metrology 4.0 trial considers two main business processes for implementation:

- **High-density metrology**: This process has developed a system capable of rapid acquiring and processing big volume of 3D point cloud data from complex

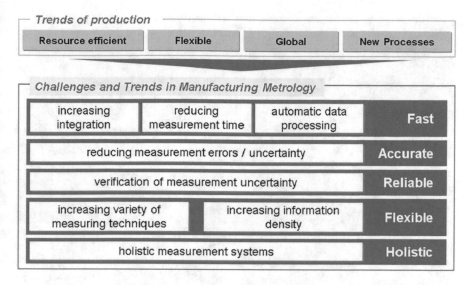

Fig. 17 Massive metrology 4.0 trial data analytics challenges

parts and analyzing massive point clouds coming from those parts by means of advanced 3D computational metrology algorithms, obtaining high-performance visualization, i.e., a realistic 3D colormap with textures.

- **Virtual massive metrology**: This business process has developed a digital metrology workflow and agile management of heterogeneous products and massive quality data, covering the whole product lifecycle management process and demonstrating an advanced semantic QIF metrology workflow (Fig. 18), enabling product design semantic interoperability with product quality control processes and advanced analysis and visualization of the quality information for decision support.

5.3 Large-Scale Trial Performance Results

The Trimek trial has allowed the implementation of advanced metrology 4.0 algorithms based on 3D computational capabilities to finally obtain a texturized mesh, instead of an annotated polygonal mesh, that is a more realistic visualization of the physical product for human-centered decision support process. Large pieces have been scanned and analyzed and the trial has demonstrated that the Boost 4.0 big data pipelines can process the whole body of a car (Fig. 19).

Table 2 summarizes the main system capacities.

These new big data capabilities have allowed the automotive industry to work fluently with 10 times larger CAD files and control beyond 600 geometrical features, which is important as having all the car modelized. The processing speed has also

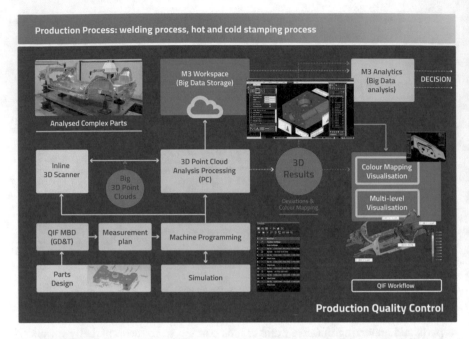

Fig. 18 QIF-ready semantic massive metrology big data analytics pipeline based on M3 platform

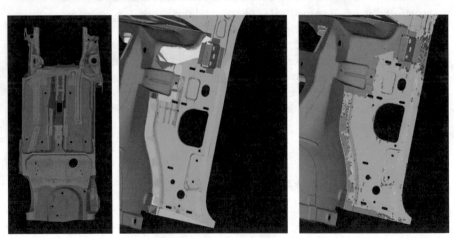

Fig. 19 Large-scale CAD file (left), annotated polygonal colormap mesh (middle), texturized colormap mesh (right)

been multiplied by 5, which allow a better performance and a more fluent analysis and visualization. Also, the time needed to program the scan of each piece has been reduced significantly, up to 80%, as having the whole car already modelized allows for a digital planification of the trajectories thus reducing the time needed to configure the scan. Overall, this derives a cost reduction of 10%.

Table 2 Trimek trial big data capabilities features for Metrology 4.0

Big data capability	M3	M3—big data
Data usage	32 bits CAD for 100 Mbs Point cloud size <six million	64 bits CAD for 400 Mbs Point cloud size >100 million
CAD file format	Mono CAD	Multi CAD
Part/CAD alignment	Mono-alignment algorithms	Multi-alignment algorithms
Region of Interest (RoI) extraction	Slow	Effective
Color mapping	Mono-core & Polygonal	Multicore & Texturized
Visualization	Monolithic	Adaptive

6 Conclusions

This chapter has discussed how Model-Based Systems Engineering (MBSE) can benefit from big data technologies to implement smarter engineering processes. The chapter has presented the Boost 4.0 testbed that has demonstrated how digital twin continuity and digital thread can be realized from service engineering, production, product performance, to behavior monitoring. The Boost 4.0 testbed has demonstrated the technical feasibility of an interconnected operation of digital twin design, ZDM subtractive manufacturing, IoT product monitoring, and spare part 3D printing services. It has shown how the IDSA reference model for data sovereignty, blockchain technologies, and FIWARE open-source technology can be jointly used for breaking silos, providing a seamless and controlled exchange of data across digital twins based on open international standards (ProStep, QIF), allowing companies to dramatically improve cost, quality, timeliness, and business results.

This closed-loop implementation allows the realization of advanced product-service processes that have been trialed by Fill, +GF+ and Trimek manufacturing equipment. The chapter has clearly presented how semantic data integration across the product and process lifecycle based on open international standards such as OPC-UA and QIF provide significant performance improvements in customization of complex machine tool installations, development of zero-defect factories 4.0, and massive product metrology 4.0. It has illustrated how QIF (Quality Information Framework) ISO-standard (ISO 23952:2020) meets MBD conditions and leverages interoperable data exchange between CAD, CAM, and other CAx systems for downstream use. This chapter has described how big data trials leverage the possibility to map data back to a single source of truth; providing traceability and validation of intended and unintended changes/outcomes. QIF with PMI (Product Manufacturing Information, aka 3D annotations) can include GD&T, Bill of Materials, Process Plans, and more for data that is unambiguous and machine-readable—able to overcome human interpretation, thereby leveraging a new generation of data-driven digital twins.

This chapter has provided additional evidence that the Boost 4.0 service development reference architecture maintained by the Digital Factory Alliance and the Big Data Value Association (BDVA) reference architecture provide a unified framework to develop high-performance big data pipeline that allow for fast transfer, replication, and adoption of product engineering and manufacturing operational optimization.

As it has become apparent in this chapter, over the last few years the role of GD&T is gaining momentum in the definition of new generations of Industry 4.0 semantic models such as Automation ML. The need for increased levels of interoperability across OPC-UA, AML, and QIF standards calls for additional research that leverage higher manufacturing autonomy levels and facilitate more efficient and effective collaborative engineering processes for highly customized and complex products. The integration and alignment of such models will be a fundamental milestone in the development of increasingly cognitive and intelligent digital twin services that seamlessly interact not only between physical and digital work but also across the product and manufacturing process lifecycle.

Acknowledgments This research work has been performed in the framework of the Boost 4.0 Big Data Lighthouse initiative; a project that has received funding from the European Union's Horizon 2020 research and innovation program under grant agreement No. 780732. This datadriven digital transformation research is also endorsed by the Digital Factory Alliance (DFA) www.digitalfactoryalliance.eu

References

1. INCOSE. https://www.incose.org/
2. INCOSE Technical Operations. (2007). *Systems Engineering Vision 2020, version 2.03*. Seattle, WA: International Council on Systems Engineering, Seattle, WA, INCOSE-TP-2004-004-02.
3. Boost 4.0. https://boost40.eu/
4. Zillner, S., Curry, E., Metzger, A., Auer, S., & Seidl, R. (2017). *european big data value strategic research & innovation agenda*. Big Data Value Association.
5. Zillner, S., Bisset, D., Milano, M., Curry, E., García Robles, A., Hahn, T., Irgens, M., Lafrenz, R., Liepert, B., O'Sullivan, B., & Smeulders, A. (Eds.) (2020) *Strategic research, innovation and deployment agenda - AI, data and robotics partnership. Third release*. September 2020, Brussels. BDVA, euRobotics, ELLIS, EurAI and CLAIRE".
6. Internet of Things Solutions World Congress. https://www.iotsworldcongress.com
7. ProStep. https://www.prostep.com
8. International Data Spaces Association (2019). *Reference architecture model*. Online. Available at: https://www.internationaldataspaces.org/wp-content/uploads/2019/03/IDS-Reference-Architecture-Model-3.0.pdf
9. FIWARE Context Broker. https://fiware-orion.readthedocs.io
10. International Organization for Standardization. *ISO 10303 Industrial Automation Systems and Integration – Product Data Representation and Exchange*. Geneva: ISO.
11. International Organization for Standardization. (2020). *ISO 23952:2020. Automation systems and integration — Quality information framework (QIF) — An integrated model for manufacturing quality information.* : ISO.

12. OPCUA. https://opcfoundation.org/about/opc-technologies/opc-ua/
13. VDI 2206. (2002). *Entwicklungsmethodik für mechatronische Systeme - Richtlinienentwurf, VDI-Richtlinienausschuß A127/VDI2206*, Paderborn.
14. Holom, R.-M., Rafetseder, K., Kritzinger, S., & Sehrschön, H. (2020). Metadata management in a big data infrastructure. *Procedia Manufacturing, 42*, 375–382. https://doi.org/10.1016/j.promfg.2020.02.060
15. Apache Spark. https://spark.apache.org/
16. Apache Avro. https://avro.apache.org/
17. SALAD. https://github.com/SaladTechnologies
18. Ackoff, R. L. (1989). From data to wisdom. *Journal of Applied Systems Analysis, 16*, 3–9.
19. CAPVIDIA's BVD Tools for Creo. https://www.capvidia.com/products/mbd-tools-for-creo
20. Innovalia Metrology's M3 Big Data Platform. https://www.innovalia-metrology.com/es/productos/metrology-software/software-metrologico

A Data Science Pipeline for Big Linked Earth Observation Data

**Manolis Koubarakis, Konstantina Bereta, Dimitris Bilidas,
Despina-Athanasia Pantazi, and George Stamoulis**

Abstract The science of Earth observation uses satellites and other sensors to monitor our planet, e.g., for mitigating the effects of climate change. Earth observation data collected by satellites is a paradigmatic case of big data. Due to programs such as Copernicus in Europe and Landsat in the United States, Earth observation data is open and free today. Users that want to develop an application using this data typically search within the relevant archives, discover the needed data, process it to extract information and knowledge and integrate this information and knowledge into their applications. In this chapter, we argue that if Earth observation data, information and knowledge are published on the Web using the linked data paradigm, then the data discovery, the information and knowledge discovery, the data integration and the development of applications become much easier. To demonstrate this, we present a data science pipeline that starts with data in a satellite archive and ends up with a complete application using this data. We show how to support the various stages of the data science pipeline using software that has been developed in various FP7 and Horizon 2020 projects. As a concrete example, our initial data comes from the Sentinel-2, Sentinel-3 and Sentinel-5P satellite archives, and they are used in developing the Green City use case.

Keywords Earth observation · Linked data · Big data · Knowledge graphs

1 Introduction

Earth observation (EO) is the science of using remote sensing technologies to monitor our planet including its land, its marine environment (seas, rivers and lakes) and its atmosphere. Satellite EO uses instruments mounted on satellite platforms to gather imaging data capturing the characteristics of our planet. These satellite

M. Koubarakis (✉) · K. Bereta · D. Bilidas · D.-A. Pantazi · G. Stamoulis
Department of Informatics and Telecommunications, National and Kapodistrian University of
Athens, Athens, Greece
e-mail: koubarak@di.uoa.gr

E. Curry et al. (eds.), *Technologies and Applications for Big Data Value*,
https://doi.org/10.1007/978-3-030-78307-5_19

431

images are then processed to extract information and knowledge that can be used in a variety of applications (e.g. in agriculture, insurance, emergency and security, or the study of climate change).

Lots of EO data are available to users at no charge today, due to the implementation of international programs such as Copernicus in Europe and Landsat in the United States. EO data is a paradigmatic case of big data bringing into play the well-known challenges of volume, velocity, variety, veracity and value. Regarding *volume*, according to the Copernicus Sentinel Data Access Annual Report of 2019 [14], the Sentinel satellites have produced *17.23 PiBs* of data from the beginning of operations until the end of 2019. Regarding *velocity*, the daily average volume of published data for the same satellites has been *18.47 TiBs* for November 2019. Regarding *variety*, EO data become useful only when analysed together with other sources of data (e.g. geospatial data or in situ data) and turned into information and knowledge. This information and knowledge is also big and similar big data challenges apply. For example, 1PB of Sentinel data may consist of about 750,000 datasets which, when processed, about 450TB of content information and knowledge (e.g. classes of objects detected) can be generated. Regarding *veracity*, EO data sources are of varying quality, and the same holds for the other data sources they are correlated with. Finally, the *economic value* of EO data is great. The Copernicus Market Report of 2019 [15] estimates that the overall investment of the European Union in the Copernicus program has been 8.2 billion Euros for the years 2008–2020. For the same period, the cumulated economic value of the program is estimated between 16.2 and 21.3 billion Euros.

Linked data is the data paradigm which studies how one can make RDF data (i.e. data that follow the Resource Description Framework[1]) available on the Web and interconnect it with other data with the aim of increasing its value. In the last few years, linked *geospatial* data has received attention as researchers and practitioners have started tapping the wealth of geospatial information available on the Web [19, 21]. As a result, the *linked open data (LOD) cloud* has been rapidly populated with geospatial data, some of it describing EO products (e.g. CORINE Land Cover and Urban Atlas published by project TELEIOS) [20]. The abundance of this data can prove useful to the new missions (e.g. the Sentinels) as a means to increase the usability of the millions of images and EO products that are expected to be produced by these missions.

However, big open EO data that are currently made available by programs such as Copernicus and Landsat are *not* easily accessible, as they are stored in different data silos (e.g. the Copernicus Open Access Hub[2]), and in most cases users have to access and combine data from these silos to get what they need. A solution to this problem would be to use Semantic Web technologies in order to publish the data contained in silos in RDF and provide semantic annotations and connections to them so that they can be easily accessible by the users. By this way, the value of the

---

[1] http://www.w3.org/TR/rdf-primer/.

[2] https://scihub.copernicus.eu/.

original data would be increased, encouraging the development of data processing applications with great environmental and processing value *even by users that are not EO experts but are proficient in Semantic Web technologies.*

The European project TELEIOS [20] was the first project internationally that has introduced the linked data paradigm to the EO domain, and developed prototype applications that are based on transforming EO products into RDF, and combining them with linked geospatial data. The ideas of TELEIOS were adopted and extended in the subsequent European projects LEO [8], MELODIES [7], BigDataEurope [2], Copernicus App Lab [3] and ExtremeEarth [18].

In this chapter, we present a data science pipeline that starts with data in a satellite archive and ends up with a complete application using this data. We show how to support the various stages of the data science pipeline using software developed by the above projects. As a concrete example, our initial data comes from the Sentinel-1 and Sentinel-5 satellite archives, and the developed application is the Green City use case we implemented in the context of project Copernicus App Lab[3].

The organization of the rest of the chapter is as follows. Section 2 introduces the Green City use case which serves the context for our application. Section 3 gives a high level of the data science pipeline and describes its various stages. Then, Sect. 4 describes how we have implemented the Green City use case using the linked geospatial data software developed in the projects mentioned above. Finally, Sect. 5 summarizes the paper.

The chapter relates to the technical priority "Data Analytics" of the European Big Data Value Strategic Research and Innovation Agenda. It addresses the horizontal concern "Data Analytics" of the BDV Technical Reference Model. It addresses the vertical concern "Standards".

The chapter relates to the "Knowledge and Learning" and "Systems, Methodologies, Hardware and Tools", cross-sectorial technology enablers of the AI, Data and Robotics Strategic Research, Innovation and Deployment Agenda.

2   The Green City Use Case

Urban areas are the source of many of today's environmental challenges – not surprisingly, since two out of three Europeans live in towns and cities. Local governments and authorities can provide the commitment and innovation needed to tackle and resolve many of these problems. The European Commission's European Green Capital Award[3] (EGCA), recognizes and rewards local efforts to improve the environment, and thereby the economy and the quality of life in cities. The EGCA is given each year to a city, which is leading the way in environmentally friendly urban living. The award encourages cities to commit to ambitious goals for further environmental improvement.

---

[3] https://ec.europa.eu/environment/europeangreencapital/about-the-award/policy-guidance.

Fig. 1 A Green City map for Paris, France

Moreover, the Green City Accord[4] is a movement of European mayors committed to making cities cleaner and healthier. It aims to improve the quality of life of all Europeans and accelerate the implementation of relevant EU environmental laws. By signing the Accord, cities commit to addressing five areas of environmental management: air, water, nature and biodiversity, circular economy and waste, and noise.

In order to define though how "green" a city is, one must combine various sources of information that would allow us to measure and illustrate the greenness of each city in Europe. In the context of the Copernicus App Lab project, we demonstrated how one can interlink heterogeneous Earth Observation data sources and combine this information with other geospatial data using Linked Data technologies to produce Green City maps [3].

In Fig. 1 we show how to determine the greenness of Paris, France, by utilizing Earth Observation data, crowd-sourced data and Linked Data technologies. To produce this map, we combined air pollution data (NO_2 concentration) with indices that measure greenness (Leaf Area Index, OpenStreetMap Parks and CORINE Land Cover-related classes). All sources were spatially interlinked using the geometries of the administrative divisions of the city. Combining these diverse datasets using Linked Data technologies allows us to produce GeoSPARQL queries that can be visualized to construct such Green City maps for cities in Europe.

---

[4] https://ec.europa.eu/environment/topics/urban-environment/green-city-accord_en.

2.1   Data Sources

Sentinel data and Copernicus Services data that are currently available are not following the linked data paradigm. They are stored in different data silos so users might need to access and combine data from more than one source to satisfy their user needs. Utilizing Semantic Web and Linked Data technologies to make Copernicus Services data available as linked data increases their usability by EO scientists but also application developers that might not be EO experts. Moreover, the interlinking of Copernicus Services data with other relevant data sources (e.g. GIS data, data from the European data portal, etc.) increases the value of this data and encourages the development of applications with great environmental and financial value.

2.2   Copernicus Sentinel Data

For the Green City use case, the most relevant Earth Observation data come from the Land Monitoring service of Copernicus and air quality indices. To detect green areas within a city, we used the Leaf Area Index (Sentinel-3) and the CORINE land cover 2018 datasets (Sentinel-2 and Landsat-8). For air quality, we used the Nitrogen Dioxide index (Sentinel-5P).

The Leaf Area Index (LAI) is defined as half the total area of green elements of the canopy per unit horizontal ground area. The satellite-derived value corresponds to the total green LAI of all the canopy layers, including the understory which may represent a very significant contribution, particularly for forests. Practically, the LAI quantifies the thickness of the vegetation cover. LAI is recognized as an Essential Climate Variable by the Global Climate Observing System. The LAI dataset is provided by the Copernicus Global Land Service[5] and is distributed in Network Common Data Form version 4 (netCDF4) file format.

The CORINE Land Cover (CLC) inventory was initiated in 1985 (reference year 1990). Updates have been produced in 2000, 2006, 2012 and 2018. This vector-based dataset includes 44 land cover and land use classes. The time-series also includes a land-change layer, highlighting changes in land cover and land use. The high-resolution layers (HRL) are raster-based datasets which provide information about different land cover characteristics and is complementary to land cover mapping (e.g. CORINE) datasets. Five HRLs describe some of the main land cover characteristics: impervious (sealed) surfaces (e.g. roads and built up areas), forest areas, (semi-) natural grasslands, wetlands and permanent water bodies. The High-Resolution Image Mosaic is a seamless pan-European ortho-rectified raster mosaic based on satellite imagery covering 39 countries. The CLC dataset is provided by

---

[5] https://land.copernicus.eu/global/products/lai.

the Copernicus Pan-European component of the Land Monitoring Service[6] and is distributed in Shapefile format.

Nitrogen dioxide (NO_2) is a gaseous air pollutant composed of nitrogen and oxygen. NO_2 forms when fossil fuels such as coal, oil, gas or diesel are burned at high temperatures. NO_2 and other nitrogen oxides in the outdoor air contribute to particle pollution and to the chemical reactions that make ozone, thus it is one of six widespread air pollutants that have national air quality standards to limit them in the outdoor air. The NO_2 index is part of the Ozone Forecast dataset provided by the LOTOS-EUROS team, consisting of the Netherlands Organisation for Applied Scientific Research (TNO), the Environmental Assessment Agency of the Dutch National Institute for Public Health and the Environment (RIVM/MNP) and the Royal Netherlands Meteorological Institute (KNMI). The NO_2 dataset was distributed through the OPeNDAP protocol.

2.3   Other Geospatial Data

In addition to the above datasets, the Green City use case utilizes data from OpenStreetMap and the global administrative divisions dataset GADM.

OpenStreetMap (OSM) is a collaborative project to create a free editable map of the world. The geodata underlying the map is considered the primary output of the project. The creation and growth of OSM has been motivated by restrictions on use or availability of map data across much of the world, and the advent of inexpensive portable satellite navigation devices. The project has a geographically diverse user-base, due to emphasis of local knowledge and ground truth in the process of data collection. Many early contributors were cyclists who survey with and for bicyclists, charting cycleroutes and navigable trails. Others are GIS professionals who contribute data with Esri tools. In this manner, OSM is an open and free map of the whole world constructed by volunteers. It is available in vector format as shapefiles from the German company Geofabrik.[7] For our use case, information about parks has been taken from this dataset.

The Global Administrative Areas (GADM) dataset is a high-resolution database of country administrative areas, with a goal of "all countries, at all levels, at any time period".[8] It is available in vector format as a shapefile, a geopackage (for SQLlite3), a format for use with the programming language R, and KMZ (compressed KML). GADM allows us to use the administrative boundaries of cities and spatially interlink it with all the information we have from the other datasets.

---

[6] https://land.copernicus.eu/pan-european/corine-land-cover.

[7] http://download.geofabrik.de/.

[8] https://gadm.org/.

3 The Data Science Pipeline

Developing a methodology and related software tools that support the complete life cycle of linked open EO data has been studied by our group in project LEO [21] following similar work for linked data, for example by project LOD2 and others [1, 27]. Capturing the life cycle of open EO data and the associated entities, roles and processes of public bodies and making available this data was the first step in achieving LEO's main objective of bringing the linked data paradigm to EO data centres, and re-engineering the life cycle of open EO data based on this paradigm. In this chapter we continue this work by presenting a data science pipeline for big linked EO data and we apply it to the development of the Green City use case presented in the previous section.

The life of EO data starts with its generation in the ground segment of a satellite mission. The management of this so-called payload data is an important activity of the ground segments of satellite missions. Figure 2 gives a high-level view of the data science pipeline for big linked EO data as we envision it in our work. Each phase of the pipeline and its associated software tools is discussed in more detail below.

3.1 *Ingestion, Processing, Cataloguing and Archiving*

Raw data, often from multiple satellite missions, is ingested, processed, catalogued and archived. Processing results in the creation of various standard products (Level 1, 2, etc., in EO jargon; raw data is Level 0) together with extensive metadata describing them.

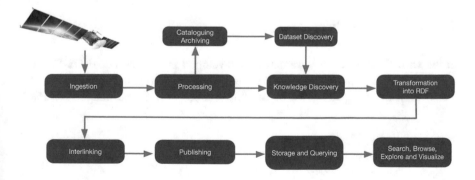

Fig. 2 The data science pipeline for big, linked EO data

3.2 Dataset Discovery

Once data become available in an archive or on the Web, they can be accessed by proprietary systems, traditional search engines or the Dataset Search service offered by Google.[9]

For example, the Copernicus Open Access Hub currently stores products from Sentinel-1, Sentinel-2, Sentinel-3 and Sentinel-5P missions[10] and offers a menu/map interface for searching for relevant data by date/time, area of interest, mission, satellite platform, etc. Similar interfaces are offered by other EO data centres hosting satellite data such as NASA[11] and the German Aerospace Center DLR.[12]

An interesting recent development in the area of dataset search is the development of the service Dataset Search by Google. This service crawls the Web retrieving metadata of datasets annotated using Schema.org vocabularies[13] following the guidelines of Google researchers.[14] Schema.org was originally founded by Google, Microsoft, Yahoo! and Yandex, and it has evolved into a community activity developing vocabularies for annotating Web resources by an open community process. Schema.org provides a unique structured data markup schema to annotate a Web page with variety of tags that can be added to HTML pages as JSON-LD, Microdata or RDFa markup. This markup allows search engines to index Web pages more effectively.

Dataset Search also offers a keyword-based search interface for discovering these datasets. For example, one can search for "CORINE land cover Copernicus App Lab" to discover the CORINE land cover dataset in linked data form published on datahub.io by our project Copernicus App Lab.[15] CORINE land cover is a dataset published by the European Environment Agency describing the land cover/land use of geographical areas in 39 European countries.

3.3 Knowledge Discovery

In the knowledge discovery frameworks developed in project TELEIOS [12, 13], traditional raw data processing has been augmented with *content extraction* methods that deal with the specificities of satellite images and derive image descriptors

---

[9] https://datasetsearch.research.google.com/.

[10] https://scihub.copernicus.eu/.

[11] https://search.earthdata.nasa.gov/.

[12] https://eoweb.dlr.de/egp/.

[13] https://schema.org/.

[14] https://support.google.com/webmasters/thread/1960710.

[15] https://datahub.ckan.io/dataset/corine-land-cover12.

(e.g. texture features, spectral characteristics of the image). Knowledge discovery techniques combine image descriptors, image metadata and auxiliary data (e.g. GIS data) to determine concepts from a domain ontology (e.g. park, forest, lake, etc.) that characterize the content of an image.

Hierarchies of domain concepts are formalized using *ontologies* encoded in the Web Ontology Language OWL2 and are used to annotate standard products. Annotations are expressed in RDF and its geospatial extension stRDF/GeoSPARQL [23, 30] and are made available as linked data so that they can be easily combined with other publicly available linked data sources (e.g. GeoNames, OpenStreetMap, DBpedia) to allow for the expression of rich user queries.

3.4 Transformation into RDF

This phase transforms vector or raster EO data from their standard formats (e.g. ESRI Shapefile or NetCDF) into RDF.

In FP7 project LEO we developed the tool GeoTriples for transforming EO data and geospatial data into RDF [26]. GeoTriples is able to deal with *vector data and their metadata* and to support natively many popular geospatial data formats (e.g. shapefiles, spatially enabled DBMS, KML, GeoJSON, etc.) The mapping generator of GeoTriples employs the mapping languages R2RML [10] and RML [11] to create mappings that dictate the method of conversion of the raw data into RDF.

R2RML is a language for expressing mappings from relational data to RDF terms, and RML is a more general language for expressing mappings from files of different formats (e.g. CSV, XML, etc.) to RDF. The mappings are enriched with subject and predicate object maps in order to properly deal with the specifics of geospatial data and represent it using an appropriate ontology.

GeoTriples is an open-source tool[16] that is distributed freely according to the Mozilla Public License v2.0.

3.5 Interlinking

This is a very important phase in the linked EO data life cycle since a lot of the value of linked data comes through connecting seemingly disparate data sources to each other.

Starting in our project LEO, we have worked on interlinking of open EO data by discovering geospatial or temporal semantic links. For example, in linked EO datasets, it is often useful to discover links involving topological relationships, for example A `geo:sfContains` F, where A is the area covered by a remotely

---

[16] http://geotriples.di.uoa.gr/.

sensed multispectral image I, F is a geographical feature of interest (field, lake, city, etc.) and geo:sfContains is a topological relationship from the topology vocabulary extension of GeoSPARQL. The existence of this link might indicate that I is an appropriate image for studying certain properties of F.

In LEO we have dealt with these issues by extending the well-known link discovery tool Silk in order to be able to discover precise geospatial and temporal links among RDF data published using the tool GeoTriples. The extension of Silk that we developed is now included in the main version. Since then other tools that carry out the same task more efficiently have been developed, for example Radon [32]. A recent comparison of geospatial interlinking systems is presented in [31].

3.6 Publishing

This phase makes linked EO data publicly available in the LOD cloud or in open data platforms such as datahub.io using well-known data repository technologies such as CKAN. In this way, others can discover and share this data and duplication of effort is avoided.

3.7 Storage and Querying

This phase deals with storing all relevant EO data and metadata on persistent storage so they can be readily available for querying in subsequent phases.

In our projects we have used our own spatiotemporal RDF store Strabon[17] which was developed especially for this purpose [24]. Strabon supports the data model stRDF and the query language stSPARQL developed by our group.

stRDF is an extension of RDF that allows the representation of geospatial data that changes over time [5, 25]. stRDF is accompanied by stSPARQL, an extension of the query language SPARQL 1.1 for querying and updating stRDF data. stRDF and stSPARQL use OGC standards (WKT and GML) for the representation of temporal and geospatial data.

Strabon extends the well-known open-source RDF store Sesame 2.6.3 and uses PostgreSQL or MonetDB as the backend spatially enabled DBMS. As shown by our experiments in [5, 16, 17, 25], Strabon is currently the most functional and performant geospatial and temporal RDF store available.

Strabon also supports the Open Geospatial Consortium (OGC) standard GeoSPARQL [30] for querying geospatial data encoded in RDF. stSPARQL and GeoSPARQL are very similar languages although they have been developed

---

[17] http://strabon.di.uoa.gr.

independently. Strictly speaking, if we omit aggregate geospatial functions from stSPARQL, the geospatial component of GeoSPARQL offers more expressive power than the corresponding component of stSPARQL. However, GeoSPARQL does not support a temporal dimension to capture the valid time of triples as stSPARQL does.

In our work stRDF has been used to represent satellite image metadata (e.g. time of acquisition, geographical coverage), knowledge extracted from satellite images (e.g. a certain area is a park) and auxiliary geospatial data sets encoded as linked data. One can then use stSPARQL to express in a single query an information request such as the following: "Find an image taken by a Meteosat second generation satellite on August 25, 2007, which covers the area of Peloponnese and contains hotspots corresponding to forest fires located within 2 km from a major archaeological site." Encoding this information request today in a typical interface to an EO data archive such as the ones discussed above is impossible, because domain-specific concepts such as "forest fires" are not included in the archive metadata, thus they cannot be used as search criteria.

With the techniques of knowledge discovery developed in our projects, we can characterize satellite image regions with concepts from appropriate ontologies (e.g. landcover ontologies with concepts such as waterbody, lake and forest, or environmental monitoring ontologies with concepts such as forest fires and flood) [13, 22]. These concepts are encoded in OWL2 ontologies and are used to annotate EO products. Thus, we attempt to close the semantic gap that exists between user requests and searchable information available explicitly in the archive.

But even if semantic information was included in the archived annotations, one would need to join it with information obtained from auxiliary data sources to answer the above query. Although such open sources of data are available to EO data centres, they are not used currently to support sophisticated ways of end-user querying in Web interfaces such as the ones discussed above under "Dataset Discovery". In our work, we have assumed that auxiliary data sources, especially geospatial ones, are encoded in stRDF and are available as linked geospatial data, thus stSPARQL can easily be used to express information requests such as the above.

In some applications it might not be a good idea to transform existing geospatial data into RDF and then store it in a triple store such as Strabon (e.g. when such data get frequently updated and/or are very large or when the data owners choose not to do so). For this case, we have developed the system Ontop-spatial,[18] which is a geospatial extension of the ontology-based data access (OBDA) system Ontop [9]. Ontop performs on-the-fly SPARQL-to-SQL translation on top of relational databases using ontologies and mappings. Ontop-spatial extends Ontop by enabling on-the-fly GeoSPARQL-to-SQL translation on top of geospatial databases [4, 6]. Ontop-spatial allows geospatial data to remain in their original

---

[18] http://ontop-spatial.di.uoa.gr.

databases (e.g. PostGIS, SpatiaLite, Oracle Spatial and Graph) and enables them to be queried effectively and efficiently using GeoSPARQL and the OBDA paradigm.

3.8 Search/Browse/Explore/Visualize

This phase enables users to find and explore the data they need and start developing interesting applications.

In FP7 project LEO, we redesigned the tool Sextant [28] for such purposes and also developed a mobile version that is distributed as an APK file for Android OS. The new version of Sextant is a web-based and mobile-ready application for exploring, interacting and visualizing time-evolving linked geospatial data.

Sextant was designed as an open-source application[19] that is flexible, portable and interoperable with other GIS tools. This allows us to use it as a core building block for creating new web or mobile applications, utilizing the provided features. The core feature of Sextant is the ability to create thematic maps by combining geospatial and temporal information that exists in a number of heterogeneous data sources ranging from standard SPARQL endpoints to SPARQL endpoints following the standard GeoSPARQL defined by the OGC, or well-adopted geospatial file formats, like KML, GML and GeoTIFF. In this manner we provide functionality to domain experts from different fields in creating thematic maps, which emphasize spatial variation of one or a small number of geographic distributions. Each thematic map is represented using a map ontology that assists on modelling these maps in RDF and allows for easy sharing, editing and search mechanisms over existing maps.

4 Implementing the Green City Use Case Using Linked Geospatial Data Software

In this section we present the implementation of the Green City use case using the pipeline of the previous section and the relevant software for each stage of the pipeline.

4.1 Ingestion

In Green City use case, access to Copernicus data and information was achieved in two ways: (1) by downloading the data via the Copernicus Open Access Hub or

---

[19] http://sextant.di.uoa.gr/.

the Websites of individual Copernicus services, and (2) via the popular OPeNDAP framework[20] for accessing scientific data.

4.2 Dataset Discovery

The Copernicus Open Access Hub[21] offers access to Sentinel data, using a simple graphical interface that enables users to specify the extent of the geographical area one is interested in. In the Green City use case though, we mainly used data from the Land Monitoring service that processes Copernicus data and produces higher-level products that are of importance in the corresponding thematic area. To detect green areas within the cities, we used the Leaf Area Index dataset, produced by Sentinel-3 data, and the CORINE land cover dataset for 2018, produced by Sentinel-2 and Landsat-8 (gap filling) data. For air quality, we used the Nitrogen Dioxide index, produced by Sentinel-5p data. Moreover, we used data from OpenStreetMap (OSM) and the Database of Global Administrative Areas (GADM).

4.3 Knowledge Discovery

Although in the Green City use case this step of the pipeline was not needed, it is a very crucial step that allows us to discover knowledge hidden in the EO images and use ontologies to describe this knowledge. Such techniques were used in the context of the projects TELEIOS and ExtremeEarth by our group in collaboration with Remote Sensing scientists. In TELEIOS, colleagues from the National Observatory of Athens developed algorithms to detect fires in SEVIRI images in the context of a fire monitoring application [20]. In ExtremeEarth, colleagues from the University of Trento perform the accurate crop type mapping needed the Food Security use case, using a deep learning architecture for Sentinel-2 images [29].

4.4 Transformation into RDF

In this stage of the pipeline, the outputs of the previous two stages are transformed into RDF, so that they can be combined with other interesting linked geospatial data. In the Green City use case, RDF is used to represent Earth Observation data produced by the Copernicus Land Monitoring service, air quality indices,

---

[20] https://www.opendap.org/.

[21] https://scihub.copernicus.eu/.

OpenStreetMap data and data from the database of Global Administrative Areas, as described in Sect. 4.2.

To transform the mentioned data into RDF, we developed INSPIRE-compliant ontologies. In the process of constructing ontologies to model Copernicus and other geospatial data, our aim is to provide standard-compliant, reusable and extensible ontologies. In this direction, we opted to follow vocabularies that have been defined in well-established standards, such as the INSPIRE directives and the OGC.

The INSPIRE directive aims to create an interoperable spatial data infrastructure for the European Union, to enable the sharing of spatial information among public sector organizations and better facilitate public access to spatial information across Europe.[22] INSPIRE-compliant ontologies are ontologies which conform to the INSPIRE requirements and recommendations. Our initial approach was to reuse existing INSPIRE-compliant ontologies, but since these efforts are not as close to the INSPIRE specifications as we would like to, we decided to construct our own INSPIRE-compliant versions, following the data specifications as closely as possible. Our aim is to reuse these ontologies for other datasets that belong to the same INSPIRE themes and also publish them so that others can reuse these ontologies for their geospatial datasets as well.

The ontologies we constructed for the Green City use case are the following:

- The ontology for the global database of *Leaf Area Index (LAI)*, as shown in the link: http://pyravlos-vm5.di.uoa.gr/laiOntology.png.
- The *CORINE Land Cover (CLC) ontology*, included in the link http://pyravlos-vm5.di.uoa.gr/corineLandCover.svg, shows the ontology constructed for the CLC dataset. The ontology is a specialization of the general ontology that we constructed to model the respective Land Cover theme of INSPIRE so that we have the first INSPIRE-compliant ontology.
- The ontology for the *Ozone Forecast* dataset, including the NO_2 index, as described in this link:
  http://pyravlos-vm5.di.uoa.gr/atmosphereTimeSeriesOntology.png.
- The *OpenStreetMap (OSM)* ontology, as shown in this figure: http://sites.pyravlos.di.uoa.gr/dragonOSM.svg.
- The ontology for the *Database of Global Administrative Areas (GADM)*, included in this link: http://pyravlos-vm5.di.uoa.gr/gadmOntology.png.

Figure 3 provides the ontology we constructed for the GADM dataset. To construct this ontology, we extended the GeoSPARQL ontology (namespaces sf and geo). For the class and properties that we introduced we use the prefix gadm.[23] The GADM ontology can be used so that a GADM dataset[24] can be either converted into RDF or queried on-the-fly.

---

[22] https://inspire.ec.europa.eu.

[23] The corresponding namespace is: http://www.app-lab.eu/gadm/.

[24] https://gadm.org/data.html.

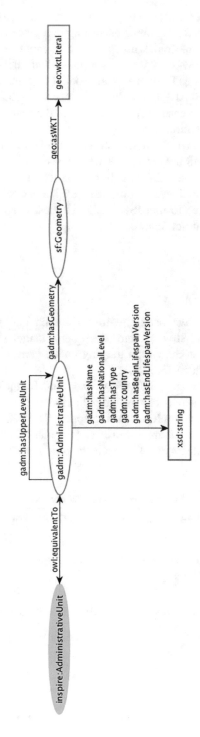

Fig. 3 The GADM ontology

For the transformation of the aforementioned datasets, we use the tool GeoTriples that automatically produces RDF graphs according to a given ontology. Shapefiles along with corresponding ontologies are provided as input to GeoTriples, which automatically creates R2RML or RML mappings that dictate the method of conversion of data into the RDF data model. Spatial information is mapped into RDF according to the GeoSPARQL vocabulary. Since GeoTriples does not support NetCDF files as input, in the case of the LAI dataset, the translation into RDF was done by writing a custom Python script.

It is very important to make the above datasets available on the Web as linked data, in order to increase their use, as, in this way, they can be made "interoperable" and more valuable when they are linked together. To achieve this goal, we followed Google Dataset Search guidelines and annotated all the datasets of the Green City use case by using the markup format JSON-LD. All these datasets can be searched and found using Google Dataset Search.

4.5 Storage/Querying

For storage and querying, we used the tools Strabon and Ontop-spatial. The spatiotemporal RDF store Strabon and the query languages stSPARQL and GeoSPARQL are used for storage and querying linked geospatial data originating from transforming EO products into RDF.

Strabon was utilized to create the SPARQL endpoints for the GADM, CLC 2018 and OSM parks data sources that are originally distributed in vector formats and are not updated frequently. For example, assuming appropriate PREFIX definitions, the GeoSPARQL query shown in Listing 1 retrieves how many CORINE areas in Paris belong to every land use category and projects the union of the geometries of these areas per category.

Listing 1 CORINE areas in Paris for every land cover category and their geometries

```
SELECT DISTINCT ?landUse (strdf:union(?w3) as ?geo) (count(?c) as
    ?instances)
WHERE{
  ?adm rdf:type gadm:AdministrativeUnit .
  ?adm gadm:hasName ?name .
  ?adm gadm:belongsToAdm2 ?adm2 .
  ?adm2 gadm:hasName
      "Paris"^^<http://www.w3.org/2001/XMLSchema#string> .
  ?adm geo:hasGeometry ?geo2 .
  ?geo2 geo:asWKT ?w2 .
  ?c corine:hasLandUse ?landUse .
  ?c geo:hasGeometry ?geo3 .
  ?geo3 geo:asWKT ?w3 .
  FILTER(geof:sfIntersects(?w2,?w3))}
GROUP BY ?landUse
```

For the rest of the data sources (LAI and NO$_2$) that are updated regularly and are distributed in raster formats, we chose to use Ontop-spatial. This solution does not require the transformation of the source data into RDF and allows us to create virtual RDF graphs on top of geospatial databases and data delivered through the OPeNDAP protocol, so they can be readily available for querying. In this case, the developer has to write R2RML mappings expressing the correspondence between a data source and classes/properties in the corresponding ontology. An example of such a mapping is provided in Listing 2, in the native mapping language of Ontop-spatial which is less verbose than R2RML.

Listing 2 Example of mappings

```
mappingId opendap_mapping
target lai:{id} rdf:type lai:Observation .
       lai:{id} lai:lai {LAI}^^xsd:float;
            time:hasTime {ts}^^xsd:dateTime .
       lai:{id} geo:hasGeometry _:g .
       _:g geo:asWKT {loc}^^geo:wktLiteral .
source SELECT id, LAI, ts, loc
       FROM (ordered opendap
       url:https://analytics.ramani.ujuizi.com/
       thredds/dodsC/Copernicus-Land-timeseries-
       global-LAI%29/readdods/LAI/, 10)
       WHERE LAI > 0
```

In the example mappings shown in Listing 2, the `source` is the LAI dataset discussed above, while the `target` part of the mapping encodes how the relational data is mapped into RDF terms. Given the mapping provided above, we can pose the GeoSPARQL query provided in Listing 3 to retrieve the LAI values and the geometries of the corresponding areas.

Listing 3 Query retrieving LAI values and locations

```
SELECT DISTINCT ?s ?wkt ?lai
WHERE { ?s lai:hasLai ?lai .
        ?s geo:hasGeometry ?g .
        ?g geo:asWKT ?wkt }
```

4.6 Publishing

Some of the RDF datasets that are used in the Green City use case have been published in the datahub https://datahub.ckan.io/organization/app-lab.

4.7 Interlinking

In the Green City use case, combining information from different geospatial sources was crucial, as we needed to spatially interlink the administrative divisions of a city with the EO data and OSM parks. We address this issue by employing the geospatial and temporal component of the framework Silk,[25] which is a component that enables users to discover a wide variety of spatial and temporal relations, such as intersects, contains, before, and during, between different sources of data.

To retrieve features for which a spatial relation holds (e.g., intersection and containment), we ask Silk to search for these relations between two RDF data sources given the relations' definitions. The outcome contains all of the entities for which the relations hold. For example, to interlink the CLC and the GADM datasets, a CLC class that intersects an administrative division is interlinked with it with the property *geo:sfIntersects*. The discovered relations are then materialized in the RDF store, resulting in a more semantically informative dataset.

Interlinking with topological and temporal relations can be used to considerably decrease the query response time by replacing the spatial and temporal functions with the respective bindings. For example, we can pose a SPARQL query by replacing the function *geof:sfIntersects* with the triple pattern *?clc geo:sfIntersects ?ad*, as the geospatial features for which the relation *geo:sfIntersects* holds have already been discovered, and the evaluation engine would simply have to retrieve the respective bindings instead of calculating the spatial filter.

4.8 Exploration and Visualization

In order to visualize the Green City use case, we used the tool Sextant to create a map for the city of Paris, France. We used Sextant to build a temporal map that shows the "greenness" of Paris, using the datasets LAI, GADM, CLC 2018, NO_2 and OSM. We show how the LAI values change over time in each administrative area of Paris and correlate these readings with the land cover of each area taken from the CORINE land cover dataset. This allows us to explain the differences in LAI values over different areas. For example, Paris areas belonging to the CORINE land cover class *clc:greenUrbanAreas* overlap with parks in OpenStreetMap and show higher LAI values over time than industrial areas.

Sextant allows us to pose GeoSPARQL/stSPARQL queries to SPARQL endpoints and visualize the results as layers on the map. Utilizing this feature, we created the thematic map for Paris[26] shown in Fig. 4, which consists of six layers:

---

[25] http://silk.wbsg.de.

[26] http://test.strabon.di.uoa.gr/SextantOL3/?mapid=mpm2tf7ha6ai5f78_.

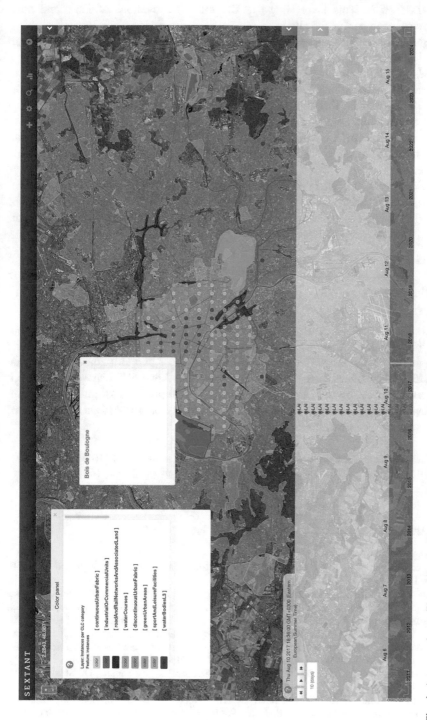

Fig. 4 A temporal map illustrating the "greenness" of Paris, created in Sextant

- *GADM Paris.* This layer shows us the different divisions of Paris and how "green" each part of the city is.
- *Instances per CLC category.* This layer shows us the different CLC classes that spatially intersect with the divisions on the city.
- *LAI.* This temporal layer consists of the different mean LAI values for area of Paris, for the months June–August 2017. The dots on the map are the centroids of 300×300 m areas that correspond to the pixel of the satellite image that contains the observation.
- *Mean LAI per Administrative Unit.* This is a statistical visualization layer, that shows us the mean LAI value for the time period of observation, for each division of Paris.
- *OSM Parks.* This layer shows us the parks that spatially intersect with the divisions of Paris.
- *NO_2.* This layer consists of the NO_2 mean concentration values for the area of Paris, for the observed time period.

5 Summary

We presented a data science pipeline for big, linked and open EO data and showed how this pipeline can be used to develop a Green City use case. The pipeline is implemented using the software developed in five FP7 and Horizon 2020 projects (TELEIOS, LEO, Melodies, Optique and Copernicus App Lab). The work presented in this chapter is now continued in the Horizon 2020 project ExtremeEarth, where we develop deep learning and big data techniques for Copernicus data in the context of two use cases: Food Security and Polar.

Acknowledgments This work has been funded by the FP7 projects TELEIOS (257662), LEO (611141), MELODIES (603525) and the H2020 project Copernicus App Lab (730124).

References

1. Auer, S., Bühmann, L., Dirschl, C., et al. (2012). Managing the life-cycle of linked data with the LOD2 stack. In *ISWC* .
2. Auer, S., Scerri, S., Versteden, A., Pauwels, E., Charalambidis, A., Konstantopoulos, S., Lehmann, J., Jabeen, H., Ermilov, I., Sejdiu, G., Ikonomopoulos, A., Andronopoulos, S., Vlachogiannis, M., Pappas, C., Davettas, A., Klampanos, I.A., Grigoropoulos, E., Karkaletsis, V., de Boer, V., Siebes, R., Mami, M.N., . . . Vidal, M. (2017). The bigdataeurope platform – supporting the variety dimension of big data. In *Web Engineering – 17th International Conference, ICWE 2017, Rome, Italy, June 5–8, 2017, Proceedings* (pp. 41–59).
3. Bereta, K., Caumont, H., Daniels, U., Goor, E., Koubarakis, M., Pantazi, D., Stamoulis, G., Ubels, S., Venus, V., & Wahyudi, F. (2019). The copernicus app lab project: Easy access to copernicus data. In *Advances in Database Technology – 22nd International Conference on Extending Database Technology, EDBT 2019, Lisbon, Portugal, March 26–29, 2019* (pp. 501–511).

4. Bereta, K., & Koubarakis, M. (2016). Ontop of geospatial databases. In *The Semantic Web – ISWC 2016 – 15th International Semantic Web Conference, Kobe, Japan, October 17–21, 2016, Proceedings, Part I* (pp. 37–52).

5. Bereta, K., Smeros, P., & Koubarakis, M. (2013). Representation and querying of valid time of triples in linked geospatial data. In *The Semantic Web: Semantics and Big Data, Lecture Notes in Computer Science* (Vol. 7882, pp. 259–274). Springer.

6. Bereta, K., Xiao, G., & Koubarakis, M. (2019). Ontop-spatial: Ontop of geospatial databases. *Journal of Web Semantics, 58*, 100514.

7. Blower, J., Clifford, D., Goncalves, P., & Koubarakis, M.: The melodies project: Integrating diverse data using linked data and cloud computing. In *Proceedings of the 2014 Conference on Big Data from Space (BiDS)* (2014)

8. Burgstaller, S., Angermair, W., Migdall, S., Bach, H., Vlachopoulos, I., Savva, D., Smeros, P., Stamoulis, G., Bereta, K., & Koubarakis, M. (2017). Leopatra: A mobile application for smart fertilization based on linked data. In *Proceedings of the 8th International Conference on Information and Communication Technologies in Agriculture, Food and Environment (HAICTA 2017), Chania, Crete Island, Greece, September 21–24, 2017* (pp. 160–171). http://ceur-ws.org/Vol-2030/HAICTA_2017_paper17.pdf

9. Calvanese, D., Cogrel, B., Komla-Ebri, S., Kontchakov, R., Lanti, D., Rezk, M., Rodriguez-Muro, M., & Xiao, G. (2017). Ontop: Answering SPARQL queries over relational databases. *Semantic Web, 8*(3), 471–487.

10. Das, S., Sundara, S., & Cyganiak, R. (2012). R2RML: RDB to RDF mapping language. http://www.w3.org/TR/r2rml/

11. Dimou, A., Vander, S., et al. (2014). RML: A generic language for integrated RDF mappings of heterogeneous data. In *Proceedings of the 7th Workshop on Linked Data on the Web*. http://events.linkeddata.org/ldow2014/papers/ldow2014_paper_01.pdf

12. Espinoza-Molina, D., & Datcu, M. (2013). Earth-observation image retrieval based on content, semantics, and metadata. *IEEE Transactions on Geoscience and Remote Sensing, 51*(11), 5145–5159.

13. Espinoza-Molina, D., Nikolaou, C., Dumitru, C.O., Bereta, K., Koubarakis, M., Schwarz, G., & Datcu, M. (2015). Very-high-resolution SAR images and linked open data analytics based on ontologies. *IEEE Journal of Selected Topics in Applied Earth Observations and Remote Sensing, 8*(4), 1696–1708.

14. European Commission, European Space Agency. (2019). Copenicus sentinel data access annual report 2019. Available from https://earth.esa.int/web/sentinel/news/-/article/copernicus-sentinel-data-access-annual-report-2019

15. GAEL, NOA, GRNET, Serco. (2019). Copernicus market report. Available from https://www.copernicus.eu/sites/default/files/2019-02/PwC_Copernicus_Market_Report_2019_PDF_version.pdf

16. Garbis, G., Kyzirakos, K., & Koubarakis, M. (2013). Geographica: A benchmark for geospatial rdf stores (long version). In *The Semantic Web – ISWC 2013, Lecture Notes in Computer Science* (Vol. 8219, pp. 343–359). Springer.

17. Ioannidis, T., Garbis, G., Kyzirakos, K., Bereta, K., & Koubarakis, M. (2019). Evaluating geospatial RDF stores using the benchmark geographica 2. CoRR abs/1906.01933

18. Koubarakis, M., Bereta, K., Bilidas, D., Giannousis, K., Ioannidis, T., Pantazi, D., Stamoulis, G., Dowling, J., Haridi, S., Vlassov, V., Bruzzone, L., Paris, C., Eltoft, T., Krämer, T., Charalambidis, A., Karkaletsis, V., Konstantopoulos, S., Kakantousis, T., Datcu, M., Dumitru, C.O., Appel, F., ... Fleming, A. (2019). From copernicus big data to extreme earth analytics. In *Advances in Database Technology – 22nd International Conference on Extending Database Technology, EDBT 2019, Lisbon, Portugal, March 26–29, 2019* (pp. 690–693).

19. Koubarakis, M., Bereta, K., Papadakis, G., Savva, D., Stamoulis, G. (2017). Big, linked geospatial data and its applications in earth observation. *IEEE Internet Computing, July/August*, 87–91.

20. Koubarakis, M., Kontoes, C., & Manegold, S. (2013). Real-time wildfire monitoring using scientific database and linked data technologies. In *Joint 2013 EDBT/ICDT Conferences, EDBT '13 Proceedings, Genoa, Italy, March 18–22, 2013* (pp. 649–660).

21. Koubarakis, M., Kyzirakos, K., Nikolaou, C., Garbis, G., Bereta, K., Dogani, R., Giannakopoulou, S., Smeros, P., Savva, D., Stamoulis, G., Vlachopoulos, G., Manegold, S., Kontoes, C., Herekakis, T., Papoutsis, I., ... Michail, D. (2016). Managing big, linked, and open earth-observation data: Using the TELEIOS/LEO software stack. *IEEE Geoscience and Remote Sensing Magazine, 4*(3), 23–37.
22. Koubarakis, M., Sioutis, M., Kyzirakos, K., Karpathiotakis, M., et al. (2012). Building virtual earth observatories using ontologies, linked geospatial data and knowledge discovery algorithms. In *ODBASE*.
23. Kyzirakos, K., Koubarakis, M., & Kaoudi, Z. (2009). Data models and languages for registries in SemsorGrid4Env. Deliverable D3.1, Dept. of Informatics and Telecommunications, University of Athens.
24. Kyzirakos, K., Karpathiotakis, M., & Koubarakis, M. (2012). Strabon: A semantic geospatial DBMS. In *The Semantic Web – ISWC 2012 – 11th International Semantic Web Conference, Boston, MA, USA, November 11–15, 2012, Proceedings, Part I* (pp. 295–311)
25. Kyzirakos, K., Karpathiotakis, M., & Koubarakis, M. (2012). Strabon: A Semantic Geospatial DBMS. In: *ISWC*.
26. Kyzirakos, K., Savva, D., Vlachopoulos, I., Vasileiou, A., Karalis, N., Koubarakis, M., & Manegold, S. (2018). Geotriples: Transforming geospatial data into RDF graphs using R2RML and RML mappings. *Journal of Web Semantics, 52–53*, 16–32.
27. Maali, F., Cyganiak, R., & Peristeras, V. (2012). A publishing pipeline for linked government data. In *ESWC*.
28. Nikolaou, C., Dogani, K., Bereta, K., Garbis, G., Karpathiotakis, M., Kyzirakos, K., & Koubarakis, M. (2015). Sextant: Visualizing time-evolving linked geospatial data. *Journal of Web Semantics, 35*, 35–52.
29. Paris, C., Weikmann, G., & Bruzzone, L. (2020). Monitoring of agricultural areas by using Sentinel 2 image time series and deep learning techniques. In L. Bruzzone, F. Bovolo, & E. Santi (Eds.) *Image and Signal Processing for Remote Sensing XXVI* (Vol. 11533, pp. 122–131). International Society for Optics and Photonics, SPIE.
30. Perry, M., & Herring, J. (2012). Geosparql – a geographic query language for RDF data. Available from https://www.ogc.org/standards/geosparql
31. Saveta, T., Fundulaki, I., Flouris, G., & Ngomo, A. N. (2018). Spgen: A benchmark generator for spatial link discovery tools. In *The Semantic Web – ISWC 2018 – 17th International Semantic Web Conference, Monterey, CA, USA, October 8–12, 2018, Proceedings, Part I* (pp. 408–423).
32. Sherif, M. A., Dreßler, K., Smeros, P., & Ngomo, A. N. (2017). Radon – rapid discovery of topological relations. In *AAAI* (pp. 175–181).

Towards Cognitive Ports of the Future

Santiago Cáceres, Francisco Valverde, Carlos E. Palau, Andreu Belsa Pellicer,
Christos A. Gizelis, Dimosthenes Krassas, Hanane Becha, Réda Khouani,
Andreas Metzger, Nikos Tzagkarakis, Anthousa Karkoglou,
Anastasios Nikolakopoulos, Achilleas Marinakis, Vrettos Moulos,
Antonios Litke, Amir Shayan Ahmadian, and Jan Jürjens

Abstract In modern societies, the rampant growth of data management technologies—that have access to data sources from a plethora of heterogeneous systems—enables data analysts to leverage their advantages to new areas and critical infrastructures. However, there is no global reference standard for data platform technology. Data platforms scenarios are characterized by a high degree of heterogeneity at all levels (middleware, application service, data/semantics, scalability, and governance), preventing deployment, federation,

S. Cáceres · F. Valverde
Instituto Tecnológico de Informática, Valencia, Spain
e-mail: scaceres@iti.es; fvalverde@iti.es

C. E. Palau · A. B. Pellicer
Universitat Politécnica de Valencia, Valencia, Spain
e-mail: cpalau@dcom.upv.es; anbelpel@upv.es

C. A. Gizelis · D. Krassas
Hellenic Telecommunications Organization S.A., Maroussi, Athens, Greece
e-mail: cgkizelis@cosmote.gr; dimkrass@ote.gr

H. Becha · R. Khouani
Traxens, Marseille, France
e-mail: h.becha@traxens.com; r.khouani@traxens.com

A. Metzger
Ruhr Institute for Software Technology, University of Duisburg-Essen, Essen, Germany
e-mail: andreas.metzger@paluno.uni-due.de

N. Tzagkarakis · A. Karkoglou · A. Nikolakopoulos · A. Marinakis · V. Moulos (✉) · A. Litke
National Technical University of Athens, Zografou, Athens, Greece
e-mail: ntzagkarakis@mail.ntua.gr; akarkoglou@mail.ntua.gr; tasosnikolakop@mail.ntua.gr;
achmarin@mail.ntua.gr; vrettos@mail.ntua.gr; litke@mail.ntua.gr

A. S. Ahmadian
Institute for Software Technology, University of Koblenz-Landau, Koblenz, Germany
e-mail: ahmadian@uni-koblenz.de

J. Jürjens
Fraunhofer-Institute for Software and Systems Engineering ISST, Dortmund, Germany
e-mail: juerjens@uni-koblenz.de

E. Curry et al. (eds.), *Technologies and Applications for Big Data Value*,
https://doi.org/10.1007/978-3-030-78307-5_20

and interoperability of existing solutions. Although many initiatives are dealing with developing data platform architectures in diversified application domains, not many projects have addressed integration in port environments with the possibility of including cognitive services. Unlike other cases, port environment is a complex system that consists of multiple heterogeneous critical infrastructures, which are connected and dependent on each other. The key pillar is to define the design of a secure interoperable system facilitating the exchange of data through standardized data models, based on common semantics, and offering advanced interconnection capabilities leading to cooperation between different IT/IoT/Objects platforms. This contribution deals with scalability, interoperability, and standardization features of data platforms from a business point of view in a smart and cognitive port case study. The main goal is to design an innovative platform, named DataPorts, which will overcome these obstacles and provide an ecosystem where port authorities, external data platforms, transportation, and logistics companies can cooperate and create the basis to offer cognitive services. The chapter relates to knowledge and learning as well as to systems, methodologies, hardware, and tools cross-sectorial technology enablers of the AI, Data and Robotics Strategic Research, Innovation & Deployment Agenda (Milano et al., Strategic research, innovation and deployment agenda - AI, data and robotics partnership. Third release. Big Data Value Association, 2020).

Keywords Industry 4.0 · Data for AI · Port authorities

1   Introduction

Serverless architecture is based on the ground that the deployment is a transparent process where the developer is not aware of the cluster/server that the stateless functions are deployed. Although the inability to specify where the functions should run seems to weaken the overall architecture caused by a lack of control, that is compensated by the benefit of performance of the application.

The digital transformation of the ports toward the Fourth Industrial Revolution is revealing opportunities related to the add-on services that can be provided. Port platforms are now integrating available data sources capturing the potential needs that arise from the increased demand for more accurate and complete information [20]. The value of that new trend is also boosted by the new wave of startups that implement over-the-top services. However, most port authorities are not technologically ready to host these services and frameworks. The necessary technology infrastructure has anti-diametric different requirements from the one that is deployed. The same drawback applies to the architecture where in most cases must be shifted to serverless-oriented solutions.

For example, port authorities have the chance to enrich their services with smart containers infrastructure. This add-on IoT infrastructure enables the online information exchange capability about the entire journey and the conditions of the cargo directly to the rest of the supply chain without human intervention. This provides greater visibility to the stakeholders within the transaction as well as to regulatory agencies who need detailed information on the consignments before they arrive at the border. This technology can be combined with other innovations such as blockchain, big data, or data pipelines to provide even more facilitation to the trading community. In all of these cases, though, we see that creating clear, unambiguous message exchange standards will allow to capitalize the full potential of the enhanced data. This shared data will enable the creation of new value-added services that the port authorities can benefit from.

Unfortunately, port authorities are unprepared nowadays to host that massive information exchange system. One reason is investments in IT that should be made, as the port systems will be upgraded to the main exchange point where companies and start-ups will connect to buy/sell data, information, and services. Another reason is the lack of expertise in new design implementations from the port authority staff since the orientation of the authority is anti-diametric different. Although a few of these needed features are addressed by serverless platforms like Openwhisk [3], OpenFaaS [33], OpenLambda [23], and Kubeless [26], some open challenges remain. Scalability, interoperability, architectural design, and standardization are the key pillars for a successful ecosystem. Definitions from a port authority view is presented in Sect. 2, where a complete analysis of these concepts is given in Sects. 3, 4, 5, and 6. Finally, we conclude with the business challenges that future systems will face after the transition to the new era.

2 Challenges for Port-Oriented Cognitive Services

A cognitive service is a software component that uses AI and big data capabilities to solve a business task. Cognitive services are ready-to-use solutions to be integrated into the context of software products for improving the decision-making process related to data. Some cloud providers offer general cognitive services such as image classification or natural text recognition/translation. However, this novel paradigm has not been applied in the ports' domain. Ports share common business tasks in which cognitive services could provide answers. Some examples are the prediction of the Expected Time of Arrival (ETA) of a container or a vessel, the truck turnaround time to deliver a container to a terminal, or the definition of a booking system to reduce environmental impact. The main goal is to build such services to be generic enough to be applied in different ports and use cases. To enable the cognitive services approach envisioned in the context of the project, the DataPorts [13] platform must address the following technical challenges:

1. One issue to address is to enable the data sharing between an undefined number of port stakeholders, such as terminal operators, port authorities, logistic carriers, and so on. The accuracy of cognitive services is directly related to the amount of available data. For building such a data ecosystem, the defined architecture must include scalability as the first design principle. To address this challenge,

we foresee the introduction of the International Data Spaces Association (IDSA) Reference Architecture Model [4], a standard solution with the required building blocks for achieving a seamless integration between organizations.

2. Scalability for AI training: Training models for cognitive services is a time-consuming task even with a powerful computing infrastructure for supporting it. As the state of the art evolves, new frameworks and techniques must be tested in order to find the optimal one, which provides better accuracy for the problem at hand. A cognitive service vision implies that not specific know-how from the data science domain is required: the end-user is who defines the training process with little manual intervention. This fact leads to the definition of several training alternatives, which must run simultaneously over a distributed infrastructure. This challenge will be addressed by the DataPorts platform, introducing the most suitable technological approaches from the ML DevOps area.

3. Heterogeneous data processing: Ports field is a domain in which several IT infrastructures, information systems like TOS or IoT sensing devices, are potential candidates to become a valuable data source. In this scenario, two main challenges arise: how to deal with the heterogeneity of the data sources (formats, schema, etc.) and with an undefined volume of data. The DataPorts platform will support the heterogeneity of schemas, applying techniques from the semantic interoperability domain and taking into account vocabularies or taxonomies from standardization bodies. Following the good practices for big data processing, such as the use of containerized application and distributed databases, we expect to provide the required tools for enabling a scalable data processing.

4. A trusted data governance: Ownership of the data is a key issue in any discussion related to data sharing between different organizations. To enhance the data sharing needed to build cognitive services, the DataPorts platform must first of all provide a trusted framework for defining data sharing rules to specific users, roles, and organizations. This framework must also enforce that the data is used following the specifications the data owner has formally defined. Data management, when this data is outside the boundaries of the organization, is a challenge that requires a set of trusted software components and clear security procedures. We foresee the use of smart contracts, in the context of a blockchain network among organizations, as the technological foundation to address this challenge.

The next sections introduce in more detail how we expect to address these overall challenges in terms of the overall architecture management, scalability, interoperability, and standardization. These challenges are in direct relation with the DaaS and FaaS overall strategic plan as well as with the horizontal concern of data management, data analytics, and data visualization of the BDV Technical Reference Model [43].

3   Scalability

In order to define a programming model and architecture where small code snippets are executed in the cloud without any control over the resources on which the code runs, the industry came with the term "Serverless Computing." [5] It is by no means an indication that no servers exist, simply that the developer should leave most operational issues to the cloud provider, such as resource provisioning, monitoring, maintenance, scalability, and fault tolerance. The platform must guarantee the scalability and elasticity of the functions of the users. In response to load, and in anticipation of potential load, this means proactively provisioning resources. This is a more daunting serverless issue because these forecasts and provisioning decisions must be made with little to no application-level information. For instance, as an indicator of the load, the system can use request queue lengths but is blind to the nature of these requests.

In a few words, serverless computing allows application developers to decompose large applications into small functions, allowing application components to scale individually [29]. The majority of Function-as-a-Service systems use Kubernetes' built-in Horizontal Pod Autoscaling (HPA) for auto-scaling, which implements compute-resource-dependent auto-scaling of function instances [6]. However, custom scaling mechanisms, such as auto-scaling based on the number of concurrent in-flight requests, can also be implemented. Generally speaking, there exist many implementations regarding scaling and auto-scaling on serverless functions.

Scalability is the ability of a system to handle a growing amount of work by adding resources to itself. This means that scalability stands as a direct solution to any workload issue that might emerge in a system. Therefore, today's frameworks should take a full advantage of scalability's benefits by implementing tools that achieve exactly that. How can a system be scalable? Solutions vary. However, the selection field narrows down a lot when it comes to serverless computing systems. That's because serverless applications are created to have the scalability issue solved in advance. What remains to be answered is how we can further improve scalability in the serverless world.

Nevertheless, a solution to the scalability issue, distinct from the majority of the available ones, is through the prism of a Data-as-a-Service Marketplace [36]. When combined with a fully working Function-as-a-Service (FaaS) platform, this approach can lead to optimum scaling results. The core idea is about creating a serverless platform as part of a Data-as-a-Service (DaaS) marketplace repository framework, which enables dynamic scaling in order to ensure business continuity, such as real-time accommodation of rapidly evolving user numbers, and fault tolerance. That framework contains a multitude of readily accessible APIs to serve the needs of a growing and changing DaaS platform marketplace, while it provides great flexibility in selecting topologies and architectures for the storage pool. In essence, any node, either located in the cloud, at the physical location of the marketplace, or even at the edge of the network, may be used to store data as part of the repository cluster. That DaaS strategy uses the cloud to deliver data storage, analytics services, and processing orchestration tools in order to offer data in a manner that new-age applications can use.

4   International Data Spaces Architecture Reference Model

However, the adopted proposal in the context of the DataPorts project [13] is the International Data Spaces Architecture (IDSA) Reference Model [4]. According to the official documentation, the International Data Spaces (IDS) is a virtual data space leveraging existing standards and technologies, as well as governance models well accepted in the data economy, in order to facilitate secure and standardized data exchange/linkage in a trusted business ecosystem (illustration of the architecture is shown later on in Fig. 1). Therefore, it provides a basis for creating smart-service scenarios and facilitating innovative cross-company business processes. At the same time, it guarantees data sovereignty for data owners. Regarding the IDSA's Reference Model, it is highly scalable. In short, the IDSA reference architecture model's high scalability is attributable to the fact that this model is a decentralized architecture ("peer-to-peer" data exchange with redundant replicated connectors and brokers) without a central bottleneck.

The two main and most important components of the IDSA Reference Model are the "Broker" and the "Connector." These components are responsible for the model's decentralized architecture and its ability to be highly scalable, as mentioned before. Additional to these two are the "Data Apps," which are data services encapsulating the functionality of data processing and/or data transformation, packaged as container images for easy installation through the application container management. The Data Apps are distributed through a secure platform, the "IDS App Store." The IDS App Store contains a list of available Data Apps. An App Store therefore facilitates the registration, release, maintenance, and query of the Data App operations, as well as the provisioning of the Data App to a Connector. The presence of an App Store, with different types and categories of applications, means that the IDSA is highly scalable, since connectors, as the main component of the IDSA Reference Model, can be modified or expanded to increase the ability and functionality of the connectors based on different requirements and domains.

To sum up, the key point (and what is worth noting) is that this architecture model is capable of achieving high scalability, in combination with an existing FaaS platform. Regarding the connection of IDSA with the DataPorts project, an architecture design for cognitive ports has to enable data sharing and data governance in a trusted ecosystem including various ports stakeholders. In the context of cognitive ports, to realize trusted data sharing and data governance, one can benefit from two main approaches:

1. **Data is stored off-chain**: Generally in this approach, we may leverage the concept of the International Data Spaces (IDS) reference architecture model

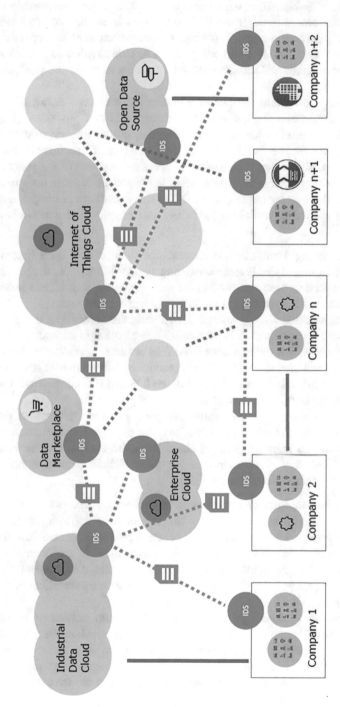

Fig. 1 International Data Spaces provides an ecosystem where various data sources are connected [4]

(RAM) [4]. In a nutshell, a peer-to-peer data exchange between the data owners and data consumers through the IDS connectors is considered. To ensure the security and privacy aspects, blockchain manages consent of access to data. Smart contracts decide if a particular access to data is allowed concerning the invoker's credentials and the specification of access rules for the particular data [1].

2. **Data is stored on chain (blockchain platform for shared data)**: Blockchain platform records transactions related to shared data and processes of all participants in a business network. For instance, coldchain temperature-alarms of a container for verification of its state such as if conditions have been compromised, allows everybody in the network to be aware of the event and act upon it. This approach in fact ensures verifiable and immutable information on shared data through the entire chain to all business network participants serving as a single source of truth and providing transparency and a not-repudiation process.

As depicted in Fig. 1 and mentioned above, the International Data Spaces (IDS) is a virtual environment that leverages existing standards and technologies, as well as governance models well-accepted in the data economy, to facilitate secure and standardized data exchange and data linkage in a trusted ecosystem [4].

As stated before and already analyzed, two main components of the IDS RAM are a broker and a connector. A brief reminder, the broker is an intermediary that stores and manages information about the data sources available in the IDS. It mainly receives and provides metadata. As mentioned before, data sharing and data exchange are the main fundamental aspect of the IDS. An IDS connector is the main technical component for this purpose.

For the IDS connector, as the central component of the architecture, different variants of implementation are available, may be deployed in various scenarios, and can be acquired from different vendors. However, each connector is able to communicate with any other connector (or other technical components) in the ecosystem of the International Data Spaces.

- Operation: Stakeholders should be able to deploy connectors in their own IT environment. Additionally, they may run a connector software on mobile or embedded devices. The operator of the connector must always be able to describe the data workflow inside the connector. Moreover, users of a connector must be identifiable and manageable. Every action, data access, data transmission, and event has to be logged. This logging data allows to draw up statistical evaluations on data usage.
- Data exchange: A connector must receive data from an enterprise backend system, either through a push mechanism or a pull mechanism. The data can be either provided via an interface or pushed directly to other participants. Hence each connector has to be uniquely identifiable. Other connectors can subscribe to data sources or pull data from these sources. Data can be written into the backend system of other participants.

In addition to what it is described, the IDS RAM benefits from an information model, which is an essential agreement shared by the participants and components of the IDS, facilitating interoperability and compatibility. The main aim of this formal model is to enable (semi-)automated exchange of digital resources in a trusted ecosystem of distributed various parties, while the sovereignty of data owners is preserved.

Data sovereignty is defined as data subject's capability of being in full control of the provided data. To this end, all the organizations attempting to access the IDS ecosystem have to be certified, and so are the core software components (for instance, IDS connector) used for trusted data exchange and data sharing. Such a certification not only ensures security and trust, but the existence of certified components guarantees compliance with technical requirements ensuring interoperability.

5 Interoperability

5.1 Introduction

Interoperability among disparate computer systems is the ability to consume services and data with one another. Each software solution provides its own infrastructure, devices, APIs, and data formats, leading to compatibility issues and therefore to the need for specifications in terms of interaction with other software systems. Interoperability, as a complex concept, entails multiple aspects that address the effective communication and coordination between components and systems that might consist of a uniform platform at a larger scale. This section focuses on interoperability from the point of view of semantic interoperability and Application Programming Interfaces. On the one hand, semantic interoperability constructs a consolidated ontology model that ensures the unambiguity of data exchanges, since it is guaranteed that the requester and provider have a common understanding of the meaning of services and data. On the other hand, APIs constitute an interoperability tool that documents all the available services that are exposed by a software system, as well as the information about the respective communication protocols. Therefore, APIs are considered a significant step toward the interoperability of a system through the standardization of the components' communication. The present section describes the evolution of the service architectural models based on the ever-changing needs of communication, as well as the most powerful state-of-the-art tool for API standardization.

5.2   Semantic Interoperability

Semantic interoperability provides the ability to computer systems to exchange data with unambiguous, shared meaning [22]. In any system focused on interoperability, it is essential to take into account the production, collection, transmission and processing of large amounts of data. An application consuming those data needs to understand its structure and meaning. The metadata is responsible for representing these aspects in a readable way by a machine. The more expressive is the language used for representing the metadata, the more accurate the description might become. The metadata provides a semantic description of the data and can be utilized for many purposes, such as resource discovery, management, and access control [21].

The concept ontology refers to a structure that provides a vocabulary for a domain of interest, together with the meaning of entities present in that vocabulary. Typically, within an ontology, the entities may be grouped, put into a hierarchy, related to each other, and subdivided according to different notions of similarity. In the last two decades, the development of the Semantic Web resulted in the creation of many ontology-related languages, standards, and tools. Ontologies give the possibility to share a common understanding of the domain, to make its assumptions explicit, and to analyze and reuse the domain knowledge. In order to achieve shared meaning of data, the platforms or systems have to use a common ontology either explicitly, or implicitly, for example, via a semantic mediator [18].

Typically, organizations in transportation and logistics, with a particular focus on port logistics, have their own local standards. Sometimes they have a poor formalization of semantics, or they don't have explicit semantics at all [19]. The development of ontologies for logistics is not a trivial task. Define and use guidelines and best practices are necessary for this domain, especially to bridge the gap between theory and practice. A proper theoretical and methodological support is required for ontology engineering to deliver precise and consistent solutions to the market, as well as to provide solutions to practical issues to be close to the real market needs [12].

DataPorts project is developing a semantic framework for describing ports data together with mappings to standard vocabularies in order to simplify the reuse of data applications for analytics and forecasting. In particular, this semantic framework will codify the domain knowledge of the domain experts, and thus can be reused and exploited by the data experts directly, thereby empowering building cognitive port applications.

The relevance of the use of data platforms and the exploitation of data sharing are boosted by the high volume of different companies and public bodies that need to collaborate among them with different degrees of digital capacities. In this aspect, a semantic interoperability framework with a currently non-existing global ontology will improve such collaboration and data representation. Regarding the target group of users from the logistics domain, DataPorts semantics components make the interpretation of data and metadata more manageable for data users, so that their discovery is straightforward according to common search criteria. On the other hand, special provisions are taken in the platform to create an active and easy to enter data marketplace for third-party developers and service providers, with minimal integration efforts, clear monetization, and business value creation.

From a technical point of view, the aim of the project is to identify the different data sources to be integrated into the DataPorts platform, including the mechanisms to store and facilitate data management. Ontologies, mechanisms and enablers must also be defined to provide semantic interoperability with the data of these digital port infrastructures. This includes IoT devices, mobile applications, and, legacy databases and systems. Finally, to develop the semantic-based tools and components needed to facilitate the generation of interfaces to interact and manage the information of these data sources through the DataPorts platform. The data platform will guarantee semantic interoperability in order to provide a unified virtualized view of the data for its use by the different data consumers and the data analytic services. Figure 2 shows the architecture designed to achieve this purpose.

The Data Access Component solves the problem of access to different data sources or origins in a secure way. Given the very different types of sources, the platform will have to cope with the variety of data sources. It is necessary to analyze the provided interfaces of each data source. These interfaces should facilitate the way to access, to analyze the format and to understand the way to receive the data.

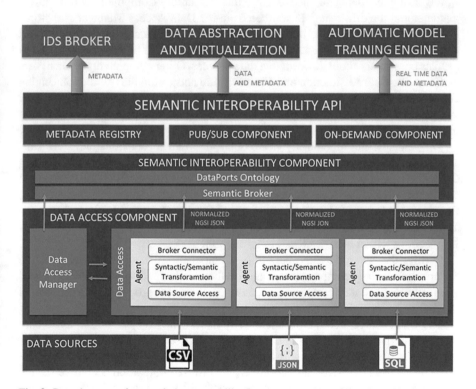

Fig. 2 Data Access and semantic interoperability Layer components of DataPorts Platform

They should recognize the volume, the velocity and the veracity of the information. In addition, to consider the existence of an ontology in the data source. To deal with these heterogeneous data sources, a data access agent is necessary for each data source integrated into DataPorts platform. An agent or adapter is the piece of software in charge of acquiring data from a source under certain conditions. Then the agents transform this data acquired from the sources to data in the DataPorts common data model. Finally, this data is sent to the other platform components. In addition, the Data Access Manager manages the metadata description of the data processed by the agents and the interaction of the agents with the other platform components. To recapitulate, the Data Access Component is the responsible for getting the sharing data and metadata description from the different data sources. It performs the processes to make the data sources understandable and available to the other platform components.

The semantic interoperability Component provides a unified API to access the data from the different data sources connected to the DataPorts platform, providing both real-time and batch data to the data consumers. In collaboration with the Data Access component, it will provide a data shared semantics solution for the interoperability of diverse data sources using existing ontologies. The output data will follow the common DataPorts ontology.

Regarding the metadata, the semantic interoperability Component obtains information about the different data sources from the Data Access Manager and stores it in the metadata registry to make it available for the other subcomponents of this layer. In addition, the metadata is sent to the IDS broker to provide information to the data consumers about the available data sources. Other components of DataPorts platform like Data Abstraction and Vitrualization and Automatic Prediction Engine can retrieve this metadata description of the data sources by asking the semantic interoperability API. The output metadata will follow the common DataPorts ontology.

The semantic interoperability Component provides a repository with the DataPorts Data Model and DataPorts Ontology. It offers an ontology definition using OWL [34] and a Data Model description using JSON schema and JSON-LD context documents. The aim is to integrate the following ontologies and data models: Fiware Data Models [14], IDSA Information Model [24], United Nations Centre for Trade Facilitation and Electronic Business (UN/CEFACT) [40] Model, and Smart Applications REFerence (SAREF) ontology [38].

Finally, the semantic interoperability Component API will interact with the Security and Privacy component to enable authentication and confidentiality, as well as to enforce the data access policies, in order to ensure proper data protection in the exchanges with the data consumers.

The open-source Fiware platform [16] has been selected to act as a core element of the Data Access and semantic interoperability components of the architecture presented, and it will be adapted, customized, and extended to fit the DataPorts project expectations. Moreover, the Fiware Foundation is involved in the development of IDS implementations and is actively cooperating with the Industrial Data Space Association (IDSA). Together, Fiware and IDSA are working on the first open-source implementation of the IDS Reference Architecture [2].

The aim behind Fiware is that technology should be accessible for everyone, with a focus on interoperability, modularity, and customizability. Data should seamlessly merge with data from other relevant sources. For that reason, Fiware components provide standardized data formats and API to simplify this integration. The platform is open source and can be easily embedded by ecosystem partners in the design of their solutions and reduce vendor lock-in risks. The standardized API means that services can operate on different vendor platforms. Fiware NGSI [17] is the API exposed by the Orion Context Broker and is used for the integration of platform components within a "Powered by Fiware" platform and by applications to update or consume context information. The Fiware NGSI (Next Generation Service Interface) API defines a data model for context information, an interface for exchanging context information, and a context availability interface for queries on how to obtain context information. The agents are the connectors that guarantee the transmission of raw data to Orion Context Broker using their own native protocols.

Some implementation decisions have been made in order to collaborate with the Fiware ecosystem. Firstly, all data and metadata formats are going to be designed to follow Fiware NGSI data models specifications [15]. Secondly, the use of the Orion Context Broker is being adopted as the semantic broker component [17]. Finally, regarding the data access agents, they are not mandatory to be implemented with a closed specific technology, but in order to provide a standardized SDK (Software Development Kit) to develop agents, the aim is to use the pyngsi Python framework [37].

The work done in previous European projects, where Fiware technology is a key element, is taken as a reference in the DataPorts implementation. For example, it is interesting to highlight projects like SynchroniCity [39] and Boost 4.0 [11]. SynchroniCity is aimed to establish a reference architecture for the envisioned IoT-enabled city market place with identified interoperability points and interfaces and data models for different verticals. The baseline of the SynchroniCity [39] data models is the FIWARE Data Models initially created by the FIWARE Community and have been expressed using the ETSI standard NGSI-LD. Regarding Boost 4.0, the aim of the project is to implement the European Data Space with FIWARE technologies.

5.3 Application Programming Interfaces for Serverless Platforms

In order for the envisioned software components of the DataPorts [13] architecture to be manifested, their interconnection through a standardized API is crucial in order to avoid a monolithic service approach that does not face the challenges presented above and causes a major drawback in the process of building a federated ecosystem. The concept of a service within a computational infrastructure has been fundamental through the evolution of different architectural designs and implementations. Application Programming Interfaces (APIs), however, constitute a powerful interoperability tool that enables communication within a heterogenous infrastructure resulting in loosely coupled components. For that reason, their usage has almost dominated the landscape of services, especially within serverless infrastructures.

Considering the WWW as an ecosystem of heterogenous services that require simplicity, Web APIs, or also known as RESTful services, have been increasingly dominating over the Web services that are based upon WSDL and SOAP. RESTful services conform to the REST architectural principles, which include constraints regarding the client-server communication, the statelessness of the request, and the use of a uniform interface. In addition, these services are characterized by resource-representation decoupling, such that the resource content can manifest in different formats (i.e., JSON, HTML, XML, etc.). Furthermore, the majority of Web APIs reliance on URIs for resource identification and interaction and HTTP protocol for message transmission result in a simple technology stack that provides access to third parties, in order for them to consume and reuse data that originate from diverse services in data-oriented service compositions named mashups [28].

The evolution of the monolithic service-oriented architectures (SOA) has commenced from the management of the complexity of distributed systems in scope of integrating different software applications and has evolved through microservices into serverless architectures. In service-oriented architectures, a service provides functionalities to other services mainly via message passing. With the modularization of these architectures into microservice ecosystems, different services are developed and scaled independently from each other according to their specific requirements and actual request stimuli, leading to the localization of decisions per service regarding programming languages, libraries, frameworks, etc. However, the rise of cloud computing has led to serverless architectures that support the dynamic resource allocation and the corresponding infrastructure management in order to enable auto-scaling based on event stimulus and to minimize operational costs [25]. In that context, Web APIs should be considered the cornerstone of the exploding evolution of service value creation, where enterprise systems are embracing the XaaS (Anything-as-a-Service) paradigm, according to which all business capabilities, products, and processes are considered an interoperable collection of services that can be accessed and leveraged across organizational boundaries [7]. Nevertheless, since the beginning of the APIs prevalence over the traditional Web service technologies, the APIs have evolved in an autonomous way, lacking an established interface definition language [28]. Hence, in terms of serverless infrastructures, the subsequent need for homogeneity in application design and development has risen.

A common denominator in the development of serverless functions is their ability to support different functionalities in a scalable and stateless manner. For instance, there might be a serverless application that is integrated with an already existing ecosystem of functions that support API calls to cloud-based storage. While the former is by definition scalable, the underlying storage system's on-demand scalability is bound to reliability and QoS guarantees. As far as these serverless implementations are concerned, two major use cases are addressed hereunder. The first one involves the composition of a number of APIs, while filtering and transforming the consumed data. A serverless function that implements this functionality mitigates the danger of network overload between the client and the invoked systems, and offloads the filtering and aggregation logic to the backend. The second serverless application involves API aggregation, not only as a composition mechanism but as a means to reduce API calls in terms of authorization, for example. This composition mechanism simplifies the client-side code that interacts with the aggregated call by disguising multiple API calls into a single one with optional authorization from an external authorization service, e.g., an API gateway [5].

Despite the commonalities among serverless platforms in terms of pricing, deployment, and programming models, the most significant difference between them is the cloud ecosystem [5]. Differences in cloud platforms lead to discrepancies in developing tools and frameworks that are available to developers for creating services native to each platform. The ever-evolving serverless APIs in combination with the corresponding frameworks and libraries represent a significant obstacle for software lifecycle management, service discovery, and brokering. The plethora of incompatible APIs in terms of serverless technology has created the need for multicloud API standardization, interoperability, and portability in order to achieve seamlessness. In this direction, informal standardization has been formed after community efforts toward addressing the lack of a common programming model that enables platform-agnostic development and interoperability of functions [42].

Following the problem identification depicted above, the solution to the lack of a standardized and programming language-agnostic interface description language is fulfilled by the OpenAPI Specification (OAS). The OpenAPI initiative was founded in November 2015 by the collaboration of SmartBear, 3Scale, Apigee, Capital One, Google, IBM, Intuit, Microsoft, PayPal, and Restlet. This initiative was formed as an open-source project under the Linux Foundation and was designed to enable both humans and computers to explore and understand the functionalities of a RESTful service without requiring access to source code, additional documentation, or inspection of network traffic. OAS enables the understanding and interaction with the remote service with a minimal amount of implementation logic according to a vendor neutral description format. The OpenAPI Specification was based on the rebranded Swagger 2.0 specification, donated by SmartBear Software in 2015 [32].

The most important advantages of the OpenAPI Specification are twofold. On the one hand, the business benefit that comes a long is the recognition of this standardization as a useful means for a lot of developers to develop open-source repositories of tools that leverage this enablement. Furthermore, OAS is supported by a group of industry leaders that contribute with their strong awareness and mindshare, while indicating stability across a diverse code base. On the other hand, OAS is registered as a powerful technical tool that is most importantly language-agnostic and provides understanding of an API without the involvement of server implementation. Its documentation is regularly updated by a broad community that provides additional example implementations, code snippets, and responses to inquiries [32].

According to the above, the significance of the standardization that OAS offers rationalizes its adoption as the proposed solution interface description language for the proposed port platform. The pluralism of different data sources that need to be integrated within this platform, in combination with the existence of different frameworks and technologies of the existing APIs that are already utilized by ports, introduces the need for a uniform description of the exposed interfaces. For instance, the IoT infrastructure that includes APIs that are exposed by smart containers enables the aggregation of information that implements the life cycle assessment (LCA) applied for port logistics operations and needs to be integrated seamlessly with the different components of the platform. Moreover, crucial role in the message exchange between the different infrastructures within the ports ecosystem, play the APIs that facilitate the communication between components. Therefore, the standardization of their interface is of utter importance for the scalability and the interoperability of the platform. The DataPorts architecture can be constructed based on the OpenAPI specification, enabling the development of add-on services and the creation of added value of the available data. Furthermore, the implementation of the platform within a serverless architecture framework underlines the significance of the OpenAPI specification as a powerful standard for the interface description of all services within and exposed by the DataPorts platform.

6 Standardization

IDSA [4] aims at open, federated data ecosystems and marketplaces ensuring data sovereignty for the creator of the data by establishing virtual space for the standardized, secure exchange and trade of data. Standards, be they national, regional, or global, are the fruit of collective efforts and guarantee interoperability. They can be revised to meet industry needs and remain relevant over time. Standards organizations, where participants from different segments of the industry gather, are among the few places where competitors work side-by-side. Standards organizations offer a safe place to do so from an antitrust perspective. Standards development participants are industry experts, tech companies, and customers representing all fields of the industry.

The adoption of global multimodal data exchange standards guarantees interoperability. In fact, smart container standardization effort [9, 10] is one of many standardization initiatives [27] supporting global trade. Standards enable stakeholders in the logistics chain to reap the maximum benefits from smart container solutions while enabling them to share data and associated costs. Standards-based data exchange usage increases the ability to collaborate, which in turn increases efficiency. Additionally, such standards reduce development and deployment costs and cut time to market for Internet of Things (IoT) solution providers.

Data exchange standards developed in an open process offer a useful aid to all parties interested in the technical applications and implementation of smart container solutions. Additionally, if solution providers find there are new data elements required to accommodate changing business requirements, it is possible to create a backward-compatible revision of the standard to accommodate their needs.

With the ramp-up of new and emerging technologies, these standards are more necessary than ever. Standards reduce the risk of developing proprietary technologies with significant deployment limitations and the lack of interoperability among systems and devices. Standards enable the parties to avoid costly and time-consuming integration and limit the risk of vendor lock-in. In this context, IDS provides a generic framework that can be leveraged by domain-specific instantiation such as UN/CEFACT smart container standard that offers transport execution and the condition under which the cargo was transported.

The United Nations Centre for Trade Facilitation and Electronic Business (UN/CEFACT) Smart Container Business Requirements Specification (BRS) ensures that the various ecosystem actors share a common understanding of smart container benefits by presenting various use cases. It also details the smart container data elements [41]. Defining the data elements that smart containers can generate accelerates integration and the use of smart container data on different platforms for the enhancement of operations. In addition, utilizing standard smart container data enables open communications channels between supply chain actors.

Standards data models and standard APIs would help stakeholders to make the necessary transformation to achieve supply chain excellence [8]. Indeed, APIs are key to ensuring simplification and acceleration of the integration of digital services from various sources.

The focus of the UN/CEFACT Smart Container project is to define the data elements via varied use cases applicable to smart container usage. Currently, the data model is being developed, which will provide the basis for the smart container standard messaging and Application Programming Interfaces (APIs). The Smart Container API catalog will be the source code-based interface specification enabling software components (services) to communicate with each other. It is crucial to first determine and align the required data elements and their semantics.

Smart containers will revolutionize the capture and timely reporting of data throughout the supply chains. Such containers are an essential building block to meet the emerging requirements for end-to-end supply chains. As leading carriers adopt smart container solutions, they gain valuable data that can be shared with shippers and other supply chain stakeholders.

However, generating and collecting data is not enough to make smart container solutions or supply chains "smart." Stakeholders already manage huge amounts of data and struggle with multiple technologies that take time away from their core businesses. A smart container solution must deliver data that matters, in a standard format for easy integration into different systems. It must enable unambiguous data interpretation and empower all involved stakeholders with actionable information. Clear semantic standards are essential for effective smart container data exchange ensuring that all stakeholders understand the same information in the same way.

Then and only then, can smart containers truly become part of digital data streams [35].

The UN/CEFACT Smart Container project aims to create multimodal communications standards that can facilitate a state-of-the-art solution in providing and exposing services. Any intermodal ecosystem stakeholder may then orchestrate and enrich these services to meet their business process needs. The availability and exposition of these services can boost the digital transformation of the transportation and logistics industry, fuelling innovation in new applications and services. Physical supply chains that move goods need a parallel digital supply chain that moves data describing the goods and their progress through the supply chain. The smart container data flows ensure that the physical flow is well synchronized with the required documents flow. Data are the raw material of Maritime Informatics. Without data streams emanating from operations, there can be no data analytics. As we digitalize, we improve operational productivity and lay the foundation, through Maritime Informatics, for another round of strategic and operational productivity based on big data analytics and machine learning.

7 Business Outcomes and Challenges

The fast-growing complexity at seaports makes data management essential, hence the optimal goal is to achieve greater efficiency. The use of large volumes of data (big data) is indisputably a major aid to this goal [31]. AI-based services available in a smart seaport is a new revenue source for many stakeholders. When such data and services are offered through a standard mechanism as is a data-driven platform, this offering acts as a leverage to improve and increase various port operations, especially those that are associated with traffic, vessel, and cargo movement, and is of high importance for third parties. Imagine the case where cargo transfer data are accessible by the shipping lines, and at the same time, all seaport's operations can be available by Port Authority's associates. Passenger mobility patterns may be available not only to the Port Authority but also to the city's decision and policy makers. Commercial or cultural associations may also be interested to access such services, especially from seaports with high passenger activity. Such services, will be used to transform the seaports into smart and cognitive, and eventually will increase the ports' stakeholders and activity boundaries. Hence, through data-driven services, the demand will also be increased. Toward this direction, the research community and the shipping-related SMEs or even the startup community may benefit from analyzing large volumes of data offered by data providers and propose additional offerings. Moreover, Analytics as a Service using data collected from shipping and freight companies, warehouses, customs brokers, and other port operations may be a key for data monetization. The opportunity of data monetization may unlock any considerations regarding data sharing, that are related to the risk of losing competitive advantages.

From a business perspective, in order for data and service sharing to be effective and useful for as many beneficiaries as possible, certain Quality of Service (QoS) characteristics should be followed. Especially in big data, among the key characteristics are considered the Volume, Velocity, Variety, Veracity, and Value. Moreover, offering data and services should match certain needs and be easily accessible for the users. Therefore, in terms of time, data should be up to date, real, and able to be authenticated. Additionally, guarantying a QoS may be difficult for heterogeneous data, especially when the competitiveness is increasing according to the demand for new data and services. Hence, a monitoring mechanism is needed to ensure the above-mentioned characteristics as well as the validity of the transferred data. A main business-related concern with data QoS is considered the regulatory compliance that today is vague, the customer satisfaction which is the goal, the validity and the accuracy of the data to allow decision making, the relevance the data should meet and their completeness for not have missing values and the consistency of data format as expected by the users.

From a technical perspective, all types of applications of the fourth paradigm of science deal with large amounts of data stored in various storage devices or systems. Distributed storage systems are often chosen for storing data of this type, as depicted in Fig. 1, framing requirements for IDS. Some of the requirements posed to those storage systems may concern Quality of Service (QoS) aspects formally expressed in a Service Level Agreement as was the traditional approach in the past. The role of QoS is to provide the necessary technical specifications that specify the system quality of features such as performance, availability, scalability, and serviceability. Within the IDS ecosystem, system qualities are closely interrelated. Requirements for one system quality might affect the requirements and design for other system qualities. For example, within connected IDS managed and framed by various companies may have different and higher levels of security policies that might affect performance, which in turn might affect availability. Adding additional servers to address availability issues affect serviceability (maintenance costs). Understanding how system qualities are interrelated and the trade-offs that must be made is the key to designing a system that successfully satisfies both business requirements and business constraints. Having these QoS attributes in mind, it's evident that the QoS management in distributed and heterogeneous environment is a challenging task given the possible storage device heterogeneity, the dynamically changing data access patterns, the client's concurrency, and storage resource sharing. The problem becomes even more complicated when distributed computing environments with virtualized and shared resources like Clouds and Blockchains are considered. Furthermore, various heterogeneous devices or objects should be integrated for transparent and seamless communication under the umbrella of Internet of Things (IoT). This would facilitate the open-access of data for the growth of various digital services. Building a general framework or selecting an approach for handling QoS becomes a complex task due to the heterogeneity in devices, technologies, platforms, and services operating in the same system. Additionally, Data's Analytics and Governance should follow an all-encompassing approach to consumer privacy and data security as opposed by Compliance Regulations that will become a benchmark for how personal data are treated in the future. In the area of cognitive ports, regulations introduce major restrictions and complexities for QoS in a technical perspective, especially those parts that address how contact data are handled—and how data quality approach can be used that involves both tools and processes as part of compliance efforts. In such heterogeneous environment, technical aspects of QoS relate to technical translation and treatment of compliance as concerning entities of rights to: "access," "be informed," "data portability," "be forgotten," "object," "restrict processing," "be notified," "rectification," and so on in addition to the aforementioned technical specifications.

The majority of technical challenges discussed above, overcame by adopting the serverless architecture approach, as described in section "Scalability". The use of APIs and the semantic interoperability in "5.2" provide vignettes for the followed approach. Therefore, utilizing serverless as well as microservices paradigms, cognitive ports constitute a PaaS and DaaS environment, in which the majority of traditional QoS aspects are dealt dynamically by inheriting system's adaptation to current needs.

Additionally, the aspects of Compliance Regulations are approached by the creation of workflows into the Blockchain and Broker infrastructure of the Cognitive Port. For example, upon the request for every provider to comply with existing regulations concerning data and identity attributes, any stakeholder that provides data is responsible for the integrity and compliance of their provided data. Furthermore, risks that might arise from analytical aspects of shared data (such as combining data with new or existing data sources within or external to Cognitive Ports environment, etc) are secured by workflows for approving data processing requests. Therefore, there are controls for either prohibit non-regulated actions, or inform consumers that upon using them, any actions needed (i.e. consents) are their responsibility. Therefore, Cognitive Ports are an ecosystem for dynamically sharing data in IDS communities.

Acknowledgments The research leading to these results has received funding from the European Commission under the H2020 Programme's project DataPorts (grant agreement No. 871493).

References

1. Ahmadian, A. S., Jürjens, J., & Strüber, D. (2018). Extending model-based privacy analysis for the industrial data space by exploiting privacy level agreements. In *Proceedings of the 33rd Annual ACM Symposium on Applied Computing - SAC '18*. ACM Press.
2. Alonso, Á., Pozo, A., Cantera, J., de la Vega, F., & Hierro, J. (2018). Industrial data space architecture implementation using FIWARE. *Sensors, 18*(7), 2226.
3. Apache OpenWhisk. https://openwhisk.apache.org/. Online. Accessed 05 January 2021.
4. Auer, S., Jürjens, J., Otto, B., Brost, G., Lange, C., Quix, C., Cirullies, J., Lohmann, S., Eitel, A., Mader, C., Schulz, D., Ernst, T., Menz, N., Schütte, J., Haas, C., Nagel, L., Spiekermann, M., Huber, M., Pettenpohl, H., Wenzel, S., Jung, C., & Pullmann, J. (2019, April). Reference architecture model for the industrial data space. https://www.internationaldataspaces.org/wp-content/uploads/2019/03/IDS-Reference-Architecture-Model-3.0.pdf. Online. Accessed 30 October 2020.

5. Baldini, I., Castro, P., Chang, K., Cheng, P., Fink, S., Ishakian, V., Mitchell, N., Muthusamy, V., & Rabbah, R. (2017). Aleksander Slominski, and Philippe Suter. Serverless computing: Current trends and open problems. In *Research Advances in Cloud Computing* (pp. 1–20). Springer.
6. Balla, D., Maliosz, M., & Simon, C. (2020). Open source FaaS performance aspects. In *2020 43rd International Conference on Telecommunications and Signal Processing (TSP)*. IEEE.
7. Basole, R. C. (2018). On the evolution of service ecosystems: A study of the emerging API economy. In *Handbook of Service Science, Volume II* (pp. 479–495). Springer International Publishing.
8. Becha, H. (2019). How standard APIs open the door to powerful digital services. https://hananebecha.home.blog/2019/11/28/the-un-cefact-smart-container-project/, November 2019. Online. Accessed 30 October 2020.
9. Becha, H. (2020). Standardization supporting global trade. *Port Technology International*, 2019. Edition 91, Shipping 2020: A Vision for Tomorrow. https://www.porttechnology.org/editions/shipping-2020-a-vision-for-tomorrow/
10. Becha, H. (2020). The UN/CEFACT smart container project. *The Report: The Magazine of the International Institute of Marine Surveying*, March 2020. Issue 91. https://www.iims.org.uk/wp-content/uploads/2020/02/The-Report-March-2020.pdf
11. Boost 4.0 – Big Data for Factories. https://boost40.eu/. Online. Accessed 05 January 2021.
12. Daniele, L., & Pires, L. F. (2013). An ontological approach to logistics. *Enterprise Interoperability, Research and Applications in the Service-Oriented Ecosystem, IWEI, 13*, 199–213.
13. DataPorts H2020 EU Project. http://dataports-project.eu. Online. Accessed 30 October 2020.
14. FireWire Data Model. https://www.fiware.org/developers/data-models/. Online. Accessed 05 January 2021.
15. Fiware Data Models. https://fiware-datamodels.readthedocs.io/en/latest/index.html. Online. Accessed 30 October 2020.
16. Fiware Open Source Platform. https://www.fiware.org/. Online. Accessed 30 October 2020.
17. Fiware Orion Context Broker. https://fiware-orion.readthedocs.io/en/master/. Online. Accessed 30 October 2020.
18. Fortino, G., Savaglio, C., Palau, C. E., de Puga, J. S., Ganzha, M., Paprzycki, M., Montesinos, M., Liotta, A., & Llop, M. (2017). Towards multi-layer interoperability of heterogeneous IoT platforms: The INTER-IoT approach. In *Internet of Things* (pp. 199–232). Springer International Publishing.
19. Ganzha, M., Paprzycki, M., Pawlowski, W., Szmeja, P., & Wasielewska, K. (2016, April). Semantic technologies for the IoT – an inter-IoT perspective. In *2016 IEEE First International Conference on Internet-of-Things Design and Implementation (IoTDI)*. IEEE.
20. Gizelis, C.-A., Mavroeidakos, T., Marinakis, A., Litke, A., & Moulos, V. (2020). Towards a smart port: The role of the telecom industry. In *Artificial Intelligence Applications and Innovations. AIAI 2020 IFIP WG 12.5 International Workshops* (pp. 128–139). Springer International Publishing.
21. Gruber, T. R. (1995). Toward principles for the design of ontologies used for knowledge sharing? *International Journal of Human-Computer Studies, 43*(5–6), 907–928.
22. Heiler, S. (1995). Semantic interoperability. *ACM Computing Surveys, 27*(2), 271–273.
23. Hendrickson, S., Sturdevant, S., Harter, T., Venkataramani, V., Arpaci-Dusseau, A. C., & Arpaci-Dusseau, R. H. (2016, June) Serverless computation with openlambda. In *8th USENIX Workshop on Hot Topics in Cloud Computing (HotCloud 16), Denver, CO*. USENIX Association.
24. IDSA Information Model. https://github.com/International-Data-Spaces-Association/InformationModel. Online. Accessed 05 January 2021.
25. Kratzke, N. (2018). A brief history of cloud application architectures. *Applied Sciences, 8*(8), 1368.
26. Kubeless. https://kubeless.io/. Online. Accessed 05 January 2021.

27. Lind, M., Simha, A., & Becha, H. (2020). Creating value for the transport buyer with digital data streams. *The Maritime Executive*. https://maritime-executive.com/editorials/creating-value-for-the-transport-buyer-with-digital-data-streams

28. Maleshkova, M., Pedrinaci, C., & Domingue, J. (2010, December). Investigating web APIs on the world wide web. In *2010 Eighth IEEE European Conference on Web Services*. IEEE.

29. McGrath, G., & Brenner, P. R. (2017, June). Serverless computing: Design, implementation, and performance. In *2017 IEEE 37th International Conference on Distributed Computing Systems Workshops (ICDCSW)*. IEEE.

30. Milano, M., Curry, E., García Robles, A., Hahn, T., Irgens, M., Lafrenz, R., Liepert, B., O'Sullivan, B., Zillner, S., Bisset, D., & Smeulders, A. (2020, September). Strategic research, innovation and deployment agenda - AI, data and robotics partnership. Third release. *Big Data Value Association*.

31. Moulos, V., Chatzikyriakos, G., Kassouras, V., Doulamis, A., Doulamis, N., Leventakis, G., Florakis, T., Varvarigou, T., Mitsokapas, E., Kioumourtzis, G., Klirodetis, P., Psychas, A., Marinakis, A., Sfetsos, T., Koniaris, A., Liapis, D., & Gatzioura, A. (2018). A robust information life cycle management framework for securing and governing critical infrastructure systems. *Inventions, 3*(4), 71.

32. OpenAPI Specification. https://www.openapis.org/faq. Online. Accessed 30 October 2020.

33. OpenFaaS. https://docs.openfaas.com/. Online. Accessed 05 January 2021.

34. OWL. https://www.w3.org/TR/owl-guide/. Online. Accessed 05 January 2021.

35. Pigni, F., Piccoli, G., & Watson, R. (2016). Digital data streams: Creating value from the real-time flow of big data. *California Management Review, 58*(3), 5–25.

36. Psomakelis, E., Nikolakopoulos, A., Marinakis, A., Psychas, A., Moulos, V., Varvarigou, T., & Christou, A. (2020). A scalable and semantic data as a service marketplace for enhancing cloud-based applications. *Future Internet, 12*(5), 77.

37. Pyngsi Python Framework. https://github.com/pixel-ports/pyngsi. Online. Accessed 30 October 2020.

38. SAREF Ontology. https://saref.etsi.org/. Online. Accessed 05 January 2021.

39. SynchroniCity. https://synchronicity-iot.eu/. Online. Accessed 05 January 2021.

40. UN/CEFACT. https://umm-dev.org/about-umm/. Online. Accessed 05 January 2021.

41. UN/CEFACT. (2019). The UN/CEFACT Smart Container Business Specifications (BRS). https://www.unece.org/fileadmin/DAM/cefact/brs/BRS-SmartContainer_v1.0.pdf. Online. Accessed 30 October 2020.

42. van Eyk, E., Toader, L., Talluri, S., Versluis, L., Uta, A., & Iosup, A. (2018). Serverless is more: From PaaS to present cloud computing. *IEEE Internet Computing, 22*(5), 8–17 (2018).

43. Zillner S., Curry E., Metzger A., Auer S., & Seidl R. (2017). European big data value strategic research & innovation agenda. Big Data Value Association.

Distributed Big Data Analytics in a Smart City

Maria A. Serrano, Erez Hadad, Roberto Cavicchioli, Rut Palmero,
Luca Chiantore, Danilo Amendola, and Eduardo Quiñones

Abstract This chapter describes an actual smart city use-case application for advanced mobility and intelligent traffic management, implemented in the city of Modena, Italy. This use case is developed in the context of the European Union's Horizon 2020 project CLASS [4]—Edge and Cloud Computation: A highly Distributed Software for Big Data Analytics. This use-case requires both real-time data processing (*data in motion*) for driving assistance and online city-wide monitoring, as well as large-scale offline processing of big data sets collected from sensors (*data at rest*). As such, it demonstrates the advanced capabilities of the CLASS software architecture to coordinate edge and cloud for big data analytics. Concretely, the CLASS smart city use case includes a range of mobility-related applications, including extended car awareness for collision avoidance, air pollution monitoring, and digital traffic sign management. These applications serve to improve the quality of road traffic in terms of safety, sustainability, and efficiency.

M. A. Serrano · E. Quiñones (✉)
Barcelona Supercomputing Center (BSC), Barcelona, Spain
e-mail: eduardo.quinones@bsc.es

E. Hadad
IBM Research, Haifa, Israel
e-mail: erezh@il.ibm.com

R. Cavicchioli
University of Modena (UNIMORE), Modena, Italy
e-mail: roberto.cavicchioli@unimore.it

R. Palmero
Atos Research and Innovation (ARI), Madrid, Spain
e-mail: ruth.palmero@atos.net

L. Chiantore
Comune di Modena, Modena, Italy
e-mail: luca.chiantore@comune.modena.it

D. Amendola
Centro Ricerche Fiat S.C.p.A. (CRF), Orbassano, Italy

© The Author(s) 2022   475
E. Curry et al. (eds.), *Technologies and Applications for Big Data Value*,
https://doi.org/10.1007/978-3-030-78307-5_21

This chapter shows the big data analytics methods and algorithms for implementing these applications efficiently.

Keywords IoT · Big data · Analytics · Distributed · Real time · Smart city

1 Introduction

Current data analytics systems are usually designed following two conflicting priorities to provide (1) a quick and reactive response (referred to as data-in-motion analysis), possibly in real-time based on continuous data flows; or (2) a thorough and more computationally intensive feedback (referred to as data-at-rest analysis), which typically implies aggregating more information into larger models [17].

These approaches have been tackled separately although they provide complementary capabilities. This is especially relevant in the context of smart cities traffic management for example, where both approaches play a fundamental role. On the one hand, delivering timely driving assistance and/or city traffic control requires real-time processing of data provided by city and smart car sensors. On the other hand, city-wide traffic data needs to be collected and processed at bigger time granularity (e.g., hourly, daily, weekly, etc.) to identify traffic issues, monitor air pollution, plan further deployment and maintenance of traffic routes, etc.

This CLASS use-case presented in this chapter demonstrates these mixed requirements. It takes place in the city of Modena (Italy), where different actors are involved. It includes a significant sensor infrastructure to collect and process real-time data across a wide urban area, supported by a private cloud and edge infrastructure, prototype cars equipped with heterogeneous sensors/actuators, and V2I connectivity. Representative applications for traffic management and advanced driving assistance domains are employed to efficiently process very large heterogeneous data streams in real-time, providing innovative services for the public sector, private companies, and citizens. The chapter relates to the technical priorities "Data Processing Architectures" and "Data Analytics" of the European Big Data Value Strategic Research & Innovation Agenda [17]. Moreover, the chapter relates to the "Sensing and Perception" and "Knowledge and Learning" cross-sectorial technology enablers of the AI, Data and Robotics Strategic Research, Innovation & Deployment Agenda [16].

1.1 Processing Distributed Data Sources

The use of combined data-in-motion and data-at-rest analysis provides cities efficient methods to exploit the massive amount of data generated from heterogeneous and geographically distributed sources including pedestrians, traffic, (autonomous) connected vehicles, city infrastructures, buildings, IoT devices, etc. Certainly, exposing city information to a dynamic, distributed, powerful, scalable, and user-friendly big data system is expected to enable the implementation of a wide range of new services and opportunities provided by analytics tools. However, big data challenges stem not only from size and heterogeneity of data but also from its geographical dispersion, making it difficult to be properly and efficiently combined, analyzed, and consumed by a single system.

The CLASS project [4], funded by the European Union's Horizon 2020 Programme, faces these challenges and proposes a novel software platform that aims to facilitate the design of advanced big data analytics workflows, incorporating data-in-motion and data-at-rest analytics methods, and efficiently collect, store, and process vast amounts of geographically distributed data sources. The software platform is meeting these needs by integrating technologies from different computing domains into a single development ecosystem, and by adopting innovative distributed architectures from the high-performance computing (HPC) domain, as well as highly parallel and energy-efficient hardware platforms from the embedded domain.

Although this chapter aims to describe the big data analytics algorithms involved in the implementation of a real smart city use-case, the next subsection briefly describes the software framework proposed in the CLASS project that supports the execution of such smart city use case and tackles the aforementioned challenges.

1.1.1 The CLASS Software Architecture

The conceptual layout of a typical big-data subsystem setup manifests the so-called compute continuum [3], in which the data is processed, transformed, and analyzed through a range of IT hardware stages, from the field devices close to the source of data (commonly referred to as edge computing) to the heavy-duty analytics in the data centers (commonly referred to as cloud computing).

Figure 1 shows the CLASS software architecture, where different components interact to distribute resources and services in a smart way, so that decision-making occurs as close as possible to where the data is originated (either at edge or cloud side), enabling faster processing time and lowering network costs. The CLASS software components include:

- **Data Analytics Platform**: This layer exposes interfaces and tools for the development and deployment of big data analytics applications for the CLASS use-cases. The analytics methods currently available include map/reduce, distributed workflow, and Deep Neural Networks (DNN). The core of the analytics platform is a serverless/FaaS (Function-as-a-Service) layer, which allows analytics to be invoked in response to invocations, or to events such as message arrival or timer. It further allows polyglot programming [6], where different application components using different analytics methods and programming languages to cooperate in computation. In CLASS, the serverless layer is further augmented for low-latency real-time computation.

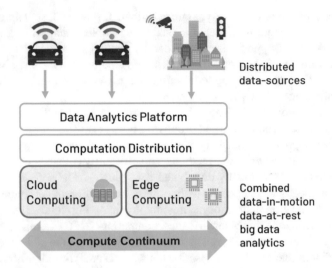

Fig. 1 Overview of the CLASS software platform to process distributed data sources

- **Computation Coordination and Distribution Framework**: This layer manages the workload distribution across the continuum, with the objective of minimizing the response time of big data analytics workflows and providing real-time guarantees. This layer also provides a shared data storage backbone among the different components of the platform.
- **Cloud Layer**: This layer provides a Container-as-a-Service (CaaS) service that abstracts the cloud infrastructure details away from the data analytics service developers. It provides not only service life cycle management but also guaranteed performance of workloads.
- **Edge Layer**: This layer provides support for the most advanced highly parallel heterogeneous embedded platforms, e.g., Nvidia GPUs, many-core fabrics, or SoC-FPGAs. On top of this hardware, the software component that supports the development of big data analytics is based on state-of-the-art Deep Neural Networks (DNN). An analytics agent deployed at this layer allows deploying various computations as part of the overall analytics layer discussed above.

Overall, a smart distribution of computing services, combined with the usage of highly parallel hardware architectures across the compute continuum, is used to significantly increase the capabilities of the data analytics solutions needed to fuel future smart systems based on big data. Multiple domains can leverage the benefits of the CLASS framework since the objective is also to provide sound real-time guarantees on end-to-end analytics responses. This ability opens the door to the use of big data into critical real-time systems, providing to them superior data analytics capabilities to implement more intelligent and autonomous control applications.

The next subsection describes the particular smart city use-case applications implemented on top of the CLASS software framework. More details of the

technologies used in this framework can be found in the CLASS project website
[4].

1.2   Big Data Analytics on Smart Cities: Applications

The CLASS software framework supports the development and execution of a
set of advanced data analytics algorithms. Interestingly, all these data analytics
engines can be optimized to execute at both, the edge and the cloud side, providing
the required flexibility needed to distribute the computation of complex data
analytics workflows composed of different analytics frameworks across the compute
continuum.

Upon the described software computing infrastructure, there is a considerable
number of city-awareness services that can be implemented. In this chapter, we
describe the combined big data analytics that provide meaningful information for
three use-case applications:

- *Digital Traffic Sign Application.* It offers the opportunity to dynamically change
  traffic conditions based on real-time traffic information collected by means of
  the distributed sensor infrastructure. In case of accidents, the traffic signals
  can advise the "best path to follow," reducing the induced traffic impact and
  improving the driver experience. For emergency vehicles (e.g., ambulances,
  firefighters, and police vehicles), it can dynamically create "green routes" by
  adjusting the frequency of the traffic lights to reduce the time of intervention.
- *Air Pollution Estimation Application.* It offers the possibility of estimating the
  pollutant emissions of vehicles in real time and segregated by areas. In particular,
  the proposed technique considers the real-time traffic conditions, e.g., as detected
  by street cameras, to model and estimate such emissions without the need of
  dedicated pollution sensors.
- *Obstacle Detection Application.* It offers the required real-time services for
  warning a driver about critical situations that may endanger the safety of the
  driver and the vulnerable road users (VRUs). The identification of potentially
  hazardous situations can be enforced at the different levels of the compute
  continuum, with different precision and latency, and considering the city cameras
  and vehicles sensors information. The implementation of this application is
  supported by the V2I communication, improving driving safety, especially in
  case of blind spots such as in intersections.

The implementation of these three applications is supported by the big data
analytics presented in the next section. The rest of the chapter is organized as
follows: Sect. 2 describes the concrete big data analytics algorithms and their
integration to implement the desired applications; Sect. 3 describes the smart city
infrastructure to execute and distribute the proposed algorithms; finally, Sect. 4
concludes the chapter.

2 Big Data Analytics Algorithms

This section describes the big-data analytics algorithms implementing the proposed smart city use-case. All the algorithms contribute to the generating of a Data Knowledge Base (DKB) from which valuable knowledge is extracted from the city and the connected cars. The final goal is to implement the three smart city applications: *digital traffic sign*, *air pollution estimation*, and *obstacle detection*.

Section 2.1 describes the considered big data analytics algorithms, each processing part of the available data. Section 2.2 presents the combined big data analytics workflow that shows the interaction and relationship between the different algorithms to cooperate toward a common objective: the generation of the DKB and the implementation of the proposed applications.

2.1 Description of Data Processing Algorithms

2.1.1 In-Vehicle Sensor Fusion

Autonomous vehicles or ADAS need to have a robust and precise perception of the surrounding environment. A precise categorization and localization of road objects, such as cars, pedestrians, cyclists, and other obstacles, is needed. In order to obtain accurate results, fusing and combining the output of several different devices has become a trend, being a good compromise to obtain good classification and 3D detection. Concretely, the solution explained in this section uses a Light Detection And Ranging (LiDAR) sensor, and multiple camera disposed to cover 360° surround the vehicle.

In the context of 2D camera object detection, Convolutional Neural Networks (CNN) are often adopted. YOLO [2] is a good example of real-time object detection and classification, based on a fully CNN. The performance in terms of precision is comparable to other methods, but the performance in terms of inference time is better because it is optimized to be used in NVIDIA Jetson products. LiDARs, instead, produce a 3D point cloud, which is then processed to place 3D bounding boxes (BBs) around the objects via clustering methods [5].

Camera detection and LiDAR clustering are merged by developing a modified parallel algorithm [13] that exploits the features of LiDAR point cloud and optimizes a YOLO CNN, to deploy an open-source real-time framework that combines camera 2D BBs with LiDAR clustering. Figure 2 shows an example of this algorithm: video frame processing and 2D BBs from four cameras installed in the vehicle (each with 120° field of view), and LiDAR 3D detections for a 360° field of view.

Moreover, GPS positions of all the detected objects are necessary to create a comprehensive picture of the area. The connected vehicle GPS sensor, if not precise

Fig. 2 Sensor fusion algorithm (vehicle's detections): output from four cameras (120° field of view each) and a LiDAR (360° field of view)

enough, could be enhanced by a previous mapping of the area, or a high-precision map externally provided, using the surrounding features detected by the LiDAR and cameras for matching. After determining a reliable position of the vehicle, the relative position of the surrounding detected objects is added using a vector sum.

As a result, the ouput of the *sensor fusion* method provides detected and classified objects in real time, their GPS position, and timestamp, as seen from a vehicle point of view. Also, the position of the connected vehicle itself is provided.

In-vehicle sensor fusion output data

Category	GPS-Position	Timestamp
ConnectedCar	(44.654540,10.933815)	1603963252
Car	(44.654550,10.933815)	1603963252
Pedestrian	(44.654534,10.933740)	1603963252
Truck	(44.654744,10.934020)	1603963252

2.1.2 Street Camera Object Detection

From cameras located on the streets, objects can also be detected and classified. This is also done with the optimized version of YOLO used in the *in-vehicle sensor fusion* method, described in the previous section. After detection and classification, the global position of the road user is also computed. To do so, each camera is manually calibrated to match known points in the image with their GPS position on a georeferenced map. As a result, this method provides detected and classified objects in real time, as seen from a street camera point of view. Figure 3 shows

Fig. 3 Object detection output from street camera

a video frame example with the bounding boxes of the objects detected, and an example of the row data generated is shown below:

Street camera object detection output data

Category	GPS-Position	Timestamp
Car	(44.655049,10.934328)	1603963252
Car	(44.654550,10.933815)	1603963252
Pedestrian	(44.655049,10.934328)	1603963252
Bike	(44.655169,10.934410)	1603963252

2.1.3   Object Tracking

The purpose of this method is to track the road users, i.e., cars, as well as pedestrians, bikes, and motorcycles, detected by both the *in-vehicle sensor fusion* and the *street cameras object detection* methods. If only detection is performed, then only different and uncorrelated detections for each video frame occur. Instead, if tracking is also performed there are two main advantages: (1) to make the detection more robust (since detection algorithms are not perfect and detection errors may occur); and (2) to perform path prediction and guess where the objects will go which in turn allows to predict possible collisions. Considering performance on edge devices, this method is based on a Kalman filter on the position points [14] (contrary to visual tracking algorithms that are computationally intensive, and not suitable for a real-time scenario).

After the object detection methods detect the bounding boxes of objects, a point of the bounding box is taken as a reference, and it is used to track each object with an

Fig. 4 Object tracking output from street camera detections

aging mechanism. It means that this algorithm is able to not only correlate objects in different frames but also to compute the speed of the objects. Figure 4 shows two video frames with the bounding boxes of the objects detected, and the lines representing the tracked trajectory. As an example, the car *entering* into the camera field of view (at the bottom of the image) in the left frame is located at the entrance of the roundabout in the right frame; the yellow line represents the tracking for this vehicle. An example of the row data generated is shown below:

Object tracking output data

Category	GPS-Position	Timestamp	ID	Speed(Km/h)
Car	(44.654550,10.933815)	1603963252	1	45
	(44.654550,10.933818)	1603963253	1	45
	(44.654552,10.933819)	1603963254	1	46
	(44.654552,10.933820)	1603963255	1	46
Car	(44.655049,10.934328)	1603963252	2	55
Pedestrian	(44.655049,10.934328)	1603963252	3	4
Bike	(44.655169,10.934410)	1603963252	4	10

In this example, the data from the *Street camera object detection* is considered, after processing multiple video frames. The car with ID 1 has been detected in four different frames; therefore, there are four entries for the same object, each with a different GPS position and timestamp, and with the computed speed.

2.1.4 Data Deduplication

When multiple objects are detected in the same area by street cameras that share part of their field of view, or by a smart connected car moving in the same area, there are duplicated road users detected by those different actors. This method manages these duplicated objects so that they appear only once in the system. A simple method deduplicates objects by searching for all nearest objects with the same category, comparing their position, and discarding the duplicated ones considering a certain threshold value. A different threshold is used depending on the category (and hence size) of the road user.

As an example, the output after the *Data deduplication* is shown below:

Data deduplication **output data**

Source	Category	GPS-Position	Timestamp	ID	Speed(Km/h)
cam_1	Car	(44.654550,10.933815)	1603963252	1	45
		(44.654550,10.933818)	1603963253	1	45
		(44.654552,10.933819)	1603963254	1	46
		(44.654552,10.933820)	1603963255	1	46
cam_1	Car	(44.655049,10.934328)	1603963252	2	55
cam_1	Pedestrian	(44.655049,10.934328)	1603963252	3	4
cam_1	Bike	(44.655169,10.934410)	1603963252	4	10
veh_1	ConnectedCar	(44.654540,10.933815)	1603963252	5	52
veh_1	~~Car~~	~~(44.654550,10.933815)~~	~~1603963252~~		
veh_1	Pedestrian	(44.654534,10.933740)	1603963252	6	4.5
veh_1	Truck	(44.654744,10.934020)	1603963252	7	32

In this example, the data from the *Street camera object detection* (source *cam* 1) and the *In-vehicle sensor fusion* (source *veh* 1) is considered. After also processing this data by the *Object tracking* algorithm, the *Data deduplication* is invoked. As a result, one of the cars, detected by the vehicle, is discarded since it has the same GPS position as Car with ID 1, detected by a street camera.

2.1.5 Trajectory Prediction

In order to foresee possible collisions between vehicles and other road users in the streets, it is necessary to predict the trajectory of all those road users. Based on the detected positions of an object and their associated timestamps, the prediction algorithm computes multiple future positions (predictions) per object, at future time points so that a complete trajectory is obtained. The trajectory prediction is calculated for multiple objects simultaneously, where input samples are not equally apart in time. This method is based on quadratic regression [12] that finds the equation of the parabola that best fits a set of data (using the detected positions). Then, the equation is used to predict the future positions.

Other known works, such as [7], employ Recurrent Neural Networks (RNNs) to compute a trajectory prediction of street objects based on their previous steps and with the environment context. This approach uses Long Short-Term Memory (LSTM) neural networks, a type of RNNs that can extract patterns from a sequence. In this case, the sequence is also a series of GPS positions along the time.

An example of the output after the *Trajectory prediction* for the Car with ID 1 is shown below:

Trajectory prediction output data

(Car ID=1)	TP_latitude	TP_longitude	TP_timestamp
	44.654553	10.933821	1603963256
	44.654553	10.933822	1603963257
	44.654555	10.933823	1603963258

The trajectory prediction function takes an object id and computes its trajectory based on its updated location history. In this example, given the four detected positions for this car (see the previous section), the algorithm is used to predict three future positions. The core trajectory prediction needs to be applied to each object covered by the cameras, only when new location data is available for the object.

2.1.6 Warning Area Filtering

According to the CLASS use-case design, connected vehicles may receive alert notifications for possible collisions with objects within each vehicle's respective *warning area*, which is defined geometrically around the vehicle's current location. This is achieved by first filtering out all objects outside a given car's warning area and then detecting a possible collision (see Sect. 2.1.7) between the car and each of the objects found to be within the warning area.

In CLASS, we employ a simple and highly efficient method of approximating a rectangular warning area around a given location, using *geohashes*. A geohash [15] is a unique string label assigned to each square in a map grid with a specific area granularity. Each GPS position can be efficiently converted to the matching geohash of the grid location containing the GPS coordinates. This is done once for each object's detected location and persisted in the shared data during the tracking phase.

The length of the geohash string matches a specific area granularity. For example, a 7-character string identifies a square shape of 153×153 m. Thus, in CLASS, once the rectangular dimensions of the warning area are determined at design time, the geohash string length is selected to be the longest string (i.e., smallest squares) such that the 3×3 grid of squares is guaranteed to cover the warning area of a car located anywhere in the central square. When coverage is guaranteed, determining if an object X is in the warning area of car C is approximated to X and C being in *neighbor* geohashes—that is, if the geohash of X is in one of the 3×3 squares surrounding the geohash of C, including that of C itself:

$$WA(C, X) \approx neighbor(C.g, X.g)$$

The *neighbor* function is highly efficient and common in most geohash libraries.

2.1.7 Collision Detection

For each given pair of objects with established trajectory predictions, it is possible to evaluate potential collisions between the objects. *Collision Detection* refers to detecting a possible collision between two objects, yielding a warning notification in case of detection. The *collision detection* algorithm computes predicted path intersections between a car and objects in its warning area as follows:

1. Quadratic regression is again used to compute the equation of the parabola for the two street objects.
2. Parabola equations are used to detect intersection points among them. An intersection indicates a potential collision between both objects, at a given GPS position.
3. If a potential collision is detected, the potential timestamps at which the intersection point occurs for both objects are computed. A threshold variable is used to determine if the intersection point is reached at the same time (within the threshold) for both objects. If this is the case, a potential collision has been detected.

The output of the *Collision detection* is very simple, as an example:

Collision detection output data

```
Potential collision detected:
   Objects IDs:   1 and 5
   GPS position:  (44.654565,10.933830)
   Timestamp:     1603963258
```

Both the *Trajectory prediction* and the *Collision detection* algorithms work in parallel, since they can be independently computed for each road user (or pairs of them). Both algorithms are suitable for map/reduce operations.

2.1.8 Vehicles Emissions Model

Based on the traffic conditions, and the information of the vehicle fleet composition in the area of study, it is possible to estimate the contamination level of such area. In this context, the interest is on the estimation of the pollution emissions of current traffic conditions in real time.

More specifically, the detected vehicles (category, timestamp, and speed) are a representation of the current traffic conditions, while the vehicle class fleet composition in the area of study is used to estimate the vehicle's engine power, based on the class of vehicle. The vehicle class compromises information on the shape, weight, and default loading of a vehicle (passenger car, heavy-duty vehicle, ...), as well as engine-related properties, for example rated power, fuel type, and most important, the actual emission class (Euro 1, Euro 2, and so on). The emissions can be then interpolated from emission curvescontaining the normalized engine's power output and vehicle data [8]. The output obtained estimates emissions of NOx, PM, CO, HC, and NO, at a configured time resolution and for road segment.

As an example, the output of the *Vehicles emissions model*, at a given time instant and for three different streets, is shown below:

Vehicles emissions model output data

Road-Segment	Emissions (g/h)					
	NOx	HC	CO	PM	PN	NO
Str. Attiraglio	85	3	59	12	2.04E+15	60
Via Manfredo Fanti	1150	33	939	148	2.60E+16	755
Via Maria Montessori	200	6	162	39	5.56E+15	129

2.1.9 Data Aggregation: Data Knowledge Base

This *Data aggregation* big data analytics method is fundamental given the dispersion nature of data collected in a smart city use-case. Concretely, this method creates and maintains a Data Knowledge Base (DKB), aggregating data into a single system, collected and processed by multiple IoT sensors and devices (located in vehicles or at city streets).

The challenging task of the DKB is to maintain (in real-time) the information anonymized, updated, and consistent across the multiple actors, i.e., the city and the vehicles, to ensure that decisions are taken based on the same updated information at all levels. A key feature of the DKB is to allow taking decisions considering information beyond the field of view of a single actor. As an example, a vehicle can receive relevant information from the city, improving the vehicle's safety. More information is provided in Sect. 3.

2.1.10 Visualization

An important feature of big data analytics is how to visualize the meaningful information. In this case, it is interesting to visualize the real-time processed information of the DKB, i.e., the traffic conditions, the emissions level, and possible alerts. There are two different final users that need to receive this information, at different levels:

- At a vehicle level, *real-time traffic conditions* is available for connected vehicle's. More specifically, a lightweight 3D user interface shows a map to the driver with its position in the city area, the position of the road users detected, and, more importantly, warnings signals about potential collisions.

- At a city level, a *dashboard* visualizes the aggregated data of the DKB, i.e., an interactive map shows the traffic conditions, based on the detected road users, and the emission levels.

2.1.11 Predictive Models

The interest of this method is to understand the potential to improve routing and travel times while responding to traffic accidents. In case of accidents, the digital traffic signals can advise the "best path to follow," reducing the induced traffic impact and improving the driver experience. For emergency vehicles (e.g., ambulances, firefighters, and police vehicles), this method can dynamically create "green routes" by adjusting the frequency of the traffic lights to reduce the time of intervention.

Trivially, the viability of modifications on a real-world urban area must be carefully planned and tested before the actual deployment. Therefore, this method is based on a well-known MATSim simulator [1] for urban transportation. The baseline version of MATSim is only capable of simulating road users that cannot communicate to nearby cars or the surrounding infrastructure. In contrast, this extended method is able to simulate the interactions among sets of connected vehicles within a smart city. It is composed of multiple modules to simulate cameras, sensors, servers, and communication systems so that connected vehicles are able to react to unexpected events such as traffic accidents.

2.2 Integration Toward a Relevant Smart City Use-Case

The big data analytics methods presented in the previous section are combined into a unique big-data analytics workflow representing the generation of the Data Knowledge Base (DKB) from which valuable knowledge is extracted from the city and the connected cars (see Sect. 2.1.9). Then, the three use-cases applications (digital traffic sign, obstacle detection, and air pollution estimation) are built upon the information included in the DKB.

Figure 5 shows the combined big data analytics workflow representation, including (1) the big data analytics methods presented in previous Sect. 2 (represented as labelled nodes), (3) the interaction between them (represented as arrows), and (3) the actors involved in the smart city system (sensors and use-cases applications).

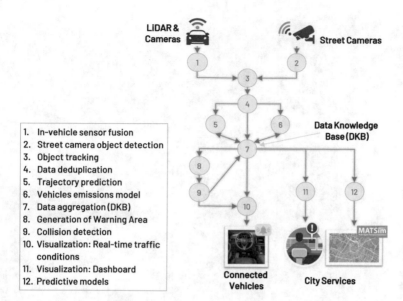

1. In-vehicle sensor fusion
2. Street camera object detection
3. Object tracking
4. Data deduplication
5. Trajectory prediction
6. Vehicles emissions model
7. Data aggregation (DKB)
8. Generation of Warning Area
9. Collision detection
10. Visualization: Real-time traffic conditions
11. Visualization: Dashboard
12. Predictive models

Fig. 5 Integration of the big data analytics

3 Distribution of Big Data Analytics Across the City Infrastructure

This section presents the actual implementation of the proposed big data analytics on a real smart city infrastructure. Section 3.1 describes the smart city infrastructure, in terms of connectivity, sensors, and computing capabilities, to support the use-cases demonstration. Concretely, the proposed big data analytics are being implemented and tested in the Modena Automotive Smart Area (MASA) [9] (see Sect. 3.1.1), using smart cars (Maserati vehicles) and connected cars (see Sect. 3.1.2). Section 3.2 describes the distribution of the big data analytics on the available city and vehicles infrastructure, and presents an example of use-case scenarios. Finally, Sect. 3.3 makes emphasis on one of the key aspects for the successful development of the considered use-cases: the real-time requirements.

3.1 Smart City Infrastructure

3.1.1 City Infrastructure: MASA

The Modena Automotive Smart Area (MASA) is a 1 km^2 area in the city of Modena (Italy), equipped with a sensing, communication, and computation infrastructure. MASA is an urban laboratory for the experimentation and certification of new technologies including autonomous driving and connecting driving technologies,

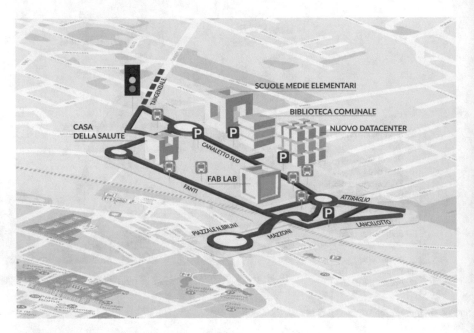

Fig. 6 Modena Automotive Smart Area (MASA)

and V2X technologies, in general. Figure 6 shows the map of the equipped streets in Modena. The tested technologies to be found in the area include interconnected traffic lights, digital signposting, cameras, and sensors, among others.

Figure 7 provides a detailed description of the sensing, communication, and computation infrastructure of MASA:

- Bullet cameras, optimized for detection or forensic purposes regardless of light conditions, connected to the fog nodes through optical fiber.
- Four-optics camera (360° overview), that provides a full detection and control of the roundabout.
- Pollution sensors, connected to the LoRa network, for the detection of air quality parameters: carbon monoxide (CO), carbon dioxide (CO2), nitrogen dioxide (NO2), and particulate matter (PM).
- Wireless communication, a 4G dedicated antenna for private local area network, 5G prototypes, and a Low-Range(LoRa) network to interconnect sensors and devices with low bandwidth needs (e.g., parking and pollution sensors).
- Optical Fiber network, connects the cameras to the available fog nodes.
- Fog Nodes, with the following computing features: an Intel®Xeon E3-1245 v.5, 32 GB of RAM, 256 GB of hard disk, and an NVIDIA Volta GPU (TitanV).

Moreover, the city of Modena provides a private cloud infrastructure in a data center, also connected to the fog nodes through optical fiber.

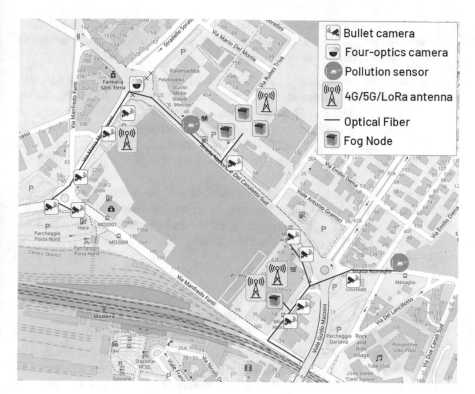

Fig. 7 Modena Automotive Smart Area (MASA) sensing, communication, and computation infrastructure

3.1.2 Vehicle Infrastructure

Maserati provides two prototype vehicles that incorporate all the sensing, communication, and computation infrastructure needed to test the use-cases described in this chapter. The vehicles are two Maserati Quattroporte (Model Year '18 and Model Year '19, respectively), a four-door full-sized luxury sports sedan vehicle (F segment). For simplicity, these are known as *smart cars*.

Each vehicle includes a set of sensor devices to obtain information about the position, speed, and typology of the objects that surround the vehicle. In addition to this data, the vehicle also provides information from the vehicle CAN network, including speed, acceleration, and collision and emergency break information. Concretely, the extra sensors installed on each vehicle are:

- Surrounding high-definition cameras, including four cameras with 120° of Field Of View (FOV), and two cameras with 60° of FOV.
- A 3D Light Detection and Ranging (LiDAR) with 360° of FOV.

- A Global Navigation Satellite System (GNSS) to increase the accuracy, redundancy, and availability of the vehicle position, compared to the positioning system already provided in the commercial vehicle.

Moreover, each vehicle is equipped with a 4G-LTE antenna receiver and a powerful embedded high-performance computing platform capable of implementing real-time bigdata analytics. Those platforms are designed to process data from camera, radar, and LiDAR sensors to perceive the surrounding environment. The first vehicle is equipped with an NVIDIA DRIVE PX2 Autochauffeur (dual TX2 SOC plus 2 discrete Pascal GPUs), and the second vehicle is equipped with the NVIDIA DRIVE AGX Pegasus [10].

For testing purposes, and to enlarge the fleet of vehicles involved in the system, without the need for a large budget, *connected cars* are also considered. A *connected car* is simply equipped with a regular laptop or an embedded device with LTE connectivity and GPS support, so that it can send its position to the system and receive alerts.

3.2 Big Data Analytics Distribution

The big data analytics methods described in Sect. 2 are executed and distributed in the available city and vehicles infrastructure, to provide the required functionality for the three use-cases: *Digital Traffic Sign*, *Obstacle Detection*, and *Air Pollution Estimation*. Concretely, the actual distribution depends on different factors: the underlying infrastructure, the source of data, and the software architecture that processes it. Moreover, the big data analytics distribution does not need to follow a static approach, but instead it can be based on the current status of the infrastructure (e.g., load or availability of computing nodes). This is one of the main challenges of the CLASS project [4].

Figure 8 shows an example of distribution of the big data analytics on the MASA and vehicles infrastructure, described in Sects. 3.1.1 and 3.1.2, respectively. This example considers three fog nodes, four street cameras, the Modena data center (private cloud), a smart car (acting as an *active* road user), and two pedestrian (*passive* vulnerable road users -VRUs-). Cameras 1 and 2 are connected to fog node 1, while cameras 3 and 4 are connected to fog node 2. Therefore, object detection and tracking are executed (independently) for each camera video feed in the corresponding fog node. Also, data deduplication is performed at fog nodes 1 and 2 with the purpose of identifying duplicated objects detected by more than one camera. This happens if two cameras (or more) have an overlapping field of view of a city area. From the smart car, the information from the sensor fusion algorithm, i.e., objects detected and the position of the car itself, is sent to fog node 3, where the tracking of these objects is performed. Then, deduplication is also executed at fog node 3, in case overlapping areas (thus, possible duplicated objects) are being recorded from cameras 1 or 2, cameras 3 or 4, and smart car.

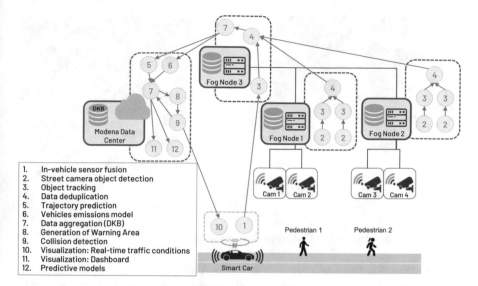

1. In-vehicle sensor fusion
2. Street camera object detection
3. Object tracking
4. Data deduplication
5. Trajectory prediction
6. Vehicles emissions model
7. Data aggregation (DKB)
8. Generation of Warning Area
9. Collision detection
10. Visualization: Real-time traffic conditions
11. Visualization: Dashboard
12. Predictive models

Fig. 8 Example of big data analytics distribution on the MASA and vehicles infrastructure

Once this data is collected and deduplicated at fog node 3, it can be aggregated into the DKB and make it also available at cloud side. In this example, the rest of big data analytics, i.e., from trajectory prediction and vehicles emission model, to the dashboard visualization and predictive models, are executed in the cloud. The visualization of the real-time traffic conditions and alerts is executed at the smart car.

The big-data distribution shown in Fig. 8 is only an example; there are many other possibilities depending on the aforementioned factors. For instance, the data deduplication can be performed only at one level, at fog node 3, but the example shows data deduplication execution at two levels (at nodes 1 and 2, and then at fog node 3) for a twofold reason: (1) to split and distribute the computation of this functionality, and (2) more importantly, to be able to obtain results (e.g., predict possible collisions) at this level, in the fog nodes 1 and 2, without the need of obtaining them in the cloud, contrary to what it is shown in this example. Another possibility is to receive the data from the smart car, directly at fog nodes 1 or 2, again with the purpose of boosting local computation at fog level. There are plenty of distribution options that rapidly increase as the number of actors involved in the use-case (e.g., smart cars, connected cars, fog nodes, etc.) increases as well. If the software framework supports a dynamic balancing of the workload, according to the different scenarios, it will promote the accomplishment of the envisioned use-case results.

(a) Scenario 1   (b) Scenario 2

Fig. 9 Real collision avoidance demonstration scenarios

3.2.1   Real Scenarios for Collision Avoidance

This section describes two possible scenarios that exploit the capabilities of the distributed big data analytics. Figure 9 shows two different scenarios to be recreated in the city of Modena. There are different actors involved: street cameras (a video frame shows its FOV), smart cars (or connected cars), a passive car and truck, and pedestrians. Arrows show the predicted trajectory of each actor.

- Scenario 1, Fig. 9a (Attiraglio Street in MASA): It evaluates the "*virtual mirror*" functionality that increases the field of view of a vehicle beyond its actual vision (or the driver's vision). In this example, there is a stationary truck that hides the view of both the smart car driver and the pedestrian who is crossing the street. This hazard situation is detected by the two cameras located at the street and processed by the combined big data analytics. As a result, an alert is sent to the smart car.
- Scenario 2, Fig. 9b (Roundabout at Via Pico della Mirandola—Parking exit of the Modena train station): It evaluates the "*two sources of attention*" functionality that aims to alert drivers when attention must be paid to two different events in opposite directions. In this example, there is a smart car exiting the parking, a regular car reaching the roundabout, and a pedestrian crossing the road. This hazard situation is detected by the camera located at the street and processed by the combined big data analytics. As a result, an alert is sent to the smart car.

3.3   Real-Time Requirements

One important aspect of the big data system presented in this chapter is the notion of real time: data is *constantly* being produced and processed, and big data analytics are highly parallelizable, e.g., new objects are detected, while tracking previous objects, or the warning area computation can be simultaneous for multiple objects. Exploiting this parallelism, along with an efficient use of the underlying infrastructure, is extremely important to guarantee that the results are meaningful by the time they are computed. This is especially relevant for the *Obstacle Detection* use-case, since alerts must raise within a time interval that is useful for the driver to react. A reasonable metric, considered in the scope of the CLASS project, is to get updated results at a rate between 10 and 100 ms. Assuming that the maximum speed of a vehicle within the city is 60 km/h, vehicles will advance between 0.17 and 1.7 m. This level of granularity is enough to implement the proposed use-cases.

4 Conclusions

One of the smart computing domains in which big data can have a larger impact on people's day-to-day life is the smart city domain. Nowadays, cities consume 70% of the world's resources, with an estimated population rate growth of 66% by 2050, according to United Nation reports [11]. Smart cities are increasingly seen as an effective technology capable of controlling the available city resources safely, sustainability, and efficiently to improve economical and societal outcomes.

This chapter described the CLASS [4] use-case, a realistic yet visionary use-case from the smart city domain, which includes the real-time elaboration of huge amounts of data coming from a large set of sensors distributed along a wide urban area, supporting intelligent traffic management and advanced driving assistant systems. The CLASS framework and use-case are currently under deployment and evaluation in the city of Modena (Italy). More details and up-to-date information can be found on the CLASS project website: https://class-project.eu/.

For the successful outcome of this research project in particular, and for addressing the big-data challenge of future smart cities in general, it is fundamental to combine multi-dimensional and multidisciplinary contexts and teams, from artificial intelligence and machine learning to data storage and engineering.

Acknowledgments The research leading to these results has received funding from the European Union's Horizon 2020 Programme under the CLASS Project (www.class-project.eu), grant agreement No. 780622.

References

1. Axhausen, K. W., Horni, A., & Nagel, K. (2016). *The multi-agent transport simulation MATSim*. Ubiquity Press.
2. Bochkovskiy, A., Wang, C.-Y., & Liao, H.-Y. M. (2020). YOLOv4: Optimal speed and accuracy of object detection. https://arxiv.org/abs/2004.10934. Online. Accessed November 2020.
3. Buchholz, D., Intel IT Principal Engineer, Dunlop, J., & Intel IT Client Architect. (2011). The future of enterprise computing: Preparing for the compute continuum. IT Intel White Paper, Intel IT.

4. Edge and Cloud Computation: A Highly Distributed Software for Big Data Analytics (CLASS). (2020). https://class-project.eu/. Online. Accessed October 2020.
5. Ester, M., Kriegel, H.-P., Sander, J., Xu, X., et al. (1996). Adensity-based algorithm for discovering clusters in large spatial databases with noise. In *Kdd* (Vol. 96, pp. 226–231).
6. Hadad, E., & Jbara, S. (2020). Using OpenWhisk as a polyglot real-time event-driven programming model in CLASS. https://class-project.eu/news/using-openwhisk-polyglot-real-time-event-driven-programming-model-class. Online. Accessed November 2020.
7. Haddad, S., Wu, M., Wei, H., & Lam, S. K. (2019). Situation-aware pedestrian trajectory prediction with spatio-temporal attention model. Preprint. arXiv:1902.05437.
8. Hausberger, S., Rexeis, M., & Dippold, M. (2017). Passenger car and heavy duty emission model (phem) light. User guide v1.
9. Modena Automotive Smart Area (MASA). (2020). https://www.automotivesmartarea.it/?lang=en. Online. Accessed October 2020.
10. NVIDIA Drive: End-to-End Solutions for Software-Defined Autonomous Vehicles. (2020). https://www.nvidia.com/en-sg/self-driving-cars/. Online. Accessed November 2020.
11. United Nations. World urbanization prospects. ISBN 978-92-1-151517-6, 2014 Revision.
12. Varsity Tutors. Quadratic Regression. https://www.varsitytutors.com/hotmath/hotmath_help/topics/quadratic-regression. Online. Accessed November 2020.
13. Verucchi, M., Bartoli, L., Bagni, F., Gatti, F., Burgio, P., & Bertogna, M. (2020). Real-time clustering and lidar-camera fusion on embedded platforms for self-driving cars. In *IEEE Robotic Computing proceedings* (2020).
14. Welch, G., Bishop, G., et al. (1995). An introduction to the Kalman filter. Chapel Hill, NC, USA.
15. Wikipedia Community. (2020). Geohash. https://en.wikipedia.org/wiki/Geohash. Online. Accessed November 2020.
16. Zillner, S., Bisset, D., Milano, M., Curry, E., Garcia Robles, A., Hahn, T., Irgens, M., Lafrenz, R., Liepert, B., O'Sullivan, B., & Smeulders, A. (Eds.). (2020). Strategic research and innovation agenda - AI, data and robotics partnership. Third Release. https://ai-data-robotics-partnership.eu/wp-content/uploads/2020/09/AI-Data-Robotics-Partnership-SRIDA-V3.0.pdf. Online. Accessed February 2021.
17. Zillner, S., Curry, E., Metzger, A., & Seidl, R. (Eds.). (2017). European Big Data Value Strategic Research and Innovation Agenda. https://bdva.eu/sites/default/files/BDVA_SRIA_v4_Ed1.1.pdf. Online. Accessed October 2020.

Processing Big Data in Motion: Core Components and System Architectures with Applications to the Maritime Domain

Nikos Giatrakos, Antonios Deligiannakis, Konstantina Bereta, Marios Vodas, Dimitris Zissis, Elias Alevizos, Charilaos Akasiadis, and Alexander Artikis

Abstract Rapidly extracting business value out of Big Data that stream in corporate data centres requires continuous analysis of massive, high-speed data while they are still in motion. So challenging a goal entails that analytics should be performed in memory with a single pass over these data. In this chapter, we outline the challenges of Big streaming Data analysis for deriving real-time, online answers to application inquiries. We review approaches, architectures and systems designed to address these challenges and report on our own progress within the scope of the EU H2020 project INFORE. We showcase INFORE into a real-world use case from the maritime domain and further discuss future research and development directions.

Keywords Big Data · Cross-platform optimisation · Data streams · Data synopses · Online machine learning · Complex event forecasting · Maritime situation awareness

N. Giatrakos (✉) · A. Deligiannakis
Athena Research Center, Athens, Greece
e-mail: ngiatrakos@athenarc.gr; adeli@athenarc.gr

K. Bereta · M. Vodas · D. Zissis
MarineTraffic, Athens, Greece
e-mail: konstantina.bereta@marinetraffic.com; marios.vodas@marinetraffic.com; dzissis@marinetraffic.com

E. Alevizos · C. Akasiadis · A. Artikis
NCSR Demokritos, Institute of Informatics and Telecommunications, Athens, Greece
e-mail: alevizos.elias@iit.demokritos.gr; cakasiadis@iit.demokritos.gr; a.artikis@iit.demokritos.gr

© The Author(s) 2022
E. Curry et al. (eds.), *Technologies and Applications for Big Data Value*,
https://doi.org/10.1007/978-3-030-78307-5_22

1   Challenges of Big Streaming Data

Today, organisations and businesses have the ability to collect, store and analyse as much data as they need, exploiting powerful computing machines in corporate data centres or the cloud. To extract value out of the raw Big Data that are accumulated, application workflows are designed and executed over these infrastructures engaging simpler (such as grouping and aggregations) or more complex (data mining and machine learning) analytics tasks. These tasks may involve data at rest or data in motion.

Data at rest are historic data stored on disks, getting retrieved and loaded for processing by some analytics workflow. Analytics tasks participating in such a workflow perform computations on massive amounts of data, lasting for hours or days. They finally deliver useful outcomes. Using a running example from the maritime domain, historic vessel position data are used to extract Patterns-of-Life (PoL) information. These are essentially collections of geometries representing normal navigational routes of vessels in various sea areas [78], used as the basis for judging anomalies.

Data in motion involve Big streaming Data which are unbounded, high-speed streams of data that need to get continuously analysed in an online, real-time fashion. Storing the data in permanent storage is not an option, since the I/O latency would prevent the real-time delivery of the analytics output. Application workflows get a single look on the streaming data tuples, which are kept in memory for a short period of time and are soon stored or discarded to process newly received data tuples.

At an increasing rate, numerous industrial and scientific institutions face such business requirements for real-time, online analytics so as to derive actionable items and timely support decision-making procedures. For instance, in the maritime domain, to pinpoint potentially illegal activities at sea [54] and allow the authorities to timely act, position streams of thousands of vessels need to be analysed online.

To handle the volume and velocity of Big streaming Data, Big Data platforms such as Apache Flink [2], Spark [5] or toolkits like Akka [1] have been designed to facilitate scaling-out, i.e., parallelising, the computation of streaming analytics tasks horizontally to a number of Virtual Machines (VM) available in corporate computer clusters or the cloud. Thus, multiple VMs simultaneously execute analytics on portions of the streaming data undertaking part of the processing load, and therefore throughput, i.e., number of tuples being processed per time unit, is increased. This aids in transforming raw data in motion to useful results delivered in real time. Big Data platforms also offer APIs with basic stream transformation operators such as filter, join, attribute selection, among others, to program and execute streaming workflows. However useful these facilities may be, they only focus on a narrow part of the challenges that business workflows need to encounter in streaming settings.

First, Big Data platforms currently provide none or suboptimal support for advanced streaming analytics tasks engaging Machine Learning (ML) or Data Mining (DM) operators. The major dedicated ML/DM APIs they provide, such as

MLlib [5] or FlinkML [2], do not focus on parallel implementations of streaming algorithms.

Second, Big Data platforms by design focus only on horizontal scalability as described above, while there are two additional types of scalability that are of essence in streaming settings. Vertical scalability, i.e., scaling the computation with the number of processed streams, is also a necessity. Federated scalability, i.e., scaling the computation one step further out, to settings composed of multiple, potentially geo-dispersed computer clusters, is another type of required scalability. For instance, in maritime applications, vessels transmit their positions to satellite or ground-based receivers. These data can be ingested in proximate data centres and communicated only on demand upon executing global workflows, i.e., involving the entire set of monitored vessels, over the fragmented set of streams.

Third, Big Data technologies are significantly fragmented. Delivering advanced analytics requires optimising the execution of workflows over a variety of Big Data platforms and tools located at a number of potentially geo-dispersed clusters or clouds [30, 34, 36]. In such cases, the challenge is to automate the selection of an optimal setup prescribing (a) which network cluster will execute each analytics operator, (b) which Big Data platform available at this cluster, and (c) how to distribute the computing resources of that cluster to the operators that are assigned to it.

Connecting the above challenges to a real-world setting from the maritime domain, on a typical day at MarineTraffic,[1] 100GB vessel position data and approximately 750M messages (volume, velocity—horizontal scalability) are processed online. This data is complemented by other data sources such as satellite image data of tens of TBs [54]. At any given time, MarineTraffic is tracking over 200K vessels in real-time (vertical scalability) over a network of approximately 5K stations (federated scalability). Additionally, the analysis engages a variety of Big Data platforms including Apache Spark, Flink, Akka and Kafka (details in Sect. 3).

Finally, applications often require an additional level of abstraction on the derived analytics results. Consider a vessel that slows down, then makes a U-turn and then starts speeding up. Such a behaviour may occur in case of an imminent piracy event where a vessel attempts to run away from pirates. The application is not interested in knowing the absolute speed, heading or direction information in the raw stream. Instead, it wants to receive continuous reports directly on a series of detected, *simple* events (`slowing down`, `U-turn`, `speeding`) and the higher level, *complex* `piracy` event or to be able to forecast such events [79]. Complex Event Processing (CEP) and Forecasting (CEF) encompass the ability to query for business rules (patterns) that match incoming streams on the basis of their content and some topological ordering on them (CEP) or to forecast the appearance of patterns (CEF) [31, 33, 35].

In this chapter, we discuss core system components required to tackle these challenges and the state of the art in their internal architectures. We further describe

---

[1] https://www.marinetraffic.com.

how we advance the state of the art within the scope of the EU H2020 project INFORE. Finally, we showcase the INFORE approach into a real-world use case from the maritime domain. We, however, stress that INFORE applies to any application domain, and we refer the interested reader to [34] for more application scenarios.

This chapter relates to the technical priorities (a) Data Management, (b) Data Processing Architectures and (c) Data Analytics of the European Big Data Value Strategic Research & Innovation Agenda [77]. It addresses the horizontal concerns Cloud, HPC and Sensor/Actuator infrastructure of the BDV Technical Reference Model and the vertical concern of Big Data Types and Semantics (Structured data, Time series data, Geospatial data). Moreover, the chapter relates to (a) Knowledge and Learning, (b) Reasoning and Decision Making, (c) Action and Interaction and (d) Systems, Hardware, Methods and Tools, cross-sectorial technology enablers of the AI, Data and Robotics Strategic Research, Innovation and Deployment Agenda [76].

2   Core Components and System Architectures

2.1   The Case for Data Synopses

Motivation There is a wide consensus in the stream processing community [25, 26, 32] that approximate but rapid answers to analytics tasks, more often than not, suffice. For instance, detecting a group of approximately 50 highly similar vessel trajectories with sub-second latency is more important than knowing minutes later that the group actually composes 55 such streams with a similarity value accurate to the last decimal. In the latter case, some vessels may have been engaged in a collision. Data synopses techniques such as samples, histograms and sketches constitute a powerful arsenal of data summarisation tools useful across the challenges discussed in the introduction of this chapter. Approximate, with tunable quality guarantees, synopses operators including, but not limited to [25, 26, 32, 46], cardinality (FM Sketches), frequency moment (CountMin, AMS Sketches, Sampling), correlation (Fourier Transforms, Locality Sensitive Hashing [37]), set membership (Bloom Filters) or quantile (GK Quantile) estimation, can replace respective exact operators in application workflows to enable or enhance all three types of required scalability as well as to reduce memory utilisation. More precisely, data summaries leave only a footprint of the stream in memory and they also enhance horizontal scalability since not only is the processing load distributed to a number of available VMs, but also it is shed by letting each VM operate on compact data summaries. Moreover, synopses enable federated scalability since only summaries, instead of the full (set of) streams, can be communicated when needed. Finally, synopses provide vertical scalability by enabling locality-aware hashing [37, 38, 46].

Related Work and State of the Art From a research viewpoint, there is a large number of related works on data synopsis techniques. Such prominent techniques are reviewed in [25, 26, 32] and have been implemented into real-world synopses libraries, such as Yahoo!DataSketch [9], Stream-lib [8], SnappyData [57] and Proteus [7]. Yahoo!DataSketch [9] and Stream-lib [8] are libraries of stochastic streaming algorithms and summarisation techniques, correspondingly, but implementations are detached from parallelisation and distributed execution aspects over streaming Big Data platforms. Apache Spark provides utilities for data synopsis via sampling operators, CountMin sketches and Bloom Filters. Moreover, SnappyData's [57] stream processing is based on Spark and its synopses engine can serve approximate, simple sum, count and average queries. Similarly, Proteus [7] extends Flink with data summarisation utilities. Spark utilities, SnappyData and Proteus combine the potential of data summarisation with horizontal scalability, i.e., parallel processing over Big Data platforms, by providing libraries of parallel versions of data synopsis techniques. However, they neither handle all types of required scalability nor cross Big Data platform execution scenarios.

INFORE Contribution In the scope of the INFORE project, we have developed a Synopses Data Engine (SDE) [46] that advances the state of the art by tackling all three types of the required scalability and also accounting for sharing synopses common to various running workflows and for cross-platform execution. INFORE SDE goes far beyond the implementation of a library of data summarisation techniques. Instead, it also implements an entire component with its own internal architecture, employing a Synopses-as-a-Service (SDEaaS) paradigm. That is, the SDE is a constantly running service (job) in one or more clusters (federated scalability) that can accept on-the-fly requests for start maintaining, updating and querying a parallel synopsis built on a single high-speed stream (e.g. vessel) of massive data proportions (horizontal scalability) or on a collection of a large number of streams (vertical scalability). The SDEaaS is customisable to specific application needs by allowing dynamic loading of code for new synopses operators at runtime, with zero downtime for the workflows that it serves.

The architecture of INFORE SDEaaS [46] is illustrated in Fig. 1a. INFORE's SDEaaS proof-of-concept implementation is based on Apache Kafka and Flink. Nevertheless, the design is generic enough to remain equally applicable to other Big Data platforms. For instance, an equally plausible alternative would be to implement the whole SDE in Kafka leveraging the Kafka Streams API. Nonetheless, Kafka Streams is simply a client library for developing micro-services, lacking a master node for global cluster management and coordination. Following Fig. 1a, when a request for maintaining a new synopsis is issued, it reaches the `RegisterRequest` and `RegisterSynopsis` `FlatMaps` which produce keys for workers (i.e., VM resources) which will handle this synopsis. Each of this pair of `FlatMaps` uses these keys for a different purpose. `RegisterRequest` uses the keys to direct queries to responsible workers, while `RegisterSynopsis` uses the keys to update the synopses on new data arrivals (blue-coloured path). In particular, when a new streaming data tuple is ingested, the `HashData` `FlatMap` looks up the

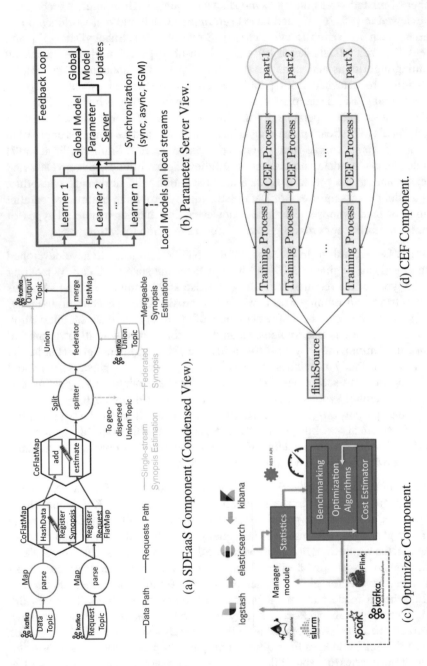

(a) SDEaaS Component (Condensed View).

(b) Parameter Server View.

(c) Optimizer Component.

(d) CEF Component.

Fig. 1 Internal architecture of key INFORE components

keys of `RegisterSynopsis` to see to which workers the tuple should be directed to update the synopsis. This update is performed by the `add FlatMap` in the blue-coloured path. The rest of the operators in Fig. 1a are used for merging partial synopses results [11] maintained across workers or even across geo-distributed computer clusters. Please refer to [46] for further details. In Sect. 3.2.3, we analyse the functionality of a domain-specific synopsis building samples of vessel positions.

2.2 Distributed Online Machine Learning and Data Mining

Motivation As discussed in Sect. 1, ML/DM APIs such as Spark's MLlib [5] or FlinkML [2] are focused on analysing data at rest. Therefore, advanced analytics tasks on data in motion call for filling the gap of a stream processing-oriented ML/DM module. ML and DM algorithms that can meet the challenges discussed in the introduction of this chapter are those that (1) are online, i.e., restricting themselves on a single pass over the data instead of requiring multiple passes, and (2) can run in a distributed fashion, i.e., they are parallelisable and thus the load can be distributed to parallel learners and parallel predictors across a number of VMs so as to provide the primitives for horizontal scalability over Big Data platforms and computer clusters. There exists a variety of algorithms that satisfy these preliminary requirements in diverse ML/DM categories, including [18, 34, 69] classification (such as (Multiclass) Passive Aggressive Classifiers, Online Support Vector Machines, Hoeffding Trees, Random Forests), clustering (BIRCH, Online k-Means, StreamKM++) and regression (Passive Aggressive Regressor, Online Ridge Regression, Polynomial Regression) tasks. These algorithms are designed or can be adapted to get executed in an online, distributed setting. The primary focus, then, is not on the algorithms themselves, but on the architecture an ML/DM module should be built upon, so that various algorithms can be incorporated and also allow for vertical scalability, federated scalability and cross-platform execution with reduced memory utilisation.

Related Work and State of the Art Towards this direction, the two most prominent approaches and modules that exist in the literature are StreamDM [19] and Apache SAMOA [48]. StreamDM is a library of ML/DM algorithms designed to be easily extensible with new algorithms, but dedicated to run on top of the Spark Streaming API [5]. Thus, it does not cover cross-platform execution scenarios, also lacking provisions for vertical and federated scalability. The only framework with a clear commitment to the cross-platform execution goals is Apache SAMOA. SAMOA is portable between Apache Flink, Storm [6] and Samza [4]. When it comes to its model of computation, the architecture of SAMOA follows the Agent-based pattern. In other words, an algorithm is a set of distributed processors that communicate with streams of messages. Little more is provided, which is intentional [48], claiming that a more structured model of computation reduces the applicability of the framework.

The state of the art in distributed ML and DM architectures is the Parameter Server (PS) distributed model [51] as illustrated in Fig. 1b, where a set of distributed learners receive portions of the training streams and extract local models in parallel. The local models are from time to time synchronised to extract a global model at the PS side. The global model is then communicated back to learners via a feedback loop (Fig. 1b). Consider for instance a set of learners each handling a subset of vessel streams within the scope of a vessel type classification task. The learners coordinate with the PS sending their locally trained classification models, while the PS responds back with an up-to-date global model. The PS paradigm enhances horizontal and federated scalability via the option of an asynchronous (besides synchronous) synchronisation policy to reduce the effect of stragglers and bandwidth consumption, respectively. In the synchronous policy, learners are communicating with the PS in predefined rounds/batches, while in the asynchronous case each learner decides individually as to when it should send updates to the PS. Performance-wise, the synchronous policy does not encourage enhanced horizontal scalability because when many learners are used, the total utilisation is usually low, should only few stragglers exist. The asynchronous one is the policy of choice in large-scale ML; the processing speed is much higher when many learners are used and the training is more scalable.

The PS paradigm has been criticised for limited training speed due to potential network congestion at the PS side and for severely getting affected by low-speed links between the learners and the PS. Under these claims, a number of decentralised ML/DM architectures have evolved which employ a more peer-to-peer alike structure, where the training rationale is based on gossiping [42, 70]. The drawback of these approaches, though, is that it is unclear how the continuously updated, but decentralised, global model can be directly deployed for real-time inference purposes. This is because knowing the network node holding the updated global model at any given time requires extra communication. Hence, in case we want to train and simultaneously deploy the updated global ML/DM models at runtime, such a decentralised architecture does not seem to mitigate low-speed issues but moves the problem to the prediction, instead of the training, stage.

INFORE Contribution In the scope of the INFORE project, we follow a PS distributed model [51]. As is the case with the SDEaaS described in the previous section, INFORE's ML/DM module includes provisions for cross-platform execution scenarios by receiving input and output streams in JSON formatted Kafka messages. Moreover, the communication between learners and the PS is performed using a lightweight middleware where a generic API for PS and learner (bidirectional) communication is provided. In that, learners can be implemented over any Big Data platform and run in any cluster, while still being able to participate in the common ML/DM task. Besides learners, INFORE's ML/DM module includes a separate pipeline of parallel predictors that can communicate with the PS in order to receive up-to-date global models continuously extracted during the training process and directly deploy them for inference purposes.

INFORE's ML/DM module accounts for vertical and boosts horizontal scalability as well. This is achieved by using INFORE's SDEaaS to partition streams to learners or to allow learners to operate on compact stream summaries, correspondingly. Remarkably, to effectively encounter congestions or low-speed links and also allow to easily and effectively deploy/update the developed models, instead of resorting to decentralised approaches [42, 70], we develop our own synchronisation policy termed FGM [67] (Fig. 1b) that improves the employed PS paradigm. The new synchronisation protocol strengthens horizontal (within a cluster) and federated scalability by bridging the gap between synchronous and asynchronous communication. Instead of having learners communicating in predefined rounds/batches (synchronous) or when each one is updated (asynchronous), FGM requires communication only when a concept drift (i.e., the global model has significantly changed based on some criterion) is likely to have occurred. This is determined based on conditions each learner can individually examine.

2.3 Distributed and Online CEF

Motivation Big Data analytics tools mine data views to extract patterns conveying insights into what has happened, and then apply those patterns to make sense of the fresh data that stream in. This only permits to react upon the detection of such patterns, which is often inadequate. In order to allow for proactive decision-making, predictive analytics tools that allow to forecast future events of interest are required. Consider, for instance, the ability to forecast and proactively respond to hazardous events, such as vessel collisions or groundings, in the maritime domain. The ability to forecast, as early as possible, a good approximation to the outcome of a time-consuming and resource-demanding computational task allows to quickly identify possible outcomes and save valuable reaction time, effort and computational resources. Diverse application domains possess different characteristics. For example, monitoring of moving entities has a strong geospatial component, whereas in stock data analysis this component is minimal. Domain-specific solutions (e.g. trajectory prediction for moving objects) cannot thus be universally applied. We need a more general Complex Event Forecasting (CEF) framework.

Related Work and State of the Art Time-series forecasting is an area with some similarities to CEF, with a significant history of contributions [56]. However, it is not possible to directly apply techniques from time-series forecasting to CEF. Time-series forecasting typically focuses on streams of (mostly) real-valued variables and the goal is to forecast relatively simple patterns. On the contrary, in CEF we are also interested in categorical values, related through complex patterns and involving multiple variables. Another related field is that of prediction of discrete sequences over finite alphabets and is closely related to the field of compression, as any compression algorithm can be used for prediction and vice versa [17, 20, 24, 63, 64, 73].

The main problem with these approaches is that they focus exclusively on next symbol prediction, i.e., they try to forecast the next symbol(s) in a stream/string of discrete symbols. This is a serious limitation for CEF. An additional limitation is that they work on single-variable discrete sequences of symbols, whereas CEF systems consume streams of events, i.e., streams of tuples with multiple variables, both numerical and categorical. Forecasting methods have also appeared in the field of temporal pattern mining [22, 50, 71, 75]. A common assumption in these methods is that patterns are usually defined either as association rules [13] or as frequent episodes [53]. From the perspective of CEF, the disadvantage of these methods is that they usually target simple patterns, defined either as strictly sequential or as sets of input events. Moreover, the input stream is composed of symbols from a finite alphabet, as is the case with the compression methods mentioned previously.

INFORE Contribution In a nutshell, the current, state-of-the-art solutions for forecasting, even when they are domain-independent, are not suitable for the kind of challenges that INFORE attempts to address. In INFORE, the streaming input can be constantly matched against a set of event patterns, i.e. arbitrarily complex combinations of time-stamped pieces of information. An event pattern can either be fully matched against the streaming data, in which case events are detected, or partially matched, in which case events are forecast with various degrees of certainty. The latter usually stems from stochastic models of future behaviour, embedded into the event processing loop, which project into the future the sequence of events that resulted to a partial event pattern match, to estimate the likelihood of a full match, i.e. the actual occurrence of a particular complex event.

Given that INFORE's input consists of a multitude of data streams, interesting events may correlate sub-events across a large number of different streams, with different attributes and different time granularities. For instance, in the maritime domain relevant streams may originate from position signals of thousands of vessels which may be fused with satellite image data [54] or even acoustic signals [40]. It is necessary to allow for a highly expressive event pattern specification language, capable of capturing complex relations between events. Moreover, the actual patterns of what constitutes an interesting event are often not known in advance, and even if they are, event patterns need to be frequently updated to cope with the drifting nature of streaming data. Not only do we need an expressive formalism in order to capture complex events in streams of data, but we also need to do so in a distributed and online manner.

Towards this direction, the CEF module of INFORE uses a highly expressive, declarative event pattern specification formalism, which combines logic, probability theory and automata theory. This formalism has a number of key advantages:

- It is capable of expressing arbitrarily complex relations and constraints between events. We are thus not limited to simple sequential patterns applied to streams with only numerical or symbolic values.
- It can be used for event forecasting and offering support for robust temporal reasoning. By converting a pattern into an automaton, we can then use historical

data to construct a probabilistic description of the automaton's behaviour and thus to estimate at any point in time its expected future behaviour.

- It offers direct connections to machine learning techniques for refining event patterns, or learning them from scratch, via tools and methods from the field of grammatical inference. In cases where we only have some historical data and some labels, we must find a way to automatically learn the interesting patterns. This is also the case when there is concept drift in the streaming data and the patterns with which we started may eventually become stale. It is therefore important to be able to infer the patterns in the data in an online manner.

INFORE's CEF module is built on top of Apache Kafka and Flink and has the ability to handle highly complex patterns in an online manner, constantly updating its probabilistic models. Figure 1d shows one possible scheme (pattern-based) for structuring multiple parallel CEF pipelines. As shown in the figure, each such pipeline processes a different CEF query [33, 35]. It is composed of a training process, which estimates the probabilities of a future event to occur, as well as a CEF process that utilises these probabilities to actually forecast complex events. Finally, one implementation detail is that each pipeline also receives a subset of the patterns (part1 to partX in Fig. 1d). The role of these loops is similar to the feedback loop of Fig. 1b. Remarkably, the CEF module can also act as a CEP one since it can not only predict but also detect occurred events of interest [14].

2.4 Geo-distributed Cross-Platform Optimisation

Motivation All the aforementioned advanced stream processing techniques and technologies will only serve their goal if they are properly used. Consider, for instance, that we perfectly tune the execution of a synopsis, ML/DM or CEF operator in a specific cluster, but we assign the execution of the downstream operator of a broader workflow to a distant cluster. The execution speed up achieved for one operator may be diminished by network latency of long network paths. Therefore, developing algorithms for optimising the execution of streaming workflows (a) over a network of many clusters located in various geographic areas, (b) across a number of Big Data platforms available in each cluster and (c) simultaneously elastically devoting VMs and resources (CPU, memory, etc.) is a prerequisite to efficiently deliver in practice real-time analytics. Within a cluster, common optimisation objectives include throughput maximisation, execution latency and memory usage minimisation, while in multi-cluster settings communication cost, bandwidth consumption and network latency are also accounted for. Quality-of-Service (QoS) and computer cluster (CPU, memory, storage) capacity constraints also apply to these objectives.

Related Work and State of the Art There are a number of works that assign the execution of operators targeting at optimising network-related metrics, such as communication cost and network latency, while executing global analytics workflows across a number of networked machines or computer clusters. The seminal work of SBON [59] seeks to optimise a quantity similar to network usage ($dataRate \times latency$), but with a squared latency, across multi-hop paths followed by communicated data. An important limitation in SBON is that by using such a blended metric, the optimisation process cannot support constrained optimisation per metric (communication cost or latency). Due to that, also other related techniques [49, 59, 62] which employ blended metrics cannot incorporate resource or QoS constraints while determining operators' assignment to clusters. Although some [49, 62] claim to support latency constraints, this comes after having determined where an operator will be executed. Finally, the approach of Geode [72] purely focuses on minimising bandwidth consumption in the presence of regulatory constraints, but it does not account for network latency.

A series of works aim at optimising the execution of analytics operators within a single computer cluster. Such works focus on optimal assignment of operators to VMs such that high performance (mainly, in terms of throughput) and load balancing among VMs is achieved; subject to multiple function, resource and QoS constraints. Related works mainly provide optimisations on load assignment and distribution, load shedding, resource provisioning and scheduling policies inside the cluster. In Medusa [16], Borealis [10], Flux [68] and Nexus [23], the focus is to primarily balance the load, choose appropriate ways to partition data streams across a number of machines and minimise the usage of available resources (CPU cycles, bandwidth, memory, etc.) while maintaining high performance.

Another category of techniques examines the optimisation of network-wide analytics, simultaneously scaling-out the computation of an operator to the VMs of the cluster that undertakes its execution. JetStream [61] trades-off network bandwidth minimisation with timely query answer and correctness, but while exploring the cluster at which an operator will be executed, it restricts itself to the MapReduce rationale (i.e. the operator is executed at the cluster where data rests), nearest site of relevant data presence or a central location. Iridium [60], basically targeting optimisation of analytics over data at rest, assumes control over where relevant data are transferred and moves these data around clusters to optimise query response latency. SQPR [45] and [21] propose more generic frameworks for the constraint-aware optimal execution of global workflows across clusters, and they also optimise resources devoted to each operator execution at each cluster. However, [21, 45] do not account for cross-platform optimisation in the presence of different Big Data technologies.

Systems such as Rheem [12], Ires [27], BigDawg [28] and Musketeer [39] are designed towards cross-platform execution of workflows, but they can only optimise the processing of data at rest,[2] instead of data in motion. Furthermore, only Rheem accounts for network-related optimisation parameters such as communication cost.

---

[2] BigDawg supports stream processing over S-Store and Rheem supports JavaStreams, but no alternatives are included to allow for optimising across different streaming platforms.

INFORE Contribution The INFORE Optimiser is the first complete solution for streaming operators [30, 34]. INFORE's Optimiser is not simply the only one which can simultaneously instruct the streaming Big Data platform, cluster and computing resources for each analytics operator, but also it does so for a wide variety of diverse operator classes including (1) synopses, (2) ML/DM, (3) CEF and (4) stream transformations. INFORE's Optimiser incorporates the richest set of optimisation criteria related to throughput, network and computational latency, communication cost, memory consumption and accuracy of SDE operators, and it also accounts for constraints per metric, fostering the notion of Pareto optimality [30, 34].

The internals of INFORE Optimiser are illustrated in Fig. 1c. We use a statistics collector to derive performance measurements from each executed workflow. Statistics are collected via JMX or Slurm[3] and are ingested in an ELK stack[4] while monitoring jobs. A Benchmarking submodule automates the acquisition of performance metrics for SDE, OMLDM and CEF/CEP operators run in different Big Data platforms. The Benchmarking submodule utilises statistics and builds performance (cost) models. Cost models are derived via a Bayesian Optimisation approach inspired by CherryPick [15]. The cost models are utilised by the optimisation algorithms [30, 34] to prescribe preferable physical execution plans.

3 Real-Life Application to a Maritime Use Case

3.1 Background on Maritime Situation Awareness (MSA)

According to the US National Concept of Operations for Maritime Domain Awareness,[5] "Global Maritime Intelligence is the product of legacy, as well as changing intelligence capabilities, policies and operational relationships used to *integrate all available data*, information, and intelligence in order to identify, locate, and track potential *maritime threats*. Global MSA results from the *persistent monitoring* of maritime activities in such a way that *trends and anomalies* can be identified".

Maritime reporting systems are distinguished into two broad categories: *cooperative* and *non-cooperative*. An example of a *cooperative* maritime reporting system is the Automatic Identification System (AIS) [43]. All commercial vessels above 300 gross tonnage are obliged to bear AIS transponders. AIS forms the basis of a lot of MSA applications, such as the MarineTraffic vessel tracking platform. Other cooperative, but not public, maritime reporting systems are the Long Range Identification and Tracking system (LRIT) [44], as well as the Vessel Monitoring

---

[3] https://docs.oracle.com/javase/tutorial/jmx/overview/, https://slurm.schedmd.com/.

[4] https://www.elastic.co/what-is/elk-stack.

[5] https://web.archive.org/web/20111004213300/http://www.gmsa.gov/twiki/bin/view/Main/MDAConOps.

System (VMS) [29] for fishing vessels. Radar on-board or ashore installations can be used as maritime surveillance systems, such as the ones installed by default in a vessel's bridge, as well as in ports. Thermal cameras and satellite imagery can also be used as additional monitoring systems for vessels. Due to the time elapsed between the actual image acquisition from a satellite and its availability on the satellite repository that can be several hours, satellite imagery data do not offer real-time snapshots of the maritime domain but can be used combined with other sources such as AIS to "fill in the gaps" of AIS coverage (e.g., identify the whereabouts of a vessel while its transponder was switched off).

Global and *continuous* monitoring of the maritime domain as well as the identification of trends and anomalies require to address the challenges pointed out throughout this chapter as well as the following generic Big Data challenges described in the scope of the maritime domain:

- *Volume*, the number of available surveillance systems and sensors increases.
- *Velocity*, applications rely on continuous monitoring (e.g., vessel tracking) and need to process high velocity streaming data in real time.
- *Variety*, data from heterogeneous surveillance systems should be combined.
- *Veracity*, most of the maritime data sources are heavily prone to noise requiring data cleaning and analysis tasks to filter out unnecessary or invalid information.
- *Value*, as the availability of more sources of maritime data as well as the advanced Big Data processing, ML and AI technologies that are now available can help to maximise the derived knowledge that can be inferred from maritime data.

3.2   Building Blocks of MSA Workflows in the Big Data Era

Figure 2b shows an example of a generic workflow, implemented in the Maritime Use Case of the INFORE project, for MSA purposes. Different applications may include a subset of operators of Fig. 2b or implement different steps. In the

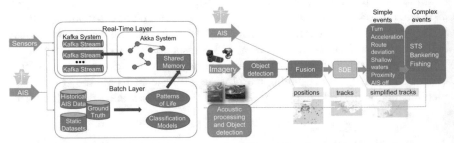

(a) Anomaly Detection at MarineTraffic.   (b) MSA workflow for maritime applications.

Fig. 2 MSA infrastructure and workflow

following, we describe the functionality of the workflow operators of Fig. 2b which serve as the building blocks of modern MSA applications.

3.2.1 Maritime Data Sources

The kinds of data sources that are provided as input in a typical MSA application (Fig. 2b) are the following:

- Vessel positions. Data about vessel positions derive from vessel reporting systems, the most popular of which is AIS. AIS forms the ground of a wide variety of MSA applications. AIS relies on VHF communication: Vessels send AIS messages that contain dynamic information (e.g., information about the current voyage, such as vessel position, speed, heading, etc.) as well as static information (e.g., vessel identifier, dimensions, etc.). For real-time applications, positional data arrive in streaming fashion to the data consumers.
- Data from other sensors. Some applications do not rely only on one source of information. For example, AIS data can be combined with acoustic data, thermal camera data and satellite data. Vessel detection algorithms are applied on this data to extract the positions of vessels. For example, AI techniques are applied on satellite imagery to extract the vessel positions which is important in the cases when a vessel is out of AIS coverage [54].
- Other datasets describing assets and activities in the maritime domain. These are datasets that describe ports, harbours, lighthouses, the boundaries of areas of interest, bathymetry datasets (e.g., for shallow waters estimation), datasets containing vessel schedules, weather data, etc. These datasets are often combined with other data (e.g., vessel positions) in order to enrich the information displayed to the end-users (e.g., the different layers of the MarineTraffic Live Map).

Kafka [3] is used at the data ingestion layer, as a fast, scalable and fault-tolerant messaging system for large data (at rest or in motion) portions.

3.2.2 Maritime Data Fusion

Data from multiple sources besides AIS, such as radars and cameras, are available in real time though in order to be used in MSA modules they must be fused together with AIS and create a unified map. This essentially translates to a need for tracking algorithms that can monitor moving objects globally and in real time using overlapping detections from multiple sensors. The Fusion operator in Fig. 2b is a custom operator with distributed implementations in order to achieve this goal. Trackers are comprised of three main components [65, 66]: (a) a method for the assignment of detections to tracks, (b) the prediction of a target's movement and (c) the architecture of the tracker that coordinates how the detections are processed.

A detection arriving to the tracker can be assigned to a track using three strategies, and each tracker implementation is based on one of them. The first way is to simply choose the track that is closest to the detection, which has the lowest computational complexity but it is not accurate in cases where two objects move very close to each other. The second method focuses on improving the accuracy in cases where a detection is close to multiple tracks by deferring the final assignment until more detections arrive, thus making a more informed decision but at the cost of significantly increasing the complexity and decreasing the responsiveness (i.e., real-time challenge). The third approach stands between the two methods and allows that a detection is assigned to multiple tracks as soon as it arrives, thus increasing the accuracy satisfactorily without increasing complexity.

Each moving object is characterised by certain physical parameters and constraints according to which several kinematic models can predict its movement under different conditions. A simple option is to choose one model, such as constant velocity that assumes the object maintains the last speed, but this affects the accuracy when an object manoeuvres. A better option is to use multiple models, such as constant turn and acceleration, at the same time so that the tracker is able to successfully detect a manoeuvring target.

3.2.3   SDE Operator For Trajectory Simplification

The plethora of incoming data from multiple overlapping sources poses a challenge for data processing workflows. A data synopsis technique with which this challenge can be tackled is trajectory simplification, i.e., reducing the amount of data (positions) so that the computational effort required is reduced as well. The ideal goal is to keep only those positions that are adequate in order to recreate the trajectory with minimal losses in the accuracy of the data processing workflow.

For that, we use INFORE's SDEaaS (Sect. 2.1) which includes an application-specific synopses, namely STSampler. The STSampler scheme resembles the concept of threshold-guided sampling in [58] but executes the sampling process in a more simplistic, yet effective in practice, way. More precisely, the sampling process is executed in a per stream fashion, i.e., for the currently monitored trajectory of each vessel separately. The core concept is that if the velocity and the direction of the movement of the vessel do not change significantly, the corresponding AIS message is not sampled. The last two reported trajectory positions are cached in the add FlatMap of Fig. 1a. When an AIS message holding information about the current status of the vessel streams in via HashData, the add FlatMap computes the change in the velocity between the lastly cached and the new AIS report, i.e., $\Delta_{vel} = |vel(prev) - vel(now)|$, and compares this value to a velocity threshold T_{vel}. Using the previously cached points, the vector describing the lastly reported direction of the vessel $dir(prev)$ is computed, while using the last cached and the newly reported positions we also compute $dir(now)$. Then, we compare $\Delta_{dir} = |dir(prev) - dir(now)|$ against a direction threshold T_{dir}. If at least one of these deltas does not exceed the corresponding threshold, the newly received

AIS message is not included in the sample by the add FlatMap. This holds, provided that a couple of additional spatiotemporal constraints are satisfied: (a) the time difference between the newly received AIS message and the last one that was included in the sample does not exceed a given time interval threshold T_{tdiff} and (b) the distance among the most recently sampled and the current position of the vessel does not surpass a distance threshold T_{dist}. SDEaaS is implemented in Flink instead of Kafka, for the reasons explained in Sect. 2.1.

3.2.4 Complex Maritime Event Processing

A very important module of the modern MSA applications is the Maritime Event Detection module. This is essentially a CEP module tailored to the maritime domain. For now, our analysis concentrates on distributed and online CEP, i.e., detecting complex events, while future work will also exploit the potential of CEF (Sect. 2.3). A description of some of the most common vessel events that can occur in the maritime domain is provided below:

- *Turn*: A vessel turns to a different direction.
- *Acceleration*: A vessel accelerates.
- *Route Deviation*: The course of a vessel deviates from "common" routes.
- *Shallow waters*: A vessel navigates in shallow waters.
- *Proximity*: A vessel is in close distance to another vessel.
- *Out of coverage*. A vessel is out of coverage with respect to one or more vessel monitoring systems such as AIS [47].

The events described above are simple events, i.e., they can be computed without depending on other events. Complex events, on the other hand, are events composed from other events. Below we provide examples of complex events:

- *Ship-to-ship*: Transfer of cargo between vessels.
- *Bunkering*: One vessel provides fuel to another vessel.
- *Tugging*: A smaller vessel (a tug) is tugging another vessel.
- *Piloting*: A smaller vessel (pilot vessel) approaches a bigger vessel so that the pilot of the vessel boards the bigger vessel in order to help it navigate into a port where special local conditions apply.
- *Fishing*: A vessel is engaged in fishing activities.

For distributed processing of streaming data in the CEP context, the Akka framework is used [1]. Akka adopts an Actor-based architecture based on message-passing communication, and it is preferred due to the fact that it is more customisable than Spark and Flink. Each Actor, run in parallel instances, is responsible for detecting a simple or complex event as those described above (Fig. 2a and b).

3.2.5 ML-Based Anomaly Detection

The ML algorithms that are relevant to the MSA workflow relate to Deep Neural Network techniques for classifying vessels according to their type (such as cargo, fishing vessel) [54]. Moreover, we are investigating ML-based techniques such as Random Forests for classifying vessel trajectories and recognise simple or complex events in them. This effort is also aided by advanced ML-based operators we have developed to extract the common routes followed by the majority of vessels for every voyage, defined as a pair of origin and destination ports [78]. At the moment, these ML tasks are performed in an offline fashion mostly using Spark's MLlib [5], which we also use to estimate sea-port area regions in [55]. The outcomes of this process performed at the batch layer of Fig. 2a can then be used as added value knowledge to the event detection or the Fusion operator of Fig. 2b. Our ongoing work focuses on incorporating INFORE's module (Sect. 2.2) to materialise ML/DM analytics in an online, real-time fashion, where possible (see restrictions on satellite images in Sect. 3.1).

3.2.6 MSA Workflow Optimisation

Across the workflow of Fig. 2b, the INFORE Optimiser is responsible for prescribing the parallelisation degree, and the provisioned resources for the maintained trajectory synopses (Sect. 3.2.3) determine the computer cluster and the number of Akka Actors devoted to MSA-related CEP tasks (Sect. 3.2.4). The Optimiser can also do the same for ML-based anomaly detection tasks (Sect. 3.2.5). An initial workflow execution plan can be re-optimised and adjusted at runtime to adapt (e.g., by increasing/decreasing the number of Akka Actors) to changing data stream distributions or to a load of concurrently executed maritime workflows. Moreover, the ongoing integration of the INFORE CEF module will allow the Optimiser to prescribe the most efficient implementation among Akka (Sect. 3.2.4) and Flink (Sect. 2.3) options for event processing tasks.

4   Future Research and Development Directions

Future research and development directions mainly lie in the synergies of ML/DM, Synopses, CEP/CEF and optimisation technologies discussed in this chapter.

Resource-Constrained ML/DM Resource-Constrained ML/DM goes beyond data processing over distributed, but computationally powerful infrastructures such as computer clusters or the cloud. The objective in resource constrained ML/DM is to bridge the gap between the very high computation and communication demands of state-of-the-art ML algorithms, such as Deep Neural Nets and Kernel Support Vector Machines, and the goal of running such algorithms (e.g. various classifiers) on a large, heavily distributed system of resource-constrained devices. Resource-constrained devices, such as sensors, pose limitations to the power supply, memory, computation and communication capacity. Fast and efficient classifiers requiring reduced power and memory should be developed, along with novel algorithms to train, apply and update the classifiers. Synergies between synopses and distributed, online ML/DM utilities are critical for such tasks.

Optimisation over Internet of Things (IoT) Platforms Optimisation over Internet of Things (IoT) platforms, since existing optimisation frameworks, should be extended to allow for planning the execution of workflows taking into consideration the whole set IoT features including: (a) resource scarcity, (b) hardware heterogeneity, (c) data heterogeneity, (d) dynamic population of devices, (e) mobility of devices, (f) security aspects over massively distributed architectures, and (g) resilience and accuracy of analytics in the presence of device failures.

CEP/CEF-Oriented Synopses CEP/CEF-Oriented Synopses techniques tailored for CEP/CEF are becoming a necessity. The work in [41] was the first to point out that load shedding schemes tailored for CEP are missing and that shedding the load in CEP significantly differentiates itself from doing so in conventional streaming settings. A few more approaches emerged since then [52, 74], but still little attention has been paid on the distributed environments and the mergeability properties of such techniques [11].

Acknowledgments This work has received funding from the EU Horizon 2020 research and innovation program INFORE under grant agreement No. 825070.

References

1. Akka v. 2.5.32. https://akka.io/. Accessed 15 September 2020.
2. Apache Flink v. 1.12. https://flink.apache.org/. Accessed 15 September 2020.
3. Apache Kafka v. 2.3. https://kafka.apache.org/. Accessed 15 September 2020.
4. Apache Samza v. 1.5.1. http://samza.apache.org/. Accessed 15 September 2020.
5. Apache Spark v. 2.4.4. https://spark.apache.org/. Accessed 15 September 2020.
6. Apache Storm v. 2.1. https://storm.apache.org/. Accessed 15 September 2020.
7. Proteus project. https://github.com/proteus-h2020/. Accessed 15 September 2020.
8. Stream-lib. https://github.com/addthis/stream-lib/. Accessed 15 September 2020.
9. Yahoo datasketch. https://datasketches.github.io/ Accessed 15 September 2020.
10. Abadi, D. J., Ahmad, Y., Balazinska, M., Cetintemel, U., Cherniack, M., Hwang, J. H., Lindner, W., Maskey, A. S., Rasin, A., Ryvkina, E., Tatbul, N., Xing, Y., Zdonik, S. (2005). The design of the borealis stream processing engine. In *CIDR*.
11. Agarwal, P. K., Cormode, G., Huang, Z., Phillips, J. M., Wei, Z., & Yi, K. (2013). Mergeable summaries. *ACM Transactions on Database Systems, 38*(4), 26:1–26:28.
12. Agrawal, D., Chawla, S., Rojas, B., et al. (2018). RHEEM: enabling cross-platform data processing - may the big data be with you! *Proceedings of the VLDB Endowment, 11*(11), 1414.
13. Agrawal, R., Imielinski, T., & Swami, A. N. (1993). Mining association rules between sets of items in large databases. In *SIGMOD*.

14. Alevizos, E., Artikis, A., Paliouras, G. (2018). Wayeb: a tool for complex event forecasting. In *LPAR*.
15. Alipourfard, O., Liu, H., Chen, J., Venkataraman, S., Yu, M., & Zhang, M. (2017). Cherrypick: Adaptively unearthing the best cloud configurations for big data analytics. In *NSDI*.
16. Balazinska, M., Balakrishnan, H., Stonebraker, M. (2004). Contract-based load management in federated distributed systems. In *NSDI*.
17. Begleiter, R., El-Yaniv, R., & Yona, G. (2004). On prediction using variable order Markov models. *Journal of Artificial Intelligence Research, 22*, 385–421.
18. Benczúr, A., Kocsis, L., & Pálovics, R. (2018). Online machine learning in big data streams. arXiv: 1802.05872.
19. Bifet, A., Maniu, S., Qian, J., Tian, G., He, C., & Fan, W. (2015). Streamdm: Advanced data mining in spark streaming. In *ICDMW*.
20. Bühlmann, P., Wyner, A. J., et al. (1999). Variable length Markov chains. *The Annals of Statistics, 27*(2), 480–513.
21. Cardellini, V., Grassi, V., Lo Presti, F., & Nardelli, M. (2016). Optimal operator placement for distributed stream processing applications. In *DEBS*.
22. Cho, C., Wu, Y., Yen, S., Zheng, Y., & Chen, A. L. P. (2011). On-line rule matching for event prediction. *VLDB Journal, 20*(3), 303–334.
23. Cipriani, N., Eissele, M., Brodt, A., Grossmann, M., & Mitschang, B. (2009). Nexusds: a flexible and extensible middleware for distributed stream processing. In *IDEAS*.
24. Cleary, J. G., & Witten, I. H. (1984). Data compression using adaptive coding and partial string matching. *IEEE Transactions on Communications, 32*(4), 396–402.
25. Cormode, G., Garofalakis, M. N., Haas, P. J., & Jermaine, C. (2012). Synopses for massive data: Samples, histograms, wavelets, sketches. *Foundations and Trends Databases, 4*(1–3), 1–294.
26. Cormode, G., & Yi, K. (2020). *Small summaries for big data.* Cambridge University Press. https://doi.org/10.1017/9781108769938
27. Doka, K., Papailiou, N., Tsoumakos, D., Mantas, C., & Koziris, N. (2015). Ires: Intelligent, multi-engine resource scheduler for big data analytics workflows. In *SIGMOD*.
28. Elmore, A. J., Duggan, J., Stonebraker, M., Balazinska, M., Çetintemel, U., Gadepally, V., Heer, J., Howe, B., Kepner, J., Kraska, T., Madden, S., Maier, D., Mattson, T. G., Papadopoulos, S., Parkhurst, J., Tatbul, N., Vartak, M., Zdonik, S. (2015). A demonstration of the bigdawg polystore system. *Proceedings of the VLDB Endowment, 8*(12), 1908.
29. FAO: VMS for fishery vessels. http://www.fao.org/fishery/topic/18103/en. Accessed 15 May 2019.
30. Flouris, I., Giatrakos, N., Deligiannakis, A., & Garofalakis, M. N. (2020). Network-wide complex event processing over geographically distributed data sources. *Information Systems, 88*, 101442.
31. Flouris, I., Giatrakos, N., Garofalakis, M. N., & Deligiannakis, A. (2015). Issues in complex event processing systems. In *IEEE TrustCom/BigDataSE/ISPA* (Vol. 2)
32. Garofalakis, M. N., Gehrke, J., & Rastogi, R. (Eds.). (2016). *Data stream management - processing high-speed data streams.* Data-centric systems and applications. Springer. https://doi.org/10.1007/978-3-540-28608-0
33. Giatrakos, N., Alevizos, E., Artikis, A., Deligiannakis, A., & Garofalakis, M. N. (2020). Complex event recognition in the big data era: a survey. *VLDB Journal, 29*(1), 313–352.
34. Giatrakos, N., Arnu, D., Bitsakis, T., Deligiannakis, A., Garofalakis, M. N., Klinkenberg, R., Konidaris, A., Kontaxakis, A., Kotidis, Y., Samoladas, V., Simitsis, A., Stamatakis, G., Temme, F., Torok, M., Yaqub, E., Montagud, A., Ponce, M., Arndt, H., Burkard, S. (2020). Infore: Interactive cross-platform analytics for everyone. In *CIKM*.
35. Giatrakos, N., Artikis, A., Deligiannakis, A., & Garofalakis, M. N. (2017). Complex event recognition in the big data era. *Proceedings of the VLDB Endowment, 10*(12), 1996.
36. Giatrakos, N., Artikis, A., Deligiannakis, A., & Garofalakis, M. N. (2019). Uncertainty-aware event analytics over distributed settings. In *DEBS*.

37. Giatrakos, N., Deligiannakis, A., Garofalakis, M. N., & Kotidis, Y. (2020). Omnibus outlier detection in sensor networks using windowed locality sensitive hashing. *Future Generation Computer Systems, 110,* 587–609.
38. Giatrakos, N., Kotidis, Y., & Deligiannakis, A. (2010). PAO: power-efficient attribution of outliers in wireless sensor networks. In *DMSN.* https://doi.org/10.1145/1858158.1858168
39. Gog, I., Schwarzkopf, M., Crooks, N., Grosvenor, M. P., Clement, A., & Hand, S. (2015). Musketeer: all for one, one for all in data processing systems. In *EuroSys.*
40. Goldhahn, R., Braca, P., Ferri, G., Munafo, A., & Lepage, K. (2014). Adaptive bayesian behaviors for AUV surveillance networks. In *UAC.*
41. He, Y., Barman, S., & Naughton, J. F. (2014). On load shedding in complex event processing. In *ICDT.*
42. Hegedüs, I., Danner, G., & Jelasity, M. (2019). Gossip learning as a decentralized alternative to federated learning. In *DAIS.*
43. IMO. (2017). Technical characteristics for an automatic identification system using time division multiple access in the VHF maritime mobile frequency band. Technical report, ITU. https://www.itu.int/dms_pubrec/itu-r/rec/m/R-REC-M.1371-5-201402-I!!PDF-E.pdf
44. IMO. (2018). Long-range identification and tracking system. Technical report, IMO. http://www.imo.org/en/OurWork/Safety/Navigation/Documents/LRIT/1259-Rev-7.pdf
45. Kalyvianaki, E., Wiesemann, W., Vu, Q. H., Kuhn, D., & Pietzuch, P. (2011). Sqpr: Stream query planning with reuse. In *ICDE.*
46. Kontaxakis, A., Giatrakos, N., & Deligiannakis, A. (2020). A synopses data engine for interactive extreme-scale analytics. In *CIKM.*
47. Kontopoulos, I., Chatzikokolakis, K., Zissis, D., Tserpes, K., & Spiliopoulos, G. (2020). Real-time maritime anomaly detection: detecting intentional AIS switch-off. *International Journal of Big Data Intelligence, 7*(2), 85–96.
48. Kourtellis, N., Morales, G. D. F., & Bifet, A. (2018). Large-scale learning from data streams with apache SAMOA. CoRR abs/1805.11477. http://arxiv.org/abs/1805.11477
49. Kumar, V., Cooper, B. F., & Schwan, K. (2005). Distributed stream management using utility-driven self-adaptive middleware. In *ICAC.*
50. Laxman, S., Tankasali, V., & White, R. W. (2008). Stream prediction using a generative model based on frequent episodes in event sequences. In *KDD.*
51. Li, M., Andersen, D., Park, J. W., et al. (2014). Scaling distributed machine learning with the parameter server. In *OSDI.*
52. Li, Z., & Ge, T. (2016). History is a mirror to the future: Best-effort approximate complex event matching with insufficient resources. *Proceedings of the VLDB Endowment, 10*(4), 85–96.
53. Mannila, H., Toivonen, H., & Verkamo, A. I. (1997). Discovery of frequent episodes in event sequences. *Data Mining and Knowledge Discovery, 1*(3), 259–289.
54. Milios, A., Bereta, K., Chatzikokolakis, K., Zissis, D., & Matwin, S. (2019). Automatic fusion of satellite imagery and AIS data for vessel detection. In *FUSION.*
55. Millefiori, L. M., Zissis, D., Cazzanti, L., & Arcieri, G. (2016). A distributed approach to estimating sea port operational regions from lots of AIS data. In *IEEE BigData.*
56. Montgomery, D. C., Jennings, C. L., & Kulahci, M. (2015). *Introduction to time series analysis and forecasting.* John Wiley & Sons.
57. Mozafari, B. (2019). Snappydata. In *Encyclopedia of Big Data Technologies.* Springer.
58. Patroumpas, K., Alevizos, E., Artikis, A., Vodas, M., Pelekis, N., & Theodoridis, Y. (2017). Online event recognition from moving vessel trajectories. *GeoInformatica, 21*(2), 389–427.
59. Pietzuch, P., Ledlie, J., Shneidman, J., Roussopoulos, M., Welsh, M., & Seltzer, M. (2006). Network-aware operator placement for stream-processing systems. In *ICDE.*
60. Pu, Q., Ananthanarayanan, G., Bodik, P., Kandula, S., Akella, A., Bahl, P., & Stoica, I. (2015). Low latency geo-distributed data analytics. In *SIGCOMM.*
61. Rabkin, A., Arye, M., Sen, S., Pai, V. S., & Freedman, M. J. (2014). Aggregation and degradation in jetstream: Streaming analytics in the wide area. In *NSDI.*
62. Rizou, S. (2013). *Concepts and algorithms for efficient distributed processing of data streams.* University of Stuttgart. https://doi.org/10.18419/opus-3209

63. Ron, D., Singer, Y., & Tishby, N. (1993). The power of amnesia. In *NIPS*.
64. Ron, D., Singer, Y., & Tishby, N. (1996). The power of amnesia: Learning probabilistic automata with variable memory length. *Machine Learning*, 25(2–3), 117–149.
65. Rong Li, X., & Jilkov, V. P. (2003). Survey of maneuvering target tracking. part i. dynamic models. *IEEE Transactions on Aerospace and Electronic Systems*, 39(4), 1333–1364.
66. Rong Li, X., & Jilkov, V. P. (2005). Survey of maneuvering target tracking. part v. multiple-model methods. *IEEE Transactions on Aerospace and Electronic Systems*, 41(4), 1255–1321.
67. Samoladas, V., & Garofalakis, M. N. (2019). Functional geometric monitoring for distributed streams. In *EDBT*.
68. Shah, M. A., Hellerstein, J. M., Chandrasekaran, S., & Franklin, M. J. (2003). Flux: An adaptive partitioning operator for continuous query systems. In *ICDE*.
69. Silva, J., Faria, E., Barros, R., Hruschka, E., Carvalho, A., Gama, J. (2013). Data stream clustering: A survey. *ACM Computing Surveys*, 46(1), 1–31.
70. Tang, H., Lian, X., Yan, M., Zhang, C., Liu, J. (2018). D^2: Decentralized training over decentralized data. In *ICML*.
71. Vilalta, R., & Ma, S. (2002). Predicting rare events in temporal domains. In *ICDM*.
72. Vulimiri, A., Curino, C., Godfrey, P. B., Jungblut, T., Padhye, J., Varghese, G. (2015). Global analytics in the face of bandwidth and regulatory constraints. In *NSDI*.
73. Willems, F. M. J., Shtarkov, Y. M., & Tjalkens, T. J. (1995). The context-tree weighting method: basic properties. *IEEE Transactions on Information Theory*, 41(3), 653–664.
74. Zhao, B., Hung, N. Q. V., & Weidlich, M. (2020). Load shedding for complex event processing: Input-based and state-based techniques. In: *ICDE*.
75. Zhou, C., Cule, B., & Goethals, B. (2015). A pattern based predictor for event streams. *Expert Systems with Applications*, 42(23), 9294–9306.
76. Zillner, S., Bisset, D., Milano, M., Curry, E., Robles, A.G., Hahn, T., Irgens, M., Lafrenz, R., Liepert, B., O'Sullivan, B., & Smeulders, A. (Eds.). (2020). Strategic research, innovation and deployment agenda - AI, data and robotics partnership. Third Release. Brussels. BDVA, euRobotics, ELLIS, EurAI and CLAIRE (September 2020).
77. Zillner, S., Curry, E., Metzger, A., Auer, S., & Seidl, R. (Eds.). (2017). *European big data value strategic research & innovation agenda*. Big Data Value Association.
78. Zissis, D., Chatzikokolakis, K., Spiliopoulos, G., & Vodas, M. (2020). A distributed spatial method for modeling maritime routes. *IEEE Access*, 8, 47556–47568.
79. Zissis, D., Chatzikokolakis, K., Vodas, M., Spiliopoulos, G., & Bereta, K.: A data driven approach to maritime anomaly detection. In *MSAW* (2019).

Knowledge Modeling and Incident Analysis for Special Cargo

Vahideh Reshadat, Tess Kolkman, Kalliopi Zervanou, Yingqian Zhang, Alp Akçay, Carlijn Snijder, Ryan McDonnell, Karel Schorer, Casper Wichers, Thomas Koch, Elenna Dugundji, and Eelco de Jong

Abstract The airfreight industry of shipping goods with special handling needs, also known as special cargo, suffers from nontransparent shipping processes, resulting in inefficiency. The LARA project (Lane Analysis and Route Advisor) aims at addressing these limitations and bringing innovation in special cargo route planning so as to improve operational deficiencies and customer services. In this chapter, we discuss the special cargo domain knowledge elicitation and modeling into an ontology. We also present research into cargo incidents, namely, automatic classification of incidents in free-text reports and experiments in detecting significant features associated with specific cargo incident types. Our work mainly addresses two of the main technical priority areas defined by the European Big Data Value (BDV) Strategic Research and Innovation Agenda, namely, the application of data analytics to improve data understanding and providing optimized architectures for analytics of data-at-rest and data-in-motion, the overall goal is to develop technologies contributing to the data value chain in the logistics sector. It addresses the horizontal concerns Data Analytics, Data Processing Architectures, and Data Management of the BDV Reference Model. It also addresses the vertical dimension Big Data Types and Semantics.

Keywords Special cargo · Knowledge acquisition · Ontology · Incident handling · Risk assessment

V. Reshadat (✉) · T. Kolkman · K. Zervanou · Y. Zhang · A. Akçay
Eindhoven University of Technology, Eindhoven, The Netherlands
e-mail: v.reshadat@tue.nl

C. Snijder · R. McDonnell · K. Schorer · C. Wichers
Vrije Universiteit Amsterdam, Amsterdam, The Netherlands

T. Koch · E. Dugundji
Vrije Universiteit Amsterdam, Amsterdam, The Netherlands

Centrum Wiskunde & Informatica (CWI), Amsterdam, The Netherlands

E. de Jong
Validaide BV, Amsterdam, The Netherlands

© The Author(s) 2022
E. Curry et al. (eds.), *Technologies and Applications for Big Data Value*,
https://doi.org/10.1007/978-3-030-78307-5_23

1   Introduction

This chapter describes ongoing work in the Lane Analysis and Route Advisor (LARA) project which aims at big data analysis and respective knowledge modeling in the logistics sector, namely, in planning shipments with special handling needs known as *special cargo*, or *special freight*, such as cargo consisting of temperature-sensitive pharmaceuticals, live animals, dangerous goods, and perishables, such as lithium batteries, flowers, and food products.

Currently, the execution of such shipments constitutes a complex process that lacks transparency and standardized knowledge resources and relies on the expert knowledge of *freight forwarders*, namely, individuals or companies organizing and planning such shipments. Freight forwarders play a key role in the special cargo industry because they possess expert knowledge on all crucial information for deciding among shipment route options, such as services provided by airlines, cargo restrictions and risks, and transport facilities. For this reason, most logistics operations are handled manually, and there is currently no transparent way of comparing and planning shipment routes, as is the case, for example, with passenger air travel planning.

Route planning for (special) cargo has significant potential for optimization with the application of advanced data analytics and artificial intelligence (AI) methods. However, an added challenge in this application lies in the acquisition and modeling of logistics and cargo knowledge from a variety of available information sources. Currently, standardization and data integration are hard not only due to the data complexity, size, and variation, but also due to cargo service providers attempting to profit from the lack of transparency and information asymmetry. Another challenge relates to processing and classifying cargo information in various types of unstructured, free-text sources, with minimal training or lexical resources. Finally, there are numerous challenges in understanding risks and constraints related to special cargo shipments [12], so as to eventually assess a candidate shipment route. In this chapter, we discuss ongoing work on addressing these challenges.

Our work addresses two of the main technical priority areas defined by the European Big Data Value (BDV) Strategic Research & Innovation Agenda [47], namely, the application of data analytics to improve data understanding and providing optimized architectures for analytics of data-at-rest and data-in-motion, the overall goal being in developing technologies contributing to the data value chain in the logistics sector. With regard to the BDV Reference Model, we address the ''vertical'' dimension: *Big Data Types and Semantics*. We also address three 'horizontal' concerns: *Data analytics*, *Data processing architectures* and *Data management*.

This chapter is organized around the three building blocks shown in Fig. 1. In Sect. 2, **Special Cargo Ontology**, we discuss the knowledge elicitation and respective research in modeling cargo knowledge into a standardized form. This work sheds more light on the design and development of a logistics knowledge base and the methodology for eliciting domain information, so as to eventually

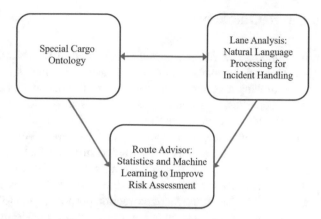

Fig. 1 An overview of the special cargo modeling system

be able to determine routing options. In Sect. 3, **Case Study: Lane Analysis and Route Advisor** we describe a test bed for future application of the knowledge modeling involving the following types of data: structured data, time series data, geospatial data, text data, network data, and metadata. [47] Subsequently, we discuss a novel palate of data analytics approaches to provide a major player in the freight forwarding industry with a set of solutions for several of their organizational issues, using this data. In Sect. 4, **Natural Language Processing for Incident Handling**, NLP and the machine learning algorithm of Random Forests are used to gain new insights on incident classification related to data quality issues in unstructured data. In Sect. 5, **Statistics and Machine to Improve Risk Assessment**, a logistic regression model is used to detect which features most profoundly influence which incident types. With regard to data management, we consider the aspect of data quality affecting the results. The analysis namely has to be considered carefully as data quality issues affect the results.

The chapter accordingly relates to three main cross-sectorial technology enablers of the **Strategic Research, Innovation & Deployment Agenda** for AI, Data, and Robotics, recently released as a joint initiative by the Big Data Value Association, CLAIRE, ELLIS, EurAI and EUrobotics [46]. These cross-sectorial technology enablers are respectively: *Knowledge and Learning* (Sect. 2), *Sensing and Perception* (Sect. 4), and *Reasoning and Decision Making* (Sect. 5). Furthermore, due to the nature of the case study involving incident analysis for special cargo, and thus digital and physical AI working together (Sect. 3), a fourth cross-sectorial technology enabler is inherently addressed: *Action and Interaction*.

2   Special Cargo Ontology

An *ontology* is defined as 'a formal, explicit specification of a shared conceptualization' [35]. One of the main advantages in using ontologies for modeling
knowledge lies in allowing a versatile representation of concepts and hierarchical
concept relations, properties, and constraints [1]. This also allows machines to
make use of the World Wide Web without any interference of humans, as an
ontology translates human concepts in machine-readable terms. For our purposes,
a special cargo ontology is intended as a knowledge structure that models special
cargo services and properties, so as to (1) have an explicit model of the domain
information requirements, (2) develop a knowledge resource for unstructured text
processing (e.g., for information retrieval or extraction purposes), and (3) eventually
use the information in the respective knowledge base for, e.g., considering important
cargo constraints when reasoning about proposing a set of possible shipment routes.

Designing and developing an ontology from scratch can be a laborious and time-
consuming process. For this reason, there are numerous approaches in learning
an ontology in an automatic or semiautomatic way, such as using automatic term
extraction and clustering, or information extraction entity and relation extraction
[8, 10, 24, 30–33, 45]. In our approach, because of the lack of existing lexical or
other knowledge resources in the special cargo domain, we have opted for a top-
down method, namely, one that relies on applying knowledge elicitation techniques
for acquiring the domain knowledge from the human experts. More details about
the size of different components of the ontology are added in Table 1 In this
section, we discuss our knowledge elicitation methodology, ontology design, and
implementation.

2.1   *Methodology and Principles for Ontology Construction*

In order to support the planning phase within the special handling cargo sector,
a knowledge structure is constructed. Based on an analysis of the ontology life
cycle, (dis)advantages, and the conformity to the nature of the special cargo domain,
different methodologies are assessed. The result of the analysis of different methodologies and techniques is the augmented UPON (Unified Process for Ontology)
methodology [9] with knowledge elicitation and evaluation tools.

The building process of special cargo ontology follows the UPON methodology
that is based on a software development process. UPON is augmented with
knowledge elicitation techniques to derive knowledge from experts and evaluation
techniques to validate the ontology. (Un)Structured interviews including the teach-
back method, laddering,[1] and document analysis techniques are implemented

---

[1] This consists of techniques consists of creating a hierarchy of the gathered knowledge, reviewing,
modifying, and validating it together with an expert.

Table 1 Different components of the special cargo ontology

Component	Size
Axiom	724
Logical Axiom	344
Declaration Axiom	197
Declaration Axiom	197
Class	129
Object Property	43
Data Property	20
Individual	7
Annotation Property	4
Class Axiom: SubClassOf	240
DisjointClasses	14
Object Property Axioms: SubObjectPropertyOf	2
InverserObjectProperties	5
FunctionalObjectProperty	8
TransitiveObjectProperty	4
ObjectPropertyDomain	4
ObjectProperyRange	3
Data Property Axioms: FunctionalDataProperty	4
DataPropertyDomain	25
DataPropertyRange	19
Individual Axioms: ClassAssertion	16
Annotation Axiom: AnnotationAssertion	183

into this methodology. The UPON methodology consists of five main workflows, namely, requirements, analysis, design, implementation, and test. In the requirements workflow, the goal is to identify the requirements and desires of the ontology users, which consists of (1) determining the domain of interest and the scope, and (2) defining the purpose, which results in the usage of knowledge elicitation techniques and an Ontology Requirement Specification (ORS) document as well as an application lexicon. In the analysis workflow, different existing ontologies are assessed, and a Unified Modeling Language (UML) use-case diagram is constructed, alongside the application lexicon. In the design workflow, the OPAL (Object, Process, Actor Modeling Language) methodology as well as justification for the relevancy of these concepts to the domain is applied to the concepts. A comprehensive explanation of concepts is defined in this step. The implementation workflow consists of implementing the lexicon and its attributes into Protege and offers performance metrics and visualization of ontology structure. The evaluation of an ontology is crucial and can be done in four strategies: gold standard, application based, data driven, and user based [20]. Due to the lack of gold standard, (technical) application, and data, human assessment is the main reference point. The final workflow is testing the ontology, and this is achieved based on the 'assessment' and 'evaluation' methods. In the assessment method, competence questions and principles are assessed. The evaluation approach consisted of a manual annotation

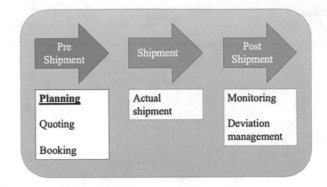

Fig. 2 Components of a cargo shipment

approach to 20 documents that are annotated by an expert. Each phase of this process is explained in more detail in the following sections.

2.2 Requirement Workflow

The application domain of the cargo ontology is the special cargo industry, with a focus on airfreight. This concerns all the processes and products that cover the interactions of special cargo airfreight forwarding within the planning phase of a shipment. Figure 2 shows a general sketch of the components of a (special) cargo shipment. This figure shows the activities that occur before the shipment planning, the actual shipment of the cargo and the activities that occur after the shipment (e.g., management of deviations).

The goal of requirement workflow is to identify the requirements of the ontology users, which consists of '(1) determining the domain of interest and the scope, and (2) defining the purpose' [9]. In this phase, the knowledge engineering techniques are applied according to the CommonKADS method [34] on top of the UPON techniques. The interviews are designed based on the guidelines and samples of CommonKADS. The knowledge elicitation is utilized in three phases, namely, knowledge identification, knowledge specification, and knowledge refinement. Knowledge identification consists of unstructured interviews and document analysis. The next step is the specification of knowledge, with structured interviews. Based on the background knowledge acquired, there are four types of experts: freight forwarders, shippers, GHAs, and support experts. While shippers play a vital role in the transportation of special cargo as it is their products being shipped, they are not concerned with the transportation jargon of the special cargo. Freight forwarders book and arrange the shipments based on the shipper's requirement. Transporters can be separated into the carriers (air carriers) and the handlers (GHA). Due to resource and time constraints, the GHAs are not consulted. The final step of the requirement workflow is knowledge refinement, and this consists of applying instances and validating the model.

To fulfill the two main goals of this workflow, an Ontology Requirement Specification (ORS) document [36] is derived. The document entails the activities of collecting the special cargo ontology requirements. The cargo that requires special handling is divided into multiple segments, namely, pharma, dangerous goods, perishables, live animals, and high value. In this regard, some information related to the purpose of the special cargo ontology, determination of the available choice set for routing options, including specific product features, capabilities, services of air carriers, and GHAs are found in the ORS document.

Along with the ORS document that includes the competency questions, an application lexicon (based on the knowledge engineering techniques) and a use-case model are the outcomes of this workflow. Applying use-case models based on the competency questions is the final step in the requirement workflow. Figure 3 shows the visualization of this use case. Laddering is conducted with a support expert and is used to elicit the UML diagram.

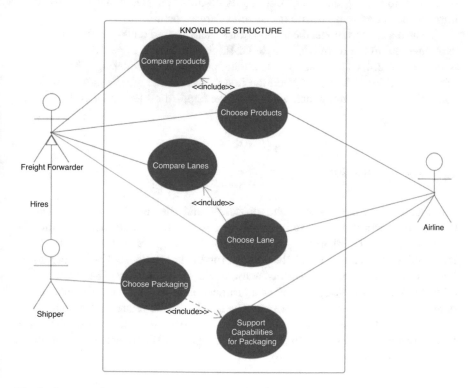

Fig. 3 Cargo ontology use case

2.3   Analysis Workflow

The analysis phase aims to refine and structure the identified requirements of the previous step. This includes reusing existing resources, modeling the application scenario using UML diagrams, and building the glossary. Considering the reuse of existing resources also entails the assessment of other domain ontologies. Existing resources or ontologies have been acquired through a search of several Ontology Libraries (OL). IATA—ONE Record,[2] the NASA Air Traffic Management Ontology,[3] and the Air Travel Booking Ontology[4] are assessed for the relevance to the domain of built cargo ontology.

The IATA—ONE Record ontology, The NASA Air Traffic Management (ATM) Ontology, and The Air Travel Booking Ontology are implemented in a message system, air traffic management, and air travel booking service, respectively. They are used in different stages of the process (i.e., planning vs booking), and the domains are not completely compatible. Although in the context of the Semantic Web, ontologies are often used with a purpose different from the original creators of the ontology [36], and these ontologies do not offer significant benefits to be implemented or associated with the Special Cargo Domain.

The next step is to model the application scenario based on the drafted UML use-case diagram, in the form of a simple UML class diagram. A part of this diagram is shown in Fig. 4, as the result of the elicitation technique laddering. The final step of the analysis workflow is to build the first version of the glossary concerning the concepts of the domain, which will merge the application lexicon and the domain lexicon.

2.4   Design Workflow

The identified entities, actors, and processes and the relations among them in the previous workflow are refined in the design phase. The steps within this workflow consist of inhabiting, categorizing the concepts according to the OPAL methodology [42], and refining the concepts and their relations. OPAL is organized into three primary modeling aspects: actor, processes, and object. The identification of the OPAL methodology, as well as a justification of why such entities exist in the ontology, is defined under the lexicon. The subclasses are related to the main class through a 'kind-of' or an 'is-a' relation. When a 'part-of' relation is defined, it is found in column 'notes'. The object, data properties, and the related explanation are found in the ontology.

---

[2] https://www.iata.org/en/programs/cargo/e/one-record.

[3] https://data.nasa.gov/ontologies/atmonto/ATM.

[4] https://www.southampton.ac.uk/~cd8e10/airtravelbookingontology.owl.

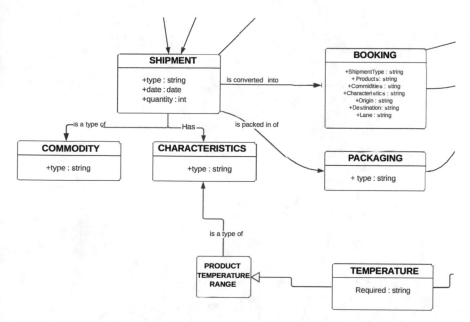

Fig. 4 A part of the Cargo Ontology class

2.5 Implementation Workflow

In this phase, ontology is formalized in a language and implemented with regard to its components. The special cargo ontology is constructed in Protege and written in RDF (Resource Description Framework) and OWL (Ontology Web Language). A part of the visualization of the special cargo ontology is shown in Fig. 5.

2.6 Test Workflow

While each ontology differs in structure and domain, testing is vital to assess the domain compliance. The goal of the test phase is to evaluate the ontology and its components and requirements. The evaluation is performed based on human-based and task-based assessment. Human-based assessment is divided into two parts: the competency questions and the principles assessment. The competency questions (CQ) are drafted in the requirement workflow, as the manual assessment will be based on the CQChecker module of Bezerra et al. [27]. The principle assessment is a subjective tool, which requires the collaboration of the ontology engineer and a

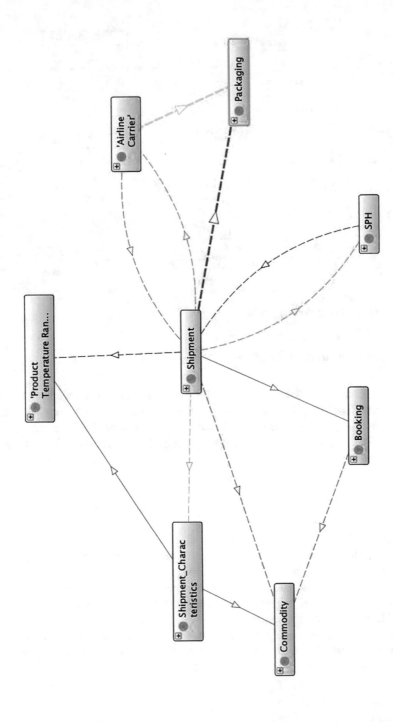

Fig. 5 'Shipment' concept in the Cargo Ontology class

Table 2 Competence Questions

Question	Real-life answer	Ontology answer	Compliance and relation
Does a pharma solution have a booked temperature range?	Yes	Yes	YES: Temperature controlled solutions 'has temperature range' some booked temperature range
Does lithium batteries transport have restrictions?	Yes	Not fully deductible	SEMI: Dangerous goods class 'has maximum capacity' (classes are not populated yet)

Table 3 Design principles

General design principles compliance	Compliance
The design should clearly state its purpose, so the user knows what the design has to offer to avoid unclear expectations	Compliant. During the extent of this research, the scope, the domain, and its purpose have been defined as well as the expectations by the LARA project
The design should remain its stability throughout time, changes, and additions	Compliant, so far. As the ontology is constructed as of late, time is hard to test on this design. However, similarly to the maintainable design principle, Protégé allows for adjustment and augmentation

domain expert. Tables 2 and 3 show some parts of these two different assessments, competence questions and design principles, respectively.

2.7 Evaluation Workflow

Task-based, data-driven evaluation is conducted by a domain expert. The evaluation is executed on two sets of ten documents concerning special cargo, collected from online cargo websites and news articles. The expert who annotated the documents has experience within the freight forwarding process as well as the risk analysis of lanes.

The final step in the testing phase is to adjust the ontology according to the result of the overall evaluation. There were three concepts ('Certification', 'Hub', 'Documentation') that were neglected in the original ontology which were implemented after the evaluation. In Table 4, a snippet from this evaluation is shown. During the analysis of the evaluation, it became clear that certain small or significant attributes were omitted in the process of creating the ontology, or in return some attributes were insignificant. In the result of this analysis, these attributes were omitted or inserted.

Table 4 A snippet from cargo ontology evaluation

Annotation	Relevance	Presence
Our products allow you to get your life-saving cargo to its destination	Yes	Yes, triple incorporated: product, pharma, and the relation
Cool Center	Yes	Yes, cool center is a synonym for temperature-controlled environment, this concept is incorporated
Highly trained experts can stand by 24/7/365 to monitor and support	Yes	Yes, trained personnel and monitoring are incorporated

2.8 Summary

In the LARA project, knowledge representation is developed for the special handling goods and services in the airfreight sector. It is designed based on the software engineering methodology with the aim of digitizing the determination of the choice set of solutions and routes for the airfreight forwarders by making data transparent and understandable to machines. For the integration of disparate knowledge sources, a special cargo domain ontology of shipping concepts is constructed for the domain of goods transported by air in a semiautomatic manner. As a structured resource, the special cargo ontology provides valuable insights into the scope of the application, the different components of the system, and the interaction between them. It can be used during the actual operation of the system [6, 22]. As an example, the fact that consumer-ready laptop computers contain a lithium battery can be modeled in the ontology means that when processing a request for shipping laptops, the system can determine that the cargo service needs to allow for lithium batteries to be shipped.

The UPON methodology is used for the construction of the cargo domain knowledge structure to get the relevant concepts and attributes. This output is evaluated based on reviewed evaluation methods and adjusted accordingly. The ontology is integrated into a software program to obtain an applicable product of the special cargo scope and domain, and subsequently, the final product is the base of an artificial intelligence route advisor based on the semantic web for the special cargo sector.

3   Case Study: Lane Analysis and Route Advisor

In the past few decades, international freight transportation has increased rapidly. This rise can be explained by technological developments, simplifying the global transport process and causing a decline in shipping costs [37]. It has led to a growing demand for freight forwarding services. Freight forwarding companies can be hired to handle the logistics of shipping goods from the customer to the consignee.

However, the process of transportation carries many risks, for which the freight forwarding company has to take responsibility. Certain types of cargo may require strict conditions during transit. For instance, some pharmaceutical products are temperature-sensitive and have to be kept at a specific temperature throughout the entire process. When constructing the route the cargo should follow, the type of packaging and possible exposure to external weather conditions need to be taken into account. Furthermore, when transporting high-value cargo (HVC) like electronics, the number of crime incidents increases. Hence, additional security measurements should be taken into consideration for HVC goods.

Freight forwarding companies aim to maintain high customer satisfaction as satisfied customers will presumably hire the company again and might help recruit other customers through positive feedback[28]. Key elements driving customer satisfaction are the service quality and the perceived value [15]. Hence, to avoid incidents and thus increase customer satisfaction, it is essential to develop a risk assessment model and to determine high-risk lanes.

Despite the aim of freight forwarders to work as carefully and efficiently as possible, incidents are inevitable. While considerable amounts of data are available regarding every incident, a lot of potential still exists to gain knowledge on factors that cause (or contribute to) incidents. Research on this matter is essential for freight forwarders; the prevention of incidents can not only contribute to keep costs as low as possible but also help forwarders maintain their reputation of a reliable forwarder.

The question arises whether factors or even combinations of factors exist that drive incident risk. A comprehensive study concerning the incident data is needed to answer this question, which is the objective of this research. This chapter focuses on incident analysis and tries to determine which factors drive risk.

For this research, high-dimensional data on incidents was provided by one of the major freight forwarding companies in the industry.

4 Natural Language Processing for Incident Handling

Logistics is defined as the process of planning, implementing, and controlling procedures for the efficient and effective transportation and storage of goods including services, and related information from the point of origin to the point of consumption for the purpose of conforming to customer requirements [25].

With regards to logistics, the data focuses on the transportation of cargo, specifically incident handling with regard to air cargo. Because this is the case, it is interesting to look at ways other chapters tackled this issue of cargo risk assessment. One of these risks is cargo loss. This is defined by Wu et al. [44] as either cargo damage or cargo theft.

According to [29], cargo damage is the most occurring problem in the logistics sector. These authors mention five main causes of cargo damage: human error (such as miscommunication); handling error (examples include incorrect placement in plane or having incorrect/missing documents); machine/tool error (such as having old or broken equipment); environment (such as temperature); and packing material.

When it comes to cargo theft, [26] mentions that employees, as well as outside offenders, may steal cargo. These authors also mention a couple of reasons why it is often difficult to detect cargo theft. One of these reasons is the fact that thefts are often under-reported. They further mention ways in which the amount of cargo theft can be decreased. These methods include, but are not limited to, placing containers with doors facing each other (so that it is more difficult to remove cargo) and minimizing waiting times for vehicles (because it is easier to steal from a still-standing vehicle).

These issues have been tackled by other authors using both predictive and descriptive analysis techniques to gain insight on cargo loss [44]. One thing they found is that high-value cargo should not be sent as land cargo, and to certain regions not as sea cargo either.

While the above-mentioned results are interesting, Hazen [18] mentions a few reasons why data analysis results regarding the logistics and supply chain industries should be considered carefully. Most of these reasons are due to data quality issues. According to these authors, data in this sector is often full of errors. They mention four key attributes in data quality that could use improvement: accuracy, timeliness, consistency, and completeness.

4.1   Random Forest Decision Trees

One of the issues that the company has been dealing with is data quality. Registration of an incident may provide text describing the incident. The classification of incidents is a subjective process in which mistakes can happen. Comparing text could, therefore, be a competent way to eliminate these mistakes. A suitable method is to process the data using Natural Language Processing (NLP) and to classify the incident using a classification model with this processed data.

NLP is a subfield of computer science that uses computational techniques to learn, understand and produce human language content [19]. The authors mention multiple reasons why NLP is useful. Among them are translating and helping the human-machine communication, which are both relevant for this research. Also, the process of analysis and learning from human language content which is available online is discussed in this chapter and might be relevant for this research as well.

One way to analyze and learn from human language content is by using machine learning algorithms. Random forests can be seen as a combination of multiple tree predictors in which each tree depends on the values of a random vector sampled independently and each following the same distribution for all trees that are included in the forest [7]. Although different types of trees exist, in this case, decision trees are used. In decision trees, the decisions are the edges of the tree and form the nodes for data classification. Decision trees are applied commonly in machine learning; one reason for their frequent usage is that they are easy to interpret [43].

However, a random forest algorithm is generally preferable over using just a decision tree; random forests improve performance by training multiple decision trees [43]. These trees are chosen randomly because, in that way, the chance of correlation between individual trees is reduced, and more accurate results are obtained.

4.2 Implementation

There are several NLP environments on the market. One of the more standard environments is the Natural Language Toolkit (NLTK) which is combined with the python `sklearn` library for the best results. A common process in NLP is tokenization. In this process, sentences are broken up into individual words, where any capital letters and interpunction are also removed. The classification model then uses a set number of most common words in the incidents. Using these most common words, the random forest classification is then trained on the training data using a set number of classification trees.

Random forest classification uses a trained classification model that is capable of classifying data based on processed text. It requires the incident data to be split up into a training and a testing set. Because of the large number of incidents, it is possible to split them up into a training set that contains 75% of the incidents and a testing set that includes 25% of the incidents. To make the random forest classification easier, the problem is reduced to a single classification problem. This means that all possible combinations of levels are given an ID, which the model tries to classify.

The accuracy of the classification model will vary based on the chosen parameters during the NLP and the classification process. Furthermore, it would be extra interesting to look at the incidents that were wrongly classified, as the original classification by the incident handlers could also be wrong.

4.3 Results and Discussion

Natural Language Toolkit (NLTK) combined with a random forest classifier provides the prediction accuracy as shown in Table 5. The NLP random forest algorithm was implemented with 1, 10, and 100 trees.

Table 5 NLP results based on NLTK with random forest classifier

#trees	Precision
1	82.0
10	82.6
100	83

An interesting result is that in all cases, the precision rounds up to being an integer, and after 100 trees, the precision does not rise much more than that integer. Since adding random forests are not prone to over-fitting and the precision flattens out at 100 trees, 100 trees seem like an adequate number of trees.

It could be interesting to look at incident types that are predicted wrong relatively often and see if it would not be better to put these under a different class.

5   Statistics and Machine Learning to Improve Risk Assessment

To be able to determine possible factors that drive incident risk, a multinomial classification model was implemented on the incident data. In this model, features are defined for every incident type that significantly predict this type. After literature research, it became clear that a Logistic Regression Model suited the data well.

5.1   Logistic Regression

A classification method known as the Logistic Regression model is used during the analysis. Regression models are used to calculate the interdependency between an outcome (response variable) and the variables thought to affect this outcome (explanatory variables). The most simple form of a regression model is the linear regression model, in which a linear function is mapped between data points. The logistic regression model is in certain ways similar to linear regression, but there are a few differences. The main difference is that with the logistic regression (logit) model used in this research, the outcome variable is discrete instead of continuous. The statistical model is typically estimated via (simulated) maximum likelihood estimation [21, 38]. Logit family models are widely applied in the transportation domain [2, 14]. Examples include: mode choice [11], route choice [23], choice of departure time [41], location choice [3, 40], and choice of products and services [13, 39].

In the logistic regression model, the relationship between the response variable and explanatory variables is expressed as a simple equation:

$$g\left(E[Y]\right) = \alpha + \beta_1 x_1 + \beta_2 x_2 + \ldots + \beta_p x_p \tag{1}$$

where g is the logit link function. In the equation above, the α represents a constant term, the x_i represent the explanatory variables, and the β_i represent a measure of the degree to which the response variable is explained by variable x_i [17]. For every explanatory variable, a t-test is performed on the corresponding β to test whether it can be statistically proven that the explanatory variable influences the response variable. To determine how useful the explanatory variables are in predicting the response variables, the ρ^2 statistic or a likelihood ratio test can be used [4].

For the implementation of the Logistic Regression in a Machine Learning fashion. a procedure is required that identifies features that are of importance to the response variable. A procedure that determines this is called Recursive Feature Elimination (RFE). After training the classifier and computing the ranking, the feature with the smallest ranking criterion is removed [16]. This step is repeated until a certain number of features n remains.

5.2 Methodology

To classify incidents into categories, these categories first had to be determined. The data that was provided for this analysis contained a feature that described what kind of incident happened. Based on this column, the incident types with the highest number of occurrences were chosen to implement in the model. Also, types that were prioritized by the company, but did not have a significantly high number of occurrences, were taken into account. In total, nine categories were obtained.

Two classification models were used, using different python packages: Biogeme and RFE. Both use the incident types described above as categories. The methods per model are described in the following subsections.

5.3 Statistical Implementation

For the implementation of the statistical Logistic Regression model, a package called Biogeme by Bierlaire [5] was used. The model performs a multinomial classification and determines significant features that predict all possible classes (the incident types). For this model, it was necessary to manually determine which variables drive the chosen incident types most and include these in the model. To determine these variables cross-tabular matrices were used, which show the proportional relation of a variable to a particular incident type. More specifically, they depict which possible values of certain variables show a connection to an incident type. The cross tab had all incident types as rows and all possible values for the variable to take into consideration as columns. For example, when taking a region variable into account, a certain incident type might have a strong correlation with a specific region. In this case, a binary variable was created, where '1' equals the situation where the incident was reported in that region, and 0 otherwise. Next, this variable was added to the regression model for this incident type. In this process, also the number of occurrences per region has to be taken into account, to avoid an unreliable view on possible predictors. If a certain region only occurred once in the data, and by coincidence an incident occurred on the shipment connected to this region, the percentage error will be 100%. Therefore, a threshold was set for the cross tabs, where every possible value taken into account in the cross tab had to have at least a minimum number of occurrences in the incident data.

The chosen features were entered for the corresponding incident type. The python package `Biogeme` was able to check for all features whether it was an important predictor for the incident type. After the first run, a base model was created that included all features for which the value of the t-test statistic was bigger than 1. This does not imply all these values are statistically significant on a 5% level, but they have enough descriptive purpose for the model. After this, all features included in the base model were analyzed for possible combinations of features with a high correlation. For instance, when the incident was caused in a particular city, the corresponding country at fault is of course always the country containing that city. This collinearity has a negative effect on the performance of the model, so for the combinations of features with high correlation, the less specific features are excluded. In the example, it would be more important to look at the city specifically, than to only take the country into consideration. Hence, in such case, the country is excluded from the model features.

Using this base, all other features were checked again for possible relevance to the incident type. This was done by performing a batch run on the base model, where every time one of the features not contained in the base model was added to the base. The rho squared and likelihood ratio test results were compared for all resulting models to determine the optimal one.

In a general logistic regression model, the predictive power of a feature cannot be determined by the beta value, since the influence on the utility function is defined as the beta times the value of the feature. Therefore, a feature with high values generally has a smaller beta than a feature with low values. However, since the features used in the regression model are all dummy variables (so only binary values are included), it is possible to compare the strength of the feature on the beta value.

5.4   Recursive Feature Elimination

For comparison with the statistical Logistic Regression model, a machine learning Logistic Regression was used as an alternative. For implementation, one of the more general packages and approaches within python, known as `sklearn`, provided the necessary tools. The main difference with the statistical approach concerns the selection of possible values per feature. All features that were determined of importance were used for the machine learning approach as well, but instead of selecting important values per feature manually using cross tabs, these values were decided using Recursive Feature Elimination (RFE). A data set was created in a 'one-hot-encoding' fashion where all features were split up into binary variables. So, for example, all possible values for the party responsible for the incident in the data got their own column, where 1 indicates that the incident was caused by that specific company and 0 indicates that another company was responsible. Since some columns have many different possible values, and some of these values only occur a few times in the data, it was decided to set a threshold on the number of occurrences. This was done to ensure predictive power and reduce the running time of the model.

It was impossible to run the model as a multinomial regression model, since then the overall best features were decided for all incident types in total, instead of per type. Therefore, the formulation of the model changed to a set of binary decisions: the model was run for every incident type separately. To achieve this, a binary column per incident type was added to use as classification feature. A drawback of this implementation is that it can also produce negative betas. Because the model is run for every incident type separately, it is now classifying features on the constraint 'is it a predictor for incident type X or not', instead of taking into account all incident types. A negative beta shows that the corresponding feature is not a good predictor for incident type X, but it is a significant predictor for some other incident type.

The machine learning implementation works in the following way. First, the model splits the data set into a training and testing set of 70% versus 30%. RFE calculates the best features to be used by the model using the training set. It was decided to let the RFE determine ten features per incident type. Following the RFE, a simple logistic regression model is constructed, using the training set to fit the model, which then provides results based on the testing set. The accuracy of the model shows the percentage of the prediction by the model that was correct.

5.5 Results

The analysis of features that could possibly predict certain incident types has led to a little over 300 different variables among the nine different regression models. So, on average around 30 possible predictors were determined per incident type. The Logistic Regression models determined the features that were the most important predictors per incident type.

The classification model implemented with the python package Biogeme gives as output an overview of all features used. A statistical test is conducted for every β, which shows whether the influence of the corresponding explanatory variable is statistically significant for the incident type it was tested for. As explained in the chapter 'Methods', a base model was created with all features for which the t-test value was bigger than 1. Per feature and value combination, the Beta gives a measure of how much this combination influences the incident type. Thus, the features with the highest Beta values for each incident type are the strongest predictors. The p-value depicts the significance of the feature in the base model. The smaller the p-value, the higher the accuracy of this feature for the model. After running the base model separately from the full model, some insignificant p-values were obtained, while they were significant when all of the features were taken into account. It would have been better to iterate over the results of the model and to exclude the insignificant features every time. However, due to the immense running time of the model, it was decided to focus on this base model and to accept the few high p-

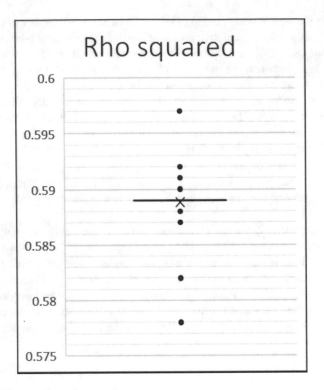

Fig. 6 All rhos of batch run

values. The rho squared for the base model was equal to 0.575 and the likelihood ratio test was equal to 93,295.63.

The batch run of the base with every single feature separately produced 220 models. Figure 6 shows a box plot of all rho squares that were obtained per model. Figure 7 shows all likelihood ratio test values. The highest rho squared and likelihood ratio are equal to 0.597 and 97,203.09, respectively.

The Logistic Regression model performed with the columns detected by the RFE gave as output ten features with most predictive power per incident type. The accuracy of each model can be found in Table 6.

5.6 Discussion

As mentioned before, the results show some insignificant p-values in the base model. These values could be explained by the fact that the corresponding features occur often for multiple incident types. Thus, they have a substantial variance as a predictor. Therefore, these features should be considered with caution. It should also be noted that some of the incident types that were implemented in the model

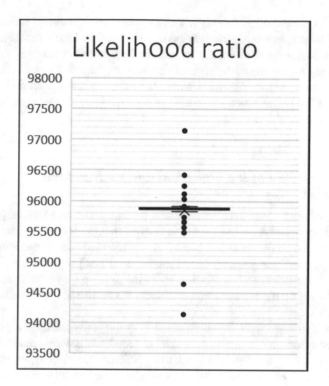

Fig. 7 All likelihood ratios of batch run

Table 6 Accuracy for the Logistic Regression model based on RFE determined features

Incident type	Accuracy
A	0.904
B	0.713
C	0.987
D	0.867
E	0.951
F	0.978
G	0.965
H	0.987
I	0.999

did not have many occurrences in the data. After running the full model, only a few predictors were determined for the base model, and they were all deemed insignificant after running the base model on its own. Therefore, a high number of occurrences is needed per incident type to acquire significant results.

The box plots with the results of the batch run (Figs. 6 and 7) show that no significant differences could be found between the resulting models. The rho squared has its mean at 0.589, and only has a few outliers. Still, the base model had a rho squared of 0.575, so adding another variable to the model generally leads to a better performing model. The highest rho squared and likelihood ratio were 0.597 and 97,203.09, respectively. However, adding the binary variable that achieved these maximum statistical measures lead to collinearity with other features. The features corresponding to the next few highest rho squares lead to the same issue. After the sixth feature, the rho squared becomes 0.59 for any feature added to the model (except for a few). Therefore, the batch run does not provide any significant results of features to add.

5.7   Comparison of the Statistical and RFE Models

The results of the two models implemented by Biogeme and RFE require special attention to compare. A reason for this is the fact that the features of the Biogeme model are judged by the statistical t-test that is performed and the resulting p-values. However, the machine learning-based model does not judge the performance of features on a statistical test. Instead, it splits the data up into a train and a test set, trains the model on the train set, and calculates the performance of the model based on the test set.

Another challenge with the comparison is the negative betas that occur for the RFE implementation. The statistical model is a multinomial implementation, which implies that it runs the model on all nine different incident types at once, and determines the appropriate features accordingly. The formulation of the machine learning model, however, is a separate set of binomial decisions. For every incident type, the model is run separately, in order to find features per incident type. This means that the model is classifying features on the constraint 'is it a predictor for incident type X or not', instead of taking into account all incident types. Because of this, the results for the RFE model contain negative betas. A negative beta shows that the corresponding feature is not a good predictor for incident type X, but it is a significant predictor for some other incident type. However, for the analysis conducted in this report, the negative betas do not add any important information. Hence, only the positive betas should be taken into account.

Still, when comparing the significant features per incident type, most of the features with a positive beta in the machine learning implementation also occurred in the results for the statistical implementation. This shows that these values are in fact important predictors for the incident types.

6   Summary, Challenges, and Conclusion

This research project is directly related to BDVA SRIA's strategic and specific goals, particularly to the topic Data Analytics to improve data understanding and providing optimized architectures for analytics of data-at-rest and data-in-motion. In this project, we conduct research into solutions based on advanced data analytics that combine the integration of various data sources ('big data'), AI-based methods such as machine learning, and natural language processing for prescriptive analytics and decision making. These methods can be applied to the optimization of route planning in global transportation and freight forwarding of sensitive products with special handling needs (e.g., COVID-19 vaccine) targeted at air freight shipment.

According to the European Big Data Value Strategic Research and Innovation Agenda (SRIA) [32], understanding data has been one of the greatest challenges for data analytics. In this regard, we use semantic and knowledge-based analysis specifically ontology engineering for Big Data sources in the special cargo domain to improve the analysis of data and provide a near-real-time interpretation of the data (i.e., accurate prediction of the lane performance). Moreover, employing Big Data analytics we develop an ontology for the products and services offered for air freight logistics providers. Based on this, a search engine can be developed to determine the available routing options for a shipment with specific features. Thus, it provides additional value in the transportation sector, leads to more efficient and accurate processes, and improves operational efficiencies and customer service.

The work has some limitations and challenges. Evaluation of the special cargo ontology is difficult and needs manual intervention, which is time-consuming and subjective. Expert intervention is required at every step of constructing ontology. Nevertheless, this work aims to make a significant contribution to the digitization of global freight forwarding, which may also pave the way toward 'no-touch' planning in airfreight transportation.

In this chapter, we also present a case study applying a novel palate of data analytics for risk assessment. A natural language processing classification model used on text in the incident handling data shows at least an 82% accuracy at identifying incident types. Furthermore, via a statistical logistic regression model for classification, it can be proven that several features are significant predictors of certain incident types. A machine learning logistic regression model also identified similar features. Focusing on these features can help the company prevent similar incidents in air cargo handling in the future.

The chapter addresses some important challenges of the airfreight industry for shipping goods with special handling needs such as vaccines. In order to design, develop, and optimize the decision making of a routing service in the special cargo domain, it is necessary to conceptualize and structure the available knowledge from different resources as a special cargo knowledge resource (ontology). This ontology is efficient for reasoning and can be used during the actual operation of the system. As an example, the fact that vaccines must be stored at an ultracold temperature can be modeled in the ontology, which means that, when processing a request for shipping vaccines, the system can determine that the cargo service needs to allow the shipment of products with special temperature needs.

Using the special cargo ontology, more heterogeneous sources of information can be automatically extracted and integrated. This information includes, for example, previous incidents and service performance. This knowledge base can be used in various downstream tasks, e.g., risk assessment model as a feature-extraction source. On the other hand, the machine learning algorithms applied in risk assessment tasks can be used for enriching the cargo domain ontology and map the extracted information to the structured knowledge source.

This research is directly related to TKI Dinalog's innovation roadmap, specifically to the topic Advanced Data Analytics in Transport Planning within the Smart ICT Roadmap.

Acknowledgments The work is supported by TKI Dinalog, the Dutch Institute for Advanced Logistics, for the LARA project—Lane Analysis & Route Advisor (Project reference number: 2018-2-171TKI). TKI Dinalog is the Knowledge and Innovation Partnership in which business, knowledge institutes, and government work together in the innovation program of the Dutch Topsector Logistics. We would also gratefully like to acknowledge the support of Daniel Lutz and Carolin Heinig in the case study application on incident handling and risk assessment.

References

1. Asim, M., Wasim, M., Khan, M. U., Mahmood, W., & Abbasi, H. M. (2018). A survey of ontology learning techniques and applications. *Database: The Journal of Biological Databases and Curation*. https://pubmed.ncbi.nlm.nih.gov/30295720/.
2. Ben-Akiva, M., & Bierlaire, M. (2003). Discrete choice models with applications to departure time and route choice. In *Handbook of Transportation Science* (pp. 7–37). Kluwer.
3. Berkelmans, G., Berkelmans, W., Piersma, N., van der Mei, R., & Dugundji, E. R. (2018). Predicting electric vehicle charging demand using mixed generalized extreme value models with panel effects. *Procedia Computer Science, 130*, 549–556.
4. Bewick, V., Cheek, L., & Ball, J. (2005). Statistics review 14: Logistic regression. *Critical Care, 9*, 112–118.
5. Bierlaire, M. (2016). *Pythonbiogeme: a short introduction*. Tech. rep., EPFL, Switzerland.
6. Black, W. J., Jowett, S., Mavroudakis, T., McNaught, J., Theodoulidis, B., Vasilakopoulos, A., Zarri, G. P., & Zervanou, K. (2004). Ontology-enablement of a system for semantic annotation of digital documents. In *SemAnnot@ ISWC*.
7. Breiman, L. (2001). Random forests. *Machine Learning, 45*(1), 5–32.
8. Buitelaar, P., Cimiano, P., & Magnini, B. (2005). *Ontology learning from text: methods, evaluation and applications* (Vol. 123). IOS Press.
9. De Nicola, A., Missikoff, M., & Navigli, R. (2005). A proposal for a unified process for ontology building. In *International Conference on Database and Expert Systems Applications* (pp. 655–664). Springer.
10. Drymonas, E., Zervanou, K., & Petrakis, E. G. M. (2010). Unsupervised ontology acquisition from plain texts: The ontogain system. In C. J. Hopfe, Y. Rezgui, E. Métais, A. Preece, & H. Li (Eds.), *Natural Language Processing and Information Systems* (pp. 277–287). Springer.
11. Dugundji, E. R., & Walker, J. L. (2005). Discrete choice with social and spatial network interdependencies: an empirical example using mixed generalized extreme value models with field and panel effects. *Transportation Research Record, 1921*(1), 70–78.
12. Faghih-Roohi, S., Akcay, A., Zhang, Y., Shekarian, E., & de Jong, E. (2020). A group risk assessment approach for the selection of pharmaceutical product shipping lanes. *International Journal of Production Economics, 229*, 107774.
13. Feilzer, J. W., Stroosnier, D., Koch, T., & Dugundji, E. R. (2021). Predicting lessee switch behavior using logit models. *Procedia Computer Science, 184*, 380–387.
14. Garrow, L. (2016). *Discrete choice modelling and air travel demand*. Routledge.

15. Gil-Saura, I., Berenguer-Contri, G., & Ruiz-Molina, E. (2018). Satisfaction and loyalty in b2b relationships in the freight forwarding industry: adding perceived value and service quality into equation. *Transport, 33*(5), 1184–1195.
16. Guyon, I., Weston, J., Barnhill, S., & Vapnik, V. (2002). Gene selection for cancer classification using support vector machines. *Machine Learning, 46*, 389–422.
17. Hall, G., & Round, A. (1994). Logistic regression – explanation and use. *Journal of the Royal College of Physicians of London, 28*(3), 242–246.
18. Hazen, B. (2014). Data quality for data science, predictive analytics, and big data in supply chain management: An introduction to the problem and suggestions for research and applications. *International Journal of Production Economics, 154*, 72–80.
19. Hirschberch, J., & Manning, C. (2015). Advances in natural language processing. *Science Magazine, 349*, 1184–1195.
20. Hlomani, H., & Stacey, D. (2014). Approaches, methods, metrics, measures, and subjectivity in ontology evaluation: A survey. *Semantic Web Journal, 1*(5), 1–11.
21. Hosmer, D., & Lemeshow, S. (2013). *Applied logistic regression*. John Wiley and Sons.
22. Klein, W., Zervanou, K., Koolen, M., van den Hooff, P., Wiering, F., Alink, W., & Pieters, T. (2017). Creating time capsules for historical research in the early modern period: Reconstructing trajectories of plant medicines. In M. Hasanuzzaman, A. Jatowt, G. Dias, M. Düring, & A. van den Bosch (Eds.), *Proceedings of the 4th International Workshop on Computational History (HistoInformatics 2017) co-located with the 26th ACM International Conference on Information and Knowledge Management (CIKM 2017), Singapore, November 6, 2017, CEUR Workshop Proceedings* (Vol. 1992, pp. 2–9). CEUR-WS.org. http://ceur-ws.org/Vol-1992/paper_2.pdf
23. Koch, T., & Dugundi, E. R. (2021). Limitations of recursive logit for inverse reinforcement learning of bicycle route choice behavior in Amsterdam. *Procedia Computer Science, 184*, 492–499.
24. Lubani, M., Noah, S. A. M., & Mahmud, R. (2019). Ontology population: approaches and design aspects. *Journal of Information Science, 45*(4), 502–515.
25. Mangan, J., & Lalwani, C. (2016). *Global logistics and supply chain management*, 3rd ed. Wiley.
26. Mayhem, C. (2001). The detection and prevention of cargo theft. *Trends and Issues in Crime and Criminal Justice, 214*, 1–6.
27. Missikoff, M., & Taglino, F. (2002). Business and enterprise ontology management with symontox. In *International Semantic Web Conference* (pp. 442–447). Springer.
28. Naumann, E., Williams, P., & Khan, M. (2009). Customer satisfaction and loyalty in b2b services: directions for future research. *The Marketing Review, 9*(4), 319—333.
29. Oktaviani, N., Yadia, Z., Nasution, N., & Veronica, V. (2017). How to reduce cargo damage. *Advances in Engineering Research, 147*, 661–670.
30. Reshadat, V., & Faili, H. (2019). A new open information extraction system using sentence difficulty estimation. *Computing and Informatics, 38*(4), 986–1008.
31. Reshadat, V., & Feizi-Derakhshi, M. R. (2012). Studying of semantic similarity methods in ontology. *Research Journal of Applied Sciences, Engineering and Technology, 4*(12), 1815–1821.
32. Reshadat, V., Hoorali, M., & Faili, H. (2016). A hybrid method for open information extraction based on shallow and deep linguistic analysis. *Interdisciplinary Information Sciences, 22*(1), 87–100.
33. Reshadat, V., HoorAli, M. & Faili, H., (2019). A new method for improving computational cost of open information extraction systems using log-linear model. *Signal and Data Processing, 16*(1), 3–20.
34. Schreiber, A. T., Schreiber, G., Akkermans, H., Anjewierden, A., Shadbolt, N., de Hoog, R., Van de Velde, W., Nigel, R., Wielinga, B., et al. (2000). *Knowledge engineering and management: the CommonKADS methodology*. MIT Press.

35. Studer, R., Benjamins, V. R., & Fensel, D. (1998). Knowledge engineering: Principles and methods. *Data & Knowledge Engineering, 25*(1–2), 161–197. http://dx.doi.org/10.1016/S0169-023X(97)00056-6

36. Suárez-Figueroa, M. C., Gómez-Pérez, A., Villazón-Terrazas, B. (2009). How to write and use the ontology requirements specification document. In *OTM Confederated International Conferences on the Move to Meaningful Internet Systems* (pp. 966–982). Springer.

37. Tester, K. (2017). The impact of technological change on the shipping industry. Technology in Shipping (pp. 11–20).

38. Train, K. (2003). *Discrete choice methods with simulation.* Cambridge University Press.

39. van Kampen, J., Pauwels, E., van der Mei, R., & Dugundji, E. R. (2019). Analyzing potential age cohort effects in car ownership and residential location in the metropolitan region of Amsterdam. *Procedia Computer Science, 151*, 543–550.

40. van Kampen, J., Pauwels, E., van der Mei, R., & Dugundji, E. R. (2021). Understanding the relation between travel duration and station choice behavior of cyclists in the metropolitan region of Amsterdam. *Journal of Ambient Intelligence and Humanized Computing, 12*(1), 137–145.

41. Vegelien, A. G., & Dugundji, E. R. (2018). A revealed preference time of day model for departure time of delivery trucks in The Netherlands. In *2018 21st International Conference on Intelligent Transportation Systems (ITSC)* (pp. 1770–1774). IEEE.

42. Vrandečić, D. (2009). Ontology evaluation. In *Handbook on ontologies* (pp. 293–313). Springer.

43. Wu, J., Feng, T., Naehrig, M., & Lauter, K. (2016). Privately evaluating decision trees and random forests. *Proceedings on Privacy Enhancing Technologies, 2016*(4), 335–355.

44. Wu, P., Chen, M., & Tsau, C. (2017). The data-driven analytics for investigating cargo loss in logistics systems. *International Journal of Physical Distribution & Logistics, 47*(1), 68–84.

45. Zervanou, K., Korkontzelos, I., van den Bosch, A., & Ananiadou, S. (2011). Enrichment and structuring of archival description metadata. In *Proceedings of the 5th ACL-HLT Workshop on Language Technology for Cultural Heritage, Social Sciences, and Humanities* (pp. 44–53). Association for Computational Linguistics, Portland, OR, USA. https://www.aclweb.org/anthology/W11-1507

46. Zillner, S., Bisset, D., Milano, M., Curry, E., García Robles, A., Hahn, T., Irgens, M., Lafrenz, R., Liepert, B., O'Sullivan, B., & Smeulders, A. (Eds.), *Strategic research, innovation and deployment agenda: AI, data and robotics partnership* (3rd ed.) BDVA, euRobotics, ELLIS, EurAI and CLAIRE (2020). https://ai-data-robotics-partnership.eu/wp-content/uploads/2020/09/AI-Data-Robotics-Partnership-SRIDA-V3.0.pdf

47. Zillner, S., Curry, E., Metzger, A., Auer, S., & Seidl, R. (2017). *European big data value strategic research & innovation agenda.* Big Data Value Association.

Printed in the United States
by Baker & Taylor Publisher Services